22nd Meeting of the International Collaboration on Advanced Neutron Sources (ICANS XXII)

Journal of Physics: Conference Series Volume 1021

Oxford, United Kingdom
27 - 31 May 2017

Editors:

Philip King
Preeti Kaur
Zoe Bowden
Sara Fletcher
Robert McGreevy
Joanne Hemstock

ISBN: 978-1-5108-6403-0
ISSN: 1742-6588

Printed from e-media with permission by:

Curran Associates, Inc.
57 Morehouse Lane
Red Hook, NY 12571

Some format issues inherent in the e-media version may also appear in this print version.

Printed by Curran Associates, Inc. (2018)

For permission requests, please contact the Institute of Physics
at the address below.

Institute of Physics
Dirac House, Temple Back
Bristol BS1 6BE UK

Phone: 44 1 17 929 7481
Fax: 44 1 17 920 0979

techtracking@iop.org

Additional copies of this publication are available from:

Curran Associates, Inc.
57 Morehouse Lane
Red Hook, NY 12571 USA
Phone: 845-758-0400
Fax: 845-758-2633
Email: curran@proceedings.com
Web: www.proceedings.com

Table of contents

Volume 1021

22nd meeting of the International Collaboration on Advanced Neutron Sources (ICANS XXII)

27–31 March 2017, Oxford, United Kingdom

Accepted papers received: 13 April 2018
Published online: 4 June 2018

Preface

Papers

Accelerators

Compact neutron sources

Data analysis and computing

Instrument design

Polarisation

Safety, compliance and radiological issues

Sample environment

Shielding

Targets and moderators

IOP Conf. Series: Journal of Physics: Conf. Series **1021** (2018) 011001 doi:10.1088/1742-6596/1021/1/011001

ICANS Proceedings Preface

Proceedings of the 22nd Meeting of the International Collaboration on Advanced Neutron Sources (ICANS-XXII); 27-31 Mar 2017, Said Business School and St Anne's College, Oxford, UK.

Philip King, Preeti Kaur, Zoe Bowden, Sara Fletcher, Robert McGreevy, Joanne Hemstock

The 22nd meeting of the International Collaboration on Advanced Neutron Sources (ICANS-XXII) was held at the Saïd Business School, University of Oxford, in March 2017. It was hosted by the ISIS Neutron and Muon Source in partnership with the UK Institute of Physics. 140 talks and 90 posters enabled discussion on a wide range of topics including neutron instrument design, neutron targets, detectors, guides and optics, neutronics, shielding, data analysis and computing, moderators, and compact neutron sources for the future. The aim was to encourage discussion and the sharing of knowledge and experience between neutron sources around the world, with the prospect of continued development of sources and instruments well into the future. The meeting was particularly timely with the European Spallation Source and Chinese Spallation Source both well into their construction phases.

This meeting was the 40th anniversary of the first ICANS workshop in 1977. John M. Carpenter (Argonne National Laboratory), Rex G. Fluharty (Los Alamos Scientific Laboratory), Leo C. W. Hobbis (Rutherford Appleton Laboratory) and Motoharu Kimura (Tohoku University and the Japanese National Laboratory for High Energy Physics, KEK) formulated the goals and structure of the collaboration, focussing on the subjects of accelerator development, source (target) and moderator development, and neutron scattering instrumentation development. ICANS now links over 15 facilities worldwide, providing a forum for scientists and engineers to openly discuss and share information on the progress and developments in accelerator-based neutron sources. ICANS has the following goals:

- to facilitate the exchange of technical information;
- to provide for review of results and designs;
- to reduce duplication of efforts;
- to identify areas for collaborative efforts, and;
- to provide means for informing participants about plans for activities relating to spallation source development.

We would like to thank all the participants of ICANS XXII for excellent presentations and discussions.

Group Photo from the conference

IOP Conf. Series: Journal of Physics: Conf. Series **1021** (2018) 011002 doi:10.1088/1742-6596/1021/1/011002

Peer review statement

All papers published in this volume of *Journal of Physics: Conference Series* have been peer reviewed through processes administered by the proceedings Editors. Reviews were conducted by expert referees to the professional and scientific standards expected of a proceedings journal published by IOP Publishing.

IOP Conf. Series: Journal of Physics: Conf. Series **1021** (2018) 012001 doi:10.1088/1742-6596/1021/1/012001

LANSCE Accelerator Update and Future Plans

Robert W. Garnett

Accelerator Operations and Technology Division
Los Alamos National Laboratory, Los Alamos NM 87545 USA

E-mail: rgarnett@lanl.gov

Abstract. Several multi-year accelerator system upgrades have recently been completed at LANSCE that will ensure the long-term future of beam operations to its users and to prepare for a planned next-generation signature facility (MaRIE) based on multi-probe capabilities. Present LANSCE accelerator operations will be discussed. A summary of the recent upgrades completed and future plans will also be presented. This work is supported by the United States Department of Energy, National Nuclear Security Agency, under contract DE-AC52-06NA25396.

1. Introduction

The Los Alamos Neutron Science Center (LANSCE) accelerator complex continues to support a diverse set of users, providing an extraordinary opportunity for U.S. science and technology programs. With its unique capability to accelerate and deliver two beam species simultaneously to multiple users and to provide peak neutron fluxes that span nearly 16 orders of magnitude in neutron energy, LANSCE supports three major Department of Energy (DOE) stakeholders: Defense Programs (DP), Office of Science (SC), and Office of Nuclear Energy, Science and Technology (NE) that support operations at LANSCE that enhance the U.S. nuclear weapons program, basic science including new scientific capabilities, and the U.S. radioisotope production capability. Unique irradiation capabilities are also provided for industry users. Recently completed RF system upgrades have enabled a return to 120-Hz accelerator operations and other multi-year system upgrades will ensure the long-term future of LANSCE beam operations to support current and future programs.

A simplified layout of the LANSCE accelerator complex comprised of an 800-MeV proton linac, a proton storage ring (PSR) and multiple experimental areas is shown in Figure 1. Simultaneous H^+ and H^- beams are delivered with flexible time structures to the experimental areas. Only the H^- beam is accelerated to 800 MeV. The H^+ beam is diverted at 100 MeV to the Isotope Production Facility (IPF) which helps satisfy the national need for biomedical radioisotopes. Other experimental user facilities at LANSCE include: the Manuel Lujan Jr. Neutron Scattering Center (Lujan Center), which exploits a moderated neutron source (meV to eV) primarily for neutron-scattering research; the Weapons Neutron Research (WNR) facility, which uses an unmoderated neutron source (keV to 800 MeV) for research in nuclear science and technology and industrial irradiations; a dedicated facility for proton radiography (pRad) at 800 MeV (Area C); Area B, which produces the most intense source of ultra-cold neutrons (UCN) in the world, and; Area A (currently inactive).

During the 2016-2017 Run Cycle, 547 unique users were supported through approved and scheduled proposals at the pRad, Lujan Center, and WNR facilities. An overall accelerator availability of 86% was achieved. The IPF facility produced enough Strontium-82 for approximately 120,000

IOP Conf. Series: Journal of Physics: Conf. Series **1021** (2018) 012001 doi:10.1088/1742-6596/1021/1/012001

cardiac images and Actinium-225 was investigated for cancer therapy. Thirty two dynamic proton radiography experiments were supported and 119 WNR and Lujan experiments were completed that advance the knowledge of total fission cross sections for uranium and plutonium. Materials science experiments at LANSCE have focused on understanding the performance of additively-manufactured materials. Experiments at the WNR "Blue Room" have contributed to developing radiation-hard electronics needed for the Large Hadron Collider (LHC) upgrade and beam delivery to the WNR "Ice House" continues to support single-event-upset measurements for the semiconductor industry.

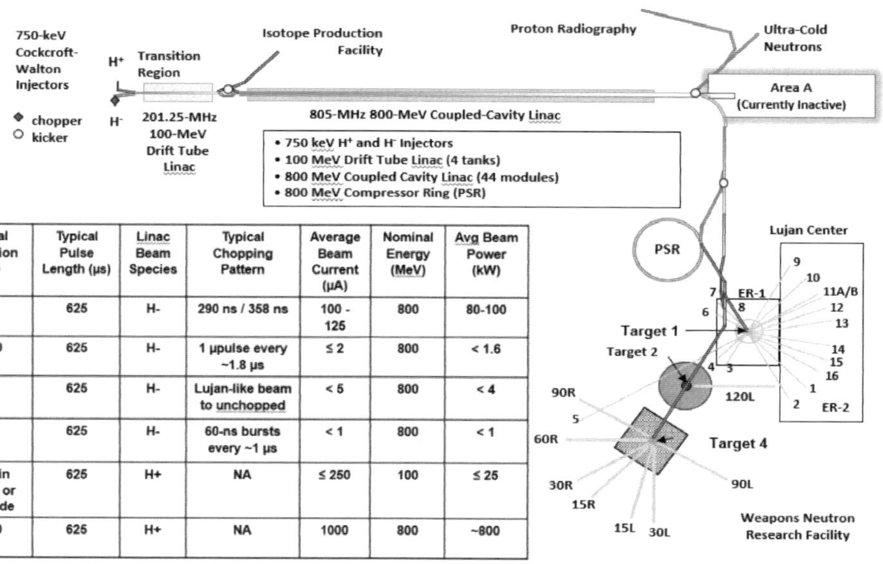

Area	Typical Repetition Rate (Hz)	Typical Pulse Length (µs)	Linac Beam Species	Typical Chopping Pattern	Average Beam Current (µA)	Nominal Energy (MeV)	Avg Beam Power (kW)
Lujan	20	625	H-	290 ns / 358 ns	100 - 125	800	80-100
WNR Tgt 4	≤ 100	625	H-	1 µpulse every ~1.8 µs	≤ 2	800	< 1.6
UCN	20	625	H-	Lujan-like beam to unchopped	< 5	800	< 4
pRad	~1	625	H-	60-ns bursts every ~1 µs	< 1	800	< 1
IPF	≤ 100 in pulsed or DC mode	625	H+	NA	≤ 250	100	≤ 25
Area A Inactive	≤ 100	625	H+	NA	1000	800	~800

Figure 1. Schematic layout of the LANSCE accelerator complex including the linac sections and the experimental areas. Also shown is a table of the possible beam parameters that can be delivered to each area.

2. LANSCE accelerator improvements

2.1. LANSCE linac risk mitigation (LRM)

Multiyear accelerator system upgrades that began in fiscal year 2009 were completed in fiscal year 2015, completing the funded LANSCE Linac Risk Mitigation project. The investments made enabled a return to 120-Hz beam operations that were previously limited to 60 Hz due to deterioration of the output power capability of the Burle 7835 power triode used in the RF amplifier system of each 201.25-MHz drift-tube linac (DTL) module. Three of four of these high-power RF systems were replaced with new diacrode-based systems [1]. Also included in the LRM project scope was the procurement of 45 new, 805-MHz klystrons to be installed as required due to failing end-of-life units, remediation of some accelerator structures and supporting equipment, magnet and other power supplies, and implementation of modern, maintainable EPICS-based instrumentation and controls. A partially funded, longer-term project goal is to replace the obsolete and aging Cockcroft-Walton-based injector systems with modern Radio-frequency Quadrupole (RFQ) accelerator based injectors. Investments made from 2009-2015 span most major LANSCE accelerator systems. The total fiscal year 2009-2015 funding for LRM upgrades was approximately $140 M (US).

Fiscal year 2014-2015 investments focused on DTL Module-2 and Module-4 diacrode amplifier installation (see Fig. 2) and procurement of components needed to replace the Module-3 system in fiscal year 2016. Also, a new computer-controlled water cooling system was installed on DTL Module 2; other modules will be upgraded in future years. New industrial controls (cRIO) were installed for the full DTL, with partial implementation in the 805-MHz coupled-cavity linac (CCL). Facility

IOP Conf. Series: Journal of Physics: Conf. Series **1021** (2018) 012001 doi:10.1088/1742-6596/1021/1/012001

improvements needed to support new digital low-level RF (dLLRF) controls systems were also completed.

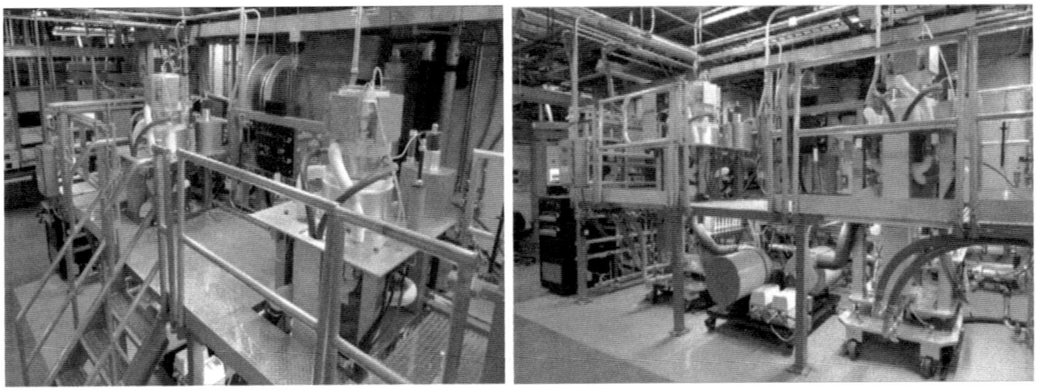

Figure 2. DTL Module-2 201.25-MHz diacode based high power RF amplifier.

2.2. Beyond LRM

Accelerator improvements are now funded on an annual basis based on impact to operations and available budget. Fiscal year 2016 efforts focused on installation of the final diacrode-based high-power RF system for DTL Module 3, completing additional controls and timing system modernization, implementation of a dLLRF system for all DTL modules, and continuing upgrading magnet power supplies and vacuum systems. Improvements to WNR Target 4 (see Fig. 1) included installation of remote pump controls to eliminate unnecessary entry into a radiation area and associated beam downtime to recover a tripped pump. Target 1 improvements included implementation of the only in-situ hydrogen ortho/para measurement capability of its kind in the neutron community [2]. A newly installed Feschenko Bunch Shape Monitor (shown in Fig. 3 below) has also been installed at 41.3 MeV in the DTL [3] and is being used to tune up the linac. Although significant progress has been made, completing the remaining scope of work in an expected flat-budget environment will require careful planning. Additional investments in out-years are needed to complete these improvements and to sustain operations and maintain high availability.

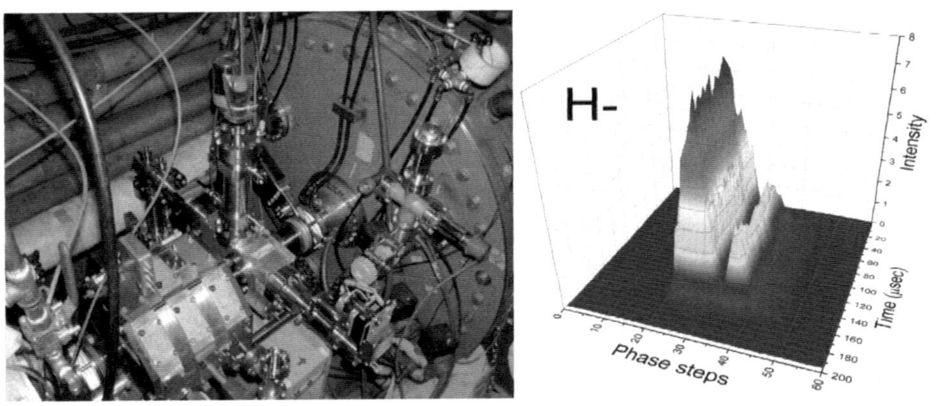

Figure 3. Left: Feschenko bunch shape monitor installed between DTL tanks 3 and 4. Right: Typical H^- bunch-shape data at 41.3 MeV, 1 Hz, 150 µs.

3. Focusing on the future

Several investments targeted at improving beam delivery to LANSCE users and to enhancing capabilities have be initiated. These are discussed in detail below.

IOP Conf. Series: Journal of Physics: Conf. Series **1021** (2018) 012001 doi:10.1088/1742-6596/1021/1/012001

3.1. H^- beam delivery enhancements
It has long been the goal to improve the LANSCE multi-cusp H^- ion source performance. A modest development program using measurements performed at the Ion Source Test Stand (ISTS) has been initiated to reach higher peak output currents and to extend the source lifetime beyond its nominal 28 days [4]. Also, presently the two LANSCE beams are combined in a common main buncher cavity whose phase and amplitude settings are a compromise for the two beams before injection into the DTL at 750 keV. Installation of a de-buncher cavity in the H^- transport line upstream of the main buncher is being considered to decouple the two-beam bunching and capture; implementation could improve the peak beam current delivered to WNR and pRad by up to 50%. Improving the H^- chopper rise time and phase control could further improve beam delivery to these areas and to Target 1.

3.2. Proton Storage Ring (PSR) and Target 1 options
Operation of the PSR at 30 Hz (normal operation is at 20 Hz) at short pulse durations (30 ns) was explored recently and is being considered as a potential routine mode of operation for the future. Pulse stacking in the PSR has also been considered recently to enable a new class of nuclear physics experiments at WNR that require very short, intense pulses to better resolve nuclear resonance phenomena. The present Target 1 will reach its end of life in the next 5 years. A new target configuration is being considered that will provide higher energy neutrons for nuclear science [5]. Preliminary calculations indicate that simple geometry changes to the present target configuration can increase neutron fluxes in the desired energy ranges. Design of the new target is expected to begin in 2018 with installation of the new target potentially as early as 2020.

3.3. 100-MeV isotope production improvement project
The LANSCE Isotope Production Facility (IPF) is the sole user of the 100-MeV H^+ beam and operates at an average beam current of up to 225 μA. During the past year a project to enhance several operational aspects of the IPF has been completed and installation is currently underway [6]. Enhancements include an improved window design capable of operation up to 300 μA with new temperature and window deflection diagnostics, an active and adjustable collimator (fixed collimation previously), shown in Fig. 4 below, with beam spill monitoring and active cooling for thermal management, new selectable raster patterns, a new beam profile and emittance measurement capability with <1 mm resolution, a 1% accuracy beam-current measurement capability over the range of 100 nA to 500 μA, and real-time beam energy monitoring (50 keV resolution). Commissioning is expected during resumed beam operations in June 2017.

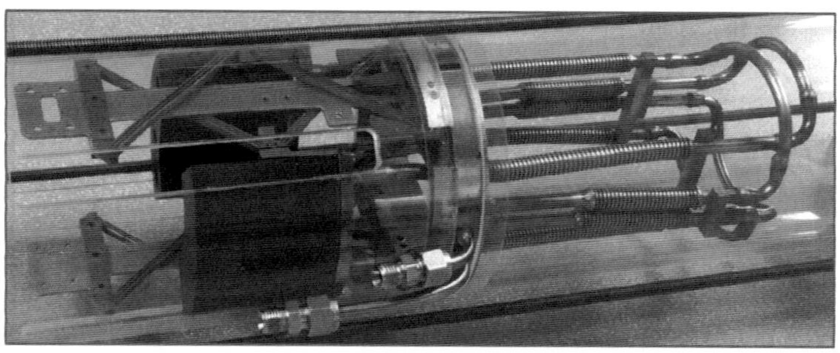

Figure 4. Adjustable collimator mock-up with complete set of cooling tubes.

3.4. H^+ RFQ to replace the Cockcroft-Walton (CW) injector
A test stand is being assembled to test a new 4-rod, 750-keV, 201.25-MHz RFQ to replace the aging and obsolete H^+ CW injector [7]. The test stand will include the ion source, low-energy beam transport, RFQ, medium-energy beam transport downstream of the RFQ, and diagnostics needed to inject beam into the LANSCE DTL. High power RF tests of the RFQ are expected to begin in the next few months.

IOP Conf. Series: Journal of Physics: Conf. Series **1021** (2018) 012001 doi:10.1088/1742-6596/1021/1/012001

3.5. Automation and enhanced diagnostics and controls

Operational efficiencies can be obtained though automation of several accelerator controls and optimization functions typically performed by trained accelerator operators. Application of automated processes and modern control algorithms using improved instrumentation and diagnostics are being explored. Interactive codes based on fast GPU processors are being developed that may soon allow automatic system and safety interlock checks, beam tuning and beam optimization, and upset recovery [8]. Automating processes at the H^- injector including ion source peaking and controlling cesium transfers is also being investigated [9], as is automated RF cavity conditioning.

3.6. Resumed 800-MeV beam operations to Area A

Since the completion of the LAMPF nuclear physics program in the mid 1990's, no 800-MeV protons have been delivered to experimental Area A. However, recently there has been renewed interest in using Area A and several viable options have been explored. At present a relatively low-cost, Low-Power Proton Source (LPPS) facility is being proposed [10]. The LPPS targets urgent needs of the national security and space science radiation effects communities that qualify components for use in space. The facility would be operated at a nominal average current of 100 nA (6.25×10^{11} protons/sec) delivered at a 1-Hz repetition rate using the 625-μs LANSCE beam macropulse length. Dual energy operation at 200 MeV and 800 MeV is being considered. The LPSS is expected to require only a modest beam transport line and experimental configuration in Area A. The project is currently under review. Implementation will require 2-3 years, including full implementation of a 100-MeV IPF kicker system to inject the proton beam into the CCL and an upstream beam chopper at 750 keV to limit the average beam current.

3.7. Matter-radiation interactions in extremes (MaRIE)

A next-generation signature facility (see Fig. 5) based on multi-probe capabilities is being planned at LANSCE that will be the first in a new generation of game-changing scientific facilities for the materials community [11]. The new Matter-Radiation Interactions in Extremes (MaRIE) facility will be used to discover and design the advanced materials needed to meet 21st-century national security and energy-security challenges by providing the tools to develop next-generation materials that will perform predictably in extreme environments. Materials samples will be synthesized and characterized at the mesoscale, and their dynamic behavior characterized in time-dependent extreme conditions by use of both imaging and diffractive scattering with multiple probes at multiple spatial and time scales. The MaRIE facility includes a new 12-GeV electron linac to drive a 42-keV X-ray free-electron laser (XFEL) coupled with the existing proton-beam capabilities of the LANSCE proton linac, new experimental halls, and materials fabrication/characterization facilities. A pre-conceptual design has been completed and performance requirements have been defined to meet a broad range of experimental needs. The project has received Critical Decision 0 (CD-0; approved mission need) and is now on the path to CD-1 (completion of an analysis of alternatives and approval of alternative selection) and is preparing for the Conceptual Design Report. Completion of the MaRIE project is expected in the 2025-2030 time frame.

4. Summary

There continue to be several challenges ahead for sustaining LANSCE operations today and building the foundation for tomorrow. Our short-term (1-5 years) efforts will be focused on maintaining stable funding for operations, on completing remaining subsystem enhancements as discussed earlier, and to continue to improve alignment of our operations with present experimental missions and to build new capabilities synergistic with the long-term MaRIE vision. In the 5-10 year timeframe, focus will be on installation of a new Target 1 optimized for nuclear science needs, enhanced uses of proton radiography, and evaluating the needs for LANSCE capabilities in the MaRIE era. Beyond these will be the transition to an integrated MaRIE facility and optimizing performance and value to the new expanded user base that is expected to result.

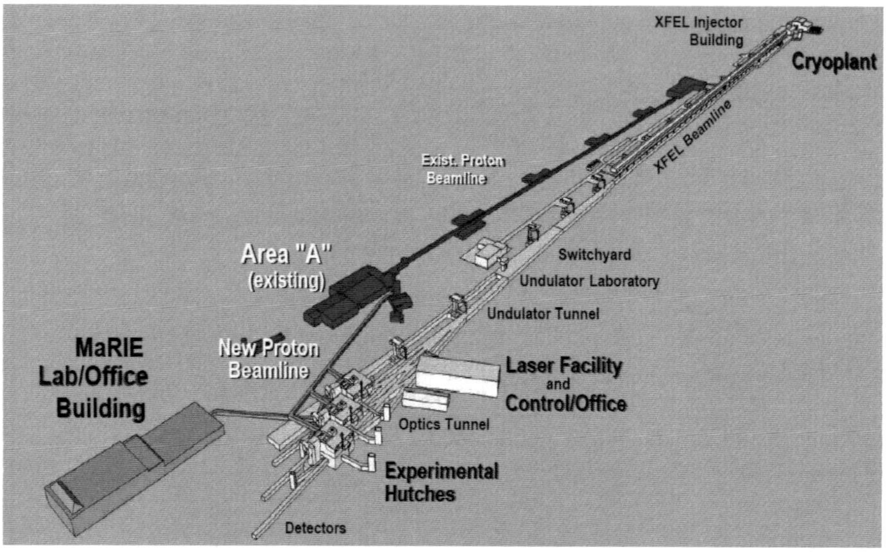

Figure 5. Pre-conceptual layout of the MaRIE facility. Note the location of the existing LANSCE proton linear accelerator (shown in blue) relative to the new MaRIE XFEL.

References

[1] J. T. M. Lyles et al 2015 Installation and operations of replacement 201 MHz high power RF system at LANSCE *Proc. 6th Int. Particle Accelerator Conf. (Richmond)* (JACoW) p 3485

[2] E. Kerstiens 2017 Status of In-Situ Hydrogen Ortho-Para Measurement Technique *this conference*

[3] I. Draganic et al 2016 Bunch shape Monitor Measurements at the LANSCE Linac *Proc. 2016 North American Particle Accelerator Conf. (Chicago)* (JACoW)

[4] I. N. Draganic et al 2016 Different approaches to modeling the LANSCE H- ion source filament performance, *Rev. of Sci. Instr.* **87**, 02B112 (2016); doi: http://dx.doi.org/10.1063/1.4932559
I. Draganic et al 2013 Progress Report of H^- Ion Beam Production at the LANL Ion Source Test Stand *Proc. of the 2013 North American Particle Accelerator Conf. (Pasadena)* (JACoW) p 667

[5] M. Mocko 2017 Redesign of the Lujan Center Neutron Spallation Source for Optimization of the Neutron Flux in the 0.001-1 MeV Energy Range, *this conference*

[6] E. R. Birnbaum et al 2014 LANL AIP Proposal: IPF Beam Transport Upgrade Project Los Alamos National Laboratory document LA-UR-14-26690
H. Watkins et al 2016 Upgrades to the LANSCE Isotope Production Facitity Beam Diagnostics *Proc. IBIC2016 (Barcelona)* (JACoW) p 1

[7] R. W. Garnett et al 2015 LANSCE H^+ RFQ Project Status *Proc. 6th Int. Particle Accelerator Conf. (Richmond)* (JACoW) p 327

[8] L. J. Rybarcyk 2016 HPSIM–Advanced Online Modeling for Proton Linacs, *Proc. HB2016 Workshop*, (JACoW) p 444
X. Pang and L. Rybarcyk 2013 Multi-Objective Particle Swarm and Genetic Algorithm for Optimization of the LANSCE Linac Operation *Nucl. Instr. Meth. A*, 741, 124, (2014)
A. Scheinker et al 2013 Model Independent Particle Accelerator Tuning *Phys. Rev. ST Accel. Beams*, 16, 102803, (2013).

[9] F. Shelley, *private communication*

[10] S. Wender and H. Quinn 2017 A Business case for a Low-Power Proton Source in LANSCE Area A, Los Alamos National Laboratory draft document; *private communication*

[11] R. W. Garnett 2016 The Los Alamos Multi-Probe Facility for Matter-Radiation Interactions in Extremes *Proc. 28th Linear Accelerator Conference (East Lansing)* (JACoW)

IOP Conf. Series: Journal of Physics: Conf. Series **1021** (2018) 012002 doi:10.1088/1742-6596/1021/1/012002

CSNS Project Construction

S N Fu[1], H S Chen[1], Y B Chen[1], L Ma[1], F W Wang[2]

[1] Dongguan Neutron Science Centre, Institute of High Energy Physics, CAS, China
[2] Institute of Physics, CAS, China

fusn@ihep.ac.cn

Abstract. The China Spallation Neutron Source (CSNS) is designed to accelerate proton beam pulses to 1.6GeV at 25Hz repetition rate, striking on a solid metal target to produce spallation neutrons for neutron scattering experiment with 20 spectrometers. The accelerator provides a beam power of 100kW for the target in the first phase and then 500kW in the second phase at the same repetition rate and in the same target station. The project construction launched in 2011 and it is planned to complete the first phase project in March 2018 for governmental acceptance. Now the accelerator construction and machine commissioning have been completed and the initial beam commissioning is in progress. Target station and three spectrometers are under final installation. This paper will update the recent construction status of the CSNS project.

1. Introduction

China Spallation Neutron Source project was proposed in 2001 and then formally approved by the Chinese central government in 2008. After several years for detailed design and key technology development, the project construction started in September 2011 and will be completed in March 2018[1]. The construction site is located at Dongguan, Guangdong Province, and the project is hosted by the Institute of High Energy Physics together with the Institute of Physics, CAS. At present, more than 350 staff works at Dongguan site for the project construction.

The CSNS accelerator provides proton beam pulses of 1.6GeV kinetic energy at 25Hz repetition rate to a solid metal target to produce spallation neutrons for neutron scattering experiment. A schematic layout of CSNS phase-1 complex is shown in Figure 1. In the phase one, an ion source produces a peak current of 25mA H^- beam. RFQ linac bunches and accelerates it to 3MeV. DTL linac raises the beam energy to 80MeV. After H^- beam is converted to proton beam via a stripping foil, the Rapid Cycling Synchrotron (RCS) accumulates and accelerates the proton beam to 1.6GeV[2]. Then the beam is extracted to the Tungsten target cladded by tantalum and cooled by water. 20 neutron channels are designed surrounding the target, but only 3 spectrometers will be built in the day-one operation due to limited budget from the centre government. More spectrometers are under planning together with users from universities, institution and local government. The beam power is 100kW on the target in the first phase. The facility design allows further upgrade to 500kW beam power at the same repetition rate and the same output energy in Phase-II by increasing the average beam current 5 times. For this purpose, we reserve 85m length in the LRBT for the installation of additional accelerating cavities, such as superconducting spoke cavity used in our ADS program or PIMS cavity used in LINAC-4 at CERN, for beam energy upgrade to more than 250MeV in the Phase-II.

IOP Conf. Series: Journal of Physics: Conf. Series **1021** (2018) 012002 doi:10.1088/1742-6596/1021/1/012002

Now the project construction will soon complete. As shown in Figure 2, all the civil constructions, including the accelerator building, the target and spectrometer hall, the utility buildings as well as the user building, are all put into usage. The accelerator has been installed. The linac is under beam commissioning and the RCS ring has completed machine commissioning. The target-moderator-reflector (TMR) has almost been installed and the construction of spectrometers is at the final stage. We can foresee the first neutrons from the target in this September. In the following sections we will report the construction status in more details.

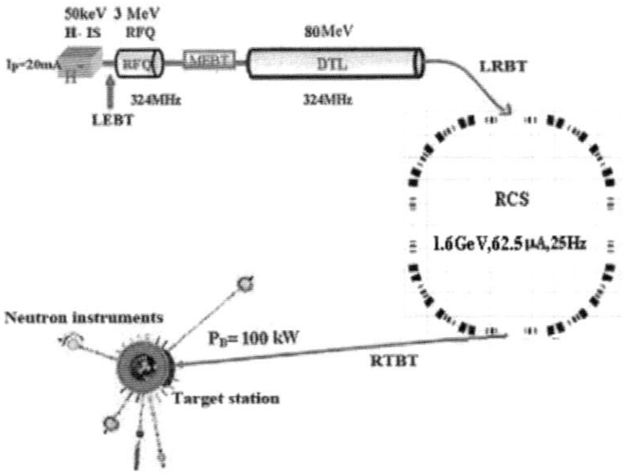

Figure 1. Schematics of the CSNS complex.

Figure 2. Bird's eye view of the CSNS campus

2. Accelerator Construction

The H^- linac is composed of a Penning H^- ion source, a LEBT with a beam chopper, an RFQ accelerator to raise beam energy to 3MeV from 50keV, a MEBT for beam match in transversal and longitudinal directions, and an 80MeV DTL with 4 tanks as the major accelerating structure of the linac. The linac has been installed in the tunnel, as shown in Figure 3.

The RCS adopts four-fold structure with triplet focusing lattice. In the four long-straight sections, we arrange injecting, collimating, extracting and accelerating elements separately. The circumference of the ring is about 228m. The RCS majorly consists of 24 dipole magnets and 48 quadrupole magnets which are all powered with DC+AC current from White resonant power supplies, 8 injection painting magnets plus 4 bump magnets, one primary collimator together with four secondary collimators, 8 kicker magnets and 8 accelerating cavities with harmonic number of two. All the ring components have been installed in the ring tunnel and Figure 4 shows an arc section of the ring.

Figure 3. CSNS linac installed in the tunnel.

Figure 4. An arc section of the RCS.

2.1. Linac beam commissioning

Beam commissioning of the frontend started in 2014. The ion source could provide maximum 40mA H^- beam, but with large emittance. Within the design emittance of 0.2 π mm.mrad, the beam current is about 15mA, which can meet the requirement of 100 kW beam power of CSNS. In order to raise further the RFQ beam transmission efficiency, a collimator at LEBT to scrape the beam beyond the acceptance of RFQ was installed. Many efforts were also put on the operation stability of the ion source, and up to now, beam trip number is around 3 per day with output H^- beam current above 50mA. The beam commissioning of RFQ was done together with MEBT. In April 2015, we got a 3MeV H^- beam of 28mA from the RFQ, which is much better than our Phase-I specification. Many beam diagnostics have been done, including BPM offset measurement, response matrix measurement, orbit measurement and correction, emittance measurement and beam chopping test. An electrostatic chopper is adopted in the LEBT, instead of a fast kicker in MEBT. The waveform measured in the MEBT shows the rise time is about 15ns(5 RF periods), as shown in the Figure 5. The rise time is comparable to the fast chopper that usually used in MEBT. In January 2016, DTL tank-1 started beam commissioning with a diagnostics plate. The commissioning started from the low peak beam current, low repetition rate and short pulse length, and then increase the average beam current to the designed peak current and pulse length. Phase scanning method is applied to searching for the RF synchrotron phase by TOF technology with two FCTs. Beam transmission rate reached 100% at 18mA output beam current, as shown in Figure 6. Same beam tuning method was also applied to the DTL tank-2 beam commissioning, which was conducted in November 2016. At present, we are conducting beam commissioning for the following tank.

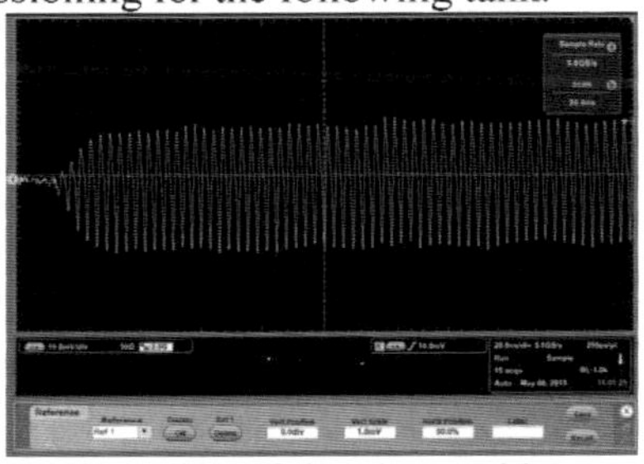

Figure 5. Rise time of the chopped beam.

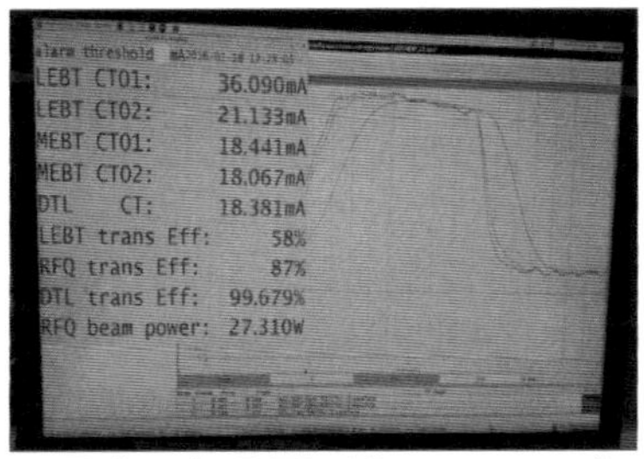

Figure 6. CT data in DTL-1 beam commissioning.

2.2. RCS machine commissioning

Machine commissioning was conducted in two steps: system commissioning and joint commissioning. Now we are coming to the second step. All dipole and quadrupole magnets were powered with their power supplies at the specified field for 7-day continuous operation. Due to 25Hz AC excitation, vibration of the magnets resulted in the break of a ceramic chamber in a quadrupole magnet. It was replaced with spare one and counter measure has been applied to all ceramic chambers to reduce the vibration. Second 7-day operation demonstrated the effectiveness of the measure. To deal with time harmonics in AC magnetic field due to nonlinearity, we proposed a new harmonic injection scheme to compensate for it, and the perfect effectiveness was confirmed with field measurement of the magnets. All 8 RF ferrite-loaded cavities and their RF power sources together the LLRF systems have been commissioned for 7-days continuous operation. LLRF control reached a satisfactory stability without beam loading, but further debugging and confirmation of LLRF operation will soon be conducted with injected beam.

3. Constructions of Target Station and Spectrometers

CSNS target station is designed to accept a proton beam power of 100 kW with the upgrade capacity to 500 kW. It means that the shielding, the embedded pipes are designed for 500 kW beam power while the TMR system is optimized for 100 kW. Figure 7 shows the engineering and technical designs of CSNS target station. Three wing-type moderator, one ambient water moderator and two cryogenic hydrogen moderators (20 K) supply diverse characteristics of pulsed neutron beams. The solid tungsten clad by tantalum is selected as CSNS target plates. The TMR neutronics design is performed and optimized by Monte Carlo simulation. The goal of radiation dose rates of 2.5 μSv/h and 25 μSv/h are set in the CSNS user area and target station high-bay area, respectively. All calculations are done for the upgrade 500 kW beam power because most shielding blocks are immovable.

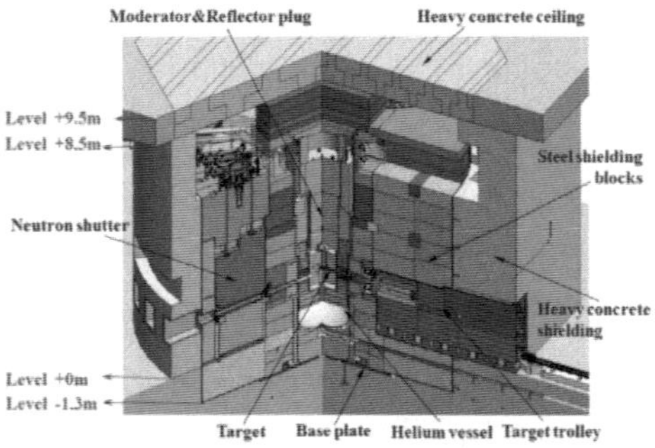

Figure 7. Engineering design of CSNS target station.

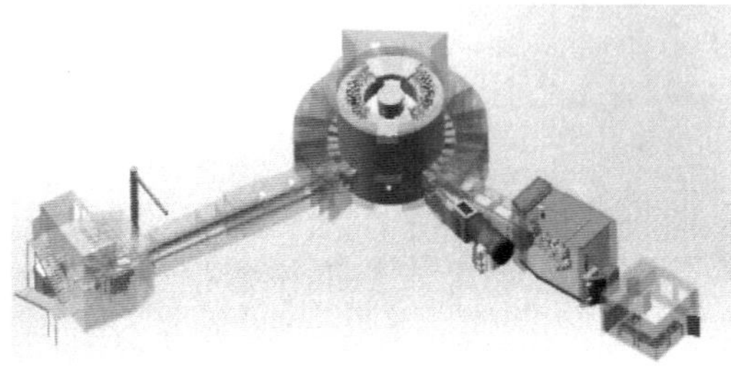

Figure 8. Engineering design of three day one instruments: GPPD, SANS and MR.

Following the characteristic of the short pulse, the scientific frontier, the demands of potential users in China and experiences of spallation source operated overseas, CSNS has planned 20 neutron scattering instruments for 20 beam lines. Because of budget limitations, only three day-one instruments: General Purpose Powder Diffractometer (GPPD), Small Angle Neutron Scattering (SANS), and Multipurpose Reflectometer (MR), are funded by the project. Figure 8 shows the engineering design of the three spectrometers. Up to now, the main equipment of target station, including of helium vessel, target trolley, shielding blocks and neutron shutters, have been installed in place, as shown in Figure 9. The water cooling system and the cryogenic system are in commissioning, and the joint commissioning with target, moderator, reflector, and proton beam window, will start in the middle of April, 2017. And then the moderator-reflector plug and beam window plug are installed. Most equipment of the three day-one neutron instruments have arrived at CSNS, and the installation is underway, as shown in Figure 10. The automatic control is debugged one by one. We expect that the first turn of joint commissioning will be completed in June, 2017, and that the first neutron beam will be produced near the end of September, 2017. Currently, all hardware of three instruments has arrived, the installation is in progress and will be completed in this July, the joint commissioning is expected to complete in this September to see the first neutron beam.

CSNS user meeting is held annually to plan the day one experiments. A few of neutron instruments, funded by local government and universities, are negotiated tightly.

Figure 9. Target station is under installation.

Figure 10. Installation of GPPD is almost completed.

Conclusions

CSNS project is going well and now at the stage of final construction and commissioning, and first neutron beam from target can be expected in the September 2017.

References

[1] Hesheng Chen, Xunli Wang, Nature Materials, VOL 15, 689-691 (2016)
[2] Shinian Fu, Hesheng Chen, et al., Proc. of IPAC2013, Shanghai, May 2013

IOP Conf. Series: Journal of Physics: Conf. Series **1021** (2018) 012003 doi:10.1088/1742-6596/1021/1/012003

Challenges in Technical Risk Management for High-Power Accelerators

R Andersson[1], S Kövecses[2], E Bargalló[3], A Nordt[4]

[1,2,3,4] European Spallation Source ERIC, Lund, Sweden
[1] University of Oslo, Norway

[1] riccard.andersson@esss.se
[2] szandra.kovecses@esss.se
[3] enric.bargallo@esss.se
[4] annika.nordt@esss.se

Abstract. The increasing demands on accelerator-based research facilities to be both reliable and cost-efficient require comprehensive risk management that takes the full scope of the facility into account. Due to the organizational and technical complexity of these facilities, a systematic risk management method is beneficial, if not necessary. However, many other aspects, aside from reliability and cost-efficiency, also call for great attention. This paper discusses the many challenges that are faced when dealing with technical risk management of accelerator facilities, as these efforts are often considered second to e.g. beam parameters and operational schedule. The paper uses the design and construction of the European Spallation Source in Sweden as a basis, but many of the contradictory aspects have been found in other similar facilities around the world. Some of the challenges for technical risk management that are discussed relate to rapid organizational expansion, pressing schedule, knowledge biases, and conflicts of interest.

1. Introduction

The construction of a large-scale research facility, such as a particle accelerator and neutron spallation source, takes on many different phases. Going from concepts and ideas to the final facility often takes decades and involves a large number of people and interest groups [1] [2] [3]. Merging these to finally achieve the end goal is far from straightforward and the approach is typically different from case to case. The ongoing construction of the European Spallation Source (ESS) in Lund, Sweden, has been and is currently undergoing such phase transitions [3].

The first ideas behind ESS dates back to 1999, where the European Neutron Scattering Association (ENSA) convinced the Organization for Economic Cooperation and Development (OECD) that neutron sources were important tools for future research, and that Europe required a new facility that could complement ISIS in the UK. The following decade included discussions and negotiations, and in 2009, Lund was chosen as the place of construction [4]. The year after, ESS was formed as a limited liability company, and in 2015 it received its status as a European Research Infrastructure Consortium (ERIC). The first spallation neutrons are expected to be created in 2020, which is then just over two decades from the initial decision for the facility [3]. ESS is then planned to be operated for 40 years before decommissioning. This shows the long time-scales for such a facility, and highlights the many decisions that need to be made to move the project forward.

IOP Conf. Series: Journal of Physics: Conf. Series **1021** (2018) 012003 doi:10.1088/1742-6596/1021/1/012003

The requirements on neutron production, both in terms of quality, or neutron brightness, and operational availability [5], implies a challenging task and all of the internal and external stakeholders need to collaborate and facilitate the design, construction, and operation of ESS. Needless to say, these challenges have been seen in prior accelerator facilities and neutron sources [6], just as they are seen for ESS. One main component in reaching the unprecedented brightness and availability is the management of technical risks. These risks encompass both damage to equipment and loss of neutron production from other aspects. In this paper, some of the challenges associated with performing a rigorous technical risk management are discussed.

2. A Systematic Approach to Risk Management

ESS has developed and is using a systematic approach to technical risk management [7] that involves a machine protection system-of-systems (MP-SoS) as well as a facility-wide reliability, availability, maintainability, and inspectability (RAMI) approach. The MP-SoS and RAMI analysts work closely together to ensure that the end goals of ESS can be fulfilled. This chapter briefly describes the method itself and its place in the ESS organization.

2.1. Method

The risk management method for MP-SoS follows the applicable ISO standards 31000 [8] and 16085 [9]. These standards outline the process of establishing a risk management context followed by identifying, analyzing, evaluating, and treating the risks in a traceable and transparent way. Additionally, the IEC 61508 [10] and 61511 [11] standards for functional safety of electric, electronic, and programmable electronic (E/E/PE) systems are guiding the lifecycle approach as well as the classification of protective functions [12]. This places the ESS risk management in a robust framework that allows for efficient and purposeful analyses and accurate treatment of the technical risks associated with running a complex high power accelerator facility. The method is currently applied at ESS and builds on close collaboration between system owners and risk analysts.

2.2. Organization

The ESS project organization is set up to address each of the three sections of the facility through a dedicated division: the proton accelerator, spallation target, and neutron science systems. Within each division, there is a set of work packages that deal with specific parts identified in the facility breakdown structure (FBS), Figure 1. The MP-SoS and RAMI analyses need to merge into the development of the different work packages at the same time as they tie back to the overall goals of ESS. MP-SoS and RAMI are therefore continuously observing and analyzing the progress of the different work packages, in order to cope with the complexity and system interaction that might lead to damage and losses that would cause ESS to not fulfill the overall goals [13].

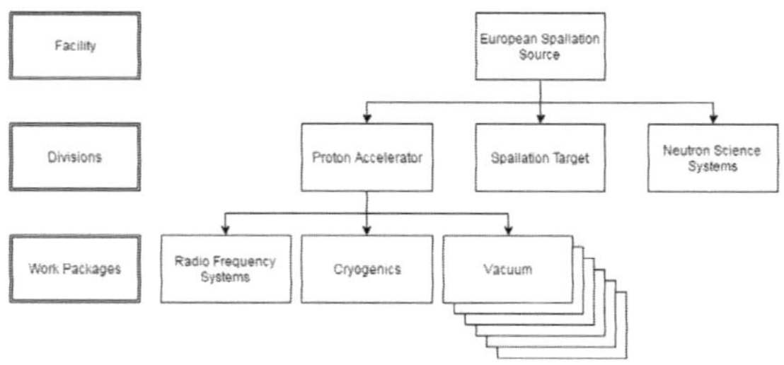

Figure 1: Part of the Facility Break Down Structure

The MP-SoS work is done by three interconnected teams, as seen in Figure 2. The protection analysis team (PAT) is in charge of translating global ESS requirements into manageable requirements and functions to be implemented into the hardware and software of the facility. The integrated protection team (IPT) analyzes interfaces between protection-related systems and the rest of the facility. Finally, the implementation and design team (IDT) makes the implementation of the requirements and serve as system experts in the discussions on possible and relevant designs and functionalities.

The RAMI work takes place through a cross-functional working group that involves stakeholders from all the divisions, where possible issues and suggested improvements can be analysed and verified. The working group meets as per request and has close discussions with external stakeholders. It has successfully derived hands-on requirements for the facility, going from high-level concepts down to applicable RAMI requirements on individual systems [13].

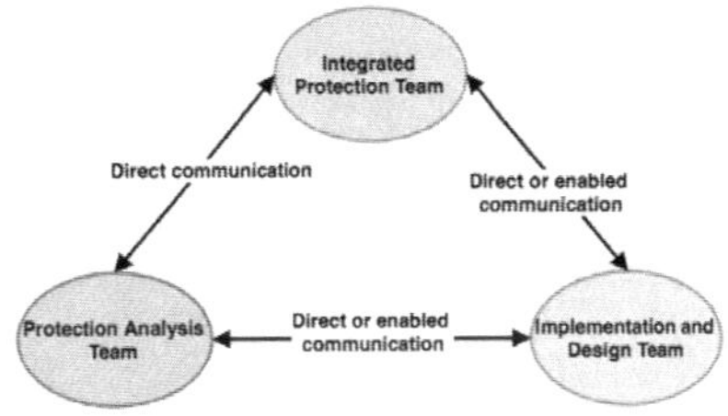

Figure 2: The three teams involved in MP-SoS work at ESS.

3. Technical and Organizational Complexities

The technical risk management in general, and the MP-SoS and RAMI approaches in particular, find certain challenges in the organization that can be attributed to the complexity of a research facility under construction. These challenges need to be highlighted and mitigated in order to avoid accidents during the commissioning and operational stages. In fact, it is claimed that basic culture and commitment to safety or protection is the most important factor in the occurrence of accidents [14].

3.1. Organizational Development

The ESS organization has and is expanding rapidly, hence the culture, communications procedures, and all of the administrative efforts, have to be developed in parallel with the design and construction of the facility. Risk management needs to cope with limited procedures for documentation as well as changing organizational structures internally and externally. This places high demands on flexibility and continuous updates of the information to be analyzed. Traceability then becomes an important aspect already in the early stages, in order to verify that the analyses are matching the latest versions of the hardware.

The rapid expansion can also be viewed from the perspective of scope, where the MP-SoS and RAMI analyses continue to include more systems as the project proceeds, having to inform and set up discussions with new stakeholders. These stakeholders are themselves required to perform concept designs of their systems at the same time as potential interfaces are changing. It is therefore important for the technical risk managers to recognize that the organizational development affects the technical developments, including systems relevant for machine protection and RAMI.

3.2. Schedule

The pressing schedule during the design, construction, and commissioning can lead to predicaments in decision-making. A too rapid decision might miss overlooked opportunities and important factors to account for, at the same time, late, or completely lacking, decisions lead to temporary solutions and workarounds that can be disadvantageous for the robustness and final protection of a system or component. Risk management on all levels needs to account for this and simplify the decisions to be taken by providing appropriate information in a timely manner.

3.3. Prior Knowledge and New Developments

The staff, contractors, and in-kind partners at ESS represent many different labs as prior work places. Each person has their own background and experience, which are to be merged with newly developed technical concepts and specialized equipment that are often not familiar to the stakeholders. Additionally, outcomes of the new organizational setup, including procedures and reporting, are not always in line with the experience from previous labs. At the same time as this work is ongoing, the staff has to ensure that interfaces between systems are in place, despite that this often requires new concepts than has not previously been used.

IOP Conf. Series: Journal of Physics: Conf. Series **1021** (2018) 012003 doi:10.1088/1742-6596/1021/1/012003

3.4. Geographical Distribution of Personnel

The process of aligning on-site personnel and in-kind contributors requires much effort in terms of setting up mutual agreements and emphasizing proper documentation and sharing of information since the different labs have different procedures and the countries have varying cultures. Refer to Figure 3. for a graphical overview of in-kind contributors. This affects the risk management process where available and up-to-date reports and documents are necessary to make appropriate analyses and suggestions.

Figure 3: Countries contributing to building ESS

3.5. Conflicting Interests

Related to the previous sections, the different experiences and methods of work across the organization sometimes lead to slowed down decision-making processes, where conflicting ideas or interests might lead to a prolonged process before system design and equipment selections are settled. If not resolved, it is possible that two separate design paths develop in parallel, which is an inefficient use of manpower and complicates the risk analyses. Often, two (or more) ideas can have different benefits and drawbacks and proper risk management is required to make the appropriate trade-offs to achieve the better outcome in the final design.

Another trade-off, or conflict of interest, is that which is involved in the risk management process itself: choosing protection or reliability. In many cases, a design emphasizing protection causes more spurious trips of the proton beam or other equipment, while a reliable design, from the operations point of view, relaxes the protection in favor of a stable operation without interruptions.

Finally, some systems benefit greatly from new developments and designs that have not been previously used. Other systems have a well-defined task that needs to be performed as reliably as possible. This conflict then touches the choice of either allowing for efforts within targeted research to optimize system designs and behavior, or to go for a defined engineering approach where the requirements are implemented with currently available equipment and processes.

4. Conclusions

This paper has highlighted some of the many challenges that appear in a rapidly growing organization that designs and constructs a complex research facility. The associated risk management processes need to be flexible enough to allow for changes and conflicting interests, and the risk managers should be aware of these challenges in order to cope with them in an efficient manner. While it is not beneficial to ignore or hide these challenges, it is not fruitful to accentuate them out of proportion either. These challenges can be difficult to tackle, but are also part of the unique environment that such a project offers.

IOP Conf. Series: Journal of Physics: Conf. Series **1021** (2018) 012003 doi:10.1088/1742-6596/1021/1/012003

5. References

[1] European Technical Working Group on ADS, "A European Roadmap for Developing Accelerator Driven Systems (ADS) for Nuclear Waste Incineration," ENEA, 2001.

[2] CERN, "The History of CERN," http://timeline.web.cern.ch/timelines/The-history-of-CERN, 2017.

[3] European Spallation Source, https://europeanspallationsource.se/ess-story/, 2017.

[4] G. Brumfiel, "Sweden snares neutron facility," *Nature News,* 2009.

[5] S. Peggs, "ESS Technical Design Report," ESS-doc-274, 2013.

[6] R. Andersson, A. Nordt, E. Bargalló and E. Adli, "Machine Protection Systems and their Impact on Beam Availability and Accelerator Reliability," *IPAC2015,* no. MOPTY044, 2015.

[7] R. Andersson, C. Hilbes, E. Bargalló and A. Nordt, "ESS Machine Protection Risk Management Process," *ESS Document (ESS-0095000),* 2017.

[8] International Organization of Standardization, "ISO 31000: Risk management - Principles and guidelines," 2009.

[9] International Organization of Standardization, "ISO 16085: Systems and software engineering - Life cycle processes - Risk management," 2006.

[10] International Electrotechnical Committee, "IEC 61508:2010," 2010.

[11] International Electrotechnical Committee, "IEC 61511: Functional safety - Safety instrumented systems for the process industry sector," 2004.

[12] C. Hilbes and A. Nordt, "ESS Machine Protection Functional Architecture Concept," *ESS Document (ESS-0035388),* 2015.

[13] E. Bargalló, "ESS reliability and availability requirements," *ESS Document (ESS-0008886),* 2015.

[14] N. Leveson, "Engineering a Safer World: Systems Thinking Applied to Safety," 2011.

The Frascati Neutron Generator: A multipurpose facility for physics and engineering

A. Pietropaolo[1], F. Andreoli[1], M. Angelone[1], U. Besi Vetrella[1], S. Fiore[1], S. Loreti[1], G. Pagano[1], R. Pilotti[1,2] and M. Pillon[1]

1- ENEA Department of Fusion and Technologies for Nuclear Safety and Security, Via Enrico Fermi 45, I-00044 Frascati (Roma), Italy
2- Department of Industrial Engineering, Università degli studi di Roma Tor Vergata Via del Politecnico 1, I-00133 Roma, Italy

*antonino.pietropaolo@enea.it

Abstract. The Frascati Neutron Generator (FNG) started operations in November 1992 making available to EU fusion community 14 MeV neutrons at an emission rate of 10^{11} s^{-1}. The fusion community strongly supported the idea of international collaboration in neutronics experiments, data base and code improvement, development of (new) experimental techniques and detectors. FNG was designed and built in ENEA that also operates the machine at its own expenses. FNG is extensively used to perform benchmarks or mock-up experiments of interest to Fusion neutronics in turn validating nuclear data and codes. A number of activities in different fields within a series of collaborations are carried out at FNG so far.

1. The Frascati Neutron Generator facility

The Frascati Neutron Generator is a compact accelerator driven neutron source installed and operating at the ENEA Frascati Research Centre [1]. It relies on both Deuteron-Deuteron (D-D) and Deuteron-Tritium (D-T) fusion reactions, in turn producing almost monochromatic 2.5 MeV and 14 MeV neutrons with a maximum neutron emission rate of 10^9 s^{-1} and 10^{11} s^{-1}, respectively. Figure 1 shows a schematic layout of FNG, where the main components are clearly indicated.

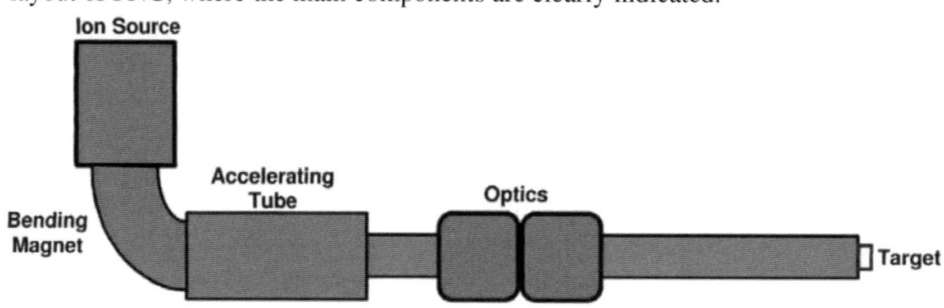

Figure 1: *Schematic layout of the FNG assembly. See Ref. [2] for details.*

Deuterons are accelerated by means of an electrostatic accelerator up to 300 kV and 1 mA current is delivered onto a Titanium target, loaded with Deuterium or Tritium, where D-D or D-T fusion reactions (namely $D+T \rightarrow n+{}^4He+17.6$ MeV and $D+D \rightarrow n+{}^3He+3.27$ MeV) take place.

In D-T mode, the source neutron emission rate is determined by means of an absolute measurement, the so-called associated alpha particle technique [2]. The D-T reaction produces in the final state a 14.1 MeV neutron and a 3.6 MeV alpha particle, respectively. A calibrated silicon detector, placed inside the drift vacuum tube at 2 m from the target and subtending a small and well defined solid angle to the interaction point, provides the absolute number of the 14.1 MeV neutrons produced at the target by counting the alpha particles from the reaction. Simulations, benchmarked by means of experimental measurements, provide a neutron yield and flux determination within an uncertainty of 3% [1]. In D-D mode the 2.5 MeV neutron emission rate is determined using foil activation technique, where a ^{115}In is irradiated and the gamma activation following the ^{115}In(n,n')^{115}In^{m} reaction channel is measured by means of an absolutely energy calibrated High-Purity Germanium detector. The D-D neutron emission is measured at ±7% uncertainty. Amongst the D-T compact sources in the world [3], FNG is one of the most intense and this feature, together with its well defined emission rate, make the facility appealing for very accurate measurements. In the following section the main FNG features are presented and discussed as well as a brief overview of the main research and development activities.

1.1. Neutron Spectra, flux and spatial distribution

Figure 2 shows the simulated neutron spectra at different neutron emission angles and the iso-flux contours, as obtained by means of MCNP Monte Carlo code, where an "ad hoc" source routine was implemented. This routine takes in to account all the physical aspects of the neutron production from the beam-target interaction, including ions scattering and fusion cross-section anisotropy. In the MCNP geometry, the target holder and the used detectors dimensions and materials are also described. The Source routine was developed by ENEA Frascati and Institut Jozef Stefan (JSI) in Ljubljana and is available in SINBAD database at NEA.

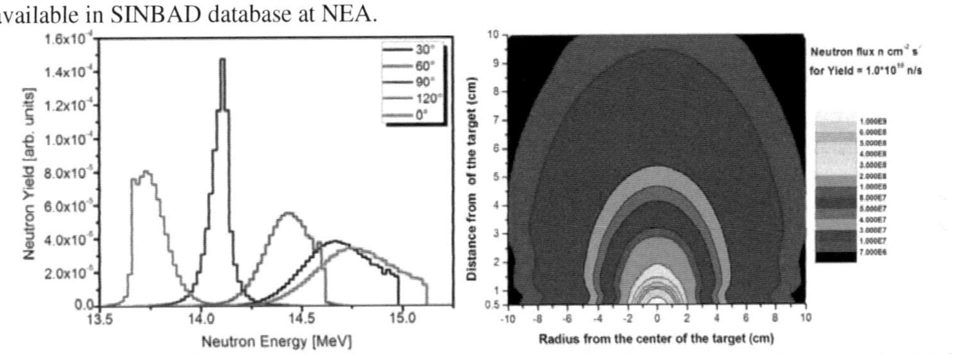

Figure 2: *(left) examples of FNG neutron spectra as a function of the neutron emission angle; (right) iso-flux contours of FNG.*

As it can be noted, the neutron spectrum is almost monochromatic, the neutron energy depending on the neutron emission angle, measured with respect to the deuteron flight direction inside the accelerator tube. Spectrum broadening is angle-dependent as well, being the narrowest at θ=90°. The overall broadening of the spectrum a the different angles depend on kinematics, on the deuteron straggling inside the tritiated/deuterated target and on the velocity distribution of the accelerated deuterons. Intensity neutron profiles were measured by means of a Single Crystal Diamond (SCD) detector. Exploiting (most effectively) the inelastic reactions ^{12}C(n,α)^{9}Be occurring in diamond when neutron energy is beyond the reaction threshold (E_{th}~5.6 MeV), it is possible to make spectroscopic measurements at different angles thank to the shift in neutron energy induced by the kinematics of the D-T reactions. Indeed, for 2.5 MeV neutrons the energy is below E_{th} so that intensities profiles are less effective with diamond. The SCD was mounted on a goniometric arm 25 cm in length and fixed on a stepper motor allowing for a fine remote control of the detector position. The SCD signal was preamplified by a ORTEC 142A and supplied with a High Voltage of 300 V by a CAEN high voltage module. Each pulse was recorded by means of a CAEN Digitizer DT5724 and both amplitude and the time of the event (featuring a resolution of 10 ns) were stored in a data file using the "list mode" acquisition. The spectra were registered for 600 s moving the SCD in steps of 5° from the position of θ= -120° to θ=120°. Left panel in figure 3, shows as a contour plot the evolution of spectra shape as a function of the detector angular position. Is evident the sharpness and the shift toward lower energy of the $^{12}C(n,\alpha)^{9}Be$ peak at high angle. The regions with depressed intensities (namely around θ =±90°) punt in evidence the effect of the materials traversed by the neutron on the detector's line of sight (mostly the metallic target). In the central panel of figure 3, is shown a 3D representation of the pulse height spectra recorded at different detector angular position in the first half of the angular scan. The peak around ADC-chn 4200 is the $^{12}C(n,\alpha)^{9}Be$ reaction signal. It can be noted as moving from θ=0° to θ=120°, the peak region is higher in intensity and narrower in energy distribution, reflecting the spectrum narrowing already shown in figure 2. In the right panel of figure 3, is shown the integrated counts (sum over the whole ADC scale) projected onto the axis labeling the detector angular position. The two region where the intensity in heavily depressed around θ=0 and 90° are the regions where neutrons intercept the materials of the cooling system around the fusion target. The intensity variation observed between θ=0 and θ=120° (almost symmetric for negative angles) is likely due to detector passing over regions (see contour plot in figure 2) characterized by different neutron fluxes.

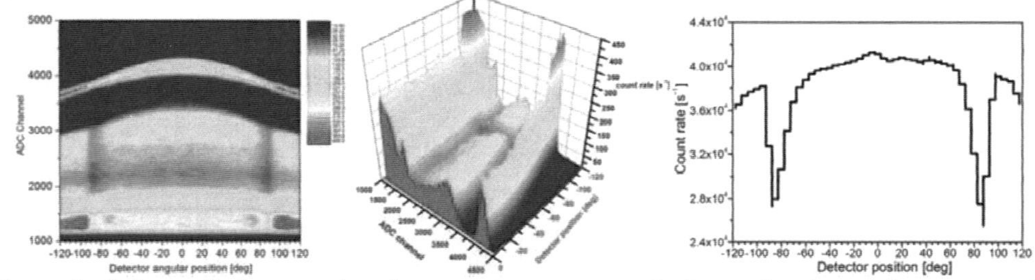

Figure 3: (left panel) *Contour plot of neutron iso-count rate loci as a function of detector angular position and ADC channel.(central panel) 3D plot representing the distribution of detector count rate as a function of detector angular position and ADC channel; (right panel) Integral count rate obtained by summing over the whole set of ADC channel as a function of the detector angular position.*

2. Overview on main FNG research activities performed and future perspectives

FNG can be considered a multipurpose compact neutron source. The activities performed so far include, referring to Fusion related activities:

1. Benchmark experiments for ITER: Stainless steel bulk shielding, Nuclear heating, Shut down dose rate, validation of nuclear data libraries and for Advanced Materials [4,5];
2. Fast neutron detectors development for the ITER experiment [6-8].

Other activities not related to fusion research

3. Electronics: testing and chip irradiation [9];
4. Engineering: Characterization of oil extraction tubes;
5. Biology: In-vivo mice irradiation to study DNA damaging [10].
6. Medicals: Radioisotopes production using 14 MeV neutrons [11];
7. Study of moderation systems for 14 MeV monochromatic beams [12]

Future activities are foreseen, such as:

8. Further development of fast and thermal neutron detectors without ^{3}He [13-15];
9. Production of innovative medical radioisotopes;
10. Improvement and upgrade of the source.

As far as radioisotopes production is concerned, it is worth mentioning that an experimental campaign [16] was devoted to the production of ^{99}Mo/^{99m}Tc exploiting the reaction channel $^{100}Mo(n,2n)^{99}Mo$ that exhibits a maximum in the cross section value around 14 MeV. This reaction channel seems exploitable for the intensive production of ^{99m}Tc radiopharmaceutical, intensively used worldwide for Single Photon Emission Computed Tomography, at present produced in research (aged) fission reactors. This study is important to assess the viability and effectiveness of this alternative route to ^{99}Mo/^{99m}Tc production.

The upgrade project under study for FNG is devoted to increase the neutron emission rate by almost an order of magnitude. This will imply a modification of the ion beam production mechanism that will likely rely on a radiofrequency ion source to obtain a more clean D^+ beam. A new HV station will be used that allow at achieving up to 10 mA D^+ current at a energy around 270 keV. In order to cope with target temperature and power dissipation a rotating target is foreseen following the experience of RTNS-I [17]. This upgrade will allow more rapid and effective irradiation experiment as well as the possibility to amply the possible applications of 14 MeV neutron beams for industrial applications using both 2.5 and 14 MeV enhanced available fluxes. These applications may include for example fast neutron radiography, and tests on radioisotopes such as ^{90}Y and ^{64}Cu, in alternative way to fission reactors.

Furthermore, following the new moderator concept developed for sealed fast neutron source (HOTNES facility [18]) FNG may produce thermal neutron beams over a large area (in the order of 10^3 cm^2) with high degree of homogeneity for testing of large device of simultaneous irradiation of small sensors. The 2.5 MeV neutron beam available at FNG may be used also as testing beam for moderator development in relation to new and upgrading large scale facilities, while the moderated beam may be useful for development/testing and performance assessment of instrumentation for neutron beam lines operating at reactors and/or accelerator driven neutron sources.

References

[1] M. Martone, M. Angelone, M. Pillon, J. Nucl. Mat. **212-215**, 1661 (1994); M. Angelone et al., Rev. Sci. Instr. **67**, 2189 (1996); www.fusione.enea.it/LABORATORIES/Tec/FNG.html.en
[2] E. Rhodes, C.E. Dickerman, A. De Volpi, C.W. Peters, IEEE Trans. Nucl. Sci. **39**, 1041 (1992).
[3] I. Anderson et al. Physics Reports **654,** 1 (2016).
[4] M. Angelone et. al., Fus. Eng. Des. **109–111**, 843 (2016).
[5] F. Moro et. Al. , Fus. Eng. Des. **84**, 1351 (2009).
[6] R. Pilotti et. al. J. Instrum., **11**, C06008 (2016).
[7] R. Pilotti et al.; Europhys. Lett. **116**, 42001 (2016).
[8] B. Esposito et al., Nucl. Instr.Meth. A **741,** 196 (2014).
[9] M. Angelone et al.; Nucl. Instr. Meth. A **623,** 921 (2010).
[10] A. Pietropaolo et al. Journal of Physics: Conference Series **746**, 012037 (2016).
[11] M. Capogni, A. Pietropaolo, L. Quintieri, Internal report **RT/2016/32/ENEA** (http://openarchive.enea.it/bitstream/handle/10840/8208/RT-2016-32-ENEA.pdf?sequence=1.)
[12] D. Flammini et al.. Appl. Rad. Isot. **125**, 129 (2017).
[13] V. Merlo et al., Appl. Phys. Lett. **106**, 113502 (2015).
[14] A. Pietropaolo et al., Nucl. Instr. Meth. A **610**, 677 (2009).
[15] C. Cazaniga et al., Nucl. Instr. Meth. A **778,** 20 (2015).
[16] M. Capogni et al.; accepted for publication on Appl. Rad. Is. (October 6th 2017).
[17] C.M. Logan, D.W. Heikkinen, Nucl. Instrum. Methods **200**, 105 (1982).
[18] R. Bedogni et al. Nucl. Instr. Meth. A **843**, 18 (2017).

IOP Conf. Series: Journal of Physics: Conf. Series **1021** (2018) 012005 doi:10.1088/1742-6596/1021/1/012005

Design of multi-pinhole collimator system for SANS based on CPHS neutron source

Zhiyuan Wang, Huarui Wu, Xuewu Wang
1.Department of Engineering Physics, Tsinghua Unversity, Beijing, China
2.Key Laboratory of Particle Radiation Imaging Tsinghua University, Ministry of Education, Beijing, China.

E-mail: wangzy_thu@aliyun.com

Abstract. A new design for Small Angle Neutron Scattering instrument based on utilization of multi-pinhole collimator is presented. The 7-pinhole structure has been proposed for 30 years by Glinka[1]. But due to the crosstalk issue, the SNR is not as high as traditional single pinhole system. Luckily, software nowadays helps us to analyze the system and optimize the parameters of the collimator. With the help of 7-pinhole collimator instead of single-pinhole, we can approach almost one magnitude higher flux than before, which is valuable especially in compact pulsed neutron source, whose flux is a big issue.

1. Introduction

Small angle neutron scattering technique has been widely used since 1970s in many fields of research including physics, biology, material science, chemistry and so on. It is used to investigate material structures in the size range of 0.5 to 500nm. In a standard SANS instrument, collimators of two or more apertures called single-pinhole system is used to get collimated neutron beam. However, in compact neutron source like CPHS (Compact Pulsed Hadron Source in Tsinghua University) whose neutron source flux is relative low compared with large facilities, as the neutron flux at sample determines whether or not the experiment's statistics are good, single-pinhole collimator is not a very good choice. To overcome the low neutron intensity from source, we are now designing a new type of collimation system called multi-pinhole collimator system, which contains several single-pinhole collimators focusing on the same position at detector. This multi-pinhole collimator idea was proposed by Glinka originally in 1986[1], and it has been used in many VSANS instruments which achieves a lower Q than traditional SANS while keeps the neutron flux not decreasing too much.[2,3,4] Besides multi-pinhole collimator structure, other types of collimator can be used to increase neutron flux at sample. For example, soller collimator, which was invented at IPNS.[5] This type of collimator uses blades of neutron absorption material to absorb high divergence neutron. However, as the surface of blades is quite long, the reflection noise is a big issue need to be overcome. Multi-channel collimator is similar with soller collimator, while it's collimated on radial direction, other than x/y direction separately.[6] In fact, soller collimator and multi-channel collimator can be regarded as a continuous multi-pinhole collimator. So it's valuable to analyze multi-pinhole collimator as a general case, while continuous one is the specific case.

There are many inherent problems that limit the multi-pinhole collimator system to replace

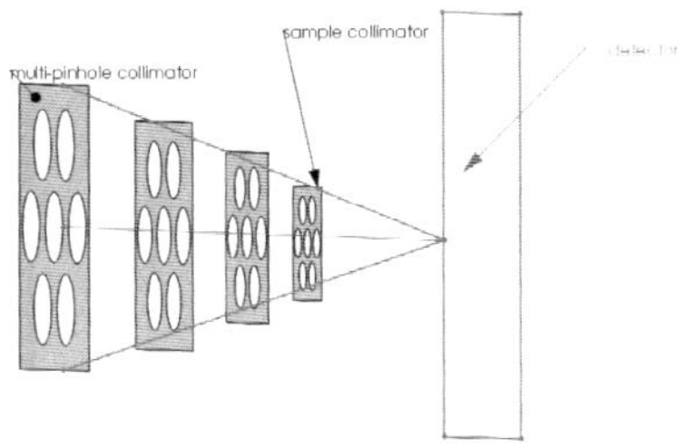

Figure 1. multi-pinhole collimator schematic diagram.

the usage of single-pinhole collimator system in traditional SANS instrument. The largest problem is cross-talk between each independent single-pinhole optic, and cross-talk can be produced by the placement of each collimator or parasitic scattering (including scattering from the thin layer, diffuse reflection, specular reflection and diffraction from the pinhole edge surface).[5] The cross-talk will lead to the uncollimated beam focus on the sample and thus reduce the SNR of the system. To make multi-pinhole system work, Glinka and Barker proposed to use beveled edge to prevent specular reflection and surface diffraction by low incident angle neutron, which can be called "open-angle" structure.[7] High neutron absorption material was used to reduce the thickness of the layer and thus the cross-talk problem can be diminished.[1]

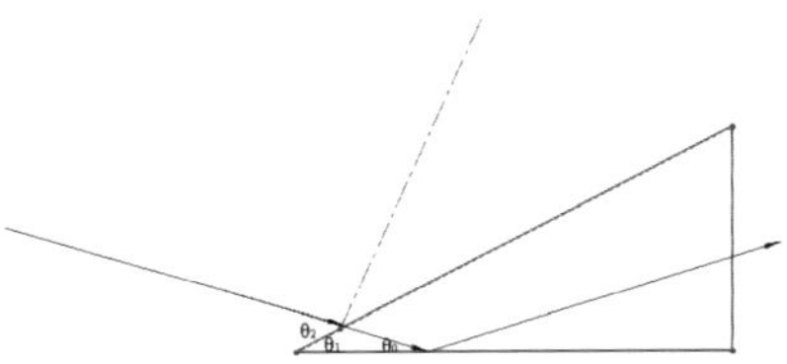

Figure 2. Open-angle structure reduces specular reflection.

2. Methods and Analysis

There are two main problems in multi-pinhole collimator system: crosstalk optics and specular reflection. For the crosstalk issue, the way to prevent is that either we can use smaller pinhole size to pinhole distance ratio, or we can use more collimator masks in the total beamline. Fig.2 shows the relationship of number of masks and pinhole size that won't have crosstalk. The tradeoff between number of masks and size of pinhole is in fact that the problem of the flux at sample and aligning work hardness.

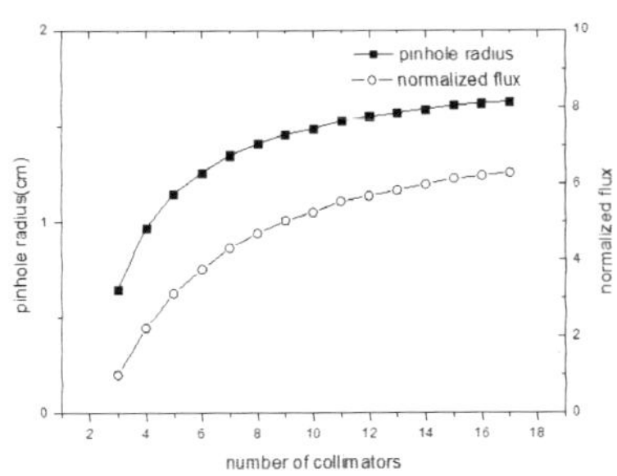

Figure 3. The crosstalk behavior for CPHS 7-pinhole collimator system.

Second problem is the specular reflection issue. The way to eliminate such noise is to add an open-angle structure to reduce reflectivity of neutron on the surface. The reflectivity expression is below:

$$k = sin^2\theta \frac{2\pi}{\rho_{SLD}\lambda^2} \tag{1}$$

$$r(\theta) = \frac{1}{(\sqrt{k}+\sqrt{k-1})^2} \tag{2}$$

Here θ is the incident angle, ρ_{SLD} is the scattering length density, λ is the neutron wavelength,$r(\theta)$ is the reflectivity.

By using the reflectivity expression, we can try to analyze the SNR of the system by adding different conditions, including absorbing material, neutron spectrum, neutron incident angle. Theoretically Signal to Noise result can be calculated by modeling. Figure 4 shows when considering specific neutron spectrum and neutron incident angle distribution, the SNR changes with the open-angle and thickness of collimator. By modeling, highest SNR can be found and thus can tell us what parameter can be used for the collimator.

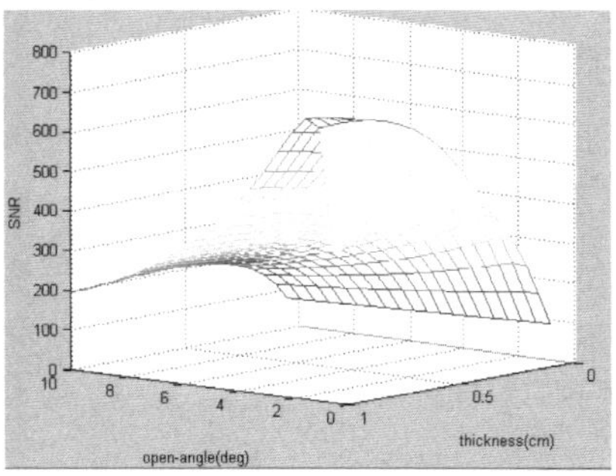

Figure 4. SNR of different open-angle and thickness for B4C material.

Mcstas simulation was also carried on to see the neutron optics for the system. Figure 5 shows when virtual source input was used, while 10 collimator masks were put in the beamline, we can see the focused neutron on the detector. By calculating the SNR from mcstas output, the correctness of modeling can also be tested.

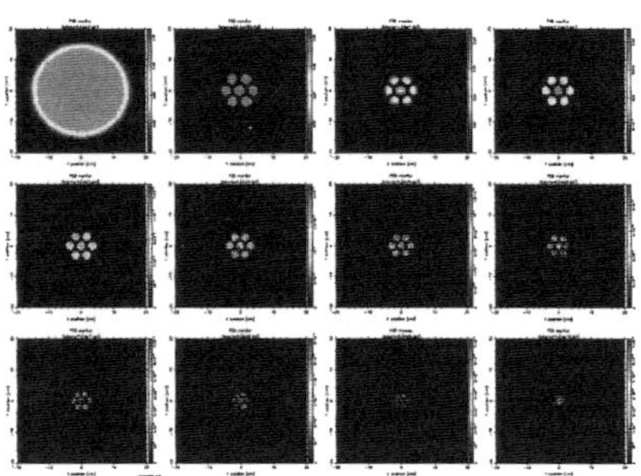

Figure 5. Mcstas simulation result.

3. Discussion

The choice of neutron absorption material is quite important. For solid material, Li-6, B-10, Cadmium and Gadolinium are most widely used. Table 1 shows the absorption cross section and gamma production of each material. It's obviously seen that gadolinium has highest absorption cross section, while it will produce high energy gamma. Oppositely, Li-6 has lowest absorption cross section, but it won't produce any gamma. As gamma can be part of noise in the beamline, it's worthy cutting down.

Table 1. The performance of different neutron absorption material.

Element	Absorption cross section for thermal neutron(barn)	Highest gamma energy(MeV)
Li-6	940	0
B-10	3836	0.478
Natural Cd	2520	9.05
Natural Gd	49700	8.89

To overcome this problem, John Barker in NCNR proposed to use "sandwich" structure to eliminate the gamma effect.[8] While using Li-6 as the first layer can reduce the effect of gamma production, gadolinium can be added as last layer to assure the neutron be absorbed, and borated aluminum alloy can be used to maintain the strength of the collimator.

4. Conclusion

Multi-pinhole collimator was used to increase neutron flux at sample while maintain the Q resolution of system, but compared with single pinhole collimator, it still has many problems need to be overcome. Luckily, Modeling and Monte Carlo simulation can be used to analyze the performance and optimize the parameters of the system. In the near future, the setup of multi-pinhole collimator for VSANS instrument will give us information to see the real performance of such system.

References

[1] Glinka C J, Rowe J M, LaRock J G. The small-angle neutron scattering spectrometer at the National Bureau of Standards[J]. Journal of applied crystallography, 1986, 19(6): 427-439.

[2] Desert S, Thevenot V, Oberdisse J, et al. The new very-small-angle neutron scattering spectrometer at Laboratoire Lon Brillouin[J]. Journal of Applied Crystallography, 2007, 40(s1): s471-s473.

[3] Brulet A, Thevenot V, Lairez D, et al. Toward a new lower limit for the minimum scattering vector on the very small angle neutron scattering spectrometer at Laboratoire Lon Brillouin[J]. Journal of Applied Crystallography, 2008, 41(1): 161-166.

[4] Grnzweig C, Hils T, Mhlbauer S, et al. Multiple small angle neutron scattering: A new two-dimensional ultrasmall angle neutron scattering technique[J]. Applied Physics Letters, 2007, 91(20): 203504.

[5] Thiyagarajan P, Epperson J E, Crawford R K, et al. The time-of-flight small-angle neutron diffractometer (SAD) at IPNS, Argonne National Laboratory[J]. Journal of applied crystallography, 1997, 30(3): 280-293.

[6] Falcao A N, Margaca F M A, Carvalho F G. Intensity and resolution effects in converging multichannel collimators for SANS by Monte Carlo simulation[J]. Journal of applied crystallography, 2003, 36(5): 1262-1265.

[7] Glinka C J, Barker J G, Mildner D F R. Comparison of pinhole collimation and focusing optics for SANS[J]. Nuclear Instruments and Methods in Physics Research Section A: Accelerators, Spectrometers, Detectors and Associated Equipment, 2015, 795: 122-127.

[8] Barker J G, (2013), unpublished memo.

IOP Conf. Series: Journal of Physics: Conf. Series **1021** (2018) 012006 doi:10.1088/1742-6596/1021/1/012006

Status Report on Accelerator and Neutron Activities of CPHS at Tsinghua University

X. Wang[1,2,*], Q. Xing[1,2], S. Zheng[1,2], Y. Yang[1,2], H. Gong[1,2], Y. Xiao[1,2], H. Wu[1,2], Z. Wang[1,2], Z. Fang[1,2], Z. Jiang[1,2], X. Guan[1,2]
[1] Key Lab of Particle & Radiation Imaging (Tsinghua Univ.), Ministry of Education
[2] Department of Engineering Physics, Tsinghua University, Beijing 100084, China
**wangxuewu@tsinghua.edu.cn*

Abstract. The CPHS project that was launched in September 2009 at Tsinghua University has reached a first commissioning stage in conjunction with ongoing activities to fulfil the eventual design goal of a ~10^{13} n/s epithermal-to-cold neutron yield for education, instrumentation development, and industrial applications. The latest progress on the commissioning and applications of 3-MeV proton and neutron beam lines, and the design, fabrication, engineering of the 13MeV/16kW proton accelerator system and the solid methane moderator system will be reported. Two prototypes of neutron detectors, based on micro-channel plate and the boron-coated straw tube design, were under evaluation and development. The research progress on neutron optics and instrumentation will also be presented.

1. Status of CPHS facility

1.1 Commissioning and applications of 3-MeV CPHS

The Compact Pulsed Hadron Source (CPHS) [1][2][3][4] facility with 3MeV RFQ had operated for about 700 hours in 2016. The present main parameters together with the designed values of the proton beam are listed in Table 1. The applications are shown in Figure 1. About half of the operation was for testing of two prototypes of neutron detectors under development, quarter of that was for neutronics performance measurement and neutron imaging beam line evaluation, and the other quarter was for the proton beam measurement.

Table 1. Main parameters of proton beam bombarding the target

Parameter	Designed Value	Present Value
Beam Energy (MeV)	13	3
Peak Current (mA)	50	28
Beam Pulse Width (μs)	500	100
Repetition Rate (Hz)	50	20

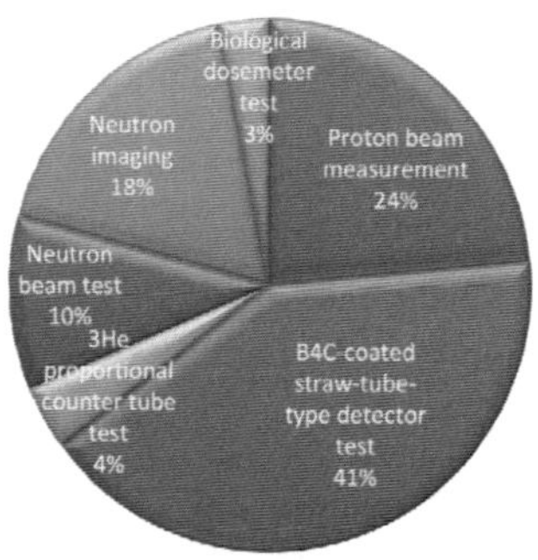

Figure 1. CPHS applications in 2016

1.2 Progress of upgrading to 13-MeV

The energy of the proton beam will be enhanced to the designed value of 13 MeV by the Drift Tube Linac (DTL) downstream the RFQ accelerator. The fabrication of the thirty-nine drift tubes and two cavities will be completed in 2017. The experiments of alignment and tuning based on one test cavity and ten test tubes have been carried out to facilitate the final assembly and cold test.

(a) (b) (c)

Figure 2. (a) The DTL cavity being fabricated, (b) a mounted PMQ and (c) a qualified drift tube.

The engineering design of solid-methane moderator of CPHS was finished by CSNS target group in March 2017, and its fabrication is expected to be finished by the end of 2017.

1.3 Beryllium target burned

The 1.2mm-thick beryllium target had been mounted with a 2mm-thick aluminium plate of high thermal conductivity and good mechanical property to avoid being broken. The aluminum was attached to the beryllium with thermal conductive adhesive. However, the neutron flux since 2016 was found to be only 18% of the simulation results. We found the target was cracked and burned after taking it out, which is shown in Figure 3. The main reason of cracking is the hydrogen deposition in target. To avoid cracking, we have designed a new beryllium target of 1.1mm backing vanadium proposed by Yamagata [5] and brazing by Ag-Cu.

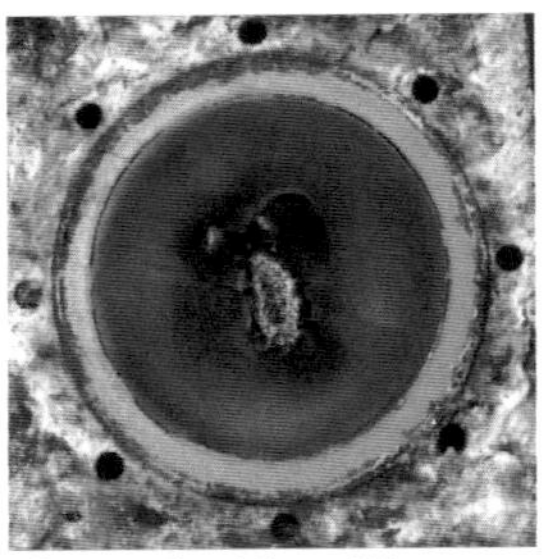

Figure 3. cracked beryllium target.

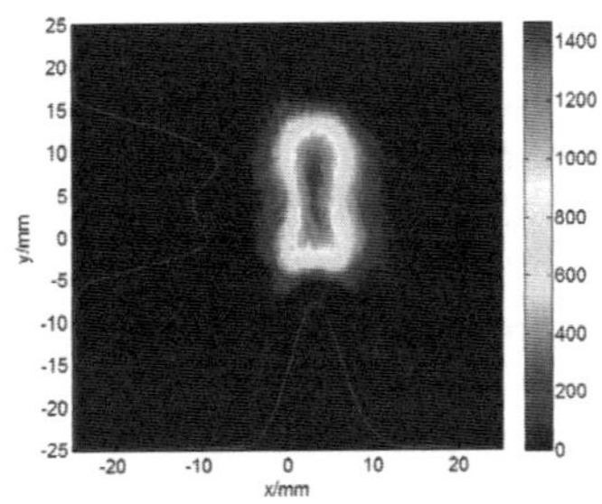

Figure 4. 2D profile of the proton beam.

1.4 Development of 2D Profile Measurement

2D profile measurement of the proton beam is under development by rotatable multi-wires based on CT algorithm, 20 carbon wires with the diameter of 30 μm are aligned and mounted on one board. With the position step of 0.1 mm and angle step of 5°, the 2D profile was acquired at the position of about six meters downstream the RFQ accelerator, as shown in Figure 4. The dynamic range of the measurement can reach $\sim 10^4$ in one-dimensional and $\sim 10^2$ in two-dimensional.

2. Development of neutron detectors

2.1 Boron coated tube array

An 800 mm × 800 mm array detector based on the 8mm-diameter boron coated tubes are expected to be built for CPHS-SANS instrument. A prototype module consisting of 64 tubes which stacked into 8 rows and 8 columns was developed as shown in Figure 5(a). Each tube is 800 mm in length and 8mm in diameter. A multiplex readout system was developed to decode the neutron hitting positions [6]. Its performance was studied on CPHS test beamline. The module was moved along its axis for 9 different positions. Position resolution for each tube of the prototype module were all measured, the average spatial resolution of 8mm × 8mm × 5.7mm was achieved and the worst axial resolution was 7.7 mm, which could still meet the CPHS-SANS' requirement.

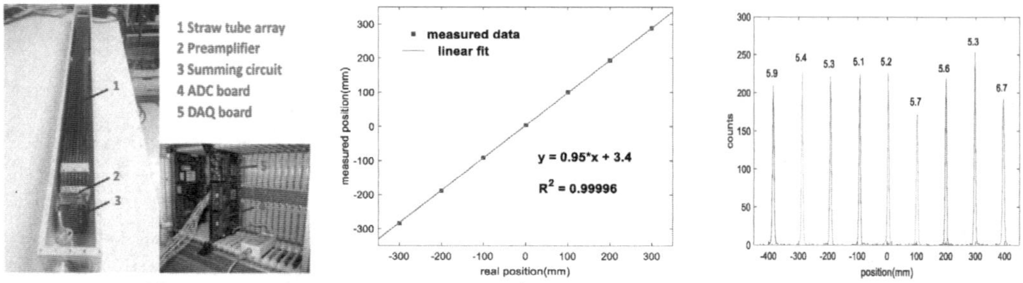

(a) (b) (c)

Figure 5. (a) Prototype module, (b) position linearity and (c) resolution of a single tube

2.2 Boron-lined honeycomb

A new detector design, as shown in figure 6(a) and proposed by Yang[7][8][9], removes the necessity of placing thousands of anode wires inside the boron lined gaseous detector, by introducing an electric field to drive the charged particles ionized electrons out to the electron multiplier, which is adjacent to the boron lined honeycomb structure. The boron lined honeycomb structure then only play the role of "neutron convertor", but not the role of "electron multiplication". The electron multiplier, which is used to replace the large number of anode wires, will collect the migrated electrons leaving from the honeycomb convertor and multiply them to form the neutron signals.

As shown in figure 6(c), this design can achieve the detection efficiency of more than 40%@25.3meV. A typical size of the detector cell is 10 cm ×10 cm ×10 cm, and many of the detector cells can be used to form a detector array to measure the neutrons scattered from the sample in the neutron scattering.

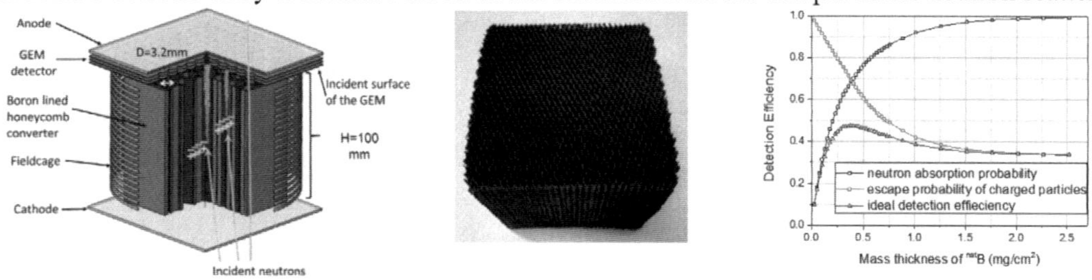

(a) (b) (c)

Figure 6. (a) The design of boron lined gaseous detector, (b) the honeycomb neutron convertor, coated with 0.9mg/cm^2 of ^{nat}B nano-powder and (c) The simulated detection efficiency of the detector.

2.3 Large area ^{n}MCP for neutron imaging

A large area (106 mm in diameter) neutron sensitive micro-channel plate (^{n}MCP) detector has been developed in Tsinghua[10][11][12][13], as shown in Figure 7. By doping 3 mole% of $^{nat}Gd_2O_3$, the ordinary MCP becomes neutron sensitive. The measured detection efficiency is 34%@25.3meV, and the spatial resolution could achieve 88 μm for thermal neutrons and 65 μm for low-energy X-rays.

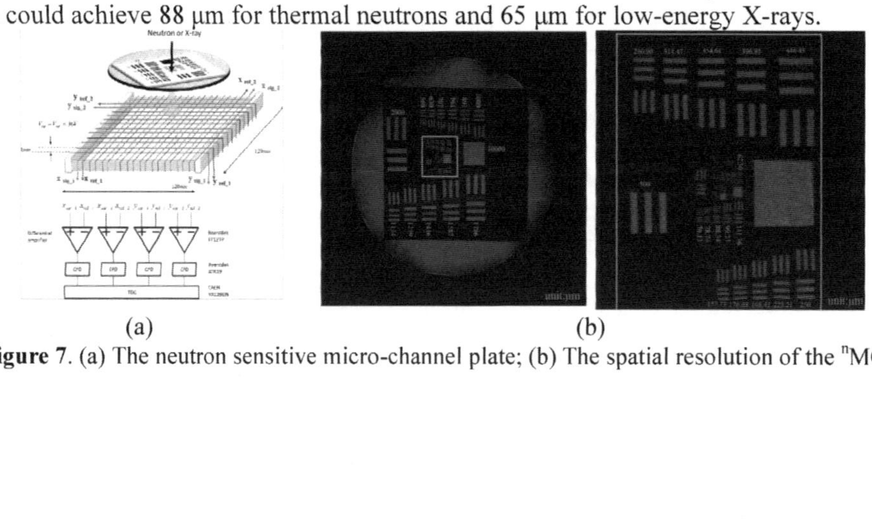

(a) (b)

Figure 7. (a) The neutron sensitive micro-channel plate; (b) The spatial resolution of the ^{n}MCP

3. Neutron optics for CPHS-SANS

3.1 Research on neutron focusing mirrors for SANS

Aiming at enabling a more practical SANS instrument, neutron focusing mirrors, as one of the potential devices to increase neutron flux, have been proposed [14][15]. A geometry of ellipsoid-shaped conical mirrors, which employs simple cones to take place of precisely curved ellipsoid, have been applied to the design of CPHS-SANS [16]. A geometrical design has been made for the test mirror as shown in Figure 8, which is a two-shell ellipsoid-shaped conical geometry with a magnification 0.6 and will be coated with m=2 supermirror. The radii of the two shells are about 42 mm and 39 mm respectively and each shell contains 4 segments with a length of 49 mm. Tongji University in China, which has the experience on X-ray telescopes [17], is fabricating the mirror.

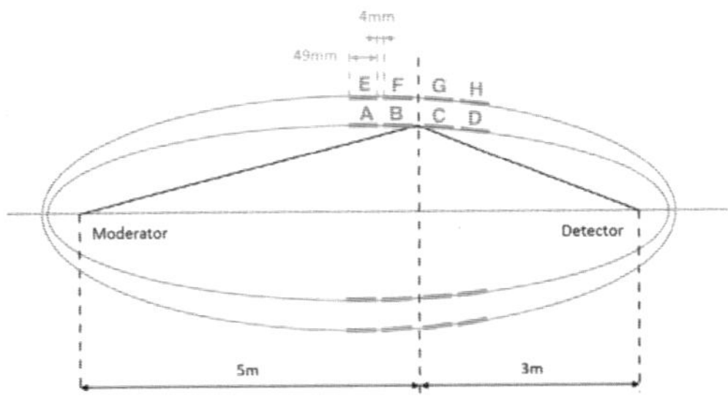

Figure 8. Schematic drawing of a two-shell ellipsoid-shaped conical mirror for test

3.2 Multi-pinhole design

There is a trade-off between flux and resolution in conventional pinhole system. Multi-pinhole collimator was proposed to increase the flux at sample while maintaining the *Q*-resolution [18]. McStas package was used to simulate and signal to noise ratio was also analyzed.

Acknowledge

This work is supported by the National Science Foundation of China (No. 11275106).

4. References

[1] Loong CK, Wei J, Guan X, and Wang X 2012 ***Physics Procedia*** 26: 8

[2] Wang X, Loong CK, Guan X and Du T 2014 ***Physics Procedia*** 60: 97

[3] Wang X, Xing Q, Loong CK, Guan X and Du T 2014 ***Physics Procedia*** 60: 186

[4] Wang X, Xing Q, Zheng S, Yang Y, et al. 2016 ***Nuovo Cimento C*** 38.

[5] Yamagata, Yutaka, et al. 2015 ***J Radioanal Nucl Chem.*** 305.3: 787-794.

[6] Yu H, Gong H, Li J, Wang Y and Wang X, et al 2015 ***Nucl Instrum Meth A*** 797:324-329

[7] Fang Z, Yang Y, Li Y, et al. 2017 ***IEEE T Nucl Sci*** DOI: 10.1109/TNS.2017.2670639

[8] Fang Z, Yang Y, Li Y and Wang X 2017. ***IEEE T Nucl Sci*** DOI: 10.1109/TNS.2017.2657655

[9] Fang Z, Yang Y and Li Y 2016 ***2015 IEEE NSS MIC.*** "The Research of High Detection Efficiency Boron Lined Detector with a Honeycomb Neutron Converter"

[10] Wang Y, Tian Y, Yang Y, Liu R, Pan J, Wang X and Zhang Z 2016 ***Chinese Phys C*** 40: 096004

[11] Wang Y, Yang Y, Wang X and Li Y 2015 ***Nucl Instrum Meth A*** 784: 226-231

[12] Tian Y, Yang Y, Pan J, Li Y and Li Y. 2014 ***Chinese Phys C*** vol 38: 086003

[13] Pan J, Yang Y, Tian Y, Zeng M, Deng T, W. Xu, et al. 2013 ***J Instrum*** 8: P01015

[14] Khaykovich B, et al. 2011 ***Nucl Instrum Meth A*** 631.1: 98-104.

[15] Furusaka M, et al. 2011 Activity Report on Neutron Scattering Research: Experimental Reports 18 (2011) Report Number: 1344

[16] Wu H and Wang W. 2016 ***2015 IEEE NSS MIC*** "Calculations on ellipsoidal and conical focusing mirrors for small angle neutron scattering".

[17] Mu B, et al. 2012 ***Chin Opt Lett*** 10.10: 103401

[18] Glinka C, Rowe J and LaRock J 1986 ***J Appl Crystallogr*** 19(6): 427-39

IOP Conf. Series: Journal of Physics: Conf. Series **1021** (2018) 012007 doi:10.1088/1742-6596/1021/1/012007

Performances of Neutron Scattering Spectrometers on a Compact Neutron Source

F Ott[1], H N Tran[3], X Fabrèges[1], A Menelle[1], N Chauvin[2], J Schwindling[2], A Letourneau[3], A Marchix[3], C Alba-Simionesco[1]

[1] Laboratoire Léon Brillouin, CEA, CNRS, Université Paris-Saclay, CEA Saclay 91191 Gif sur Yvette France
[2] IRFU/SACM, Université Paris-Saclay, CEA Saclay 91191 Gif sur Yvette France
[3] IRFU/SPhN, CEA, CNRS, Université Paris-Saclay, CEA Saclay 91191 Gif sur Yvette France

Abstract. There is currently a big effort put into the operation and construction of world class neutron scattering facilities (SNS and SNS-TS2 in the US, J-PARC in Japan, ESS in Europe, CSS in China, PIK in Russia). On the other hand, there exists a network of smaller neutron scattering facilities which play a key role in creating a large neutron scattering community who is able to efficiently use the existing facilities. With the foreseen closure of the ageing nuclear research reactors, especially in Europe, there is a risk of seeing a shrinking of the community able to use efficiently the world class facilities. There is thus a reflection being conducted in several countries for the replacement of smaller research reactors with low energy accelerator based sources. We consider here a reference design for a compact neutron source based on existing accelerator components. We estimate the performances of various types of neutron scattering instruments built around such a source. The results suggest that nowadays state of the art neutron scattering experiments could be successfully performed on such a compact source and that it is thus a viable replacement solution for neutron research reactors.

1. Introduction

Considering the political situation in Europe, it is very unlikely that new nuclear research reactors will be built in replacement of the old ones. However, currently aging facilities are providing a broad user base to the most performing ones. It is thus necessary to try to find a solution to ensure that the broad user base (6000 users [1]) can be maintained in Europe in order to make the best use of the most powerful sources. Any possibility to build neutron scattering facilities which could replace existing nuclear reactors would be welcome provided the investment is in the 100-200 M€ range which would make it affordable to a single country on par with a synchrotron or a power laser facility. During the last decade a number of groups have independently considered the possibility of operating a high current / low energy proton accelerator to produce thermal and cold neutrons [2-7]. A few facilities have actually been built and are operating scattering instruments [8]. A UCANS network gathering these groups has been created [9].

In this communication we consider the design of a compact source based on existing components (proton source, accelerator, target, moderator, neutron optics) and evaluate the performances of a range of neutron scattering instruments which could operate on such a source. We focus mainly on elastic scattering instruments since the flux requirements are less stringent and the modelling of elastic instruments is significantly simpler than inelastic instruments.

As a starting point we consider a source design based on existing components: a proton source (100 keV, 100mA), a RFQ accelerator able to operate at proton currents up to 100mA bringing the proton energy to 3.6 MeV and a DTL section boosting the proton energy to 20 MeV. We also consider a pulsed operation (coupled with Time-of-Flight spectrometers) since this operation mode allows optimizing the (neutron production / energy input) ratio. For the pulsed operation we consider a duty cycle of 4% which corresponds to a beam energy on the target of 80kW and an average proton current of 4 mA. We do not make any assumption about the repetition rate and pulse length. We consider Beryllium as the target material.

Monte-Carlo simulations as well as a few experimental data show that rather simple moderator combining polyethylene and a beryllium reflector can allows to achieve a thermal neutron brilliance on the order of $3x10^7$ n/s/cm²/sr/μA at the exit of the moderator. Note that these values are coherent with the values which can be extrapolated from the measurements of Allen et al [10] at 10 MeV – 30μA on the Birmingham Nuffield Cyclotron using a PE moderator. Assuming an average proton current of 4 mA this leads to a source brilliance of $1.2x10^{11}$ n/s/cm²/sr. A source which the above brilliance was used as an input for the simulation of instruments using the McStas software [11] which is a tool dedicated to the modelling of neutron scattering instruments.

2. Instruments designs

For each class of instruments we assume that the repetition rate and the pulse length can be freely optimized (within the 4% duty cycle power envelope). We consider that this makes sense since we are trying to evaluate the optimal performances of different types of instruments.

2.1. Reflectivity

It can be considered that wavelength resolutions $\delta\lambda/\lambda$ up to 10% are perfectly usable for reflectivity. Let us assume a source operating with pulses of length 2 ms and with an operation frequency of 20 Hz. This corresponds to a duty cycle of 4% and is very close to the ESS neutron pulse structure. On a pulsed source, the wavelength resolution degrades very quickly for short wavelengths. For pulse widths of w=2 ms and a flight path L=16m, the usable wavelength band ranges from 3 to 12 Å which matches a cold neutron spectrum, with a wavelength resolution of 12% at the peak flux (4 Å). The resolution improves quickly to 7% at 8 Å. A shorter instrument would degrade the wavelength resolution but would also increase the cut-off wavelength to 16Å which might be usable in reflectivity experiments. A longer instrument (24 m) would provide a better wavelength resolution at the expense of a narrower wavelength bandwidth (λ_{max} = 8 Å). While it seems acceptable to work in the (L=16 m, w=2 ms) configuration even though the resolution is as low as 15% at 3 Å, there are a number of situations where there is an interest in improving the resolution. Besides it is somewhat unsatisfactory to have a resolution varying from 15% to 4% over the bandwidth. A simple improvement such as a double disk-chopper [12] right at the moderator exit would (i) improve the wavelength resolution at short wavelengths (2) cut-off neutron pulse tails. Note that if the double disk chopper solution is used, due to the versatility of the system, a shorter instrument could be built (typically 12m) so that the resolution could be set at 10% below 6Å with the double disk chopper and improved down to 5% at 13Å. This would roughly match the POLREF@ISIS settings [13]. In order to avoid a direct view from the moderator, since the instrument would be very short, a neutron mirror (e.g. m=6, length 500mm, height 20mm) could be used. The detector position (12m from the source) would be offset by more than 40cm from the direct view of the moderator.

We performed Monte Carlo simulations using McStas and the following design. From the moderator: straight guide of length 8m with m = 4, cross section 100x50mm²; a 2 m long collimator with F1 = 2 mm and F2 = 2 mm and a side guide with m = 4; a detector at 2 m from the sample position. The neutron flux at the sample position is $6x10^6$ n/cm²/s which is on the order of CRISP@ISIS and HERMES@LLB. Hence it should be possible to operate a reflectometer with sufficient flux to perform useful science on the reference source design.

IOP Conf. Series: Journal of Physics: Conf. Series **1021** (2018) 012007 doi:10.1088/1742-6596/1021/1/012007

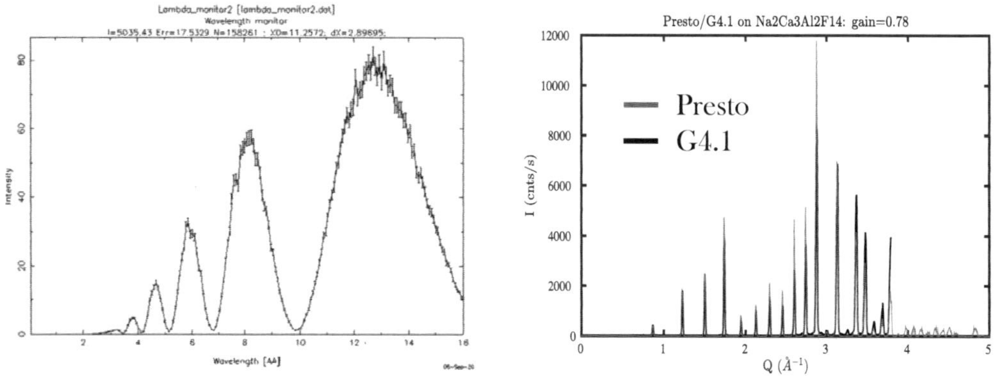

Figure 1: (a) McStas simulation of a 12m neutron reflectometer with a 10x5 cm² sample (Ni 20nm//Si) at an incidence angle of 3°. Wavelength spectrum measured in the specular reflectivity direction. (b) Comparison of the diffraction spectra measured on G41@LLB (λ = 2.43Å) and PRESTO (ToF). A wider Q range is accessible in ToF (above 4Å^{-1}).

2.2. Small Angle Scattering

SANS scattering instruments are workhouse instruments for neutron scattering in soft matter and metallurgy. Considering the implementation of a SANS instrument on a CNS, we may use as a starting point the fact that SANS instruments are low resolution instruments which can be nicely operated with a wavelength resolution $\delta\lambda/\lambda$ ranging from 10 up to 20%. For SANS measurements, a wide lambda range is desirable to cover a Q range as large as possible. Hence a source with long pulses and a slow repetition rate is the most efficient way to operate. Let us again assume a source operating with pulses of length 2ms and with an operation frequency of 20 Hz (4% duty cycle). The wavelength resolution function is identical to the case of the reflectometer. Again, the configuration (L=12 m, w=2 ms) is rather favorable: (i) The wavelength band ranges from 3 to 16Å and matches perfectly a cold neutron spectrum with a wavelength resolution of 16% at the peak flux (4Å). The resolution improves quickly to 8% at 8 Å; (ii) for metallurgical studies, which require wavelengths above 5Å, the resolution is below 12% which is optimal. A shorter instrument would degrade the wavelength resolution and the wavelength band extension beyond 16 Å is realistically not usable. A longer instrument would provide a better wavelength resolution at the expense of a narrower wavelength bandwidth. Note that in general the total length of a SANS instrument can be varied by moving the detector closer or further from the sample position.

We suggest that an instrument with the following specifications would be perfectly suitable for SANS studies: (i) cold source; (ii) Source-Sample distance of 8 m. Sample – detector distance variable from 1 to 7m, total flight path from 9 to 15m; (iii) useful bandwidth from 3 Å to 16 Å. If one compares with the SANS2D@ISIS (f = 10 Hz, Source-sample = 19 m, Total length = 21 – 31m / Q_{min}~0.002 Å^{-1}, Q_{max}~3 Å^{-1}), the above specifications scale exactly as the operating frequency. At 20 Hz, an instrument should be twice as short as an instrument operating at 10 Hz for the same Q range. Monte-Carlo simulations for various configurations were performed. The results are summarized in Table 1. The flux are also on par with the PAXE instrument at the LLB [14]. The calculated flux are also on the order of the flux expected on SANS2D at ISIS TS2 (~10^6 cm^{-2} s^{-1}) [15].

Configuration	**L_g (m)**	**L_1 (m)**	**L_2 (m)**	**L_{tot} (m)**	**D_1 (mm)**	**D_2 (mm)**	**Flux (n/cm²/s)**
Small Q	1	7	7	15	20	16	0.7x10^6
Medium Q	4	4	4	12	20	16	2.2x10^6
High Q	6	2	1	9	20	16	6.7x10^6
PAXE [36]	6	2.5	2.5	11	12	8	0.7x10^6

Table 1: Flux at the sample position for various SANS configurations (small, medium, high Q). L_g being the length of the guide from the source to the first pin-hole, L_1 being the length of the collimation and L_2 being the sample-detector distance. D_1 and D_2 are the sizes of the holes at the entrance and exit of the collimator.

2.3. Powder diffraction

Power diffraction instruments are also workhorses in neutron diffraction. High resolution instruments are used for structure determination. High flux instruments are used to study phase transitions. Efficient powder diffraction in Time-of-Flight mode requires a large coverage of the space with detectors which turns into a rather expensive instrument (contrary to monochromatic instruments).
We consider a design for a powder spectrometer aiming at replacing G41@LLB (low resolution - high flux). The G41 instrument [16] is operating with a graphite monochromator on a cold neutron guide. The horizontal divergence is 0.3° and the vertical divergence is 3°. The detector is covering a solid angle of 80° x 3°. The neutron flux at $\lambda = 2.43$ Å is $4x10^6$ n/cm²/s at the sample position. We consider the following design for a powder diffractometer: (i) cold source, (ii) pulse width w = 250μs, f = 40 Hz (1% duty cycle), (iii) Source-sample distance 52m, (iv) sample-detector distance of 1m, total flight path of 53m. This leads to a usable bandwidth ranging from 1.4Å to 3.3Å with $\Delta\lambda/\lambda$ ranging from 1.3% to 0.5%. The horizontal and vertical divergence are set to 0.6°. The flux on the sample would be $1.5x10^6$ n/cm²/s (~70% of G41. By using the 7C2@LLB detector which is covering a solid angle of 120°x20° with a detection efficiency of 90% a gain of a factor 20 would be obtained. This shows that a powder instrument on a CNS can easily outperform existing very productive instruments at neutron reactor facilities.

3. Discussion

From our simulations, we think that it should be possible to build a compact neutron source with performances in the range of ORPHEE / ISIS TS2. While the performances of such a source would not match the performances of ESS, it could still serve the need of about 50-75% of the users who do not need extreme neutron flux. Such sources could successfully replace aging neutron reactors as well as lower the barrier for new entrant countries in the field of neutron scattering.

Technique	Flux on sample	Reference spectrometer	Potential gains
SANS	$1x10^6$ n/s/cm² (low Q) $3x10^6$ n/s/cm² (med Q) $9x10^6$ n/s/cm² (high Q)	PAXE@LLB (low Q) $0.7x10^6$ n/s/cm² SANS2D@ISIS $1x10^6$ n/s/cm²	Slit setup x10 Focusing optics for VSANS (small Q) x10
Reflectivity	$6x10^6$ n/s/cm²	HERMES@LLB $8x10^6$ n/s/cm² POLREF@ISIS ~$1x10^7$ n/s/cm²	SELENE concept x10 [17] Advanced Deconvolution x3
Low resolution powder diffraction	$1.5x10^6$ n/s/cm²	G41@LLB $2x10^6$ n/s/cm²	Large solid angle detector (7C2 type) x20

Table 2: Comparison of the performances of various types of instruments on a compact source with existing instruments.

References

[1] ESFRI Report, Neutron scattering facilities in Europe, Present status and future perspectives
[2] Lavelle C M 2008 Nuclear Instr. And Methods A **587** 324
[3] Halfon S, Arenshtam A, Kijel D, et al 2015 Appl. Radiat. Isotopes **106** Special Issue 57-62.
[4] Rücker U, Cronert T, Voigt T et al 2016 Eur. Phys. J. Plus **131** 19.
[5] Wang X et al 2014 Physics Procedia **60** 186-192.
[6] Yamagata Y, Hirota K, Ju J et al. 2015 J Radioanal. Nucl. Chem. **305** 787-794.
[7] Sordo F et al, 2013 Nuclear Instruments and Methods in Physics Research A **707** 1–8
[8] LENS, www.indiana.edu/~lens/sans.html
[9] UCANS, www.ucans.org/facilities.html
[10] Allen D A, Hawkesworth M R, Beynon T D et al, 1994 Nucl. Instr. Methods A **353** 128-133.
[11] McStas, www.mcstas.org/
[12] A.A. van Well, Physica B 180&181 (1992) 959-961.
[13] POLREF, http://www.isis.stfc.ac.uk/instruments/polref/polref3024.html
[14] PAXE@LLB, http://www-llb.cea.fr/spectros/spectro/g5-4.html
[15] SANS2D@ISIS, http://www.isis.stfc.ac.uk/instruments/sans2d/sans2d3000.html
[16] G41, http://www-llb.cea.fr/spectros/spectro/g4-1.html
[17] ESS Instrument Construction Proposal, ESTIA focusing reflectometer https://europeanspallationsource.se/sites/default/files/estia_proposal.pdf

IOP Conf. Series: Journal of Physics: Conf. Series **1021** (2018) 012008 doi:10.1088/1742-6596/1021/1/012008

Validation of Geant4 simulation tool for low energy proton induced reactions

H.N. Tran[1], A. Marchix[1], A. Letourneau[1], N. Chauvin[1], A. Menelle[2], F. Ott[2], J. Schwindling[1]

[1]IRFU, CEA, Université Paris-Saclay, F-91191 Gif-sur-Yvette, France
[2]LLB, CEA, CNRS, Université Paris-Saclay, F-91191 Gif-sur-Yvette, France

Email: ngoc-hoang.tran@cea.fr

Abstract. With the development of high-intensity accelerators, it is now envisaged to build compact high-intensity neutron sources for various applications. At the CEA Saclay, we have started a project to promote such technology based on the existing 100 mA IPHI proton injector. We are focusing our interest on low-energy proton-induced reactions (less than 20 MeV) to provide an economical way to produce neutrons and also to limit the radiological constraints. Optimization studies of the target-moderator assembly require validated simulation tools in order to maximize the neutron flux. In this document, the development status of the GEANT4-based simulation tool is presented, as well as its validation by comparison with gold foil activation measurements on the IPHI proton injector. The experimental setup and the simulation procedure are described. A good agreement is obtained between experimental data and simulation results when considering angular distribution of primary neutrons provided by Howard and Yanch's measurements.

1. Introduction. The coupling of high-intensity accelerator technology and the concept of compact neutron source permits to envisage a large panel of applications, such as solid state physics studies with neutron scattering experiments, nuclear data measurements for nuclear industry, neutron radiography for industry materials qualification and also medical purposes like isotope production and neutron-capture therapy. Nevertheless, when dealing with high-intensity proton beams, many challenges appear. First from the accelerator side, strong space charges can induce emittance growth, beam halo and beam losses that have to be controlled. Then from the target/moderator side, heat removal is the main challenge for the target and the limiting parameter for beam intensity. At CEA Saclay, in France, we have started a project to promote such technology based on the existing 100 mA IPHI proton injector developed at IRFU within a CEA – CERN – CNRS collaboration [1]. We are focusing our interest on low-energy proton-induced reactions (less than 20 MeV) on beryllium (Be), aiming to provide an economical way to produce neutrons and also to limit radiological constraints. As most applications need moderated neutrons (less than 100 meV), the moderator/reflector assembly has to be optimized, in order to slow down primary fast-neutrons and to maximize the neutron flux on the sample, by using validated and predictive Monte-Carlo simulation tools (Geant4 version 10.02 [3]). In the following parts, we will present the development status of the GEANT4-based simulation tool and its validation by comparison with gold foil activation measurements.

2. Experimental validations of GEANT4 code. It is well-known that the efficiency of a moderator depends on the large neutron elastic cross sections of its constituents, a reduced neutron capture cross

IOP Conf. Series: Journal of Physics: Conf. Series **1021** (2018) 012008 doi:10.1088/1742-6596/1021/1/012008

section and a large energy loss per collision for the scattered neutrons. For neutron energies at a few MeV, hydrogenous materials are the best suited moderator, slowing down neutrons in a few collisions (about 20). With its high hydrogen concentration and easy shaping, polyethylene (PE) was considered as a potential material for a neutron moderator. In this section, we describe an experimental validation of simulation neutron transport based on a simple geometry of PE moderator.

2.1 Experimental setup. To validate the simulation tool in terms of neutron transportation and moderation in PE moderator, we have performed an experiment using the high-intensity proton injector (IPHI) which was installed at CEA-Saclay. It consists of a proton source emitting 100 keV protons (the SILHI source) [1]. The protons are then accelerated in a Radio Frequency Quadrupole up to an energy of 3 MeV. The accelerator was designed to operate in continuous mode with proton currents up to 100 mA, corresponding to a maximum power of 300 kW. For tests, the accelerator was operated in pulsed mode with pulses of length 100 μs and a repetition rate of 1 Hz to minimize the number of fast neutron produced and be able to perform time-of-flight measurements. Figure 1 shows the scheme of the experimental setup with a simple moderator made of polyethylene. The proton beam was impinging on a thin beryllium target of thickness 0.5 mm. The support was air cooled to remove the heat deposit, limited to a maximum of 10 W for this test. The proton beam size was limited to 16 mm in diameter. The proton current incident on the target ranged from 2.6 to 3.1 μA.

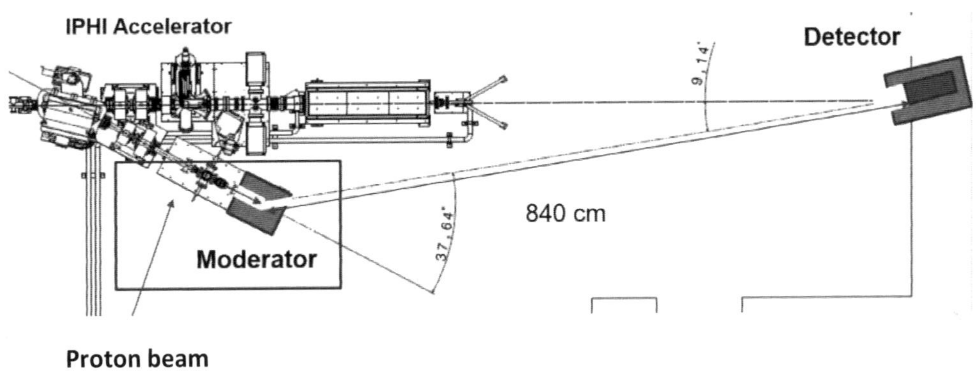

Figure 1: Scheme of the experimental set-up.

The beryllium target was located close to the centre of a polyethylene (PE) moderator box (300x300x400 mm^3) (see Figure 2). A 20 mm diameter hole was drilled through the moderator at 38 degree with proton beam axis (see Figure 2). To measure the neutron flux inside the moderator, 0.112 g gold foils, in the form of 6 mm diameter discs, were set at two positions inside the moderator (at 70 mm and 210 mm from the intersection of the proton beam axis with the neutron guide one). The foils were then irradiated for durations ranging from 15 to 22 minutes and a proton beam current of about 3 μA. ^{198}Au isotopes were formed from neutron capture on ^{197}Au and decayed by the emission of gamma-rays. The measuring of these latter leads to the number of ^{198}Au isotopes present in the foil after activation.

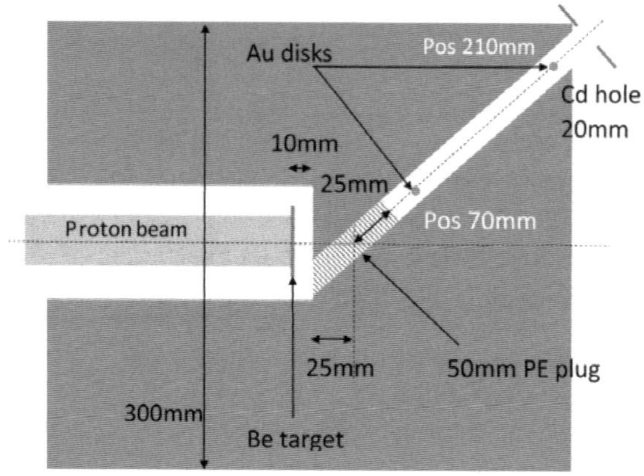

Figure 2: Polyethylene (blue) moderator.

2.2 GEANT4 simulations. In the simulations of the experiment, we compared the effect of the energy and angular distributions by defining the neutron source term at the target position either from nuclear data using the *particlehp* package (ENDF/B-VII nuclear data library [4]) or from the energy and angular distributions measured by Howard and Yanch [5] at 3.7 MeV or the angular distribution measured by Marion [6] at 3.06 MeV combined with the energy distribution of Howard and Yanch at 3.0 MeV (see details in [2] and [7]). The lack of experimental data for 3 MeV proton beam does not allow a direct comparison with simulations, as only the total neutron yield and the angular distribution are available. The experimental distributions were normalized to the total neutron yield measured by Porges *et al.* [8] at 3.05 MeV, i.e. $3.0x10^{11}$ n/mA. For the *particlehp* calculation, the calculated neutron yield gives $2.19x10^{11}$ n/mA, i.e. 27% less than the experimental data. The neutron transport was realized using the neutron-data library G4NDL4.5 based on the evaluated ENDF/B-VII nuclear data library [4]. The thermal scattering library was also included for neutron energies less than 4 eV, to take into account for molecular effects. The density of PE was set to 0.94 g/cm^3.

Gold foil activations were calculated along the cylinder hole with Geant4 and compared with experimental data (see Figure 3). The density of the gold discs was set to 19.320 g/cm^3. The calculation using the angular distribution of Marion shows an underestimation for the whole range of measurements (roughly a factor 2) whereas using Howard and Yanch's data a better agreement with experimental data is found but still below those ones. For this latter, an underestimation of about 30 % is found close to the target while the discrepancy reduces to about 20% close to the exit hole. It clearly shows the effect of the angular distribution as data from Howard and Yanch are mainly focused in the forward direction whereas Marion's one present a maximum near 120°. One may point out that the experimental uncertainties are of 20 %, leading to a fair agreement in between experimental data and simulation results. However, a determination of angular distribution and neutron yield at 3 MeV could confirm the results and lead to useful information for moderator geometry optimization.

From these results we can extrapolate the thermal neutron flux for an installation based on 100 mA and 20 MeV protons on beryllium target. A gain in the neutron yield of a factor 200 can be achieved by increasing the proton energy from 3 MeV to 20 MeV (see details in [2]). The flux of the simple PE moderator would be of about $5.7x10^6$ $n/cm^2/s$ (in pulsed mode with a typical duty cycle of the order of 4%) at a distance of 4 meters and a pinhole of size $5x5cm^2$, the geometry corresponding to an L/D radio of about 80 as routinely used on the radiography station G45 at the Orphée reactor. With an optimized moderator-reflector geometry, a flux of about $1.0x10^7$ $n/cm^2/s$ should be achievable. For neutron imaging applications, the ICON@SINQ beamline [10] provides a flux of about $1.3x10^7$ $n/cm^2/s$ and the IMAGINE station at Laboratoire Léon Brillouin [11] provides a flux of about $1.0x10^7$ $n/cm^2/s$. These latter flux values are obtained for high L/D collimation ratio of 350. However, for industrial screening applications, low L/D ratios (~80) are not detrimental to the spatial resolution of the measurements. In absolute values, an imaging station on a pulsed compact source would be about 20 times less bright for high L/D ratios than a continuous source. This is due to the fact that

IOP Conf. Series: Journal of Physics: Conf. Series **1021** (2018) 012008 doi:10.1088/1742-6596/1021/1/012008

radiography techniques cannot benefit from the neutron pulse time structure. On the other hand, these figures suggest that techniques which can benefit from the time structure should be as efficient on a compact neutron source as on a medium power reactor or a continuous spallation source.

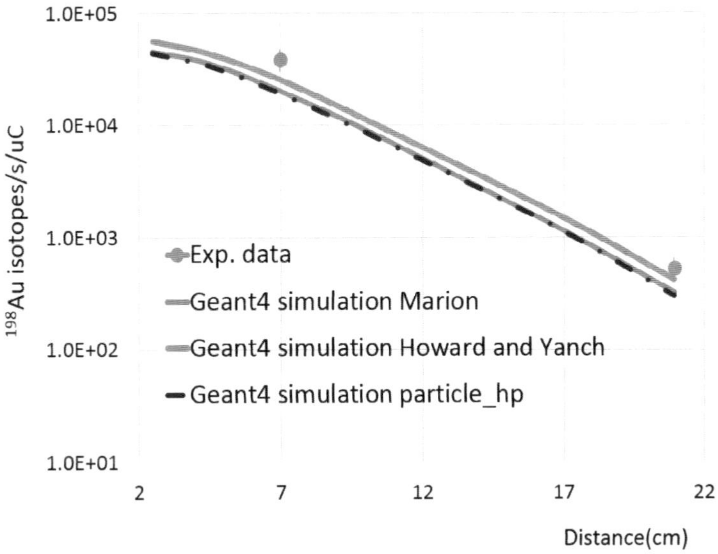

Figure 3: Comparison of gold activation as a function of distance from ^{9}Be target.

3. Conclusion. A good agreement was observed between experimental data and simulation results for the transportation and the moderation of low-energy neutrons in polyethylene moderator. This permits to validate the transportation and the moderation process performed in GEANT4 using G4NDL library. Simulations put into light a wide influence of the angular distribution of emitted neutrons in the ^{9}Be(p,n) reaction on the thermal neutron distribution inside the moderator. New measurements will be performed to confirm the angular distribution of the emitted neutrons for 3 MeV protons.

References

[1] R. Gobin, at al., (2002),"High intensity ECR ion source (H+, D+, H−)(H+, D+, H−) developments at CEA/Saclay" Rev. Sci. Instrum. 73 pp 922.
[2] A. Letourneau, et al., (2016), "Development of Compact Accelerator Neutron Source" , ND2016, 11-16/09/2016, Bruges, Belgium, accepted for publication in EPJ Web of Conference.
[3] J. Allison, et al. (2016), "Recent developments in Geant4", Nucl. Instr. and Meth. A, 835 pp 186–225.
[4] M.B. Chadwick, et al., (2006) "ENDF/B-VII.0: Next generation evaluated nuclear data library for nuclear science and technology", Nucl. Data Sheets 107 pp2931.
[5] W.B. Howard and J.C Yanch, (2001) "Measurement of the Thick-Target 9Be(p,n) Neutron Energy Spectra" Nuc. Sci. and Eng. 138 pp 145.
[6] J.B. Marion, (1956) "Excited States in B10", Phys. ReV. 103 pp 713.
[7] H.N. Tran, et al., (2016), "Validation of physics models in Geant4 for compact neutron source at low-energy proton beam" , Proceeding of SATIF-13 at Helmholtz-Zentrum Dresden-Rossendorf, 10-12 October 2016
[8] K. Porges, et al. (1971), "Argonne National Laboratory Progress Report", ANL7910 pp361.
[9] C.M.Lavelle et al., (2008) "Neutronic design and measured performance of the Low Energy Neutron Source (LENS) target moderator reflector assembly", Nucl. Instr. and Meth. A 587 324–341.
[10] A.P. Kacstner et al., (2011) "The ICON beamline–A facility for cold neutron imaging at SINQ", Nucl. Instr. and Meth. A 659 387–393
[11] A. Lagorce-Tachon et al., (2015) "The cork viewed from the inside", Journal of Food Engineering 149 pp 214–221

IOP Conf. Series: Journal of Physics: Conf. Series **1021** (2018) 012009 doi:10.1088/1742-6596/1021/1/012009

General use of low-dimensional moderators in neutron sources

L. Zanini[1], F. Mezei[1], K. Batkov[1], E. Klinkby[1,2] and A. Takibayev[1]

[1]European Spallation Source ESS ERIC, Box 176, S-221 00 Lund, Sweden

[2]Technical University of Denmark, DTU Nutech, Risø Campus, Frederiksborgvej 399, Bldg. 201, 4000 Roskilde, DK

E-mail: luca.zanini@esss.se

Abstract.
The European Spallation Source (ESS) will use low-dimensional moderators for cold and thermal neutron production. Low-dimensional moderators deliver higher slow neutron brightness compared to conventional volume moderators. The concepts developed at ESS could be used in reactors and compact neutron sources to reach large increases in neutron flux to the experiments, compared to common practice. For reactors, the smaller volume of the moderators allows positioning them at the optimum flux region. For compact neutron sources, the reduced heat deposition and lower radiation environment allows for particularly efficient implementation of the ESS concepts. Using tube moderators in a compact source would allow reaching a higher brightness per unit yield at least a factor of 4, with respect to a spallation source.

1. Introduction

High-brightness low-dimensional moderators are the baseline choice at the European Spallation Source [1, 2, 3, 4]. The figure of merit used in the design of the neutron moderators was the brightness of the source. Compared to conventional volume moderators, low-dimensional moderators have higher brightness; by carefully designing the optics system for neutron transport, this will usually result in large gains in flux at samples [5].

One of the reasons for the higher brightness of low-dimensional moderators is the fact that thermal neutrons are effectively moderated in about 1 cm of liquid hydrogen; the optimal thickness of a moderator is of the same order [3]. Another key aspect is the directionality of neutron emission of low-dimensional moderators, which can be flat (quasi 2D) or tube (quasi 1D) shaped: in such moderators, neutrons that are not going in the "right" direction along the moderator walls have a chance to be scattered back into the flat or tube moderator and have a second possibility to exit the moderator. This works particularly well in a tube moderator because of the long extension of the tube, comparable to the mean free path of cold neutrons in pure parahydrogen, which is about 10 cm. Therefore, while low-dimensional moderators can be made of any hydrogenous material, they work particularly well if pure parahydrogen is used, due to the moderator-shape determined directionality of the cold neutron escape in this particular case. Other aspects that contribute to the increased brightness of low-dimensional moderators are: *i*) the lower parasitic neutron absorption from the hydrogen itself, due to the lower amount of moderator material used; *ii*) the fact that less reflector material is removed; *iii*) the closer

IOP Conf. Series: Journal of Physics: Conf. Series **1021** (2018) 012009 doi:10.1088/1742-6596/1021/1/012009

position of the moderator to the target. The latter aspect can be of particular interest for compact sources.

Besides the use of low-dimensional moderators for ESS, in Ref. [2] we suggested the application of low-dimensional moderators at reactors, and suggested in this case the possible use of tube moderators. Beam extraction channels at reactors are commonly spread over a small angular range, which is the ideal application for the very directional tube moderators. Moreover, the compact size of these moderators, with respect to the large D_2 moderators, allows for placing them in the position of maximum thermal flux in the reflector. Tube moderators are also considered for the Jülich high-brilliance neutron source project [6].

In this paper we focus on applications of low-dimensional moderators to compact sources, giving some examples of possible configurations.

2. Application to compact sources

2.1. Possible configurations

Low-energy proton and deuteron beams can be used for fast neutron production in compact sources. For energies below about 30 MeV, proton and deuteron reactions in lithium and beryllium are the most efficient for neutron production [7]. Compared to spallation, these reactions are much less efficient, with yields per accelerated particle of the order of 10^{-2} n/p compared to typical yields achieved in spallation sources, of 30 n/p or more. Of course, with up to 3 orders of magnitude difference in particle energies, there is much less difference between yields per unit beam energy. Despite this large difference in neutron production, compact sources are of interest for reasons of cost; the lower energy of the proton beam minimizes the activation of the target system, and lowers the shielding and cooling requirements. The cost per useful slow neutron produced is optimal for high power spallation sources, in particular for ESS. The low facility costs make compact sources attractive for a host of applications which only require moderate neutron beam flux. The reduced sizes of the targets and incoming beams in compact sources are expected to add more flexibility to the design of the target-moderator-reflector system with respect to spallation sources. We considered two possible configurations using Be targets and proton beams of 13 MeV.

The first configuration is shown in Fig. 1. It consists of a two-target system, with two Be targets, 2 mm thick, 5 × 5 cm^2 area, cooled by a flow of water. Having two separate proton beams would allow reducing the heat load in each of the targets. Moreover, a configuration with the moderator placed between the targets, can increase the neutron flux at the moderator, with respect to a configuration with only one target and the moderator on its side. The cold moderator consists of a tube-shaped Al container filled with pure parahydrogen. It is wider at the center to allow for an increased angular range of directional emission [2] in the beam extraction channels. The length of the moderator is of 20 cm. On the side, a water moderator has been placed, which can deliver thermal neutrons, and also serve as premoderator for the cold moderator. Neutrons are extracted from the moderator faces with an area 1.5 cm wide and 3 cm high. A Be reflector surrounds the moderators.

An alternative configuration is shown in Fig. 2 [9]. In this second configuration, a proton beam is hitting the Be target from the top. The cold moderator consists of three tubes, of 3 cm height and 3 cm width, surrounding the target and forming a triangular shape. Three openings in the Be reflectors of 60° would allow providing cold beams to more beam lines than in the previous example.

2.2. Performance

The calculated neutron yield for a 2 mm thick Be target, for a 13 MeV proton beam, using MCNPX [8] with ENDF-B/VII proton libraries, is of 6.2×10^{-3} n/p.

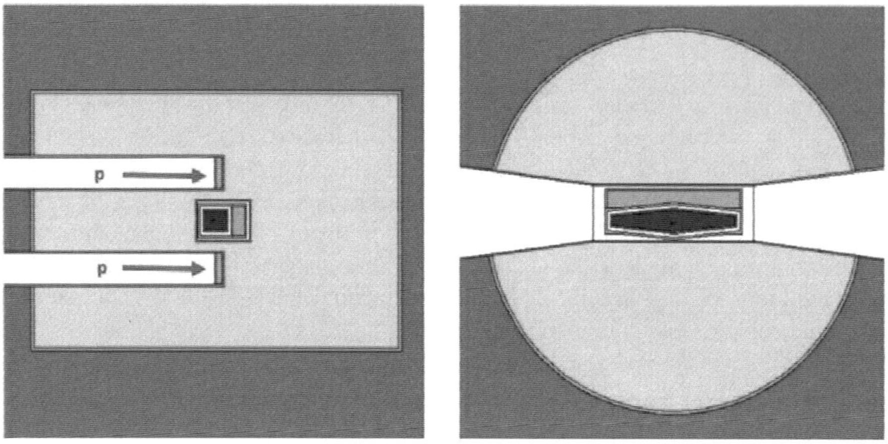

Figure 1. Geometry of the first proposed configuration of compact source using tube moderators [9]. Cold moderator (blue), and thermal moderator (light blue) are surrounded by the Be reflector (green) and outer steel reflector (red). *Left*: side view showing the two proton beam channels, with proton beams hitting the thin, water-cooled Be targets. *Right*: top view, showing cold and thermal moderators, with neutron beam extraction channels.

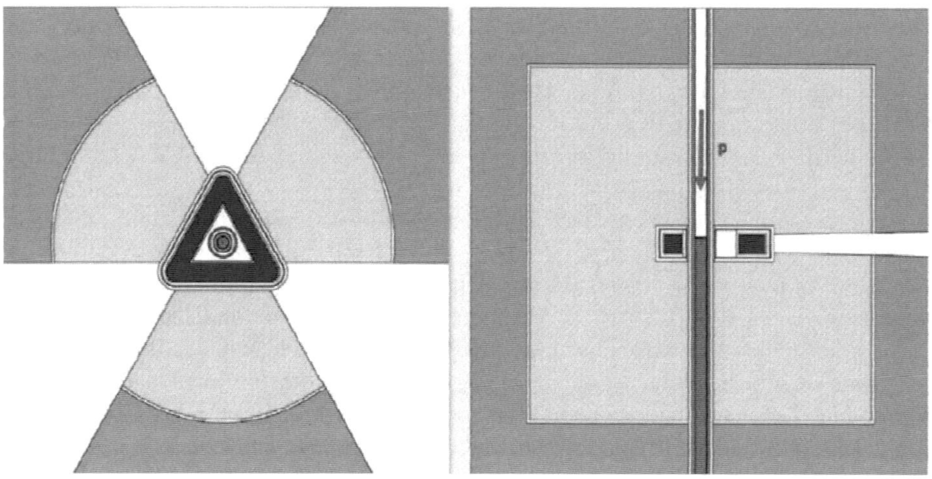

Figure 2. Geometry of the second proposed configuration of compact source using tube moderators. The cold moderator (blue) consists of three tubes filled with parahydrogen, surrounding the Be target (red). The Be reflector is in green, outer steel reflector/shielding in orange. *Left*: top view (proton beam is perpendicular to the figure), showing neutron beam extraction openings in the reflector. *Right*: side view, showing proton beam channel, with protons entering vertically from the top.

Time-average, integrated (between 0 and 5 meV) cold brightness was calculated using F5 tallies, for the configurations of Fig. 1 and 2. Only basic optimizations of the geometry were performed in this preliminary study. The cold neutron brightness delivered by both designs is

IOP Conf. Series: Journal of Physics: Conf. Series **1021** (2018) 012009 doi:10.1088/1742-6596/1021/1/012009

similar. For the first configuration, the integrated time-average cold brightness for 100 kW power is of about 3.2×10^{10} n/cm^2/s/sr. An integrated (between 20 meV and 100 meV) time-average thermal brightness of 3.3×10^{10} n/cm^2/s/sr was calculated for the thermal moderator.

The second configuration (Fig. 2) was ran with two different proton beam energies: 13 MeV and 40 MeV. At 13 MeV the performance is very similar to the first case. With 40 MeV protons, a time-average brightness of about 1×10^{11} n/cm^2/s/sr at 100 kW is obtained. The integrated cold brightness below 5 meV from the ILL yellow book is of 3.3×10^{12} n/cm^2/s/sr. This gives an indication that in a pulsed compact source with the same duty cycle of ESS (4%), the peak brightness could reach values comparable to high-flux research reactors.

2.3. Comparison with spallation reactions

In any neutron source, the brightness of the source (moderator) is proportional to the yield of neutrons produced by the primary reaction (e.g. spallation of protons on target, nuclear fission, proton or deuteron reaction on Li/Be targets). These reactions produce fast neutrons with rather similar center of gravity around a few MeV. Therefore, a possible figure of merit to determine how efficient is a neutron source, is the ratio between the moderator brightness and the neutron yield. The neutron yield Y of 2 GeV protons in tungsten is of about 38 n/p, according to the semi-empirical formula in [7]. The ESS brightness per proton (always integrated between 0 and 5 meV) is B=1×10^{-3} n/cm^2/s/sr/p. For a 13 MeV proton on Be, the calculated neutron yield is Y=6.2×10^{-3} n/p. The cold brightness for the model in Fig. 1 is of 6.7×10^{-7} n/cm^2/s/sr/p. Therefore, by considering as figure of merit for slow neutron production efficiency the ratio B/Y, we get a B/Y ratio four times higher in the case of the small source. This gain is probably due in part to the use of tube moderators, in part to the advantages given by the more compact geometry of a small neutron source.

3. Conclusions

We have shown two examples of application of low-dimensional moderators to compact neutron sources. The above results are preliminary, and more in-depth studies are needed for a better exploitation of the properties of these moderators to the case of compact sources. More calculations should be performed with optimization of the modelled sources, including engineering details. Nevertheless, these results indicate that low-dimensional moderators can be used at compact sources to deliver cold neutron beams, perhaps even more efficiently than at spallation sources.

References

[1] K. Batkov, A. Takibayev, L. Zanini and F. Mezei, *Unperturbed moderator brightness in pulsed neutron sources*, Nuclear Instruments and Methods in Physics Research Section A: Accelerators, Spectrometers, Detectors and Associated Equipment 729 (2013) 500 - 505. http://dx.doi.org/10.1016/j.nima.2013.07.031.

[2] F. Mezei, L. Zanini, A. Takibayev, K. Batkov, E. Klinkby, E. Pitcher, and T. Schönfeldt, *Low dimensional neutron moderators for enhanced source brightness*, Journal of Neutron Research 17 (2014) 101 – 105.

[3] L. Zanini, K. Batkov, E. Klinkby, F. Mezei, T. Schönfeldt, and A. Takibayev, *The neutron moderators for the European Spallation Source*, these proceedings.

[4] L. Zanini, K. Andersen, K. Batkov, E. Klinkby, F. Mezei, T. Schönfeldt, and A. Takibayev, *Design of the cold and thermal neutron moderators for the European Spallation Source*, in course of publication, 2017.

[5] K. Andersen, M. Bertelsen, L. Zanini, E. Klinkby, T.Schönfeldt, P. Bentley, J. Saroun, *Optimisation of moderators and beam extraction at ESS*, in course of publication, 2017.

[6] U. Rücker, T. Cronert, J. Voigt *et al.* Eur. Phys. J. Plus (2016) 131: 19. https://doi.org/10.1140/epjp/i2016-16019-5.

[7] J. M. Carpenter, C.-K. Loong, *Elements of slow-neutron scattering*, Cambridge University Press, 2015, and references therein.

[8] D. Pelowitz, editor, MCNPX Users Manual, Version 2.7.0, Number LA-CP-11-0438, 2011.

[9] A. Takibayev, Presentation at the UCANS-VI conference, Xian, China, 2016.

IOP Conf. Series: Journal of Physics: Conf. Series **1021** (2018) 012010 doi:10.1088/1742-6596/1021/1/012010

LINUS, the Integrated LNL Neutron Source facility

D. Bisello[1], E. Fagotti[2], J. Esposito[2], C-K. Loong[3], M. Maggiore[2], P. Mastinu[2], G. Prete[2], L. Silvestrin[1,3], J. Wyss[1,4]

[1] INFN, Padova, via Marzolo 8, 35131 Padova, Italy
[2] INFN, LNL, Viale dell'Università 2, 35020 Legnaro (PD), Italy
[3] DFA, University of Padova, via Marzolo 8, 35131 Padova, Italy
[4] DICeM, University of Cassino, Viale dell'Università, 03043 Cassino (FR), Italy

luca.silvestrin@unipd.it

Abstract. LINUS is a project at the INFN Legnaro National Laboratories (LNL, Italy) to create a suite of different neutron sources (LSNS, NEPIR, SLOWNE) driven by existing accelerators. LSNS, driven by a 40 mA, 5 MeV proton RFQ, will use Li and Be targets to deliver cold, thermal, epithermal and fast neutrons. The SPES high current (0.75 mA), 70 MeV proton cyclotron will drive the NEPIR and SLOWNE facilities. NEPIR will alternatively deliver quasi mono-energetic neutrons with energy peak down to 20 MeV, and a neutron beam with a continuous energy distribution similar to that of neutrons present in the Earth atmosphere in the accessible energy range. SLOWNE is an intense neutron source for applications outside the LSNS range.

1. LINUS: a multipurpose neutron facility from cold to fast neutrons

Reactors play an essential role in delivering intense neutron beams but many are being retired and will not be replaced. On the other hand, present and future large, high-energy, high-neutron-yield spallation source facilities (ISIS at RAL, ESS…) cannot possibly fulfill the future demands of all the users from Europe and elsewhere. [1]

Lower-energy non-spallation-based Compact Accelerator-driven Neutron Sources (CANS) can bridge the capacity insufficiency and the widening expanse of neutron applications. However, the majority of CANS are not new and many do not allow a broad range of applications under variant configurations. Therefore, the realization of new Accelerator-driven, Brilliant, and Compact (ABC) Neutron Sources is an imminent goal to ensure an expansive neutron R&D landscape.

The Legnaro Integrated NeUtron Source facility (LINUS) wants to be an important contributor to this effort. LINUS consists of three parts: the Legnaro Slow Neutron Source (LSNS) at a high current RFQ 5 MeV accelerator, and a fast (E_n > 1 MeV) Neutron and Proton Irradiation (NEPIR) facility and an intense source of slower neutrons (SLOWNE) at the 70 MeV SPES cyclotron facility.

2. LSNS

LSNS is based upon state-of-the-art accelerator and neutron production technologies already developed at LNL: a high-power, high-current radio-frequency-quadrupole (RFQ) proton accelerator [2] and Li and Be neutron producing targets [3]. The LSNS design will deliver high-brilliance neutron beams over the cold, thermal and epithermal energy ranges for independent utilization to serve diverse users across different fields ranging from basic nuclear science, to experimental nuclear astrophysics, materials research, life science, medical therapy, nuclear instrumentation, cultural heritage and education of the next generation of neutron scientists and engineers. Significant savings in construction and operation costs are expected due to the low proton energy, compactness and modularity of the accelerator and target systems.

2.1. Proton beam Pulse Selection and Distribution Station (PSDS)

The LSNS is conceived as modular in structure, see Fig. 1a, and a schematic of the layout is shown in Fig. 1b. A Pulse and Selection Distribution Station (PSDS) will deliver an intense low energy (≤ 5 MeV) proton beam from the RFQ to the various parts of LSNS. Continuous Wave (CW), Short Pulse (SP) and Long Pulse (LP) modes will be all available. The LSNS suite shown schematically in Fig. 1b includes a cryogenic Cold Neutron Target Station (CNTS) and a Thermal Neutron Target Station (TNTS). The TNTS system is identical in design to CNTS, but it will operate at ambient temperature. The PSDS system also delivers beams to the LENOS [4] (astrophysics) and BNCT [5] (oncology) targets stations.

Legnaro Slow Neutron Source (LSNS) proton RFQ (30-50 mA, 150-250 kW, CW or P at	Cold Neutrons (Be-target, 25 Hz, 0.5 ms LP)	Imaging and radiography	Small-angle scattering	Neutron reflectometry	Instrumentation R&D (neutronics, polarization...)
	Thermal neutrons (Be-targets,50 Hz, 0.01 ms SP) Duty cycle 1/2000	Diffraction	Neutron interrogation	Instrumentation R&D	
	LENOS (Li-targets,125 kHz, 200 ns bunched 2 ns SP) Duty cycle 1/40	Nuclear astrophysics TOF	Nuclear astrophysics activation		
	Medical therapy (Be-targets, CW) duty cycle = 1	BNCT (thermal)	BNCT (epithermal)		

Figure 1a. LSNS beam characteristics for different applications

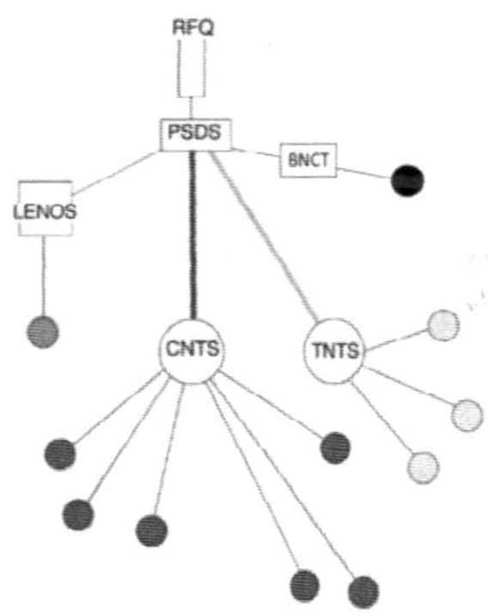

Figure 1b. A sketch of the LSNS multiplexed structure.

All the neutron experiments at the CNTS, TNTS and the LENOS systems are conducted in the time-of-flight (TOF) mode. However, each system entails different time structures for the various epithermal, thermal and cold neutron applications. The PSDS makes the best use of the RFQ proton beam for concurrent service to neutron production at all the facilities with high efficiency.

A chopper upstream of the RFQ will condition the proton beam to be pulsed at 125 kHz with a pulse width (FWHM) of ~200 ns. These short pulses are shortened further to 1-2 ns by a series of choppers and bunchers: a clock starts the accumulation of the short pulses to achieve a pulse width of ~10 μs in conjunction with the defocusing of the proton beam to a cross-section of 10 cm × 10 cm. This proton pulse is then delivered to TNTS and the process is then repeated every 20 ms to enable a short pulse (SP) neutron source running at 50 Hz. After each SP a second clock starts at to accumulate a defocused long pulse (LP) of ~500 μs width to feed the CNTS. This process repeats every 40 ms for a LP cold neutron source running at 25 Hz.

2.2. The Cold and Thermal Neutron Target Stations

The core of the CNTS consists of a Be target and a cryogenic moderator to contain solid methane at ~10 K, and a reflector (typically graphite or water as employed by other neutron sources). It is set on a concrete base and surrounded by shielding materials (alternating shells of lead and borated polyethylene) with overall typical dimensions: 4 m (diameter of the pile) by 3 m (height). The configuration of the TNTS is the same as that of CNTS except that the moderator contains water at ambient temperature. The identical design of the TNTS and CNTS results in substantial saving in shielding fabrication and instrumentation design costs. Up to 6 neutron exit ports with individually operated beam gates are installed at TNTS. Three beam lines are planned at the initial stage.

IOP Conf. Series: Journal of Physics: Conf. Series **1021** (2018) 012010 doi:10.1088/1742-6596/1021/1/012010

2.3. Instruments System

Behind the target station is a biological shield wall through which neutron guides deliver the neutron pulses with minimal intensity loss along incident flight paths that define the neutron beamlines. Each beamline serves a neutron instrument. The beamlines and instruments to be installed in the initial phase of LSNS are four cold-neutron instruments at CNTS: a Small-Angle Neutron Scattering, a General-Purpose Neutron Reflectometer, a Cold-Neutron Imaging & Radiology, and a Cold-Neutron Instrumentation Development beamline, and three thermal-neutron instruments at TNTS: a High-Intensity Powder Diffractometer, a Multiple-Purpose Materials Interrogation, and Neutron Device and Detector Development in the TNTS hall.

3. NEPIR and SLOWNE at the SPES cyclotron

The fast Neutron and Proton Irradiation (NEPIR) and the SLOWNE facilities will be driven by the SPES accelerator, a proton cyclotron able to produce proton beams in the 35-70 MeV energy range with currents up to 700 μA [6]. Both will be housed in the Hall 9 of the SPES building.

3.1. NEPIR

The original purpose of NEPIR is to study radiation damage effects in electronics induced by fast neutrons present at flight-altitudes and sea level and by solar protons. It will deliver two complementary neutron beams: Quasi Mono-energetic Neutron (QMN) beams in the 30-65 MeV energy range and a neutron beam (ANEM) with a continuous energy distribution, up to 65 MeV, similar to that of neutrons found in the Earth atmosphere (an extension to emulate the Mars atmosphere is foreseen). A direct variable energy proton beam line, not described here, will be also available.

The QMN and ANEM targets will share the same proton beam line. The ANEM target is installed upstream of the QMN target: an aperture on the ANEM target will allow alternate operation with the QMN source and direct proton beam. The energy range of both QMN and direct protons will be extended down to 20 MeV by proton-energy degraders.

QMN reference fields allow one to study energy dependent neutron interaction mechanisms with matter [7]. Many diverse research fields can take advantage of this facility: nuclear physics, both pure and applied, neutron detector development and calibration, shielding experiments, neutron effects on electronics and biological matter, etc.

QMNs will be produced by using thin (1-4 mm) Li or Be targets, the emerging protons being magnetically deflected towards a beam dump. In the forward direction (0°), the neutron energy spectrum presents a peak close to the energy of the impinging proton beam and a broad low energy tail coming from nuclear breakup [8], each of which contains about half of the total neutron intensity. The calculated flux of neutrons in the energy peak is $\sim 5\times10^5$ n cm^{-2} s^{-1}, for a 10 μA 70 MeV proton current on a 4 mm Li target, at 3 m from the target. The unwanted effects due to the low energy tail can be assessed by taking data concurrently at larger angles (15°-30°), where the high energy peak is absent [9] and subsequently corrected in data treatment. To this end, a static multi-angle collimator system, as at iTHEMBA and CYRIC [10][11], will be initially used.

The QMN at LNL [12] will be an important part of the landscape of European neutron facilities. The recently closed TSL [13] was capable of operating in the 22-175 MeV energy range and did not allow the correction for data collected at forward direction. NFS, under construction at GANIL, is limited to energies below 40 MeV [14].

The Atmospheric Neutron Emulator (ANEM) system will produce an atmospheric-like neutron spectrum by using a weighted convolution of neutrons coming from different target materials. This is achieved by using a rotating composite target (a 5 mm thick W disk and a 24 mm thick Be sector). Rotation allows to distribute the deposited energy over a large area of target elements (Fig. 2a). The target is housed in a cylindrical vacuum chamber connected to the beam pipe; heat transfer is ensured by a water cooling system. The proton beam is not stopped by the Be sector (to avoid excessive damage) and the protons that emerge from it are stopped by a W disk [12]. With 10 μA of 70 MeV protons, a neutron flux of 3.5×10^6 n cm^{-2} s^{-1} is delivered at a downstream distance of 6 m. That allows accelerated test (acceleration factor F = 10^9) of the effects of atmospheric neutrons on electronic devices and systems and on biological matter.

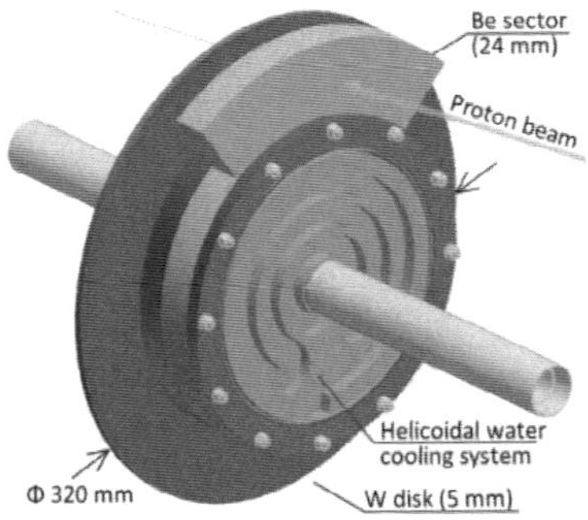

Figure 2a. The ANEM rotating target.

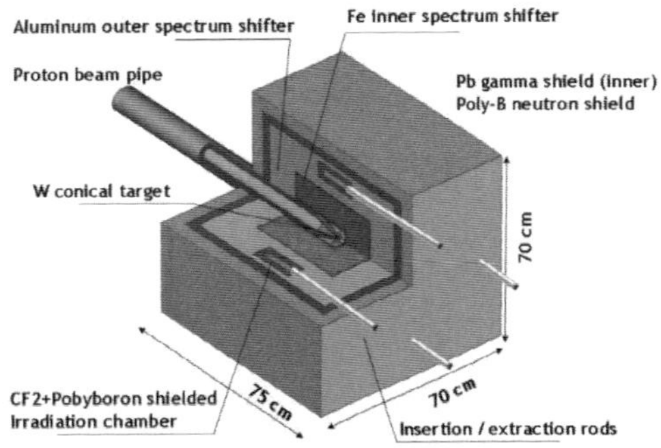

Figure 2b. The self-shielding neutron source developed for the FARETRA project.

3.2. SLOWNE

SLOWNE is an independent high flux neutron line for those special applications that require neutron energies outside the energy range of LSNS, such as transmutation studies and radiation damage in electronics. It is based on a high-power conical water cooled 5 mm thick W target housed inside a neutron reflector and energy spectrum shifter with related elements for radiation shielding (Fig. 2b). With a 500 μA current of 70 MeV protons the source flux of neutrons is 5.6×10^{14} n/s.

The target was originally developed for FARETRA, a Fast Reactor Emulator for TRAnsmutation studies, the purpose of which is to obtain new data about neutron induced reactions on actinides, fission fragments and structural materials. Application to BNCT is also under study [5].

References

[1] I. Anderson et al. Research opportunities with compact accelerator-driven neutron sources, Phys. Rept, 2016, 654: 1
[2] E. Fagotti et al., Completion of the fabrication of TRASCO RFQ, Proc. 25th Linear Acc. Conf. LINAC 2010 Tsukuba, Japan, TUP057
[3] J. Esposito et al., Be target development for the accelerator-based SPES-BNCT facility at INFN Legnaro, Appl Radiat Isot. 67 (7-8 Suppl): S270-3, Jul. 2009
[4] P.F. Masinu et al., A method to obtain a Maxwell-Boltzmann neutron spectrum at kT=30 keV for nuclear astrophysics studies NIST A 601 (2009) 333–338
[5] C. Ceballos et al., Towards the final BSA modeling for the accelerator-driven BNCT facility at INFN LNL, Appl. Radiat. and Isotopes, V. 69, I. 12, Dec. 2011, 1660–1663
[6] G. Prete et al., The SPES project at the INFN- Laboratori Nazionali di Legnaro, EPJ Web of Conferences 66, 11030 (2014)
[7] S. Pomp et al., High-energy quasi-monoenergetic neutron fields: existing facilities and future needs, EURADOS Report 2013-02
[8] M. Baba et al., Characterization of a 40-90 MeV ^{7}Li(p,n) neutron source at TIARA using a proton recoil telescope and a TOF method, Nucl. Instr. Meth. A 428, 1999, 454-465
[9] S. Kamada, et al., Measurement of energy-angular neutron distribution for ^{7}Li, ^{9}Be(p,xn) reaction at Ep=70 MeV and 11 MeV, J. of Korean Phys. Soc. 59, No. 2, Aug. 2011, 1676
[10] M. Mosconi et al., Characterisation of the high-energy neutron beam at iThemba LABS, Rad. Meas. 45, I 10, Dec. 2010, 1342-1345
[11] A. Terakawa et al., New fast-neutron time-of-flight facilities at CYRIC, NIST A 491, I. 3, 2002
[12] L. Silvestrin et al., SPES and the neutron facilities at Laboratori Nazionali di Legnaro, Eur. Phys. J. Plus (2016) 131: 72
[13] A. V. Prokofiev et al., The TSL neutron beam facility, Rad. Prot. Dosim.126(1-4), 18-22, 2007
[14] X. Ledoux et al., The Neutrons for Science Facility at SPIRAL-2, Nucl. Data Sheets 119, 353 (2014).

IOP Conf. Series: Journal of Physics: Conf. Series **1021** (2018) 012011 doi:10.1088/1742-6596/1021/1/012011

Towards a compact Laser based Neutron source

A Cianchi[1;2], C. Andreani[1], R. Bedogni[3], G. Festa[4], O. Sans Planell[3], R. Senesi[1]

[1] University of Rome Tor Vergata, Via della Ricerca Scientifica, 1-00133 Rome, Italy

[2] INFN-Roma Tor Vergata, Via della Ricerca Scientifica, 1-00133 Rome, Italy

[3] INFN-LNF, Via E. Fermi, 40 - 00044 Frascati, Rome, Italy

[4] Centro Fermi, Piazza del Viminale, 1, 00184, Rome, Italy

E-mail: alessandro.cianchi@uniroma2.it

Abstract. Several experiments for neutron generation using high intensity laser sources, with a power exceeding 10^{19} W/cm^2 via TNSA (Target Normal Sheath Acceleration) or other similar methods, have been performed in recent years in different laboratories. However, so far there is no one running neutron source based on such a technology. We have just started a study of the advantages and limits of this novel technique, trying to identify the critical points in order to pave the way for future experiments.

1. Introduction

The need for accelerating gradients orders of magnitude larger than existing ones drives the research in particle accelerators towards plasma based accelerators [1],[2]. These machines can sustain fields greater than tens of GV/m paving the way for the realization of table top accelerators. In every considered scheme under investigation a PW class laser is used.

Likely these lasers will be not considered uniquely to drive the main accelerator but they will also be devoted to other several activities. The interaction of such a laser with matter produces a large number of electrons, ions, positrons, protons via several different mechanisms depending on the laser intensity and the target material compositions and dimensions [3]. While a lot of effort is dedicated nowadays to improving the quality of these charged or neutral beams of particles to use them, up to now there is not yet a user facility using these kind of beams while there are several studies concerning neutron production via laser-matter interaction (for instance among others [4],[5], [6]).

2. Methodological analysis

Our interest is to focus mainly on a very compact neutron source. In order to achieve it we consider a moderator with the Bedogni design [7], implemented already in a thermal neutron irradiation facility with extended and very uniform irradiation area (HOTNES at ENEA-INFN, Frascati). A neutron source is located on the bottom of a large cylindrical cavity (30 cm diameter, 70 cm in height) delimited by polyethylene walls. Owing on a polyethylene shadow-bar, only multiple-scattered neutrons can reach the irradiation volume. The resulting neutron spectrum is highly thermalized (Cd-ratio about 10). Irradiation planes are disks (30 cm in diameter) showing very uniform thermal field

IOP Conf. Series: Journal of Physics: Conf. Series **1021** (2018) 012011 doi:10.1088/1742-6596/1021/1/012011

(1-2%) over their whole surface. The moderating efficiency (thermal fluence per primary neutron) is about 2 10^{-4} cm^{-2}.

HOTNES was designed using the Monte Carlo MCNPX 2.7 Monte Carlo code [8] using the ENDF/B-VII cross section library [9] for neutrons with energies below 20 MeV and the room temperature cross section tables for thermal neutrons in polyethylene, $S(\alpha,\beta)$. In our case we replace the radionuclide neutron source with the laser compact source.

3. Conventional source

We use as a master reference the parameter of the Eupraxia collaboration [10], a European design study to produce the first conceptual design report of a plasma based accelerator. Eupraxia is still considering several alternatives, and we focus on the case where the beam is produced in a conventional accelerator and later plasma-accelerated. An electron beam of 1 GeV energy, with bunch charge 100 pC, a repetition rate of 10 Hz and an average current of 1 nA is foreseen. We consider the production of neutrons via bremsstrahlung, using a tungsten target of 5x5x9 cm^3, being 9 cm the maximum yield. We can obtain about 0.4 neutrons for each primary electron. With the Eupraxia design values we can have about 2.5 10^9 neutrons/s. This number sets the lower limit of any laser based neutron compact source that can be consider interesting at such a facility.

Using a moderator like [7] we conclude that we can have about 2.32 10^{-4} neutrons per primary particle. We are not considering the beam collimation because it has an impact less than an order of magnitude in terms of lost neutron flux, and it does not change the results of our analysis. With these figures we estimate about 9.3 10^{-5} n/cm^2 per primary on the sample.

4. Laser based source

When a high intensity laser (10^{19} W/cm^2 or larger) is focused on a solid target several effects can be triggered, like for instance Target Normal Sheath Acceleration (TNSA) [11], Radiation Pressure Acceleration (RPA) [12], collisionless shock acceleration [13] and Break Out Afterburner (BOA) [14], depending on target material, thickness and surface contaminations for instance. Let us focus on TNSA. Fast electrons are accelerated through the material by the laser. These electrons penetrate the target, ionizing other particles and escaping from the other side. In this moment they build up a very strong electric field, in the order of TV/m. This field extracts protons and ions from the rear surface, producing an intense beam of particles. A sketch describing the physics of the interaction can be find in [15]. While there are scaling laws of the process, it is very difficult to define the energy spectrum, the flux intensity, and the particle geometrical distribution in a general case, as the emission is strongly linked to the target material, surface contamination, laser energy and intensity, and laser contrast. To better underline this argument the reader can have a look to the spectra of Figure 1 produced from very different lasers. While their energies are quite different, the intensities are much more similar. The increasing of the energy increases both proton number and average energy. However this dependence is not followed very strictly, mainly due to some particular experimental arrangement of the target, as well as the use of different kind of targets triggering different mechanisms.

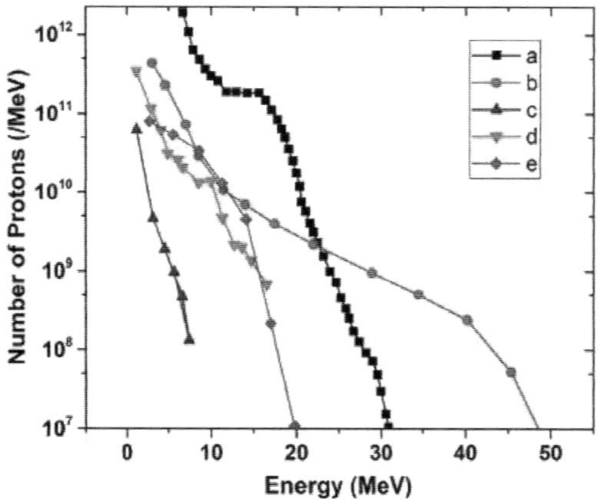

Figure 1. Comparison of proton spectrum emerging in different experiments. For the symbol caption see the Table 1.

Table 1. Symbol caption for spectrum in Fig.1.

Label	Name	Intensity(W/ cm^2)	Energy(J)	Reference
a)	Vulcan	2.0 10^{20}	200	[16]
b)	Trident	1.5 10^{20}	80	[17]
c)	Arcturus	1.0 10^{20}	3	[18]
d)	Vulcan	1.0 10^{20}	42	[19]
e)	Astra Gemini	1.0 10^{21}	10	[20]

What is really important is the number of protons in an energy range below 10-20 MeV, because even a laser with energy around or below 40 J can already produce a proton number in the order of 10^{11} in such energy range. Once that the primary beam is produced, the protons/ions hit a material like for instance LiF or Be, in order to produce a neutron flux. This scheme is usually called pitcher-catcher scheme. In [6] there is a comparison between results obtained with different lasers and target materials. This kind of comparison is very difficult because the properties of the proton or ion beams are not completely and uniquely linked to the laser but also to the target material, geometry and thickness.

5. Expected flux

We focus on a typical catcher target of lithium fluoride. Different yields are found in the literature for the same reaction. For this simulation and for the following, the neutron production yield was generated using a custom Labview^{TM} based software based on a continuous projectile slowing down in the target. The stopping power data were generated from PSTAR (NIST) [21] for protons and SRIM 2011 [22] for deuterons. Cross sections data were taken from ENDFB VII [9] and EXFOR (protons) [23] or TENDL2009 (deuterons) [24]. The differences in the yield in the range of the tens-few tens of MeV are only about a factor 2-4, much less than an order of magnitude, making appealing also protons of lower energy that can be produced by smaller laser with higher repetition rate. The thickness of the target was also optimized in order to maximize the neutron flux. With 0.2 cm of LiF target, the typical

IOP Conf. Series: Journal of Physics: Conf. Series **1021** (2018) 012011 doi:10.1088/1742-6596/1021/1/012011

moderation yields in thermal neutron flux per primary fast neutron is about 1.55 10^{-4} n/p per proton of 5 MeV.

We also considered the case of a Beryllium target (0.3 mm thick) with a deuteron beam, because usually the targets are hydrogenated on the surface to increase the deuteron number. Considering 7 MeV deuterons we can have a moderation yield in thermal neutron flux per primary fast neutron of about 1.24 10^{-4} n/d.

6. Conclusions

Looking at Table 2 we can compare the results of the several cases examined.

Table 2. Neutrons from different scenarios.

Primary	Energy	Yield (n/prim)	m(moderation efficiency)	Y x m
Electrons	1 GeV	4.0 10^{-1}	2.32 10^{-4}	9.3 10^{-5}
Protons	5 MeV	8.67 10^{-4}	1.55 10^{-4}	1.4 10^{-7}
Ions	7 MeV	7.64 10^{-4}	1.24 10^{-4}	9.4 10^{-8}

It is clear that to be competitive, in a lab where there is already an accelerator with GeV energy beam, this kind of source must have a flux of primary particles at least 3 orders of magnitude greater. Let us consider now some specific applications for such a laser based neutron source such as Prompt Gamma Activation Analysis (PGAA) and neutron radiography. Usually a series of many techniques is used to study the objects in this field like THz, IR, X-ray radiation and neutron based techniques. There is a particular interest in considering new plasma accelerators for cultural heritage, because for instance in the Eupraxia project, all of these sources are foreseen. Adding also, with the same instrumentation, a neutron source could be very interesting, giving the possibility of having on the same site all these techniques together. Neutron radiography requires parallel beam or divergent beam of low energy neutrons having intensity in the range of only 10^4 - 10^6 neutrons/cm^2/s to avoid formation of significant amount of long-lived radioactive isotope from neutron absorption within the specimen. PGAA are less demanding considering that even conventional portable sources are used for this end, giving flux on the sample in the order of 10^3 n/cm^2/s.

These kind of numbers are in the some order of several CANS [25]. For a laser based neutron source, with such a high efficiency and compact moderator the required initial flux of primary particles is in the order of 10^{11}-10^{12} particles/s in an energy range around tens of MeV. This number is really very close to the existing results of several experiments. In particular, while it is already achieved in a single shot, it must be demonstrated for lasers with lower energy, let's say less than 30 J, that can operate with a repetition rate of 1-10 Hz. The gap can be filled in the next years with improvements in the targets and laser technologies.

Laser plasma based neutron sources are very promising CANS. The reduced dimensions, the absence of RF devices, the small number of personnel needed to operate them, are all in favor of further development in this field. The state of the art is really close to consider interesting these kind of sources in some particular applications.

References

[1] J. Faure,et al.,Nature 431 (2004) 541.
[2] W. P. Leemans ,et al.,Physical Review Letters 113 (2014) 245002.
[3] Ledingham, K. W. D., and Wilfried Galster , New Journal of Physics 12.4 (2010): 045005.
[4] Roth, M., et al., Physical review letters 110.4 (2013): 044802.
[5] Kar, S., et al., New Journal of Physics 18.5 (2016): 053002.

IOP Conf. Series: Journal of Physics: Conf. Series **1021** (2018) 012011 doi:10.1088/1742-6596/1021/1/012011

[6] Alejo, A., et al., Nuovo Cimento C 38 (2016): 188.
[7] Bedogni, R., et al., NIM A 843 (2017): 18-21.
[8] Pelowitz, D. B., et al, Los Alamos National Laboratory, LA-UR-11-01502(2011)
[9] Chadwick, M.B., al., Nucl. Data Sheets 107, 29313060, (2006)
[10] http://www.eupraxia-project.eu/
[11] Hatchett, S.P. et al., Phys. Plasmas 2000, 7, 20762082
[12] Esirkepov, T. et al., Phys. Rev. Lett. 2004, 92
[13] Haberberger, D. et al., Nat. Phys. 2012, 8, 9599
[14] Henig, A. et al., Phys. Rev. Lett. 2009, 103, 045002
[15] H. Schwoerer, et al., Nature 439, 445-448(26 January 2006)
[16] Yang, J. M., et al., Journal of applied physics 96.11 (2004): 6912-6918.
[17] Gaillard, S. A., et al., Physics of Plasmas 18.5 (2011): 056710.
[18] Kar, Satyabrata, et al., Nature communications 7 (2016).
[19] Clark, Eugene Laurence. PhD Thesis, (2002).
[20] Green, J. S., et al., Applied Physics Letters 104.21 (2014): 214101.
[21] PSTAR, NIST. URL http://physics.nist.gov/PhysRefData/Star/Text
[22] Ziegler, J. F., et al., SRIM 2011 Code, 19842011.
[23] https://www-nds.iaea.org/exfor/exfor.htm
[24] Koning, A. J., and D. Rochman., Nuclear Research and Consultancy Group, Petten (2009).
[25] Anderson, I. S., et al., Physics Reports 654 (2016): 1-58.

Bayesian Inference in MANTID – An Update

B. Llopis Vidal,[1] E. Oram,[2] R. Alvarez Baños,[2] L.C. Pardo,[1] S. Mukhopadhyay,[2,3] and F. Fernandez-Alonso[2,4]

[1]Grup de Caracterització de Materials, Departament de Física, EEBE, Universitat Politècnica de Catalunya, C/ Eduard Maristany 10, E-08019 Barcelona, Catalonia, Spain.
[2]ISIS Facility, Rutherford Appleton Laboratory, Chilton, Didcot, Oxfordshire OX11 0QX, United Kingdom
[3]Imperial College London, Exhibition Road, London SW7 2AZ, United Kingdom.
[4]Department of Physics and Astronomy, University College London, Gower Street, London WC1E 6BT, United Kingdom.

Abstract. In the context of neutron science, Bayesian inference methods have been recently implemented within the MANTID framework [Monserrat D *et al.* 2015 *J. Phys. Conf. Ser.* **663** 012009 (2015)]. In this contribution, we highlight the advantages of this software package for robust data analysis and subsequent model selection. To this end, we use the celebrated Rosenbrock function to illustrate its merits and strengths relative to *classical* fitting algorithms. We also introduce the latest additions implemented in MANTID, with a view to increasing its user friendliness as well as stimulating wider use. These include simulated-annealing schemes to reduce the need for initial guesses, as well as new options for multidimensional fitting.

1. Reverend Bayes to the Rescue

Data fitting is very often a key step in the scientific method. In the context of neutron science, the analysis of Quasielastic Neutron Scattering (hereafter QENS) data is a good case in point, as the measured response almost invariably consists of the cumulative contribution of partially overlapping signals convolved with the instrument resolution in the presence of (typically) unwanted features such as backgrounds and/or so-called 'spurions.' In this situation, a robust analytical description of a given experimental data set requires the use of several and often complex spectral functions, whose precise number and associated line shapes carry vital information on the nature of the underlying properties of the material under investigation. In the language of data analysis and statistics, we are faced with a problem in *model selection*.

Using the conceptual framework originally introduced by Sivia and Skilling [1], previous work has focused on the development of a Fitting Algorithm for Bayesian Analysis of DAta (FABADA) [2, 3]. This algorithm uses an adaptive Markov Chain Monte Carlo (MCMC) method to effect Bayesian model selection, and it is applicable to any type of data. More recently, FABADA has been implemented within the MANTID framework [4], as described in more detail in Refs. [5, 6, 7]. Beyond its well-established applicability to analyse QENS experiments [8, 9, 10], FABADA in MANTID (hereafter FIM) has also found recent use in the interpretation of Compton and mass-selective neutron-scattering data, techniques which in many respects suffer from a similar degree of spectral congestion as QENS [11, 12].

In what follows, we outline the primary advantages of FIM in its current incarnation, as well as give a brief description of the most recent additions to this software package.

IOP Conf. Series: Journal of Physics: Conf. Series **1021** (2018) 012012 doi:10.1088/1742-6596/1021/1/012012

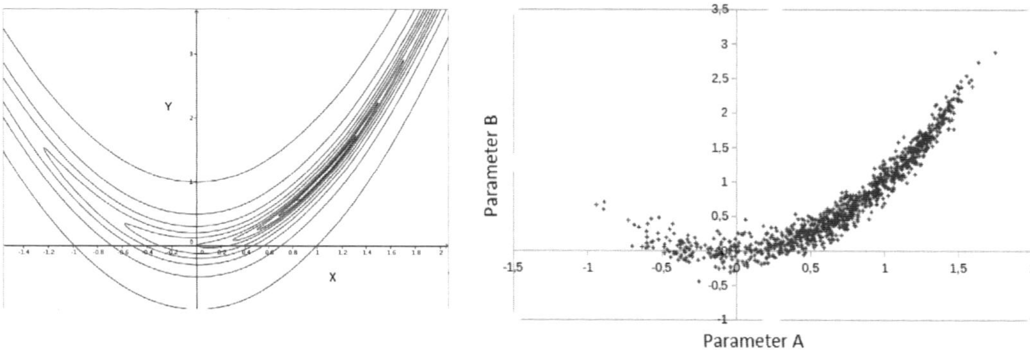

Figure 1. Contours of Rosenbrock function (**left**) and FIM converged-chain points (**right**) for that function described in the main text (for $a = 1$ and $b = 100$). A and B are fitting parameters of fitting function f(x|A,B).

2. Advantages of FABADA in MANTID – A Quick Dive

FIM aims to provide a user-friendly platform to implement Bayesian analysis with an emphasis on neutron-scattering data of any kind. In addition, the MANTID framework contains a plethora of features for the further visualisation and analysis of FIM output, including its much-needed integration with other neutron-focused algorithms and data workflows.

On a more technical front, FIM uses the natural tendency of adaptive MCMC methods to explore the entire parameter space with a (mathematically proven!) [2, 13] tendency to reach the global minimum irrespective of dimensionality. The method also yields a histogram for the *a posteriori* Bayesian parameter distributions, not just the optimal values available in widely used minimisation packages based on linear or non-linear methods, *e.g.*, Levenberg-Marquardt algorithms. This feature is a crucial one, as it enables rigorous model selection via an assumption-free analysis of the underlying Probability Density Functions (PDFs). Such an analysis can deal quite naturally with any general χ^2 landscape, including multimodal and asymmetric PDFs. These PDFs typically challenge the realm of applicability of *classical* fitting algorithms, particularly when dealing with sparse data as those arising from low-count-rate techniques. FIM can deal with these cases by simply replacing the associated cost function by suitable alternatives, and work is underway to automate these possibilities.

Figure 1 illustrates the current capabilities of FIM using the Rosenbrock function $R_{a,b}(x, y) = (x - a)^2 + b(x^2 - y)^2$. This seemingly simple function is widely used to test the robustness and reliability of minimisation algorithms to reach global minima. Standard minimisers including non-linear ones like Levenberg-Marquardt can struggle with this case owing to the non-convex nature of the χ^2 landscape, whereas FIM easily samples parameter space as well as can identify the valley minimum via the use of simulated annealing techniques, as explained in more detail below.

3. Latest Additions

3.1. Simulated Annealing

All fitting algorithms require an initial guess of all parameters, and this condition can cause serious convergence issues. FIM depends far less strongly on this choice, as MCMC methods have the tendency to explore the entire parameter space with little bias – *i.e.*, parameter sets characterised by a higher cost function (typically the χ^2) have a non-zero probability, thus enabling escape from local minima. In this situation, any χ^2 barrier towards the global minimum may be overcome provided enough time is spent exploring parameter space.

IOP Conf. Series: Journal of Physics: Conf. Series **1021** (2018) 012012 doi:10.1088/1742-6596/1021/1/012012

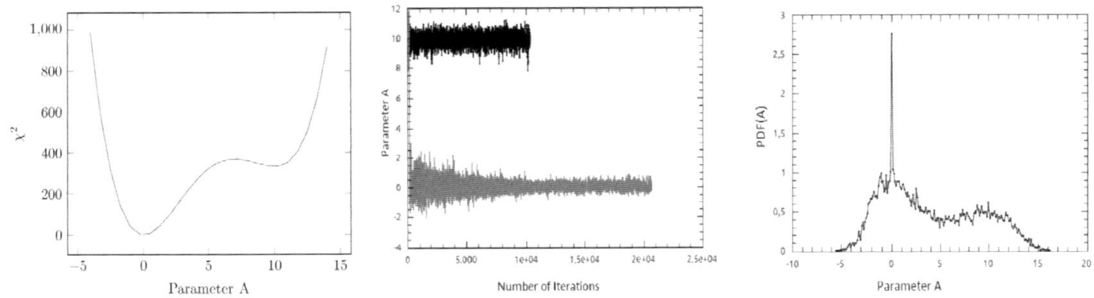

Figure 2. Simulated annealing in FIM: (**left**) χ^2 landscape; (**center**) convergence of *classical* fitting algorithms (black, local minimum) and FIM (red, absolute minimum) using an annealing temperature of 10; and (**right**) resulting PDF from FIM, showing its clear multimodal character.

To accelerate this process, FIM has been modified to include a simulated annealing option that allows for efficient jumps across χ^2 barriers of arbitrary height. In essence, this procedure works by multiplying the cost function by a constant that serves to flatten χ^2 features, therefore enabling an easier exploration of the landscape. This procedure is particularly useful when an initial guess is unavailable. In addition, it can quantify how robust parameter estimates for the location of the minimum are using the associated PDFs. These notions find a familiar analogy in thermal physics, where such a constant may be considered the temperature of the system dictating the total number of accessible states. In this vein, the *fusion* temperature of a given minimum in χ^2 space can be used to define a quantitative criterion for the robustness of a fit – in many respects, this is quite a profound connection between energy landscapes (the realm of physics) and χ^2 distributions (the realm of statistics). Figure 2 provides a simple illustration of simulated annealing in FIM. The model has been constructed to have the χ^2 landscape shown on the left, characterised by the presence of both a global and a metastable basin. The center panel shows how a sophisticated *classical* fitting algorithm such as Levenberg-Marquard gets stuck in the latter, whereas FIM can reach the global minimum without much difficulty. The figure on the right shows the associated PDF obtained with FIM, illustrating its intrinsic multimodal character.

3.2. Multidimensional Fitting

The ability to perform multidimensional fits with FIM has also been developed and preliminary tests have been performed. Using this option, it is possible to fit data collected as a function of several variables in a simultaneous fashion, *e.g.*, spatial dimensions or momentum- and energy-transfers, or parametrically as a function of external stimuli such as temperature or pressure. As example, Fig. 3 shows the uncertainties from a FIM fit to a two-dimensional Gaussian. Amongst a number of applications, we anticipate that this fresh addition to FIM will enable the use of the entire dynamic structure factor to implement model selection, as recently demonstrated in Ref. [8].

4. Outlook

Following its introduction in 2015, FIM continues to grow and evolve. Simulated annealing and multidimensional fitting constitute the latest additions to this software package and extensive testing of these capabilities is currently underway. Looking further into the future, extensions to FIM include: dealing with sparse, low-count-rate data in a statistically rigorous fashion; the definition of quantitative criteria for a systematic analysis of the information content (and associated correlation) of the underlying PDFs; the development of a library of models starting with those typically used for the analysis of QENS and Compton data; or the use of genetic

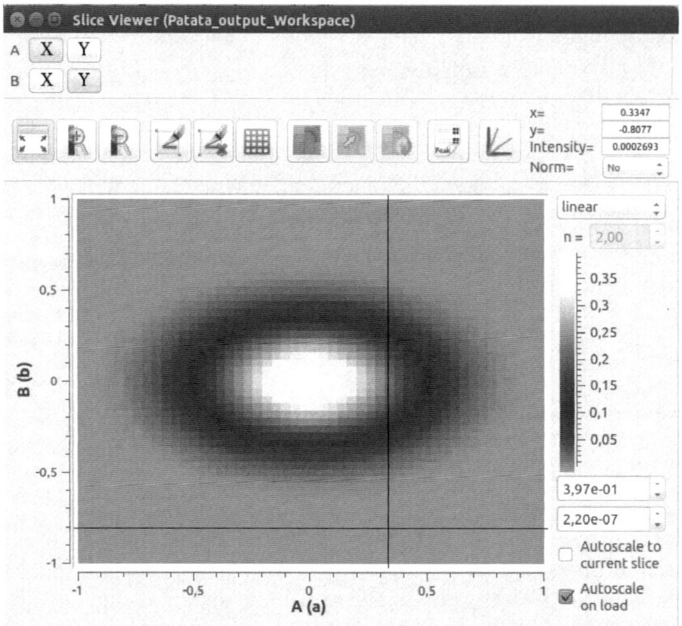

Figure 3. A snap shot of the uncertainties of two fitting parameters, A and B, obtained after multidimensional fitting with FIM to a two dimensional function. The value of χ^2 is presented by colour map. See text for more details.

algorithms to increase overall efficiency. All in all, the primary objective behind these efforts is to develop a multi-purpose state-of-the-art tool for the intuitive and robust analysis of complex neutron data.

Acknowledgements

We gratefully acknowledge the *Universitat Politècnica de Catalunya* and the *UK Science & Technology Facilities Council* for their continued support.

References

[1] Sivia D and Skilling D 2006 *Data Analysis: A Bayesian Tutorial* (Oxford University Press).
[2] Pardo LC, Rovira-Esteva M, Busch S, Moulin JF and Tamarit JLl 2011 *Phys. Rev. E* **84** 046711.
[3] Pardo LC, Rovira-Esteva M, Busch S, Ruíz-Martín MD and Tamarit JLl 2011 *J. Phys. Conf. Ser.* **325** 012006.
[4] The latest version of the MANTID software is available from `dx.doi.org/10.5286/SOFTWARE/MANTID`
[5] For an up-to-date tutorial of FIM to analyse QENS data, see `www.mantidproject.org/Fitting_QENS_using_Bayesian_Routine_FABADA`
[6] Monserrat D, Vispa A, Pardo LC, Tolchenov R, Mukhopadhyay S and Fernandez-Alonso F 2015 *J. Phys. Conf. Ser.* **663** 012009.
[7] D. Monserrat, *BSc Thesis* (Universitat Politècnica de Catalunya, Spain, 2016). WEB: `epubs.stfc.ac.uk/work/24224303`
[8] Vispa A, Busch S, Tamarit JLl, Unruh T, Fernandez-Alonso F and Pardo LC 2016 *Phys. Chem. Chem. Phys.* **18** 3975.
[9] Fernandez-Alonso F, McLain SE, Taylor JW, Bermejo FJ, Bustinduy I, Ruiz-Martin MD and Turner JFC 2007 *J. Chem. Phys.* **126** 234509.
[10] Fernandez-Alonso F, Bermejo FJ, McLain SE, Turner JFC, Molaison JJ and Herwig KW 2007 *Phys. Rev. Lett.* **98** 077801.
[11] Krzystyniak M, Drużbicki D, Romanelli G, Gutmann MJ, S Rudić, Imberti S and Fernandez-Alonso F 2017 *Phys. Chem. Chem. Phys.* `dx.doi.org/10.1039/C7CP00997F`
[12] Romanelli G and Krzystyniak M 2016 *Nucl. Instr. Meth. Phys. Res. A* **819** 84.
[13] Soren Fiig Jarner, Ernst Hansen, 2000 *Stoch. Proc. Applic.* **85** 341.

IOP Conf. Series: Journal of Physics: Conf. Series **1021** (2018) 012013 doi:10.1088/1742-6596/1021/1/012013

A New Design for Live Neutron Event Data Visualisation for ISIS and the ESS

M D Jones[1], F A Akeroyd[2], O Arnold[1], M J Clarke[2], N Draper[1], M Gigg[1], L Moore[2] and T Richter[3]
[1] Tessella, Abingdon, OX14 3YS, UK
[2] ISIS Pulsed Neutron and Muon Source, Rutherford Appleton Laboratory, Science & Technology Facilities Council STFC, Harwell Oxford, OX11 0QX, UK
[3] European Spallation Source, P.O Box 176, SE-22100 Lund, Sweden

E-mail: `matt.clarke@stfc.ac.uk`

Abstract. Viewing and reducing event data *live* during an experiment is an important component of contemporary experiment control systems. At ISIS a simple TCP streaming mechanism for neutron event data allows data to be processed by Mantid during an experiment; however, this existing solution is likely to encounter performance issues with the new generation of acquisition hardware. As part of an ESS in-kind collaboration, ISIS is working on developing a new data streaming system based on Apache Kafka for managing and delivering live experiment data to multiple clients. This work provides an ideal opportunity to replace the existing ISIS streaming mechanism with this new system, which not only provides enhanced functionality for ISIS but also functions as a prototype for developing and testing requirements of the ESS.

1. Introduction

A range of different data must be aggregated during an experiment on a beamline at a pulsed neutron source. These include measurements from sensors monitoring the sample environment, details of the proton pulse from the accelerator and experiment metadata, such as the names of the current users of the beamline. The majority of data by frequency and volume are typically neutron detection events, which comprise information about the location on a detector that a neutron was detected, its time of flight from source to detector and the wall-clock time of the neutron pulse which the neutron is associated with.

Live visualisation of neutron events and other data is increasingly seen as a critical component of experiment control. Visualisation during an experiment allows the scientist to optimise the experiment configuration via the control system based on this continuous feedback. This feedback loop has the highly valuable result of making more efficient use of beamtime compared to the alternative of loading and visualising the data from a file written at the end of each experiment run.

Transporting data from the detectors and other data sources to where it is processed and visualised during an experiment requires a robust and high-throughput data streaming system. Contemporary streaming systems must handle the increasing data rate requirements of facilities which exhibit a trend towards higher intensity neutron sources [1], detectors with larger detector areas, higher spatial resolutions and increased detection efficiencies [2]. Such systems must also be reliable in the sense that they should neither waste beamtime due to system downtime nor compromise scientific validity of results due to data loss.

By working with the European Spallation Source (ESS) Data Management Group in addition to their BrightnESS [3] partners Paul Scherrer Institute and ELETTRA Sincrotrone Trieste in an in-kind project, ISIS has an excellent opportunity to improve on current systems in use and to prototype the future ESS system. It will be invaluable to the ESS to have proven software with well understood operational requirements ahead of the facility's start of operations in 2019 [4]. The ESS will use the Experimental Physics and Industrial Control System (EPICS) [5] for instrument control and has chosen to use the Mantid Framework [6] for live data processing and visualisation. Both of these technologies are in use at ISIS and therefore sharing knowledge and implemented software across the institutions is mutually beneficial.

2. Existing system

The *Instrument Control Program* (ICP) aggregates data on a single machine for each beamline at ISIS. It is connected to the detector electronics and has access to other data sources such as metadata databases and sample environment measurements via a local area network. The ICP is responsible for writing a file in the NeXus format [7] which is then copied to network attached storage from where it can be read for analysis and visualisation using the Mantid Framework. It also serves data to several existing visualisation clients, including Mantid's *Live Listener*, LabVIEW clients and the IBEX control system [8].

The current data streaming solution has several limitations. Firstly, each new connection places additional strain on the instrument server. Since this server is responsible for aggregating and writing data to file it is crucial that it is not overloaded. Secondly, the clients are only able to capture data from the moment they join the stream and will lose data if they cannot process and display data at the same rate as the stream. Data can also be lost on the wire without the knowledge of the client, which is of critical concern when streaming data for analysis or to a file writer. Finally, there are some data types, such as images from the EPICS's *areaDetector* module [9], which are not yet supported.

For the ESS, the facility's high beam intensity and cutting edge detectors are expected to result in neutron detection rates one to two orders of magnitude higher than existing institutions [10]. In the past, the volume of data would be reduced early in the pipeline by histogramming events, however modern analysis requires retaining full event data [11]. Therefore, detection rates of up to 10^8 neutrons per second per beamline create throughput requirements of up to 10 Gbps for neutron data from a single beamline. Implementing a system without the aforementioned limitations is essential since having a single instrument server responsible for aggregation, file writing and serving data with this bandwidth is beyond current hardware capabilities. Furthermore, robustness and data delivery guarantees are of particular importance due to the intention to carry out live data reduction at the ESS [12].

3. The new design

3.1. Apache Kafka

The key technology on which the new design for ISIS and the ESS is based is Apache Kafka. Kafka is an open-source, high throughput, message streaming platform with a configurable level of message delivery guarantees [13]. Kafka implements a publish-subscribe architecture, which is depicted in Figure 1. *Producer* clients publish messages to named data streams, called *topics*, on a cluster of Kafka *brokers*. *Consumer* clients subscribe to one or more topics to receive messages.

Topics are persisted on disk on brokers in one or more *partitions*. Each newly produced message is appended to the end of a partition, thus partitions are log files of messages. Partitions are distributed across brokers in the cluster to allow throughput for a topic to exceed the capacity of a single broker. Topics can also be configured to replicate their partitions on different brokers, with priority given to brokers in different server racks. Broker machines or entire racks can therefore fail without the system losing data or clients experiencing loss of service. The oldest messages, at the start of the partition, are removed once their age reaches a configurable *retention time* or to ensure a specified limit on data volume is not exceeded.

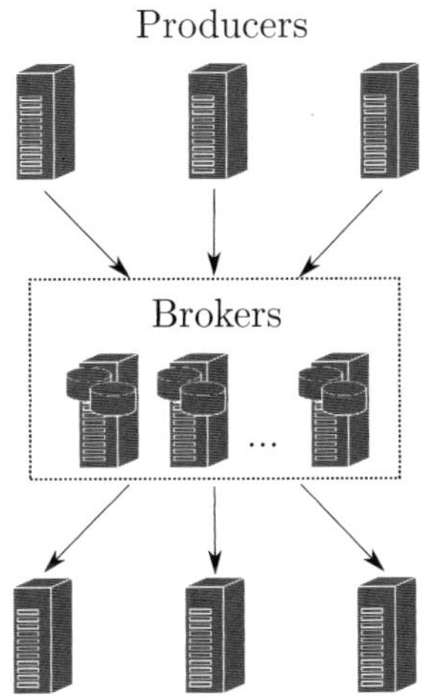

Figure 1. Basic architecture of Apache Kafka.

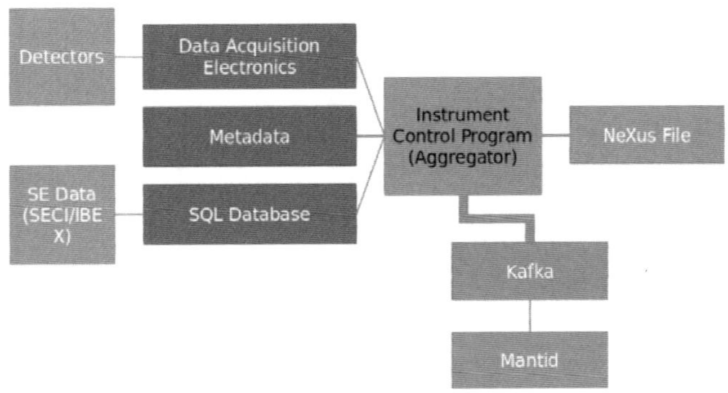

Figure 2. Architecture of the initial prototype of the new neutron data streaming system at ISIS.

3.2. Initial prototype

Figure 2 shows the initial prototype of the new system at ISIS. In this design the ICP functions as a producer and streams aggregated data into the Kafka cluster. Since the scalable Kafka cluster is then responsible for serving data to any consumers this approach solves the issue of additional clients placing strain on the ICP. As Kafka stores messages in partitions on the cluster, clients do not have to miss data if their processing cannot keep up with data production. An additional benefit is that clients can request historical data, for example data from the start of the current experimental run even if they join the stream part way through.

Mantid's Live Listener interface was updated to allow it to function as a consumer client for the Kafka-based system. The Live Listener subscribes to a topic containing neutron event data and updates a data structure, a *workspace*, at a specifiable interval with information that has been streamed to it. At each interval, processing algorithms can be configured to run on the new, or cumulative, data and the instrument view visualisation can be updated. An example of Mantid's *Instrument View* is shown in Figure 3 and displays accumulated counts of detected neutrons on a three-dimensional view of the detectors of the WISH beamline at ISIS during an experiment run.

3.3. Future

In the prototype system the ICP remains responsible for aggregating data and writing the NeXus file. As previously mentioned, the expected throughput requirements of the ESS make this approach impossible for the new facility. Instead, various producer clients are being implemented to publish data from different sources, such as EPICS and databases providing metadata. Similarly a separate, NeXus file writing, consumer client is in development by colleagues at the Paul Scherrer Institute as part of the BrightnESS project [3]. The system at ISIS will be

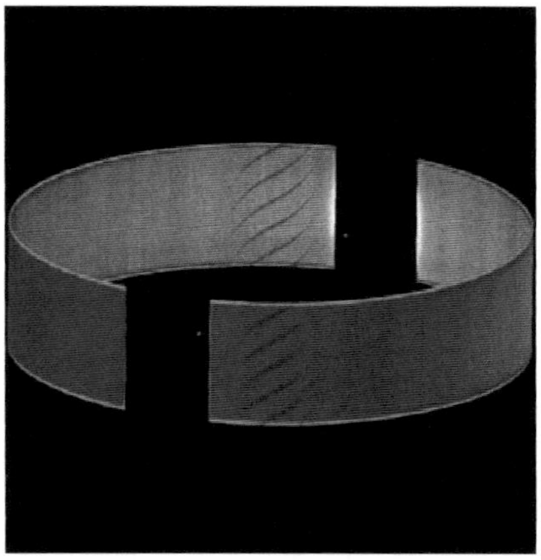

Figure 3. MantidPlot's Instrument View interface. A colour map of cumulative, detected neutron count is displayed on a three-dimensional view of the detectors of the WISH instrument.

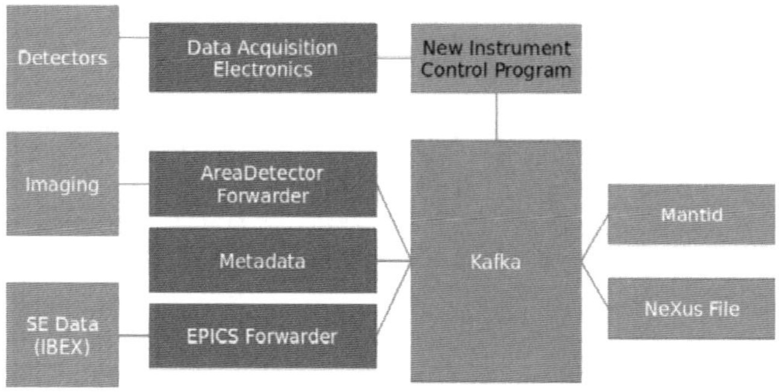

Figure 4. Planned architecture for future development of the data streaming system at ISIS.

updated to make use of these clients and mimic the planned architecture for the ESS, as shown in Figure 4. Not only will this benefit ISIS by adding support for datatypes not streamed by the current system, but also make the system a true prototype for the ESS.

Acknowledgments
The authors wish to acknowledge the contributions of M Brambilla, M Könnecke and D Werder at the Paul Scherrer Institute, and M Christensen, A Mukai and J Nilsson at the European Spallation Source.

References

[1] Peggs S *et al* 2013 *ESS technical design report* ISBN 9789198017328 URL `http://eval.esss.lu.se/cgi-bin/public/DocDB/ShowDocument?docid=274`
[2] Hall-Wilton R, Höglund C, Imam M, Kanaki K, Khaplanov A, Kirstein O, Kittelmann T, Nilsson B and Scherzinger J 2012 *IEEE Nuclear Science Symposium Conference Record* 4283–4289 ISSN 10957863
[3] BrightnESS URL `https://brightness.esss.se/`
[4] European Spallation Source 2015 Activity Report 2015 Tech. rep. URL `http://europeanspallationsource.se/sites/default/files/ess_activity_report_2015_printer_version.pdf`
[5] EPICS: Experimental Physics and Industrial Control System URL `http://www.aps.anl.gov/epics/`
[6] Arnold O *et al* 2014 *Nuclear Instruments and Methods in Physics Research, Section A: Accelerators, Spectrometers, Detectors and Associated Equipment* **764** 156–166 ISSN 01689002 (*Preprint* 1407.5860) URL `http://dx.doi.org/10.1016/j.nima.2014.07.029`
[7] Könnecke M *et al* 2015 *Journal of Applied Crystallography* **48** 301–305 ISSN 16005767
[8] Clarke M J *et al* 2015 *ICALEPCS 2015 Conference Proccedings* 0–3 URL `http://icalepcs.synchrotron.org.au/papers/mopgf048.pdf`
[9] areaDetector: EPICS Area Detector Support URL `http://cars9.uchicago.edu/software/epics/areaDetectorDoc.html`
[10] Klinkby E and Soldner T 2016 *Journal of Physics: Conference Series* **746** 012051 ISSN 1742-6588
[11] Peterson P F, Campbell S I, Reuter M A, Taylor R J and Zikovsky J 2015 *Nuclear Instruments and Methods in Physics Research, Section A: Accelerators, Spectrometers, Detectors and Associated Equipment* **803** 24–28 ISSN 01689002 URL `http://dx.doi.org/10.1016/j.nima.2015.09.016`
[12] European Spallation Source URL `https://europeanspallationsource.se/`
[13] Apache Kafka: A distributed streaming platform URL `https://kafka.apache.org`

IOP Conf. Series: Journal of Physics: Conf. Series **1021** (2018) 012014 doi:10.1088/1742-6596/1021/1/012014

YUI and HANA: control and visualization programs for HRC in J-PARC

Daichi Kawana[1], Minoru Soda[1,3], Masahiro Yoshida[1], Yoichi Ikeda[1,4], Toshio Asami[1], Ryosuke Sugiura[1], Hideki Yoshizawa[1], Takatsugu Masuda[1], Takafumi Hawai[2], Soshi Ibuka[2], Tetsuya Yokoo[2], and Shinichi Itoh[2]

[1] Neutron Science Laboratory, Institute for Solid State Physics, the University of Tokyo, 106-1 Shirakata Tokai, Ibaraki 319-1106, Japan

[2] Neutron Science Division, Institute of Materials Structure Science, High Energy Accelerator Research Organization, 1-1 Oho Tsukuba, Ibaraki 305-0801, Japan

E-mail: `kawana@issp.u-tokyo.ac.jp`

Abstract. We developed control and visualization programs, YUI and HANA, for High-Resolution Chopper spectrometer (HRC). YUI is a comprehensive program to control DAQ-middleware, the accessories, and sample environment devices. HANA is a program for the data transformation and visualization of inelastic neutron scattering spectra. We describe the basic system structures and unique functions of these programs from the viewpoint of users.

1. Introduction

High Resolution Chopper spectrometer (HRC) is an inelastic neutron chopper spectrometer to investigate elementary excitations of spin and orbital fluctuations in strongly correlated electron systems with high energy resolution. Recently it is used also for the study of hydrogen atoms in a solid. HRC is installed at BL12, in Materials and Life Science Experimental Facility (MLF), J-PARC [1, 2]. Basic accessories including vacuum chamber, T0 chopper, Fermi chopper and Soller collimator, and minimal sample-environmental devices had been prepared until 2010, and then we have constructed computing system and developed software environment to control these devices and to handle neutron signals obtained from detectors. Comprehensive-control program YUI and visualization program HANA have central roles in the software environment. In this paper, we firstly explain the configuration of the computing-system which underlies the two programs. Next, we explain basic constructs and unique and characteristic functions in YUI and HANA.

2. System configuration

The HRC computing system consists of six Linux-based machines (PC), three network-attached storage (NAS) machines, and four switching (SW) hubs, as shown in figure 1. For every PC

[3] Present address: Center for Emergent Matter Science, RIKEN, 2-1 Hirosawa Wako, Saitama 351-0198, Japan.
[4] Present address: Institute for Materials Research, Tohoku University, 2-1-1 Katahira Aoba Sendai, Miyagi 980-8577, Japan.

except the FL-net gateway, which is only communicated with the vacuum-chamber controller, DAQ-middleware (DAQ-MW) [3] was installed.

The roles of these machines are as follows. Two DAQ-CPU machines collect the event data generated from the position-sensitive detectors (PSD) via the NeuNET modules [4]. The number of the machines is determined by taking into account the total number of PSDs (324 PSDs as in Mar. 2017). The obtained event data are stored to one NAS machine. This sequent process is controlled by the DAQ-operator machine, on which the program YUI simultaneously runs. The versatile server machine aggregates and logs statuses from the devices. Its communication mechanism is described in the following section. The analysis server machine behaves as a CPU of the program HANA.

These machines are linked via the three independent HRC-intranet networks, i.e., storage LAN, control LAN, and analysis LAN, and the other J-PARC network infrastructure, called JLAN. They play roles in handling the obtained event data, intervening between the computing system and all the controllable devices, processing the data visualization, and connecting to the internet, respectively.

In our computing system, a client-server model is adopted for the convenience of users. YUI and HANA are respectively available as the terminal in two iMac PCs in the user's room of the HRC cabin. Users do not need to operate the Linux machines directly.

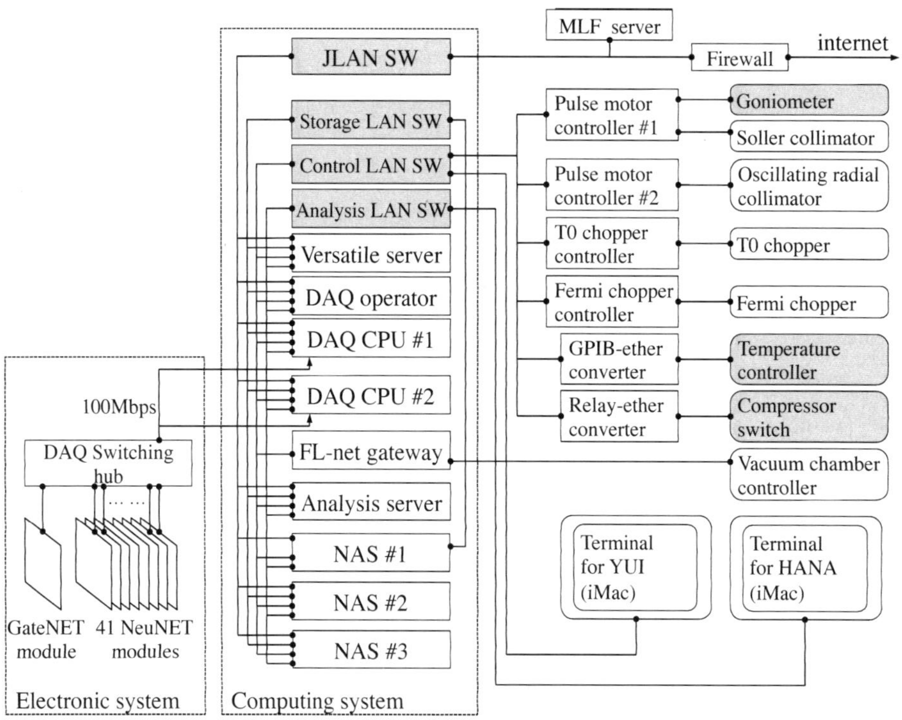

Figure 1. Schematic diagram of the computing system in HRC. Open and hatched round-squares denote common-use accessories and sample-environment devices, respectively.

3. YUI

The comprehensive control program YUI is the abbreviation of *Yasashii* or *Yoku-dekita* User Interface. Here, these italic words are Japanese which means friendly or easy, and well-constructed, respectively. YUI allows users to perform the measurement by sequential operation of DAQ-MW, to control the devices, and to log their status. The main window of YUI is shown in figure 2.

The most basic function is a measurement by DAQ-MW, so-called "DAQ Run". By sending "begin" and "end" commands to DAQ-MW, one measurement is completed. On YUI, DAQ Run is implemented as the built-in macro, called "user-command". Other than DAQ Run, several user-commands to control the accessories such as the Fermi Chopper have been prepared. Users can perform their experiment by editing and executing a sequence of appropriate user-commands.

To control the accessories, of which their communication protocols are different, we developed a "device server," a virtual machine which intervenes between YUI and the corresponding device. The installed device servers work for DAQ-MW, MLF server, Fermi chopper, Soller collimator, oscillating radial collimator, goniometer and temperature controller. Here, the MLF server is regarded as the accessory, to obtain the number of protons injected to the neutron production target between the beginning and completion times of DAQ Run. The number of protons is used to normalize the observed neutron intensity on HANA, as below. The device servers of the last two are prepared for each set of sample-environment devices, such as the GM-type refrigerator and the ^{3}He circulation-type refrigerator [5]. Users activate and connect their required device servers on the device-server list subwindow as shown in figure 2.

Set and current values of the devices are continuously monitored and recorded to the log files every 5 seconds. The obtained log data can be graphically displayed on a browser page named "Log-visualizer". This helps users to check the time variation of the current values of the devices, such as stability of the phase of the Fermi chopper. The details are described elsewhere [6].

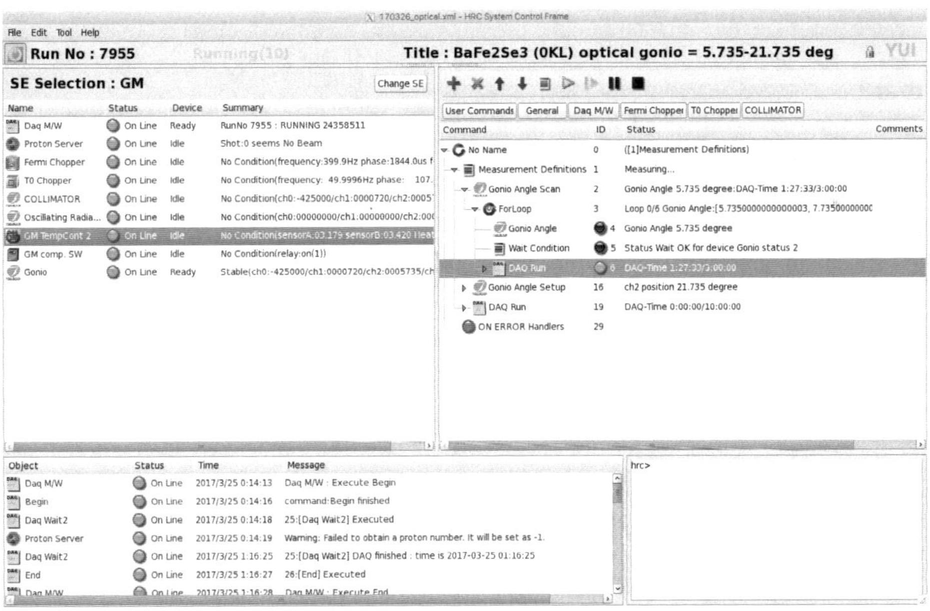

Figure 2. The main window of YUI. It consists of three subwindows: list of activated device servers (top-left), sequence edited by users and its progress status (top-right), and messages communicated with devices, called system log (bottom).

4. HANA

HANA (HRC ANAlyzer) is a data-reduction program for the initial data handling and the data visualization. The basic flow is creation of intensity-histogram data from the event data, sequent reduction of the histogram data, and visualization of the dynamical structure factor in four-dimensional space.

One detected neutron is treated as a data set labeled with three parameters of (d, p, t), where d, p, and t denote the PSD number, position on the PSD, and time of flight (ToF) of the neutron, respectively. Here, p is given from the ratio of the electrical pulse heights at the upper and lower ends [3]. The parameters (d, p) can be regarded as the geometric coordinate on the PSD array. From the data set (d, p, t) and the incident energy E_i, the energy transfer ω is calculated. This calculation is executed for each neutron and then the intensity-histogram data of $I(d, p, \omega)$ is created. Here, the observed value of E_i can be determined by a simple fitting procedure on HANA.

The dynamical structure factor in four-dimensional space can be visualized in two patterns. One is for a single-crystal. The momentum transfer $\boldsymbol{Q}$ has the reciprocal-lattice coordinates of (Q_a, Q_b, Q_c). Their directions are defined from the given lattice parameters. Users can visualize a two-dimensional intensity map, on which two of the four parameters Q_a, Q_b, Q_c, and ω are selected as vertical and horizontal directions. The other pattern is for a powder sample. The momentum transfer is simply regarded as $|\boldsymbol{Q}| = Q$. Users can visualize an intensity map of the Q-ω plane.

As unique functions of HANA, the histogram data can be transformed to a data format of the visualization program DAVE MSlice [7]. The alignment support function is convenient and indispensable in the beginning of single crystalline experiments. From a Laue profile measured by a white beam, users can obtain the information of the relation between the coordinate of the crystal and that of the spectrometer with an easy procedure. The details of the procedures are descrived elsewhere [6].

5. Summary

We have developed the control program YUI and the data-analysis program HANA for HRC spectrometer. They have been improved day-by-day for convenience of users and instrument staff members. The hardware of HRC is always updated; for example a new Soller collimator has been installed for more neutron flux, and a new device for sample environment will be installed in future. YUI also evolves with the progress of the hardware. On the other hand in HANA, we are planning to implement a functions of the analysis for four-dimensional data obtained from the continuously rotation scan of a single crystal.

Acknowledgments

The development of the computing system and software environment for HRC are approved and progressed by the Neutron Scattering Program Advisory Committee of IMSS, KEK (Proposal No. 2010S01-2016S01). We would like to appreciate Bee Beans Technologies Co.,Ltd. (Tsukuba, Japan) for gratefully supporting our works as partnership.

References

[1] Itoh S, Yokoo T, Satoh S, Yano S, Kawana D, Suzuki J and Sato T J 2011 *Nucl. Instr. and Meth. Phys. Res. A* **631** 90.
[2] Itoh S, Yokoo T, Kawana D, Yoshizawa H, Masuda T, Soda M, Sato T J, Satoh S, Sakaguchi M and Muto S 2013 *J. Phys. Soc. Jpn.* **82** SA033.
[3] Satoh S *et al.* 2009 *Nucl. Instr. and Meth. Phys. Res. A* **600** 103.
[4] Yano S, Itoh S, Satoh S, Yokoo T, Kawana D and Sato T J 2011 *Nucl. Instr. and Meth. Phys. Res. A* **654** 421.
[5] Itoh S, *et al.* 2016 *proc. of ICANS-XXI (JAEA-Conf 2015-002/KEK Proceedings 2015-7)* 298.
[6] Kawana D *et al.* arXiv:1710.00487.
[7] Azuah R T, Kneller L R, Qiu Y, Tregenna-Piggott P L W, Brown C M, Copley J R D and Dimeo R M 2009 *J. Res. Natl. Inst. Stand. Technol.* **114** 341.

IOP Conf. Series: Journal of Physics: Conf. Series **1021** (2018) 012015 doi:10.1088/1742-6596/1021/1/012015

Applications of the differential events reading method at MLF, J-PARC

Y. Inamura[1], T. Ito[2] and J. Suzuki[1,3]

[1] Materials and Life Science Division (MLF), J-PARC Center, Tokai, Ibaraki 319-1195, Japan
[2] Neutron Science and Technology Center, Comprehensive Research Organization for Science and Society (CROSS), Tokai, Ibaraki 319-1106, Japan
[3]High Energy Accelerator Research Organization (KEK), Tsukuba Ibaraki, 305-0801, Japan
E-mail: yasuhiro.inamura@j-parc.jp

Abstract. An event recording method for data acquisition for neutron scattering and its analysis provides a lot of benefits for the measurement of time-resolved and time transient phenomena. Almost all instruments installed in the Materials and Life Science Experimental Facility (MLF) at J-PARC have adopted this method since the first beam came and have achieved good results. On the other hand, our treatment of event recorded data leaves room for improvements and developments to achieve more effective utilization. This paper introduces one of the applications of the event recording method to achieve the pseudo real-time data treatment in MLF in a simple way.

1. Introduction

An event recording method for neutron scattering measurements has been adopted widely at a lot of neutron instruments in the world because this method has advantage for time-resolved measurements and response phenomena in a sample under an external field. The Materials and Life Science Experimental Facility (MLF) also has adopted an event recording method as a standard data acquisition system (DAQ), which is realized by DAQ middleware [1], since the first neutron beam was produced. The DAQ middleware system can gather neutron detection events from the board connected to each detector, i.e. ^{3}He-position sensitive detector (PSD), 2D- or 1D- scintillation counter, N_2 beam monitor and so on, to store the data as binary files. Now the event recorded information comprises not only a neutron or muon detection but also conditions of sample environment and instrument devices in MLF, for example an orientation of sample stage goniometer, temperature, choppers, beam narrower (beam slits) and so on. The neutron detection event in MLF includes the time of flight and the position of the neutron detection. This method gives MLF users higher flexibility and efficiency on the measurement and analysis.

The first step to treat event recorded data is to convert events to histogram data with a unit conversion for easy understanding and to do a variety of corrections on the data, so-called data reduction. At almost all instruments in MLF, "Manyo Library" [2,3] and "Utsusemi" software [4] are utilized to do such data reduction steps. The Manyo Library is a software framework for handling data containers with operations and developing data reduction software. The Utsusemi, which is based on the Manyo Library, is a set of software to analyze and visualize event recorded data with a graphical interface [Fig.1]. Especially, the Utsusemi successfully contributed to enable the multi-Ei method in inelastic scattering for the first time in the world by its flexible data reduction [5].

The data reduction functions in Utsusemi are developed under the assumption that they are to be carried out with a static data file after its data acquisition finishes. This means that the histogram conversion code reads data files from the beginning to end for each time. On the other hand, users require strongly a "real-time analysis and visualization" to see and check the obtained data during measurements, because users want to confirm the current measurement conditions frequently and to decide the next condition as soon as possible. In this case, as the data files are accessed from both the DAQ middleware and the data reduction code for a long time, the risk of trouble such as file destruction, becomes higher. In addition, it takes a long time to read data files from the beginning. To resolve these issues, we produced a kind of monitoring system which observes the current event recorded data files and periodically executes data reduction only on differential events.

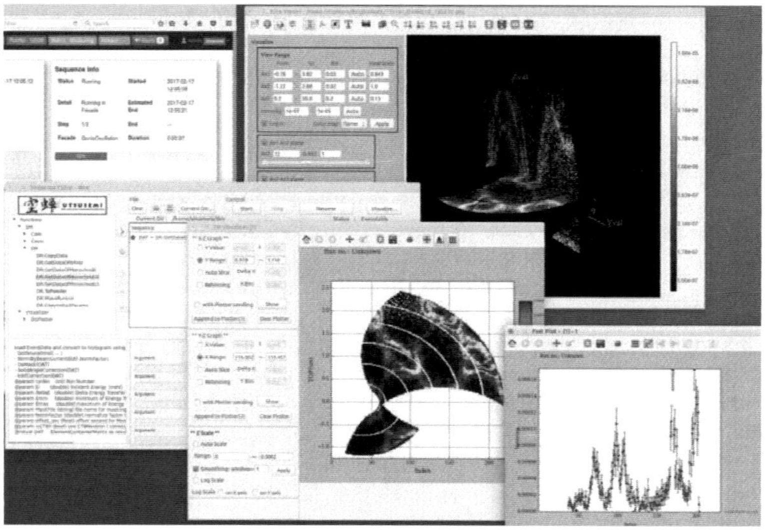

Figure 1. Screen shot of the desktop that the Utsusemi software is working on. Utsusemi is the set of software to analyse and visualize the neutron scattering data observed at MLF.

2. The event treatment and the concept for developments

To realize 'real-time' data reduction and visualization, we have considered that we should make the event recording method practical to use. So we decided to utilize the differential events in the event data files produced continuously by a DAQ system. Therefore, this system is not a true 'real-time' treatment but a pseudo one. In this section, we show the format and the treatment of event data in MLF and the concept of the development of the differential events reading method.

2.1. Event recorded data format and treatments

The events used in MLF are classified into several types, i.e. a neutron detection event, T0 event which indicates the origin of time-of-flight with its identification number, an absolute time event and a common electrical signal event. A neutron detection event includes the internal position information on the detector and time-of-flight, but event format for each detector, for example between ^{3}He-PSD and a scintillation counter, has a slightly different characteristic. Especially since ^{3}He-PSD event includes the pulse height values of amplified electrical signals from detectors as position information, we distinguish events from electrical noise by pulse height filtering on histogram conversion. This function is very effective for analysis of the data including gamma ray background from instrument devices and sample by incident neutrons. Neutron detection events are produced from an electronics board connected with detectors, which are developed by the DAQ group in MLF. One board groups all events of several detectors into one event stream to be gathered by the DAQ middleware to save as the event data files. Each event data file includes identical T0 events and absolute time events for keeping time synchronization between files.

On converting event data to histogram format in Utsusemi, the time-of-flight values included in events can be converted to another unit, i.e. energy, wavelength, momentum transfer and energy transfer, using detector positions with any range given by users. In addition, software enables extraction of events in the given absolute time region to make a histogram simply. A period of the neutron production at the source of MLF is a typical unit used for a time slicing, 25 Hz, which is called a frame. We generally adopt a frame as the time unit because wide energy neutrons are usable in this period. For speeding up reading and treating event data files, the Utsusemi code divides the event stream from the data file into the unit of frames to be used for data reduction with multiple processes. To divide the event stream effectively, we prepare a bookmark list for T0 events inside a file, called T0 index information, which records the number of events from the beginning of a file to each T0 event and its absolute time as shown in Fig.2. These bookmark T0 indexes are used not only to distribute each divided data stream to multiple processes for off-line reading of data files but also to slice by a given time range with high speed. T0 index information can be made from each event data file and can be reused once it is made. For instruments using Utsusemi in MLF, their DAQ system also includes the software module to make T0 index information directly from streaming of events and save them as a file.

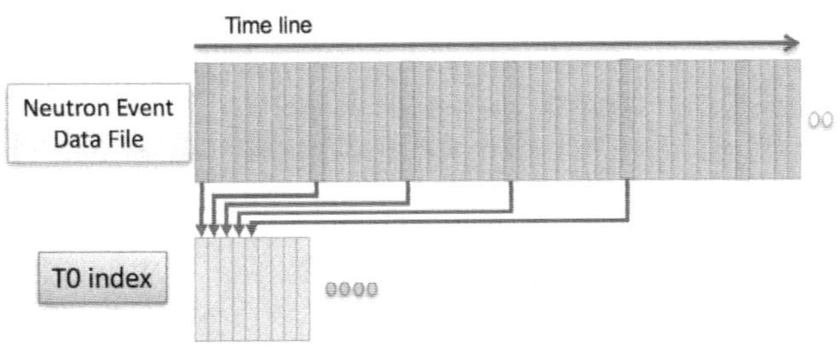

Figure 2. Diagram showing the meaning of T0 index. Each event data file includes T0 events (red box) and neutron events (grey box). T0 index events (blue box) is made by the analysis of the event data file and includes the T0 event position in the file used as a bookmark. T0 index list is saved as a binary file.

2.2. The concept to realize a pseudo 'real-time' data reduction

To realize a pseudo 'real-time' data reduction, we produced the differential events reading method which observes the event data files periodically to read the differential part of the files updated and convert them to histogram format containers incrementally [Fig.3]. Reading differential events from the file is easy to introduce into the existing software at beamlines because it has little effect on the DAQ system and software environment already in use.

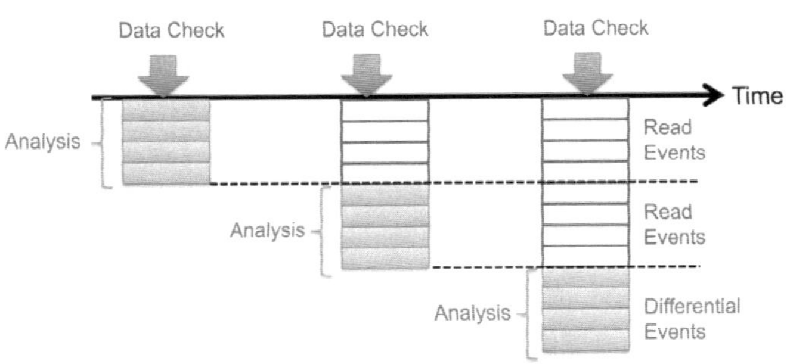

Figure 3. Diagram for the differential events reading method. Analysing only the differential events reduces the time cost of data treatment for the data reduction process.

At first, the software allocates and keeps the histogram area with given information by a user on the memory in PC and starts the observation on the updated data file. The target files of the observation are not event data files but T0 index files to reduce the access to the storage. If updated T0 index files are found, the software reads the updated part of T0 index to extract the differential events stream from the files. If there is a disagreement between the T0 information and the actual event data file by a different output timing from the DAQ system, the software adjusts the T0 information for the actual situation. The extracted differential events are converted incrementally to histogram format containers in the memory to be used for the data reduction steps. As this histogram format is the same as that used at the off-line data reduction and visualization, the existing code can be utilized directly. Users can clear the incremental counts of histogram in the memory for each observation to visualize the differential data like a current data monitoring. This method is very convenient but does not contribute to the time cost for the data reduction steps. In addition, if users want to change the axis unit or the ranges of the histogram, the software must delete the histogram in the memory area to read event data files from the beginning again. We understand that there are still risks that the data on a the storage is damaged but the risk becomes small.

3. Applications in MLF

Our differential event reading method has several features. One is the processed speed become higher because of the decrease of the amount of events to be treated at a time. Second is the flexibility of the data observation, which enables users to decide an interval time to obtain the differential events and to convert events to histograms with counting not only incrementally but also differentially, i.e. counts summation from beginning or only differential events in the interval period. The third is that it is a simple task to introduce this differential data reduction by only replacing the program to read event data with this software module.

This method has already been introduced in several instruments in MLF. One example is a neutron beam monitor with visualization. This simple software creates incremental histogram data by reading the differential events and plots it with a 1D plotter. This system is installed at BL15, BL17 in MLF [Fig.4(a)]. The other case is where many detectors must be treated. At BL11, in which over 300 PSDs are installed, users want to watch the 'real-time' incremental data which the data reduction process finishes [Fig.4(b)]. The third case is to be used in the inelastic beamlines. In the inelastic scattering measurements on a single crystal, we have many opportunities for measuring a crystal rotated over a wide angle to obtain the whole excitation in the energy-momentum space as a map. Recently we successfully produced the system of 4D mapping measurement not by a step-by-step change of a single crystal sample orientation but by a continuous rotation of a sample with a pseudo real-time visualization for the incremental data during the measurement, as shown in Fig.4(c). In this system, orientation information is recorded as events with a time stamp the same as one of the neutron detections events. We added the new function to the differential data reduction software to divide neutron events by orientation information at the detection time for each angle. We do not describe the details of this result in this paper.

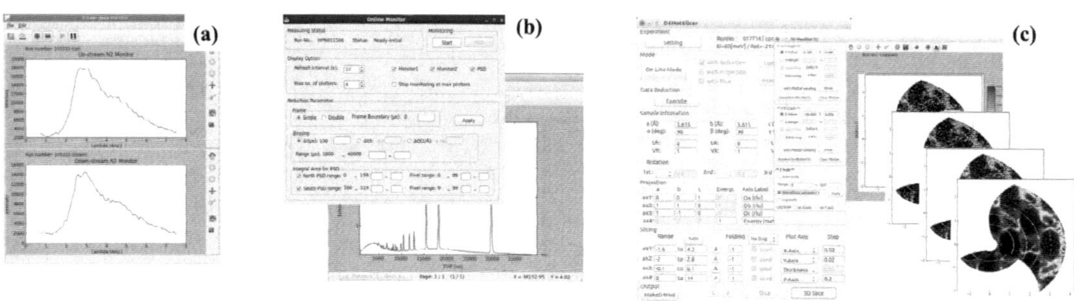

Figure 4. The screen shots of GUI for applications using the differential event reading method. (a) is the beam monitor, (b) is 'real-time' data reduction of many detectors and (c) is the 4D space mapping measurement by the continuous rotation of a single crystal.

4. Future plans

The basic system for the differential event reading method is ready, however this is introduced in only a few beamlines because of the lack of easy-to-use graphical interface to execute each data reduction step with this method. Therefore, we continue to develop these. On the other hand, there is an intense progress of the technology treating a message queue recently. We also consider that this technology of an asynchronous communications protocol should be used for our data acquisition system and the data reduction, especially for monitoring current measurement data and device conditions. We have already started an attempt to import the message queue system into our DAQ system and to apply it to the data reduction part in Utsusemi. It is not difficult to replace the differential event conversion with a queue protocol as mentioned above.

5. Summary

We introduced, in this paper, how to treat event recorded data in MLF and successful use for differential event reading method as a common data monitoring system in MLF. We achieved the situation that users can get full analyzed data just after finishing their measurement. However, this system leaves room to be upgraded with the new technology and should be utilized for more beamlines. We will continue the developments for these purposes.

References

[1] Y. Yasu, K. Nakayoshi, H. Sendai, E. Inoue, M. Tanaka, S. Suzuki, S. Satoh, S. Muto, T. Otomo, T. Nakatani, T. Uchida, N. Ando, T. Kotoku, S. Hirano, International Conference on Computing in High Energy and Nuclear Physics (CHEP09), 2009.

[2] J. Suzuki et. al., Nuclear Inst. and Meth. Phys. Res., A534 175 (2004).

[3] J. Suzuki et. al., Nuclear Inst. and Meth. Phys. Res., A600 123 (2009)

[4] Y. Inamura, T. Nakatani, J. Suzuki, and T. Otomo, J. Phys. Soc. Jpn. 82, (2013), SA031-1 – SA031-9

[5] M. Nakamura, R. Kajimoto, Y. Inamura, F. Mizuno, M. Fujita, J. Phys. Soc. Jpn. 78(2009) 093002-1 – 093002-4

IOP Conf. Series: Journal of Physics: Conf. Series **1021** (2018) 012016 doi:10.1088/1742-6596/1021/1/012016

Measurement of activation cross sections of the target and the proton beam window materials at J-PARC

H Matsuda, S Meigo, H Iwamoto
JAEA, 2-4, Shirakata, Tokai-mura, Naka-gun, Ibaraki, Japan
E-mail: matsuda.hiroki@jaea.go.jp

Abstract. To improve the accuracy of the nuclear design for spallation neutron sources and transmutation systems, nuclear reaction cross sections are required. Considering decommissioning of the accelerator facilities, accurate cross sections are mandatory. The activation cross sections have been measured in several facilities. However, they have low accuracy and precision. Especially, since there is almost no data for 3 GeV protons which are used for the spallation neutron source (MLF) in J-PARC, the experimental data is required for in order to improve the target materials. For the sake of the forthcoming full-time measurement of the activation cross sections for various nuclei, we measured the cross sections of aluminium with 0.4 GeV, 1.3 GeV, 2.2 GeV, and 3.0 GeV protons as a test case. It was found that more accurate data than current ones would be measured by using precise beam controls and highly accurate beam monitoring. We compared the experimental data, the evaluated data (JENDL-HE/2007), and the calculations with several intranuclear cascade models using the PHITS code. Although the experimental data agreed with JENDL-HE/2007, the calculations underestimated by about 40%. This arises from the evaporation model (GEM) being included in PHITS code. We found that the calculations could be made to agree with the experimental data by upgrading the PHITS code.

1. Introduction

In the Japan Proton Accelerator Research Complex (J-PARC) [1], a MW-class pulsed neutron source, the Japan Spallation Neutron Source [2], and the Muon Science facility (MUSE) [3] will be installed in the Materials and Life Science Experimental Facility (MLF). Since 2008, this source has produced a high-power proton beam of 300 kW. In 2015, we successfully ramped up beam power to 500 kW and delivered a 1-MW beam to the targets. To produce a neutron source, a 3 GeV proton beam collides with a mercury target, and to produce a muon source, the 3 GeV proton beam collides with a 2-cmthick carbon graphite target. To efficiently use the proton beam for particle production, both targets are aligned in a cascade scheme, with the graphite target placed 33 m upstream of the neutron target. For both sources, the 3 GeV proton beam is delivered from a rapid cycling synchrotron (RCS) to the targets by the 3NBT (3 GeV RCS to Neutron facility Beam Transport) [4, 5] Before injection into the RCS, the proton beam is accelerated up to 0.4 GeV by a LINAC. The beam is accumulated in two short bunches and accelerated up to 3 GeV in the RCS. The extracted 3 GeV proton beam, with a 150 ns bunch width and a spacing of 600 ns, is transferred to the muon production target and the spallation neutron source.

At J-PARC, another facility called the TEF is planned for development of the Accelerator Driven System (ADS) target system as well. A similar proton beam window will be applied at the TEF. For the decommissioning of the accelerator and TEF facilities, the activation cross sections known with high accuracy are mandatory for various materials. As a first step for further forthcoming measurements, the cross sections for aluminium with 0.4 GeV, 1.3 GeV, 2.2 GeV, and 3.0 GeV protons have been measured.

2. Experiment

2.1. Setup

In order to obtain activation cross sections, the experiment was carried out at the beam transport to the RCS and the MLF. A sample foil was placed at the beam dump line used for tuning the RCS. To estimate background radiation, some of the foils were placed at the outside of the stage. A liner stage guide was utilised to control the irradiation status. Thin aluminium foils with 0.1 mm thickness and 25 mm $\times$ 25 mm square were used as irradiation targets. Each of them was sandwiched by aluminium foils of size 38 mm $\times$ 60 mm and with the same thickness as the target. Four sets were placed at the entrance of the beam dump. The dump is made of iron, where is placed downstream at about 12 m from the foil. For the measurement of the cross section, it is important to control the beam irradiation conditions since the dump is utilised for beam tuning of the accelerator. To control the irradiation conditions, the foil was placed on a movable stage in the vacuum chamber. After irradiation, the sample holder was extracted from the vacuum chamber and then retracted from the holder. During the foil extraction and placement into the vacuum chamber, the sample was slowly depressed and pressed with slow leak valve to avoid rupture and deformation of the sample.

2.2. Irradiation

Each foil was irradiated by 0.4 GeV, 1.3 GeV, 2.2 GeV, and 3.0 GeV protons. The beam width was observed with three sets of Multi Wire Profile Monitors (MWPM), which are placed to measure the beam profile at the beam dump. In order to avoid unnecessary irradiation of the wires, the frame can retract from the beam and moves with a pendulum motion. It was shown that the beam width at the foil was smaller than 2.5 mm to 6.5 mm. It should be noted that the beam position was very stable since the RCS and the beam transport has quite good stability to avoid beam loss to deliver the beam to the spallation neutron source.

The repetition of the shots was chosen to be0.4 Hz to avoid the samples melting. The number of protons per shot was about 6×10^{12}, which was monitored by current transformers shots by shots. The total number of protons irradiated was 1.77×10^{14} for 0.4 GeV, 1.72×10^{14} for 1.3 GeV, 1.78×10^{14} for 2.2 GeV, and 2.20×10^{14} for 3.0 GeV.

2.3. Analysis

Decay gamma rays from irradiated samples were measured by the HPGe immediately to survey short-lived nuclei. The main activation products in aluminium are ^{7}Be (life-time: 53.22 d), ^{22}Na (life-time: 2.6018 y), and ^{24}Na (life-time: 14.997 h). Disintegration data were taken from the latest IAEA Nuclear Data Services [6].

Since there was only one HPGe, each of samples were periodically measured at intervals from two hours to a day. The samples were mounted on an acrylic spacer to keep the detector-to-sample geometry rigidly. Radioactivity for each product was deduced using the following equation:

$$A = \frac{\lambda N}{(1 - e^{-\lambda t_m})\epsilon I} \exp(\lambda t_m) \quad (1)$$

IOP Conf. Series: Journal of Physics: Conf. Series **1021** (2018) 012016 doi:10.1088/1742-6596/1021/1/012016

Table 1: The activation cross sections

Proton energy GeV	Cross sections mb. ^{7}Be	^{22}Na	^{24}Na
0.4	3.55 ± 0.45	16.2 ± 1.0	11.5 ± 0.8
1.3	7.10 ± 0.66	11.9 ± 0.9	10.4 ± 0.8
2.2	8.77 ± 0.73	11.7 ± 0.8	9.80 ± 0.77
3.0	8.76 ± 0.73	11.0 ± 0.8	10.4 ± 0.8

where N: count of γ-rays peak, λ: decay constant, t_m: measurement time, ϵ: detection efficiency, and I: absolute γ-rays intensity, respectively. The production cross section was derived from the following equation:

$$\sigma = \frac{A}{\lambda N_p n} \quad \text{barn} \tag{2}$$

where N_p is the number of protons irradiated, and n is the number density of the sample.

The obtained cross sections are shown in table 1. Both statistical and systematic uncertainties are included. The main contribution to the uncertainty is the number density which is 5%. The next one is detection efficienty (5%). The others (the number of protons, self-absorption of gamma-rays, decay gammas) are about 1%. In total, the systematic uncertainty is estimated as 6.1%.

It is estimated that the contribution of sodium production by neutron and other charged particles is negligible since there was no gamma rays observed in the samples used for background measurement.

3. Discussion

In figure 1, the measured, calculated, and evaluated activation cross sections for ^{27}Al$(p, X)^{7}$Be, ^{27}Al$(p, 3p3n)^{22}$Na , and ^{27}Al$(p, 3p1n)^{24}$Na are shown. Comparing with other experimental data, the present data have smaller uncertainties than others thanks to the highly stabilised and controlled proton beam. Their values are also consistent with others. Thus the irradiation experiment at the 3NBT beam dump line produces valid results and is ready for full-time measurements with various materials.

It was found that the GEM code in the PHITS code slightly differs from the original one [9]. Thus we implemented the original one in the PHITS code, then calculated the activation cross sections. They are superposed in figures 1b, 1d and 1f respectively. It is clearly shown that especially for ^{7}Be production the calculated ones are drastically improved by about 40% (see figures 1a and 1b) though the one with INCL still underestimates above 2 GeV. On the other hand, the ones for ^{22}Na and ^{24}Na production are less affected since Na comes from fission products rather evaporation products. This indicates that improvement of the intranuclear cascade model is required.

4. Conclusion

In order to improve the accuracy of the nuclear design for spallation neutron sources and transmutation systems, nuclear cross sections are important. Prior to the forthcoming full-time measurement with various materials, the activation cross sections of aluminium for ^{7}Be, ^{22}Na, and ^{24}Na with 0.4 GeV, 1.3 GeV, 2.2 GeV, and 3.0 GeV protons were measured to validate the availability of the irradiation place at the 3NBT beam dump line in J-PARC. The present experimental results are in good agreement with other experimental ones and have less uncertainty than other experimental ones.

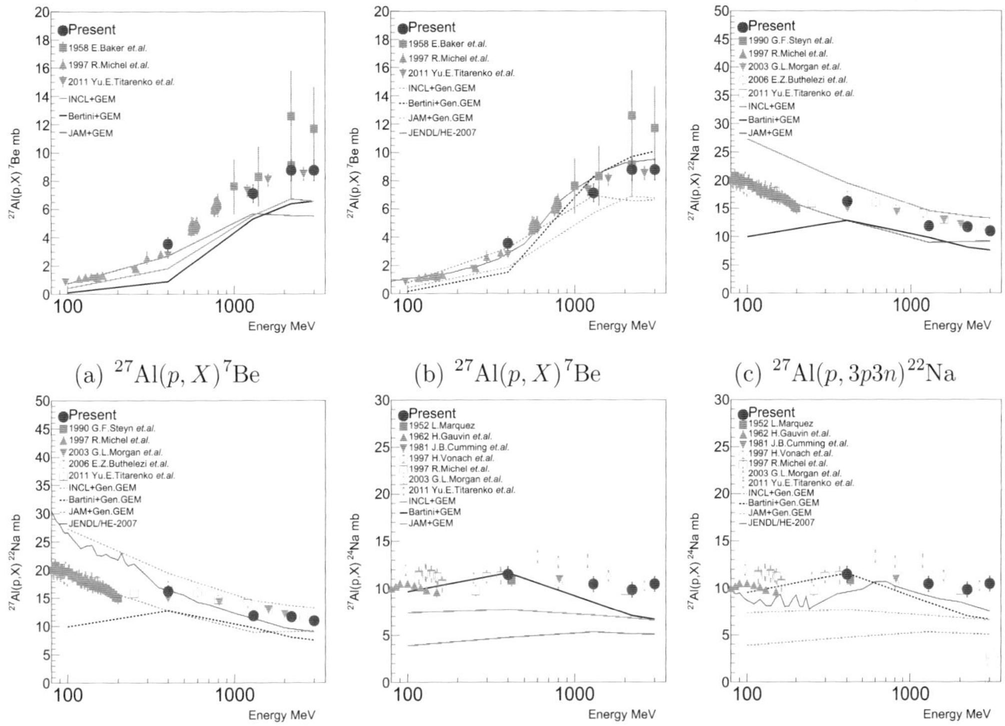

(a) $^{27}Al(p,X)^{7}Be$ (b) $^{27}Al(p,X)^{7}Be$ (c) $^{27}Al(p,3p3n)^{22}Na$

(d) $^{27}Al(p,3p3n)^{22}Na$ (e) $^{27}Al(p,3p1n)^{24}Na$ (f) $^{27}Al(p,3p1n)^{24}Na$

Figure 1: Comparison of the present production cross sections with other experiments, the evaluated data, and the calculations of the original PHITS [7] code and the upgraded one. Figures 1a, 1c and 1e are compared to the calculation done by the original PHITS. Others (figures 1b, 1d and 1f) are to the one done by the upgraded PHITS (labeled by *Gen.GEM*). The present results (full circle) and other experiments (other symbols) taken from EXFOR [10] data are plotted.

It was shown that the calculation by the PHITS code underestimates the present measurements. By implementing the original GEM code to the PHITS code, ^{7}Be production cross section were improved drastically. However, this is not so for sodium production which is mainly produced by direct reaction.

In future works, measurement of the activation cross sections for noble gases (especially Xe) with Pb and Bi materials for parameter determination of TEF facilities is planned.

Acknowledgments

The authors are grateful to M Nishikawa and the members of Nippon Advanced Technology (NAT) for technical assistance with the experiments.

References

[1] The Joint Project Team of JAERI and KEK, *JAERI-Tech* **99-56**, 1999
[2] Y. Ikeda, *Nucl. Instrum. Meth.* A**600**, (2009) 1
[3] Y. Miyake, *et al.*, *Physica* B **404** (2009) 957
[4] S. Meigo, *et al.*, *Nucl. Instrum. Meth.* A**562**, (2006) 596
[5] S. Meigo, *et al.*, *Nucl. Instrum. Meth.* A**600**, (2009) 41
[6] IAEA Nuclear Data Services, `www-nds.iaea.org`, 01.Jun.2017 visited
[7] T. Sato, *et al.*, *J. Nucl. Sci. Technol.* **50**:9, 913-923 (2013)
[8] A. Boudard, *et al.*, *Phys. Rev.* C**80**, 014606 (2013)
[9] S. Furihata, *Nucl. Instrum. Meth* B**171**, (2000) 251
[10] IAEA, `www-nds.iaea.org/exfor/exfor.html`, 17/March/2017 visited

IOP Conf. Series: Journal of Physics: Conf. Series **1021** (2018) 012017 doi:10.1088/1742-6596/1021/1/012017

The measurements of neutron energy spectrum at 180 degrees with the mercury target at J-PARC

H Matsuda, S Meigo, H Iwamoto
JAEA, 2-4, Shirakata, Tokai-mura, Naka-gun, Ibaraki, Japan
E-mail: matsuda.hiroki@jaea.go.jp

Abstract. Spallation neutron production at 180 degrees is of importance for an evaluation of radiation protection for ADS (Accelerator-Driven System) and nuclear physics. It was, however, quite difficult to measure this. We measured the energy spectrum of spallation neutron at 180 degrees at the proton transport beam line (3NBT) to MLF (Materials and Life Science Experimental Facility) on J-PARC using a NE213 liquid scintillator. The irradiated proton energy was 3 GeV, and the intensity was 1×10^{10} protons. The neutron energy was determined by a Time-Of-Flight method with n-gamma discrimination. We also simulated the energy spectrum by using the PHITS code and compared this with measured spectrum. In this paper, the overview of the experiment and the results are described.

1. Introduction

In the Japan Proton Accelerator Research Complex (J-PARC) [1], a MW-class pulsed neutron source, the Japan Spallation Neutron Source [2], and the Muon Science facility (MUSE) [3] are installed in the Materials and Life Science Experimental Facility (MLF). Since 2008, this source has produced a high-power proton beam of 300 kW. In 2015, we successfully ramped up the beam power to 500 kW and delivered a 1 MW beam to the targets. To produce a neutron source, a 3 GeV proton beam collides with a mercury target, and to produce a muon source, the 3 GeV proton beam collides with a 2 cm thick carbon graphite target. To efficiently use the proton beam for particle production, both targets are aligned in a cascade scheme, with the graphite target placed 33 m upstream of the neutron target. For both sources, the 3 GeV proton beam is delivered from a rapid cycling synchrotron (RCS) to the targets by the 3NBT (3 GeV RCS to Neutron facility Beam Transport) [4, 5] Before injection into the RCS, the proton beam is accelerated up to 0.4 GeV by a LINAC. The beam is accumulated in two short bunches and accelerated up to 3 GeV in the RCS.

To evaluate the ADS system parameters, data for fast neutrons measured with low uncertainty is mandatory since ADS is a fast reactor system. However, uncertainties mostly affect ADS parameters [6]. Thus we tried to develop an irradiation field of the neutron at 3NBT. As a first step, the actual neutron spectrum of the spallation target was measured. On the entrance of a magnet at the most downstream side, significant radioactivation was observed. It is also our motivation of this measurement to make establish that whether this activation was caused by beam loss or spallation neutrons.

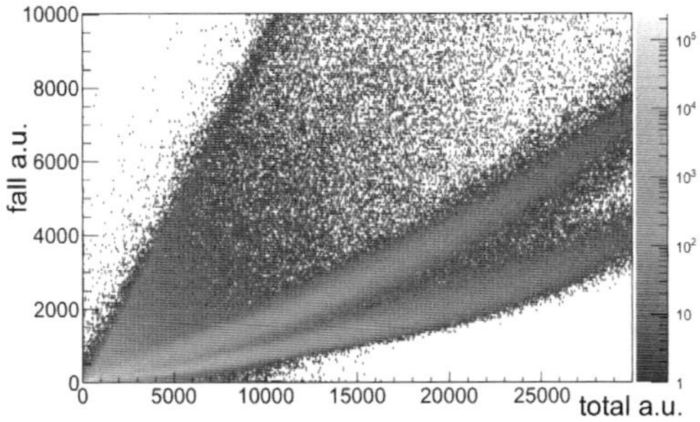

Figure 1. The two dimensional histogram of total and fall components of NE213 signals.

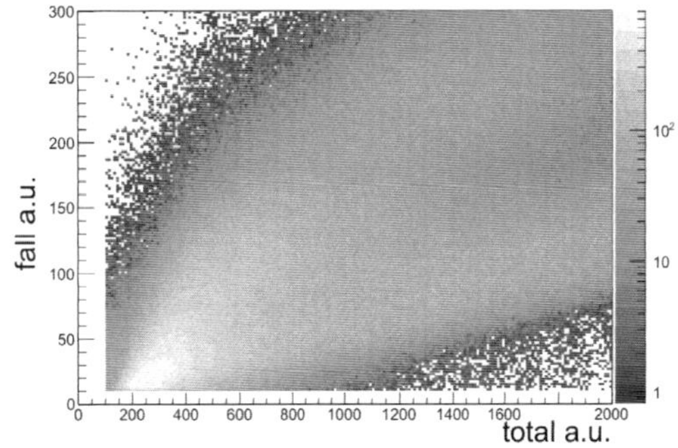

Figure 2. The zoomed histogram for the low energy region.

2. Experiment

2.1. Setup

NE213 liquid scintillator (ϕ8 mm × 20 mm) was employed, whose response and characterisation can be found in [7]. The scintillator was directly connected onto a photomultiplier (Hamamatsu H3164-10) surface. A magnetic shield made of FINEMET® surrounded it to avoid magnetic field leakage from bending magnets. To concentrate on the low energy range, a high voltage of −1400 V was applied. The photomultiplier was fixed on the beam pipe directly to minimise neutron attenuation. A low-attenuation signal cable (SUCOFEED) whose length was about 100 m was used to achieve n-γ discrimination. At full power operation there are unavoidable pile-up events. Thus the beam power is kept lower than 150 W to 200 W.The proton intensity was monitored by the current transformer shot by shot with 1% uncertainty. The typical intensity was 1×10^{10} protons per shot. The nominal repetition rate was 25 Hz. A VME digitiser SIS3316-125-16 [8] was employed in this measurement. The start signal for the Time-Of-Flight (TOF) was the beam-synchronised pulse. The time resolution was 4 ns per channel.

We measured the accumulated neutron flux by using the irradiation foil method [9]. $^{115}\mathrm{In}(n,n')^{115m}\mathrm{In}$ reaction was employed to evaluate the fast neutron flux. The threshold neutron energy is 0.34 MeV for this reaction. The shape of indium sheet was 25 mm square, and the weight was 8.7 g. The sheet was placed onto the beam pipe flange near the upstream side of the scintillator. The sheet was irradiated for one week, which is a typical run cycle of the MLF.

2.2. Analysis

NE213 Scintillator The distribution of PSD is shown in figure 1. In figure 2, signals for neutron and gamma are clearly discriminated at low energy. Below channel 1000 which is equivalent to ^{241}Am 60 keV total absorption peak, the signals are clearly separated.

A TOF spectrum is shown in figure 3. The first peak around channel 475 is the flash-gamma signal in the carbon target for the muon source [1]. By fitting with a Gaussian to the second peak, it is found that the time resolution is 5.3 ns. The distance from the mercury target to the scintillator is 126.4 m. Thus the solid angle is 3.15×10^{-9} sr. The response function for NE213 in the energy range from 0.1 MeV to 80 MeV was calculated by SCINFUL-R [10].

The obtained spectrum for 180 degrees spallation neutrons is shown in figure 4 in blue. The

[1] The peak around 580 ns was caused by ringing.

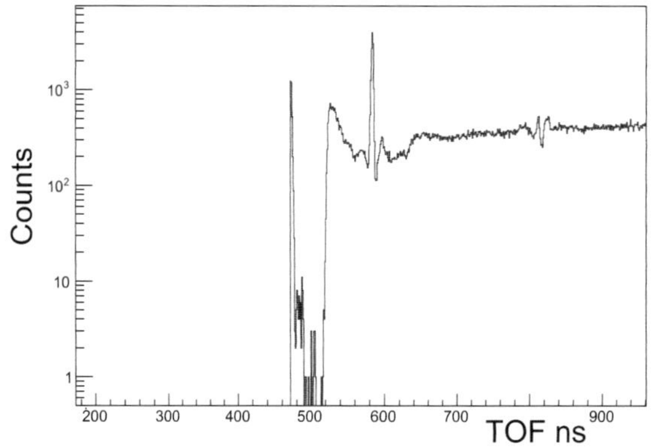

Figure 3. TOF spectrum.

statistical uncertainty is less than 1%. The systematic uncertainty, however, still needs to be estimated: it is expected to less than 10%. The energy resolutions, which are determined by the TOF resolution, are 0.15% at 0.1 MeV, 0.5% at 1 MeV, and 1.5% at 10 MeV.

When proton beam loss occurs, beam loss monitors would detect it and immediately the proton beam is stopped within the next proton bunch. If there is loss, TOF signals would be observed before the flash-gamma signal in figure 3. Thus it is concluded that almost no neutrons from beam loss were observed.

Indium sheet During one week of irradiation, the beam condition was quite stable. The actual operating rate was 98.8%, and the beam power was 150.4 kW on average. From equation 1 in Ref. [9], the reaction rate was evaluated as 8.29×10^{-34} /proton with 10% uncertainty.

3. Discussion

3.1. Comparison of measured data with calculations

Simulations were performed by using PHITS code [11]. The structures of the mercury target, the muon target, the beam pipe line, and the beam tunnel are described in detail in the code. We simulated the TOF spectrum to be the same as the real measurement. The calculated TOF spectrum is converted to an energy scale.

The measured and calculated spectra are shown in figure 4. The one from the bare mercury target is also superposed. The shape of the measured spectrum is consistent with the calculated one. The resonance peaks for the total inelastic scattering from the structure materials (Al, Ti, and C) appears in the same neutron energy channel. In the energy range below 1 MeV, the measured flux is lower than the calculated one. Above 10 MeV range, it seems that the bare, the measured, and the calculated spectra converge.

3.2. Comparison of absolute flux measurement

To calculate the reaction rate from the measured spectrum, the total cross section used in [9] was employed, obtaining 8.85×10^{-34} /proton. Thus the reaction rate from the indium irradiation method showed a good agreement with the one from the measured flux within the uncertainties.

3.3. Comparison of the total inelastic scattering of carbon with the evaluated data

The 2 cm thick carbon graphite target is used as a muon source at MUSE. To confirm the validity of this measurement, the neutron spectrum with and without the muon target was measured to

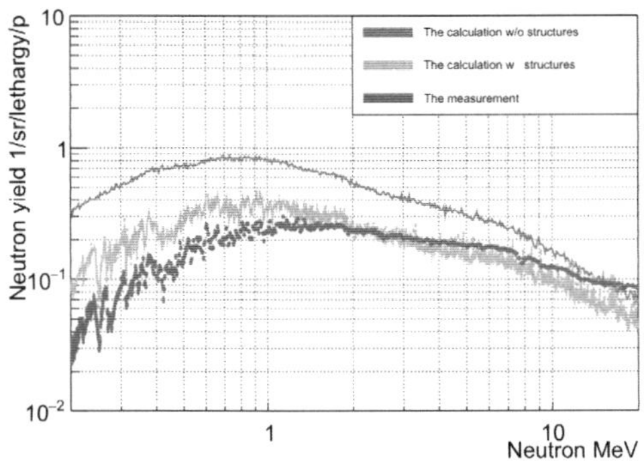

Figure 4. Neutron spectra (Red: the calculated spectrum of the bare target, Green: the calculated one, Blue: the measured one).

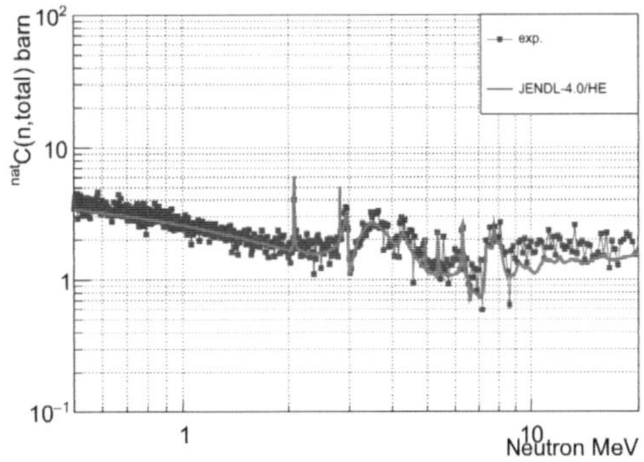

Figure 5. Comparison of the total inelastic scatterings of carbon. The red line for JENDL-4.0/HE. The blue filled square with line for the measured results.

evaluate the total inelastic scattering of carbon, which is defined as $N = N_0 \exp(-N_{\mathrm{carbon}}\sigma t)$ where N (N_0) is the number of neutrons with (without) the carbon target, N_{carbon} is the number of carbon atoms in the muon target, σ is the cross section of $C(n, n')$, and the t is the target thickness. The obtrained spectrum is seen in figure 5 with JENDL-4.0/HE data [12]. It is obviously seen that the obtained spectrum shape follows the evaluated data and a sharp peak around 2.1 MeV is observed above the statistical uncertainty. Thus it is indicated that our measurement is certainly valid.

4. Conclusion

We measured the energy spectrum of spallation neutrons at 180 degrees in J-PARC for the first time. NE213 liquid scintillator was employed to discriminate neutron signals from photons. We also used the indium irradiation method to measure the absolute flux above 0.34 MeV neutron energy range. The neutron spectrum was successfully obtained with a low power beam of about 150 W. The calculated spectrum using the PHITS code is consistent with the measured one. Resonance "dips" appeared in same energy channel. The absolute flux measured by the irradiation method is also consistent with the measured flux within the uncertainties. By comparing the total inelastic scattering of carbon using the graphite target used for the muon source, it showed that this measurement is valid. To measure neutron flux at higher energy, the irradiation method using Bismuth should be employed.

References

[1] The Joint Project Team of JAERI and KEK, *JAERI-Tech* **99-56**, 1999
[2] Y. Ikeda, *Nucl. Instrum. Meth.* A**600**, (2009) 1
[3] Y. Miyake, *et al.*, *Physica* B **404** (2009) 957
[4] S. Meigo, *et al.*, *Nucl. Instrum. Meth.* A**562**, (2006) 596
[5] S. Meigo, *et al.*, *Nucl. Instrum. Meth.* A**600**, (2009) 41
[6] H. Iwamoto, *et al.*, *J. Nucl. Sci. Technol.* **8** (2013) 856
[7] CRYSTALS SAINT-GOBAIN, `www.crystals.saint-gobain.com`, 10.October 2017 visited
[8] struck innovative systeme, `www.struck.de/sis3316.html`, 10.October 2017 visited
[9] Y. Kasugai, *et al.*, *JAEA-data-code* **2015-033**, (2015)
[10] S. Meigo, *Nucl. Instrum. Meth.* A**401**, (1997) 365
[11] T. Sato, *et al.*, *J. Nucl. Sci. Technol.* **50**:9, 913-923 (2013)
[12] K. Shibata, *et al.*, *J. Nucl. Sci. Technol.* **48**(1), 1-30 (2011)

IOP Conf. Series: Journal of Physics: Conf. Series **1021** (2018) 012018 doi:10.1088/1742-6596/1021/1/012018

High performance controls architecture for real-time non-Boolean interlocks using PLCs for ESS Target system

M Zaera-Sanz, A Nordt and S Kövecses

European Spallation Source ERIC, Lund, Sweden

Manuel.Zaera-Sanz@esss.se Annika.Nordt@esss.se Szandra.Kovecses@esss.se

Abstract. The 600m long linear accelerator of the European Spallation Source ERIC, transporting a 5MWatt proton beam from the source to the target station, is currently under construction in Lund (Sweden) and plans to start neutron production in 2020. In order to protect the equipment from damage and to take the appropriate actions required to minimise recovery time, a dedicated set of Programmable Logic Controller (PLC) based interlock systems is being designed. The target protection system represents one part of the machine protection in the overall ESS protection architecture. Its main purpose is to provide dependable and, if possible, predictive interlocking for the target station and the associated subsystems (i.e. monolith, wheel, moderator-reflector system, beam extraction systems and cooling systems) in case of malfunctioning, by taking the appropriate actions to bring the machine into a safe state. This goal implies non-Boolean complex computations in real-time with critical constraints. To achieve this goal, an initial framework based on the last generation of PLCs and networking components (cluster solution) is presented. This framework is based on Siemens Totally Integrated Automation (TIA) portal with Step 7 programming environment, PROFINET fieldbus networking and intelligent devices (I-devices).

1. Introduction and requirements

Machine protection at ESS focuses on the protection of the investment (i.e. hardware) and beam (i.e. performance), hence aiming for very high beam availability towards users.

The target protection system, as part of the ESS machine protection, should fulfill a set of protection functions that are currently being defined for each subsystem: monolith, wheel, moderator-reflector system, beam extraction and cooling systems. Besides, high availability figures (95%) must be fulfilled to provide neutrons to the users.

In order to fulfill such availability figures, investment protection must be provided, and a new technique called predictive interlocking [1] is under development. Its main goal is to avoid stopping the machine by being predictive and taking the proper actions (instead of just being reactive), minimizing machine downtime and therefore saving cost in terms of equipment and labor.

Predictive interlocking will make use of artificial intelligence software techniques which may require high performance computation platforms. There are basically two problems to solve: one related to the complexity of the predictive interlocking technique to be applied, and the second one is the complexity of the set of protection functions. Implementing such complexity would require a robust platform, with real-time processing and communications constraints, and possibly SIL (Safety Integrity Level) rated. We believe that nowadays PLCs can fulfill these requirements.

The Target Protection System has to fulfil the following main requirements:

- Protect the components: In case of failure (coming from the wheel, moderator-reflector, beam extraction or cooling system), the necessary steps have to be taken leading the system to a safe state.
- Protect the beam: the system should not generate beam stops if this is not strictly necessary. Faulty trigger signals leading to neutron production stops must be kept to a strict minimum. Predictive interlocking plays a key role to provide this requirement.
- Provide the evidence: In case of failure, the operators should be given means of easy fault tracking and understanding. The system must support the identification of the initial failure, also in case of multiple alarms (one initial failure that causes subsequent failures).
- Assist improving the operation: The diagnostics for failures should be easy. The status of the system must be clearly presented in the control room and should be transparent to the operator.

To fulfil the above set of requirements we are implementing an interlock system based on three high performance PLC CPUs making use of PROFINET intelligent devices (I-devices) and shared I-devices under Siemens TIA portal [2].

2. Predictive Interlocks for ESS target local protection system

The mission statement is to minimize machine downtime by predicting failures in the equipment and taking the appropriate action, increasing the availability and saving costs. For this purpose, PLC technology provides the needed robustness, safety and real-time performance, executing artificial intelligence software in a cluster configuration.

It is important to stress that predictive interlocks are not substituting the traditional interlocks. It is a technique applied in an above layer which purpose it is to prevent triggering the interlocks action as much as possible.

Figure 1 shows the ESS target system and Figure 2 shows an example of predictive interlocking applied to the rotating target wheel. Artificial intelligence software is monitoring the temperature of the wheel. This software is also receiving data coming from cooling, alignment and timing systems. When it detects that the wheel temperature is rising, a probability of failure associated to the cooling, alignment and timing systems is being computed. According to that probability, the health status of the related system is being evaluated; valve positions, motor speeds or timing drifts are being adjusted and the outcome is being observed. If there is no feasible solution, either no action is taken and the traditional interlocks keep operating normally, or in case the traditional interlocks are part of the predictive interlocks, the beam is being stopped.

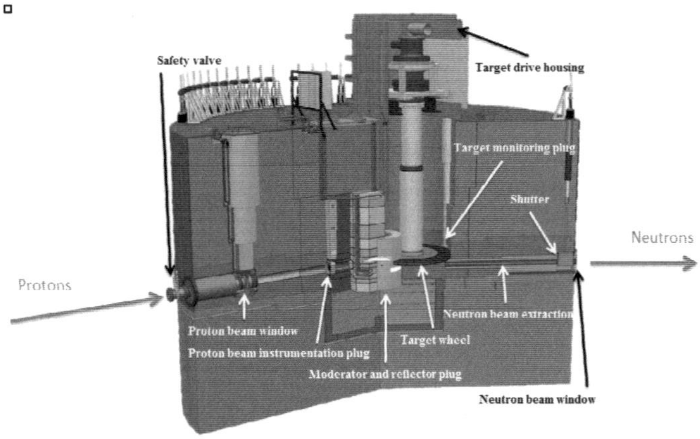

Figure 1. ESS target system

IOP Conf. Series: Journal of Physics: Conf. Series **1021** (2018) 012018 doi:10.1088/1742-6596/1021/1/012018

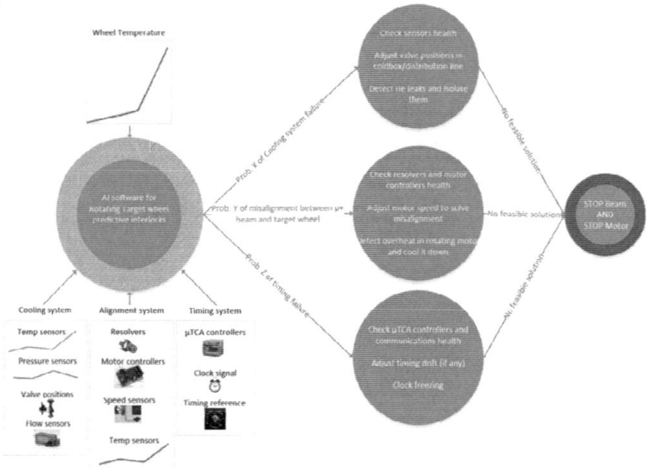

Figure 2. Predictive interlocks applied to the target wheel

3. High performance computing using PLCs: concept and implementation prototype

Our intention is to build a computer cluster using PLCs as a platform for predictive interlocking instead of using traditional computers or industrial computers.

As Figure 3 shows, any cluster solution consists of a parallel programming environment and a cluster middleware which provides a single system image. With PLCs, the parallel programming environment can be built by using STEP 7 and TIA portal tools like PROFINET I-devices and shared I-devices. The application middleware is difficult to develop, but initially we plan to implement it using a specific problem oriented approach (communicating, distributing and synchronising tasks for each specific algorithm), and in the future we plan to build a library implementing this as a single PLC system image.

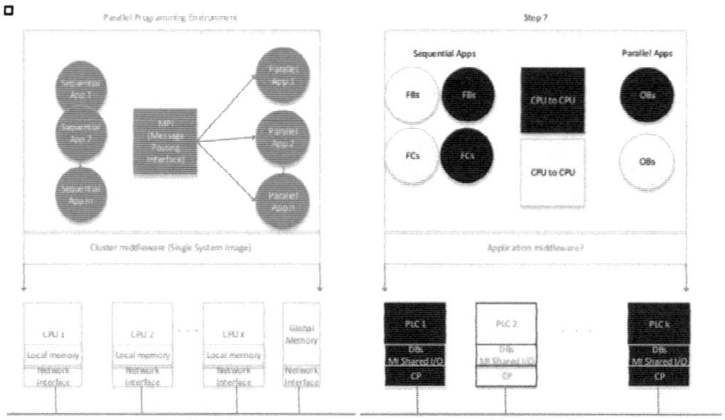

Figure 3. Typical cluster solution vs. PLC based one

We have currently built a prototype of our cluster solution based on PLCs. Figure 4 shows the main components: three PLC CPUs S7-1518, SCALANCE IRT switch, one CPU S71516, several analog and digital I/O modules in ET200MP format and SITOP redundant powering system. This cluster solution is classified as a scalable multiprocessor with physically distributed memory but logically shared. It follows a programming model based on shared memory in I-devices. Each processor acts as I/O controller and shared I-device being able to address all the memory areas. The processing information comes from an additional PLC configured also as shared I-device. We have built bus and star network topologies in order to connect our PLCs using a SCALANCE switch under real-time PROFINET profile for communications.

IOP Conf. Series: Journal of Physics: Conf. Series **1021** (2018) 012018 doi:10.1088/1742-6596/1021/1/012018

Figure 4. PLC cluster

4. Conclusions and further work

Particle accelerators and accelerator driven facilities are time and cost critical, and predictive interlocks could play a key role:

- Avoid to stop the machine providing high availability
- Minimize machine downtime and therefore saving cost
- Speed-up machine tuning and provide advanced fault detection and fault recovery
- Perform preventive maintenance (structural health)

As further work we plan to:

- Define communications and synchronization primitives for distributed processing
- Parallelize protection functions implementation to run in the PLC cluster
- Evaluate the response time for the interlocks reaction
- In case of failsafe requirements, failsafe I-devices are possible but not "shared devices" which implies a different approach for cluster construction.

High performance computing platforms using PLCs are promising candidates for complex interlocks applications in which non-Boolean protection functions are required. Real-time high performance computing platforms to run artificial intelligence predictive interlocks software for machine protection is perhaps a good way to increase dependability.

References

[1] Zaera-Sanz M December 2014 *Predictive interlocks and high performance computing using PLCs* 2nd PLC based interlock systems workshop ITER International Fusion Organisation https://indico03.esss.lu.se/indico/event/218/overview

[2] Zaera-Sanz M and Nordt A February 2016 *High performance computing platform for predictive interlocks using PLCs at ESS* 3rd PLC based interlock systems workshop CERN European Laboratory for Particle Physics https://indico.cern.ch/event/461080/

IBEX - an EPICS based control system for the ISIS pulsed neutron and muon source

F A Akeroyd[1], K V L Baker[1], M J Clarke[1], J R Holt[2], G D Howells[1], D P Keymer[1], T Löhnert[1], C M Moreton-Smith[1], D E Oram[1], A Potter[2], I H Rey[2], T A Willemsen[2] and K Woods[2]

[1] ISIS Pulsed Neutron and Muon Source, Rutherford Appleton Laboratory, Science & Technology Facilities Council, Harwell Campus, DIDCOT, OX11 0QX, UK
[2] Tessella, 26 The Quadrant, Abingdon, OX14 3YS, UK

E-mail: freddie.akeroyd@stfc.ac.uk

Abstract. The ISIS pulsed neutron and muon source is currently migrating its instrument control system from a mainly locally written system, called SECI, to one based on EPICS for device control and using Eclipse RCP and Control System Studio for the user interface.

The new control system, called IBEX, was initially deployed on two new instruments, but recently the first existing instruments have been converted from SECI to IBEX. For compatibility with our existing business infrastructure, to ease migration and also to allow dual running, we have continued to use Microsoft Windows as our underlying operating system. We will present an overview of IBEX and its architecture, current status and our future development and rollout plans.

1. Introduction

The ISIS pulsed neutron and muon source [1] is a world-leading centre for research in the physical and life sciences and currently has over thirty beamline instruments. A key component in the successful delivery of experiments is the software and hardware used to control these beamline instruments.

As detailed previously [2] we are moving from a locally developed LabVIEW [3] based system, known as SECI (Sample Environment Control Interface), to one based on EPICS [4]. EPICS is used at many large facilities around the world, including the Diamond Light Source (on the same site as ISIS), the European Spallation Source (ESS) and the Spallation Neutron Source (SNS) at ORNL. As well as providing a solid software framework, adopting EPICS provides opportunities for collaboration and sharing that were not possible with the old control system.

2. General approach

As ISIS has many established beamlines, introducing a new control system must be done carefully to avoid lost beam time. Instrument scientists must also feel that the new system is both reliable and tested. To make this transition as smooth as possible the following approach has been taken:

- Continue to use Microsoft Windows as the underlying operating system [2]. Linux is more common at other facilities, but there is less expertise and familiarity at ISIS and we would run the risk of changing too much at once. We also maintain direct compatibility with existing business services (such as archiving, backup and neutron data acquisition), more easily use existing

LabVIEW drivers when necessary, and have the option of reverting back to SECI if needed. The IBEX system has, however, been developed without excluding a move to Linux at a future date.
- Develop the software using Continuous Integration (CI) managed via Jenkins, which runs both system and unit tests and creates a deployment package.
- Make use of device emulators for testing, we are using the LeWIS [5] Python framework which was developed separately as part of ISIS-ESS in-kind work.
- Make use of EPICS within the existing SECI system, both as a way to avoid duplication when new devices software is required, but also to increase user confidence. The CALab [6] package allows easy creation of a LabVIEW EPICS client. An upgrade of the muon front end magnet control was carried out this way, the ultimate requirement for shared control and monitoring across several beamlines was a natural fit for the EPICS client-server model.
- Provide LabVIEW integration tools. We have a great deal of existing LabVIEW controlled equipment, for much an EPICS driver either already exists or can easily be created using e.g. StreamDevice [7]. We do, however, have third party equipment with vendor supplied LabVIEW control, in these cases creating a clean interface by incorporating the equipment rather than rewriting control in EPICS is better. For such situations we have developed LVDCOM [8] (for LabVIEW front panel control) and NETSHRVAR [9] (for Network Shared Variable [10] control).
- Allow scientists to quickly create a simple device driver. Sometimes a new piece of equipment needs to be connected at short notice, the SDTEST package has been used by them to quickly implement an EPICS StreamDevice driver by defining simple EPICS macros in a text file.

3. Hardware architecture

This has been covered in more detail previously [2]. Though EPICS allows greater potential for distributed control with IOCs running on multiple computers, most IBEX systems currently run all IOCs on a single Virtual Machine (VM). The exception is the IMAT beamline where two neutron imaging cameras have their own dedicated PCs for camera control, these PCs run a “cut down” version of the IBEX server to allow control from the central VM instance.

4. Software architecture

4.1. IBEX server

Individual devices are controlled via EPICS IOCs running on the loopback network interface, each wrapped within a procServ [11] process harness instance. Two CS-Studio [12] archive engines record values of interest from IOCs, and are backed by a MySQL instance which is also used for storing log messages generated from IOCs [2].

Each IOC also has an associated config.xml file specifying configuration parameters (EPICS macros) it supports. These files are assembled at system build time and served to the IBEX GUI, they describe both available parameters and parameter validation rules.

On starting each IOC writes a list of its Process Variables (PVs) and associated EPICS DB info records to the instrument MySQL instance, this is then used by the GUI for PV browsing/lookup. To help with user selection, a high/medium/low “interest” level can be assigned to PVs via the DB info record and used for filtering. The info records are also used for configuring the archiving of PVs.

The neutron data acquisition is controlled by an IOC that communicates with the same program used by the SECI system. This IOC also runs an instance of the EPICS PCAS software so is able to dynamically create process variables to serve detector spectra. We are thus able to serve data from any detector pixel as an EPICS waveform record without having to pre-create a potentially large number of EPICS process variables. On instruments using event mode neutron data collection the IOC can now make use of plugins from the EPICS areaDetector framework [13] to serve integrated images to clients for visualisation during an experiment.

Running alongside the IOCs is the BlockServer process, which is the main manager process of IBEX. Its job is to manage which IOCs should be running at any particular time, and it does this by allowing the user to create instrument configurations. The BlockServer is based on the EPICS PCASpy package [14], which allows it to serve values to EPICS clients via Channel Access.

4.2. Instrument configurations and blocks

The equipment used on a beamline can vary, so it is important to specify which IOCs should be active. Leaving IOCs running when equipment is not present can lead to large log files or unnecessary errors being flagged to users. This collection of IOCs (and any necessary parameters, such as EPICS macro and default PV values) is known as a configuration and represented as XML. An aggregation style model is supported, where a configuration can reference smaller parts called components. Thus an "instrument base" component can be created for the fixed beamline equipment and various smaller components for transient sample environment equipment. A configuration can then be assembled using relevant components for the particular experiment type being performed.

A configuration also specifies "Blocks", which are scientist friendly names for particular EPICS values important to the experiment e.g. a SampleTemp block representing the "SOME:READBACK" IOC device PV. Blocks can have a logging frequency set and values are automatically recorded in the NeXus [15] format data file created at experiment end. A block may also be given a valid range and put under "run control", which means that neutron data collection will automatically be suspended while it is out of range.

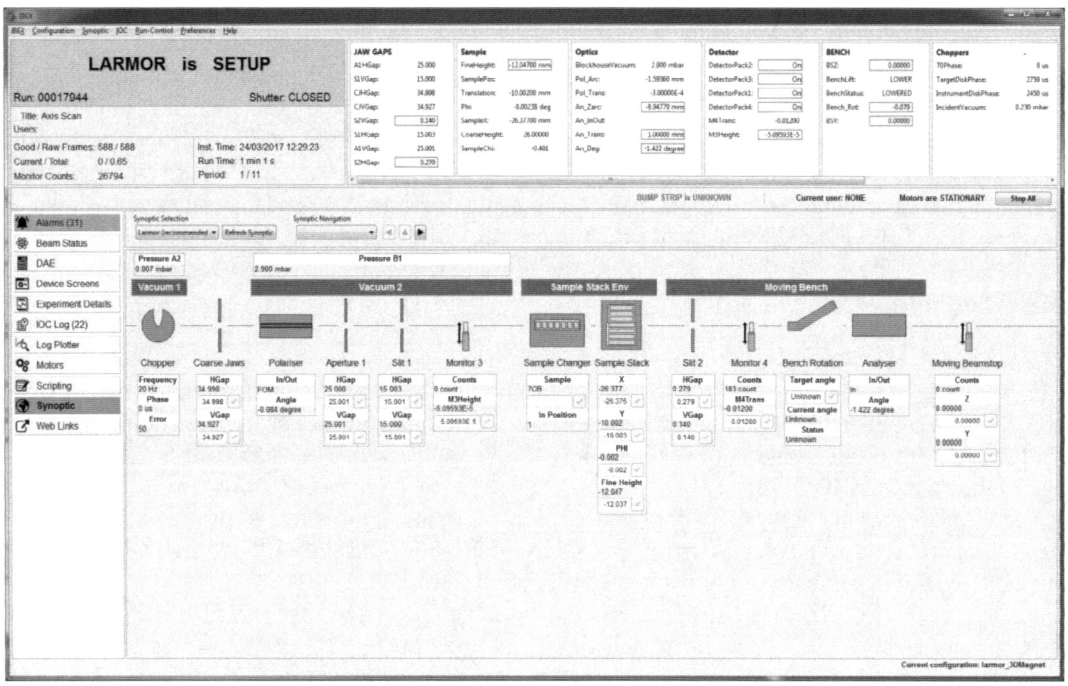

Figure 1. The main IBEX client window.

4.3. IBEX client

The main GUI is the "IBEX Client" (Figure 1) which is an Eclipse RCP based program. It has a similar layout to SECI, but also incorporates a new central synoptic display. A Python scripting window running the ISIS "GENIE Python" command set for instrument control is also provided.

The synoptic provides an overall view of the instrument state and key values can be displayed along with the icons. It also supports "drill down" to further screens that contain CS-Studio BOY OPIs [12] for individual devices. Multiple synoptic views can be defined, providing different views of the instrument – these are created using a simple drag and drop editor. It is also possible to create e.g. a synoptic displaying sample environment values across several beamlines.

A recent addition to the GUI is the concept of "device screens". Previously it was not possible to open a CS-Studio BOY OPI unless it was connected to a synoptic icon. It is now possible to define these views and associated OPI parameters separately.

As the client is configured from the server, it supports switching to view a different instrument – this has proved useful for checking the status of multiple instruments from a technician's desktop.

5. Status and future plans

IBEX has successfully been used to commission two new instruments, and is now also running on two established instruments. We are in the process of rolling out IBEX to more instruments, and hope to complete the majority of beamlines within the next two years. As with all new software development programmes, we have experienced challenges with new hardware types and commissioning existing hardware already operational on beamlines. To mitigate these issues, we have employed device simulation as well as ensuring close communication with instrument scientists at the facility.

Currently user control scripts are run from a Python terminal on the main instrument computer. A table based script generation tool (MkScript) is available, but there is currently no direct feedback from the script itself on how far it has progressed. We are currently investigating using the nicos-daemon part of the NICOS instrument control package [16] as a script server. Work on this is progressing well, an ActiveMQ [17] proxy has been created to allow easier integration of NICOS with the IBEX control system. NICOS provides direct feedback to the client GUI as to where in a script is currently executing, as well as options to modify future operations.

As part of separate ESS in-kind work, Apache Kafka has been used for live event streaming [18]. The latest ISIS data acquisition hardware supports a network based interface and Linux drivers for this are available, so plans are being developed to update the ISIS neutron acquisition software chain which could lead to a fully Linux based IBEX instrument prototype in the near future.

References

[1] The ISIS Pulsed Neutron and Muon Source URL http://www.isis.stfc.ac.uk/
[2] Clarke M J *et al* 2015 ICALEPCS 2015 Conference Proceedings 0-3 URL http://accelconf.web.cern.ch/AccelConf/ICALEPCS2015/papers/mopgf048.pdf
[3] National Instruments LabVIEW Software URL http://www.ni.com/labview/
[4] EPICS: Experimental Physics and Industrial Control System URL http://www.aps.anl.gov/epics/
[5] LeWIS stateful device simulator URL http://lewis.readthedocs.io/en/latest/
[6] CALab software package URL https://www.helmholtz-berlin.de/zentrum/locations/it/software/exsteuer/calab/index_en.html
[7] EPICS StreamDevice software URL http://epics.web.psi.ch/software/streamdevice/
[8] LVDCOM software URL https://github.com/ISISComputingGroup/EPICS-lvDCOM/
[9] NETSHRVAR software URL https://github.com/ISISComputingGroup/EPICS-NetShrVar/
[10] National Instruments Network Shared Variables URL http://www.ni.com/white-paper/5484/en/
[11] Procserv Process Server URL https://sourceforge.net/projects/procserv/
[12] Control System Studio URL http://controlsystemstudio.org/
[13] EPICS areaDetector URL http://cars9.uchicago.edu/software/epics/areaDetector.html
[14] PCASpy: Portable Channel Access Server in Python URL https://pypi.python.org/pypi/pcaspy/
[15] Könnecke M *et al* 2015 *J. Appl. Cryst.* **48** 301-305 ISSN 16005767
[16] NICOS Control System URL http://cdn.frm2.tum.de/fileadmin/stuff/services/ITServices/nicos-2.0/dirhtml/
[17] Apache ActiveMQ URL http://activemq.apache.org/
[18] Jones M D, Akeroyd F A, Arnold O, Clarke M J, Draper N, Gigg M and Moore L 2017 ICANS XXII *J. Phys.: Conf. Ser.* (in press)

IOP Conf. Series: Journal of Physics: Conf. Series **1021** (2018) 012020 doi:10.1088/1742-6596/1021/1/012020

Development of a target imaging system for the European Spallation Source

Nicolò Borghi[1], Esben B Klinkby[1], Bent Lauritzen[1], Eric Pitcher[2], Nigel Poolton[1] and Luca Zanini[2]

[1]Technical University of Denmark, DTU Nutech, Risø Campus, Frederiksborgvej 399, Bldg. 201, 4000 Roskilde, DK

[2]European Spallation Source ERIC, Tunavägen 24, 223 63 Lund, SE

E-mail: nicbo@dtu.dk

Abstract. At the European Spallation Source (ESS) neutrons will be produced by a proton beam impinging on a rotating target wheel. The technology of the target wheel, which comprises a large number of closely spaced tungsten bricks and is cooled by helium, is largely untested. The durability of the target wheel and hence the overall ESS neutronic performance depend on the integrity of the tungsten bricks. In order to monitor whether the target geometry is preserved over the expected 5 year lifetime of the target wheel, we propose a Target Imaging System (TIS). The TIS consists of a scintillator array detecting the collimated single photon emission (decay gammas) from the activated tungsten bricks. Preliminary Monte Carlo simulations support the feasibility of this imaging system. As a proof-of-principle, an experimental test-rig is being constructed allowing to test the main aspects of the imaging system under conditions relevant to ESS.

1. Introduction

The European Spallation Source (ESS) will be the most powerful spallation neutron source in the world. Neutron production is initiated by a 2 GeV, 5 MW proton beam impinging on a rotating target wheel, comprising a large number of closely spaced tungsten bricks encased in a steel vessel [1]. The technology of the helium-cooled tungsten wheel, however, is largely untested and the durability of the target wheel depends *inter alia* on the integrity of the tungsten bricks. The tungsten bricks will operate in a brittle regime after exposure to radiation and thermal stresses, both of which may induce erosion or cracking of the bricks, eventually resulting in the blocking of the coolant channels, unwarranted heating and loss of mechanical stability of the wheel.

2. The Target Imaging System

Among several diagnostic systems currently under development, we propose a Target Imaging System (TIS) in order to investigate whether the target geometry is preserved over the expected 5 year lifetime of the target wheel. The TIS is intended to provide a means for a high spatial resolution recording of the decay gammas from the target wheel.

During operation, the target with the tungsten bricks is exposed to intense particle irradiation, especially protons and neutrons, giving rise to highly radioactive bricks, encased in a steel vessel of somewhat lower activity. By recording the decay gammas from the tungsten

IOP Conf. Series: Journal of Physics: Conf. Series **1021** (2018) 012020 doi:10.1088/1742-6596/1021/1/012020

bricks, the TIS will be able to detect changes to the brick layout, especially the presence of brick fragments in the cooling channels between the bricks.

The TIS is not a safety system but is intended as a monitoring instrument capable of providing support to issues of target maintenance and operation, as well as future design development.

As shown in figure 1(a), the TIS is positioned in the vertical target diagnostic plug that lies above the target wheel, on the side opposite to the incoming proton beam and forming a 15° angle (azimuthal) with respect to the proton beam axis.

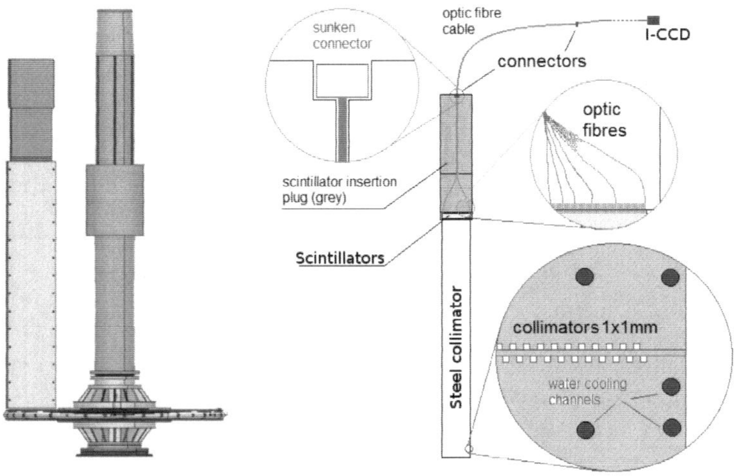

(a) (b)

Figure 1. (a) Target diagnostic plug positioning relative to the target system; (b) schematic representation of the main components of the TIS.

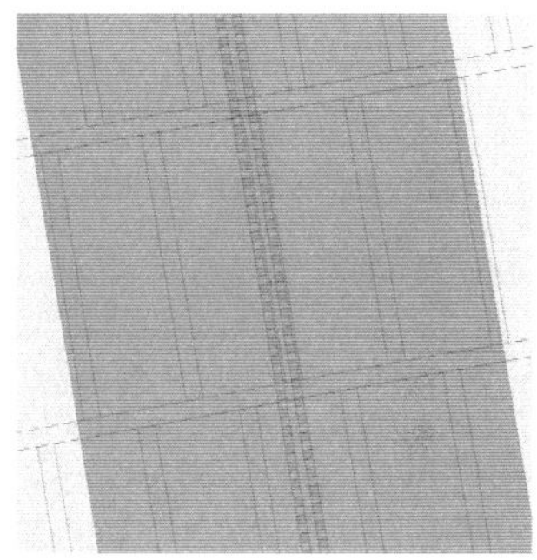

Figure 2. MCNPX model of the horizontal cross section of tungsten bricks superimposed with collimator grooves.

The TIS is composed of four main components, outlined in figure 1(b):

- a 2.8 m steel collimator suspended 2 cm above the target wheel and normal to its plane. A total of 500 1×1 mm^2 grooves are machined into the collimator structure, aligned along two rows, offset by 1 mm with respect to each other and separated by a 1 mm septum;
- on top of the collimator, a scintillator cartridge housing 500 $1 \times 1 \times 10$ mm^3 scintillators. Each scintillator is precisely coupled to a single collimator groove;
- an optical fiber bundle, where each fiber is coupled to a scintillator. The fiber bundle is $10 - 15$ m long and will be routed outside the ESS target monolith. To comply with ESS requirements, extra shielding material is surrounding the optical fiber bundle;
- a fast-switching (3 ns) image-intensified CCD camera coupled to the optical fiber bundle, to register the scintillation photons.

The TIS can be seen as the collection of 500 independent read-out channels, each composed of a collimator, a scintillator, a fiber and a CCD detector. Each of these channels will detect decay gammas emitted by the tungsten bricks, allowing for a high spatial resolution image.

In figure 2, details of the MCNPX geometry are shown, where the horizontal cross section of the tungsten bricks has been superimposed on the collimator grooves. The two rows of collimators allow the desired spatial resolution (1 mm) to be achieved while ensuring radial coverage of the target wheel; the imaging along the azimuthal direction is achieved through rotation of the wheel.

3. Data acquisition and imaging

With the high spatial resolution required for proper monitoring of the target wheel, each of the scintillators will have a count rate of about 10 γ/s. For each of the collimators, the passing time for a collimator across the gap between the bricks (that act as cooling channels) is of the order of 300 μs during normal operation, hence real-time image reconstruction is not possible. However, by taking advantage of fast image-intensified CCD cameras, several accumulation techniques may be envisaged to gather sufficient statistical significance in a relatively short time during ESS operation. This includes sector-averaging, in which only high-spatial resolution is maintained in the radial direction, supplemented by intelligent gating procedures allowing for high-spatial azimuthal angular resolution covering smaller sections of the wheel.

Two main sources of noise will affect the measurements:

- prompt γ background, resulting from the spallation process. This can effectively be avoided by limiting the acquisition time to the beam-off intervals;
- delayed gamma background, resulting from the activation of all the components within the monolith, including the scintillators themselves. This background can be limited with specific shielding.

4. Neutronic simulations

Extensive Monte Carlo simulations and activation calculations, performed with MCNPX [2] and CINDER'90 [3], respectively, are carried out to investigate the signal-to-noise ratio of the TIS.

The MCNPX and CINDER'90 simulations are carried out in three phases: (i) MCNPX simulation of a 2 GeV proton beam impinging on the target wheel to calculate the neutron fluxes in one sector of the wheel; (ii) CINDER'90 activation calculation. For the latter, we employed the following time structure:

- 1 year irradiation at 5 MW i.e., average power of the proton beam;
- 30 irradiation cycles comprising 2.86 ms irradiation at peak power (125 MW) interleaved with 2.57 s cooling time;
- cooling time of 1 s.

The 1 s cooling time corresponds to the time needed for the wheel to move from the irradiation position to the imaging position, i.e. the measurement of a sector of the target wheel is performed 1 second after the proton pulse. Finally, an MCNPX simulation is performed to calculate the instantaneous gamma flux at the scintillators.

In figure 3, the result of the simulation is shown, corresponding to the layout of figure 2. The figure shows the gamma flux at the position of the scintillators; each point corresponding to one of 113 scintillators included in the simulation. The cooling time is 1 s.

The blue dots correspond to the left row of scintillators in figure 2, imaging three bricks, whereas the red dots are for the right row, imaging two bricks only. The cooling channels in between the bricks are associated with the reduced gamma fluxes at around 300 $\gamma/\text{cm}^2/\text{s}$. Note that the overall gamma intensity grows with the distance from the target shaft, since the primary proton beam intensity is larger at the wheel edge. The constant values of the gamma flux across each of the bricks is an artifact of the simulation where each brick is treated as a single cell.

5. The test-rig

As a proof-of-principle for the TIS, a test-rig is being constructed at DTU. A 3D view of the setup is provided in figure 4(a). The test-rig will be equipped with 60 collimator grooves; 30 with 1 mm^2 cross section and 30 with 0.5 mm^2 cross section, allowing for different spatial resolutions. A scaled model of the wheel, shown in figure 4(b), will be fitted with several resolution targets

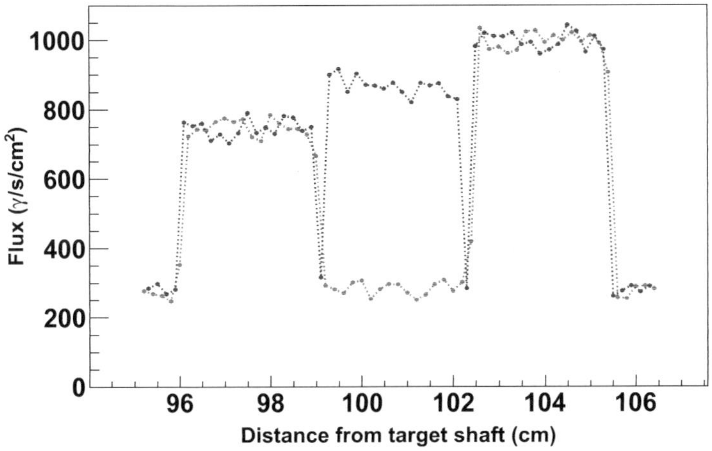

Figure 3. Simulated gamma flux at the scintillators 1 s after irradiation: left row (blue), right row (red). Background due to activation of the monolith has not been included. Dotted lines are to guide the eye.

made of lead. While the lead targets will not be activated themselves, measurements will be performed using the DTU Nutech medical radiation source (^{60}Co, 450 TBq).

The test-rig will allow assessment of the detector performance of the proposed TIS in terms of spatial resolution, temperature-dependent signal degradation, radiation damage of the components, etc. In addition, data acquired from the measurements of the test-rig will be used for the development of the control and data analysis software.

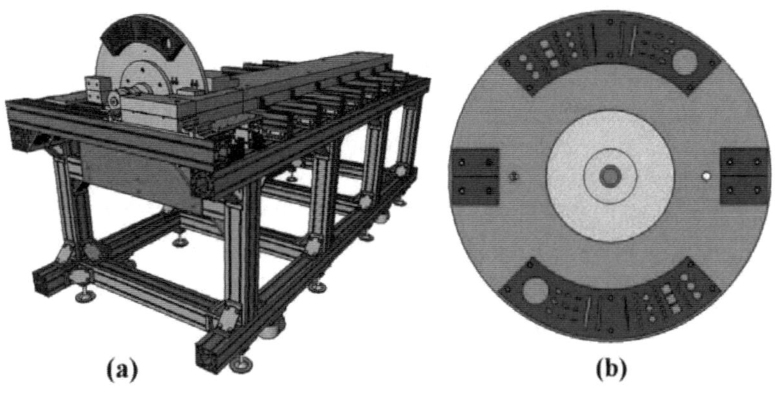

Figure 4. (a) The DTU test-rig: 3D rendering of the entire setup; (b) detail of the model wheel with resolution targets.

References

[1] Peggs S (editor) 2013 *ESS Technical Design Report* ESS-doc-274
[2] Pelowitz D (editor) 2011 *MCNPX User's Manual, Version 2.7.0* Los Alamos National Laboratory LA-CP-11-00438
[3] Wilson W B *et al* 2007 *RSICC Computer Code Collection - CINDER 1.05 - Code system for actanide transmutation calculations* Oak Ridge National Laboratory LA-UR-07-8412

IOP Conf. Series: Journal of Physics: Conf. Series **1021** (2018) 012021 doi:10.1088/1742-6596/1021/1/012021

The detector systems of the IBR-2M spectrometers.

A V Churakov, A V Belushkin, A A Bogdzel, V A Drozdov, V V Kruglov, S A Kulikov, F V Levtchanovski, E I Litvinenko, V M Milkov, S M Murashkevich, Ts Ts Panteleev, V I Prikhodko, V N Shvetsov, V V Zhuravlev

Frank Laboratory of Neutron Physics, Joint Institute for Nuclear Research, 141980 Joliot-Curie, 6, Dubna, Moscow region, Russia

E-mail: churakov@nf.jinr.ru

Abstract. . The variety of research being conducted at the instruments on the external beams of the IBR-2M pulsed fast reactor in Frank Laboratory of Neutron Physics Joint Institute for Nuclear Research (FLNP JINR) is the reason of the differences in the requirements of the detectors for these instruments. This leads to the necessity of developing a variety of detectors in the Laboratory that are used in experiments. This report reviews the neutron detection systems developed and used at the instruments on the external beams of the IBR-2M pulsed research reactor, the current status and operating features of which have been considered.

1. Introduction

The pulsed high-intensity neutron reactor IBR-2 [1] is the main facility of the Frank Laboratory of Neutron Physics (FLNP) at the Joint Institute for Nuclear Research (JINR). It was put into operation in 1984 and was modernized in 2007-2011. The unique complex of neutron instruments of the reactor allows a wide range of cross-disciplinary research to be carried out in the fields of condensed matter physics, material science, chemistry, biology, geophysics, pharmacology, medicine, nuclear physics, ecology, etc. The essential parts of the experimental equipment are the detectors of neutrons and the detector systems on which they are based.

2. Data Acquisition and software

Electronic and software systems used on the reactor [2] are unified as much as possible. For multi-counter detector systems consisting of helium or scintillation counters, as well as for annular ("ring") detectors DAQ systems based on MPD16 and MPD32 [3] units are used. The maximum number of connected counters is 16 for MPD 16 and 240 for MPD32. For 1D and 2D position-sensitive detectors PCI the system DeLiDAQ [4] or a new DeLiDAQ2 [5] NIM unit is used. All the electronic systems can operate in the Time-Of-Flight (TOF) mode. The electronics are standardized and the specific features of each spectrometer are taken into account when programming the internal FPGA. All the electronic components have been developed in FLNP, JINR (DeLiDAQ developed in collaboration with HZB, Berlin).

Software includes SONIX+ [6] [7] software complex and DeLiDAQ [8] software system. SONIX+ is a software for the automatization of experiments. It is extensible, based on the Python script language and allows fully automated experiments to be carried out. Software system DeLiDAQ has been designed for real-time data observation and detector testing. It is based on the ROOT library. The parts of DeLiDAQ can be embedded in other systems.

3. Detector systems of the IBR-2 spectrometers

The list of scientific instruments at IBR-2 includes 3 reflectometers, 2 inelastic spectrometers, 1 small-angle spectrometer, 7 diffractometers, 1 spectrometer for neutron radiography and tomography, 2 installations for nuclear physics and installations for radiation materials science and neutron activation analysis. Here is some information about the neutron detector systems on these installations.

3.1. Reflectometers

The REMUR reflectometer (beam №8) is intended for the study of properties of thin films, limits of the section in multilayered structures, various types of streamlining, self-organization in biological systems, etc. by the method of specular and not specular reflection of neutrons from the surface of films. The reflectometer with polarized neutrons REFLEX (beam №9) is designed for the study of structural properties of thin films and layered nanostructures. Reconstruction has added a chance to operate this instrument as a spin echo small-angle neutron scattering spectrometer. The GRAINS instrument at beam №10 is a multifunctional neutron reflectometer with a horizontal sample plane, intended for the determination of the reflectivity of thermal neutrons in various dimensions from surfaces and interfaces. It has the possibility to measure both the solid-solid and the liquid-solid interfaces, liquid surfaces. Each of these three instruments is equipped with a 2D PSD [9] and a ^{3}He proportional counter on the moving platform.

Figure 1. 2D Position-Sensitive Detectors

The position-sensitive detector (figure 1) is a general purpose detector developed in FLNP JINR. It is designed for neutron diffraction and neutron reflectometry. The sensitive area of 2D PSD is 200x200 mm^2, the spatial resolution is 2x2,3 mm^2, efficiency for the thermal neutrons (2A) is 65%. The detector is made on the principle of the multiwire proportional chamber. An anode, a cathode and drift planes are located in the hermetically sealed internal volume of the detector. The time of neutron detection is determined by the anode signal, coordinate information received using delay lines. Detectors of this type are rather universal, and at the same time cheap tools for the neutron measurements what the experience of use on the installations at FLNP JINR testifies to. The same 2D PSD is operated at some other research centers in Russia, as well as at the research reactor LVR-15 of Nuclear Physics Institute (NPI ASCR) in Rez near Prague, the Czech Republic.

3.2. Inelastic scattering

There are two inelastic instruments on IBR-2, the spectrometer DIN-2PI on beam №2 and NERA-PR on beam 7b. Both spectrometers are equipped with multi-counter systems based on ^{3}He counters, covering large area. The DIN-2PI inelastic neutron scattering spectrometer is a "direct geometry", double time-of-flight, multi-counter system with a set of curved-slit and background choppers, and a cadmium periodic shutter for background measurements. The spectrometer offers a wide range of resolutions, energy transfers, momentum transfers, scattering angles and sample environments. The NERA-PR spectrometer is an inverse geometry time-of-flight multi-purpose instrument. It was designed to study inelastic and quasielastic neutron scattering with simultaneous monitoring of the phase of the sample by neutron diffraction. The detector system of the spectrometer NERA-PR [10] consists of two symmetrical sections. Each section includes nine chambers, with multi-counter arrays. One chamber intended for diffraction measurements includes blocks of counters with collimators and eight chambers for inelastic measurements are equipped with counters, beryllium filters and analyzers

IOP Conf. Series: Journal of Physics: Conf. Series **1021** (2018) 012021 doi:10.1088/1742-6596/1021/1/012021

(crystals or pyrolytic graphite). Filters and analyzers are used for switching between working regimes with various luminosity and energy resolution.

3.3. Small-angle scattering

The YuMO small angle scattering (SANS) spectrometer operates on the beam №4 of the IBR-2M reactor. It can obtain information about sizes, spatial correlations and shapes of objects in a range of 1 nm to several hundred nanometers. It is suitable for the study of polymers, biological objects, colloidal solutions and surfactants. The YuMO detector system consists of two ring detectors [11] placed one after another and a monitor of the direct beam. The appearance of the ring detector is shown in the figure 2a. All detectors work in the vacuum casing of the spectrometer. Each ring detector is divided into 8 rings, which allows the angular distribution of the neutron flux to be measured. All the rings of a detector are located in a single gas volume. The ring detectors have a central hole for the passage of the direct beam through the detector to the direct beam monitor. The direct beam monitor is a low-efficiency multiwire proportional chamber with a ^{10}B neutron converter. It is designed to calibrate the intensity of the incident beam.

3.4. Diffractometers

The biggest classes of instruments on the IBR-2 reactor are the diffractometers. The DN-12 neutron diffractometer is intended for the investigations of micro-samples at high pressure. It is located at beam №12. The sapphire/diamond anvils in combination with low-background neutron diffractometry have allowed the pressure range of the experiments to be extended to several GPa. The detector system of the DN-12 diffractometer consists of a multi-counter detector system, covering 0.125 sr.

The DN-6 spectrometer for microsample investigations is the direct development of the existing spectrometer DN-12. The detector system includes a set of proportional counters and a 16-section ring detector [12]. The ring sectioned detector (figure 2b) covers an angle of 360 degrees in the circular plane. The detector is divided into 16 sectors, 22.5 degrees each, located in the same hermetical volume. Each of the radial sectors is additionally divided into 6 sections along the cylindrical surface, thus the total number of measuring channels is 96. Sections and sectors are mechanically limited with laminated fiberglass plates with a thickness of 1 mm. The width of the entrance window of the detector is 7 mm, the thickness of the sensitive volume is 40 mm. The radius of the internal part of the detector is 320 mm. The efficiency to thermal neutrons (4A) reaches ~70%.

a b

Figure 2 8-section SANS ring detector.(a) 16-section ring detector (b).

The EPSILON (beam № 7a-1) strain/stress diffractometer allows bulk samples to be investigated and strain in space and time to be resolved, separately for each phase [13]. Its 9 banks of ^{3}He counters mounted on a ring at 2θ=90°. The SKAT [14] (beam № 7a-2) detector system is a multi-counter diffractometer consisting of three independent detector rings. Each detector system is characterized by a unique scattering geometry of the detectors, allowing texture experiments to be carried out with high and very high resolution. The RTD (former DN-2) spectrometer is a general purpose time-of-flight diffractometer with medium interplanar spacing resolution. RTD is designed for research processes in real time mode. The detector system of RTD consists of ring detectors (eight rings like the YUMO detector) in reverse and small angles of dispersion in axial geometry, blocks of 8 proportional counters on intermediate corners (15-135 degrees) and 2D PSD 200x200 on a rotary platform on the angles of dispersion from 2 to 140 degrees.

Specifics for the operation of Fourier diffractometers is the method of time-focusing. This method has very strict requirements for determining the time of registration of neutrons and that is why these diffractometers use mostly scintillation-based detector systems. At IBR-2 there are three Fourier diffractometers: HRFD at beam №5, FSD at beam № 11 and FSS at beam № 13 (under construction).

The HRFD [15] is a neutron reverse time-of-flight High Resolution Fourier Diffractometer designed for precision analysis of polycrystalline structures. HRFD combines high neutron flux at the sample position with high resolution over a wide range of d-spacing. On the HRFD spectrometer three scintillation detector systems with ^{6}Li glasses are supplemented by a ^{3}He 1D PSD. The aperture of the scintillation detectors is 0,2 sr, the active area of 1D PSD - 200x80 mm^2.

The high resolution Fourier Stress Diffractometer FSD [16] is a neutron reverse time-of-flight diffractometer intended for precise residual stress investigations in bulk samples and advanced materials. The main specific feature of FSD is the capability of analysis of triple correlations of the signals from the neutron source, the Fourier-chopper and the detector. This leads to a substantial decrease of the correlation background, better quality of diffraction patterns, and permits the useful wavelength interval to be extended. The spectrometer FSD has a scintillation system ASTRA [17] with combined electronic and geometric focusing, and a backscattering detector with 16 ^{6}Li-glass scintillator elements (figure 3). ASTRA consists of two multielement sections with ^{6}LiF+ZnS neutron converter and wavelength shifting fibers-type readout.

a b

Figure 3. The FSD detector system. Part of ASTRA scintillator detector system (a), backscattering detector (b).

FSS is a time-of-flight Fourier spectrometer for non-destructive characterization of residual stresses in various polycrystalline materials. The equipment for this spectrometer was received from the research center GKSS (Geesthacht, Germany). The detector system of the FSS spectrometer will include 2 banks of scintillator detectors with ^{6}Li-glass.

3.5. Neutron radiography and nuclear physics

Neutron radiography is a powerful non-destructive testing method. The experimental station of neutron radiography and tomography, the NRT is located on beam №14. A collimated beam with a size of 200x200 mm^2 goes through the sample. A scintillator detector with ^{6}LiF/ZnS scintillator

IOP Conf. Series: Journal of Physics: Conf. Series **1021** (2018) 012021 doi:10.1088/1742-6596/1021/1/012021

screen and CCD-matrix is used for neutron detection. The overall spatial resolution of the system is about 230 μm.

Various types of detectors are used at nuclear physic experiments. The experimental complex KOLKHIDA [18] operates on IBR-2 beam №1a. It is designed for the investigation of nuclear neutron precession and nuclear magnetic structure. It consists of a polarized neutron spectrometer and a polarized nuclear target installation. A ^{3}He counter is used as a neutron detector and a ^{235}U fission chamber is used to monitor the primary neutron beam.

For experimental studies with fast neutrons a new device, "neutron telescope", was developed [19]. It operates on the principle of recoil protons from a hydrogen-containing gas target. This device makes it possible to measure the kinetic energies of neutrons with a good resolution (~2%) in the range from 100 keV to 14 MeV. It can be used to determine the spectra of fast and resonant neutrons of all the neutron-producing targets of IBR-2, IREN and EG-5 FLNP, as well as other installations.

References

[1] Dragunov Y G et. al. , 2012, At. Energy, **113,** 29–38

[2] Kulikov S A, Prikhodko V I, 2016, *Phys. Part. Nucl.*, **47**, 702-710

[3] Levchanovskiy F V, Murashkevich S M, 2013, *Proc. XXIV Int. Symp. on Nuclear Electronics and Computing NEC`2013 (Varna,Bulgaria)*, (Dubna, JINR), pp176-179.

[4] Levchanovsk F V, Gebauer B, Litvinenko E I, Nikiforov A S, Prikhodko V I, Schulz C and Wilpert T, 2004, *Nucl. Instr. Meth. Phys. Res.* A, **529**, 413–416

[5] Levchanovskiy F, Murashkevich S, 2016, , *Phys. Part. Nucl. Letters*, 2016, **13**, 591–594

[6] http://sonix.jinr.ru/wiki/doku.php?id=en:index, Sonix User Guide.

[7] Kirillov A S, 2016, The unified software package Sonix+. Analysis of development experience, *Proc. 11 Conf. New Opportunities for Better User Group Software NOBUGS2016 (Copenhagen, Denmark),* p35

[8] Levchanovsky F V, Litvinenko E I, Nikiforov A S, Gebauer B, Schulz C and Wilpert T, 2006 *Nucl. Instr. Meth. Phys. Res.* A, **569**, 900–904

[9] Belushkin A V et al, 2010, *Phyisics. of Solid State,* **52**, 1025-28

[10] Natkaniec I, Chudoba D, Hetmańczyk Ł, Kazimirov V Yu, Krawczyk J, Sashin I L and Zalewski S, 2014, *J. Phys.: Conf. Ser.,* **554** 012002

[11] Kuklin A I, Islamov A Kh, and Gordeliy V I, 2005, *Neutron News*, **16**, 16-18

[12] Belushkin A V et al, 2013, *Phys. Part. Nucl. Letters*, **10**, 436–441

[13] Walther K, Scheffzuek C, Frischbutter A, 2000, *Physica Condens. Matter* B, **276-278**, 130-131

[14] Ullemeyer K, Spalthoff P, Heinitz J, Isakov N N, Nikitin A N, Weber K, 1998, *Nucl. Instr. Meth. Phys. Res.* A, **412**, 80-88

[15] Balagurov A M, 2005, *Neutron News*, **16** , 8-12

[16] G.D. Bokuchava G D, Balagurov A M, Sumin V V, Papushkin I V,2010, *J. Surface Investigation. Xray, Synchrotron and Neutron Techniques,* **4**, 879-890

[17] Kuzmin E S, Balagurov A M, Bokuchava G D, Zhuk V V, Kudryashev V A,2002, *J. of Neutron Res.*, **10**, 31-41

[18] Abov Yu G, Alfimenkov V P, Lason L, Mareev Yu D, Pikelner L B, Tsulaia V M, Tsulaia M I, Salamatin I M, 2009, *Nucl. Instr. Meth. Phys. Res.*, **601**, 317-325

[19] Milkov V M, Panteleev Ts Ts, Bogdzel A A, Shvetsov V N, Kutuzov S A, Borzakov S B and Sedyshev P V, 2012, *Phys. Part. Nucl. Letters*, **9**, 508–51

IOP Conf. Series: Journal of Physics: Conf. Series **1021** (2018) 012022 doi:10.1088/1742-6596/1021/1/012022

Optical Observation of Single Neutron Detection

Marita Mosconi[1], Jon Ortega[1], Estefanía Abad[1], Paolo Mutti[2], Patrick Van Esch[2] and Emilio Ruiz-Martinez[2]

[1] ESS Bilbao, Poligono Ugaldeguren III, Pol A -7B, 48170 Zamudio, Spain.
[2] Institut Laue Langevin, 71 avenue de Martyrs, 38000 Grenoble, France.

E-mail: mmosconi@essbilbao.org

Abstract. The feasibility of using highly pixelated light sensors as neutron detectors is the subject of a collaboration between ILL and ESS Bilbao. The fast paced development of camera sensors can meet the future challenges of neutron detection if an optical coupling able to increase the signal to noise ratio and enlarge the detection area is identified. Fiber optics tapers seem the best candidates, but issues about their radiation resistance and price make photographic objectives a more sensible choice. The performances of photographic optics have been tested as well as the light output of several thermal neutron scintillators. Thanks to the use of a highly transmissive optics, single neutrons could be observed in ^{6}LiF/ZnS(Ag) for the first time.

1. Introduction

The planned ESS instrument suite challenges the state of the art of the position sensitive neutron detection by requesting detectors able to sustain count rates which may reach the MHz range [1] (a ^{3}He proportional counter can sustain up to 150 kHz [2]) at a finer spatial resolution (from the micrometric range to some mm) over relatively large areas (up to tens of m^2).
Current scientific CMOS camera sensors, considerably more robust and cheaper than ECCD, have reached sensitivities of the order of a few photons. As a consequence, the observation of single neutron detection in scintillators coupled to these sensors with lenses may be possible. This technique would allow a single sensor to cover an area large enough to make this detection profitable for relatively large areas and the high pixelation would assure a competitive spatial resolution. Optical observation of the light emitted by ZnS based scintillators interacting with charge particles played an important role in the physical research at the beginning of the 20th century (see, for example, the so-called "Rutherford gold foil experiment" [3]). Those observation were done using microscopes with a magnification at least of the order of x20 [4] and Regener first stressed the importance of maximizing the light collection [5]. Nowadays, Miller et. al reviewed several scintillators and proposed a system able to take pictures of any scintillator by transporting its light to an intensifying plate through a fiber optic taper [6]. Being ZnS massively employed in thermal neutron detection, this kind of scintillators seems the best candidate to prove the concept and investigate how to fine tune the technique. Apart ^{6}LiF/ZnS based scintillators which are cheap, bright, with low γ-sensitivity [7], but limited in efficiency [8], mainly ^{6}Li glasses are used when detection efficiency is the main requirement and other scintillators can be considered once a highly transmittive optical system has been designed. Because coupling high speed photography and sensitivity is still a problematic issue, the design of the most suitable electronics able to

detect spots of light in a highly pixelated silicon sensor has to be investigated. ILL is tackling the problem and a developments of a board able to count single neutrons is presented in [9].

2. Tests of transmission in photographic objectives

The potentiality of photographic objectives have been tested using four samples: two reference lenses whose transmission value T has been already measured and two lenses of higher acceptance. The brightest lens is a 50 mm f/0.75 branded by Canon under the name of Canon XI. The other lenses are a C-mount Tamron 8mm f/1.4, which can focus from 8 cm to infinity, a Nikon-mount AF Nikkor 50mm f/1.8D t/2.0 [10], which can focus from 45 cm to infinity and a Nikon-mount AF-S DX Micro Nikkor 40mm f/2.8G t/3.3 [10], which can focus from 16 cm to infinity. Pictures of a homogeneous white surface under controlled illuminations are used to measure the brightness of the objectives. By different amount of time exposures, the effective transmission of an objective can be deduced by the slope of the relation between exposure time and luminosity of the pictures. Pictures are taken in focus on the subject and focusing at infinity, if possible. In figure 1 the luminosity captured at the different exposure times, normalized over the field of view, is shown as the averaged gray level of the picture. The Canon XI lens has been identified as the brightest lens of the group with a transmission not better than t/1.4.

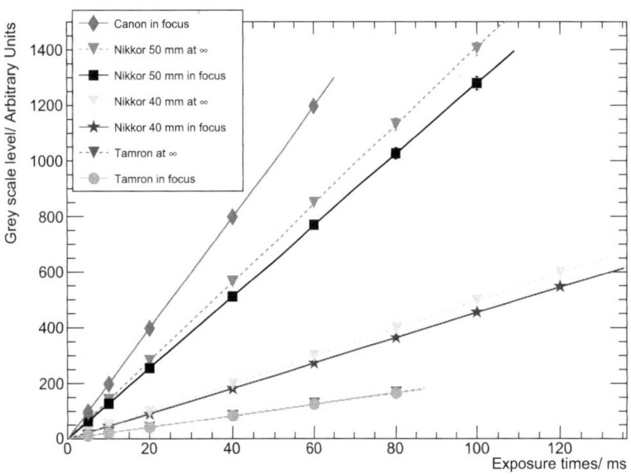

Figure 1. Above: tested optics, from right: Canon XI, Tamron, Nikkor f/1.8, Nikkor f/2.8.
Right: Luminosity captured by the tested optics at different exposure times.

3. Tests on scintillators

Several thermal neutron scintillators have been tested at the same conditions at the beam line CT1 of the Institut Laue-Langevin: the Scintacor ^{6}LiF/ZnS(Ag) (yielding blue light, in the following quoted also as blue scintillator) and ^{6}LiF/ZnS(Cu,Al,Au) (yielding green light, in the following quoted also as green scintillator) screens in two formulations (2:1 and 4:1, ZnS:^{6}LiF) and thicknesses (250 μm and 450 μm), two Eljen ^{6}LiF/ZnS(Ag) scintillators (thicknesses 320 μm and 500 μm) and two ^{6}Li glasses doped with increasing amount of ^{6}Li (GS20 and KG2) in the thickness of 2 mm. The scintillators were placed in beam facing a mirror inside a dark box and their light output recorded by a Hamamatsu ORCA-Flash4.0 camera mounting a f/0.95 MVL50HS objective for tests with full neutron beam and the Canon XI for imaging single neutrons. The use of the mirror allows to avoid to expose the camera at the direct neutron beam. The interior of the dark box during the experiments can be seen in figure 2. The unattenuated beam was scarcely visible by Li glass scintillators so no further analysis has been undertaken. The absorption in the other scintillators is evaluated by measuring the beam flux with a ^{3}He proportional counter without scintillator and after it. The absorption in ^{6}LiF/ZnS(Ag) 2:1

450 μm is the largest, but the its total light output is inferior to a green scintillator of the same thickness and even to the ^{6}LiF/ZnS(Ag) 4:1 450 μm. Indeed, the luminosity relative to the absorption confirms the thinner scintillators, green scintillators and formulations containing more ZnS with respect to ^{6}LiF brighter than their counterpart.

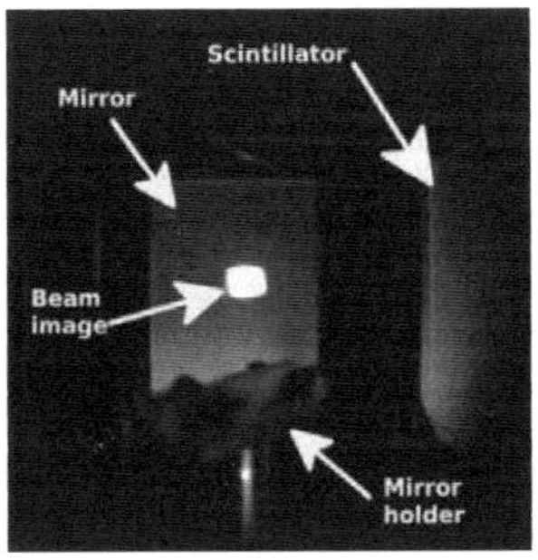

Figure 2. The experimental setup as pictured by the 8 mm Tamron objective on the camera of the experimental setup.

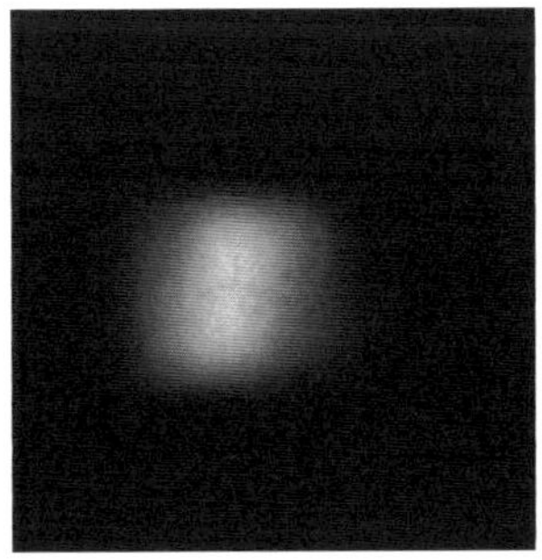

Figure 3. A picture of the light emitted by the unattenuated beam. The localization to the beam position is identifying the are where single neutron detections may be found.

In order to observe single neutrons, the flux of the beam line CT1 has been reduced by inserting plastic scatterers. Pictures were taken with exposures ranging from 10 ms to 100 ms in steps of 10 ms and from 200 ms to 500 ms in steps of 100 ms. The identification of single neutrons rely on the localization of the spots of light at the beam position which is a small portion of the total area of the pictures as can be visible in figure 3 and in the comparison with the events identified in the picture taken during the campaign [11] where an area free of neutrons was defined by inserting a cadmium slab. Such events appear as a cluster of pixels having gray level of the order of the highest value of the dark noise or slightly above as can be seen in figures 6 and 8. This events are identified by filtering the pictures using the standard Non-Local Mean Denoising tool of ImageJ [12] as it is shown in figure 6 and 8. The possibility

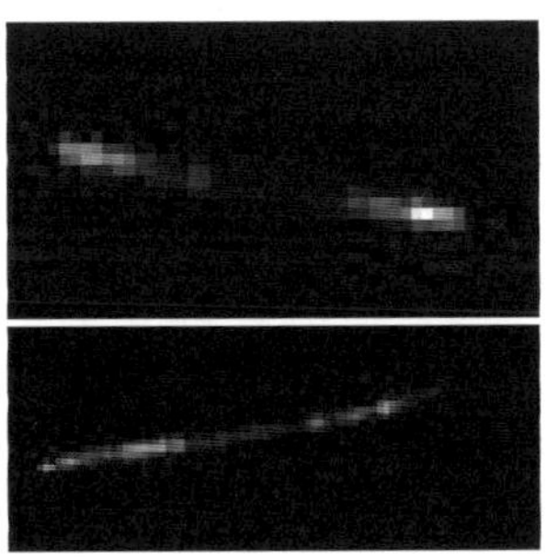

Figure 4. Detections on the silicon sensor of γ-rays or charge particles.

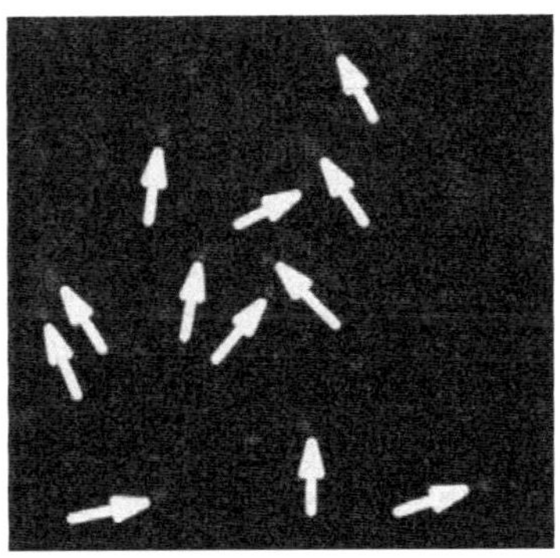

Figure 5. Neutron detection in a green scintillator, raw picture, 200 ms exposure. Identified detection events are highlighted.

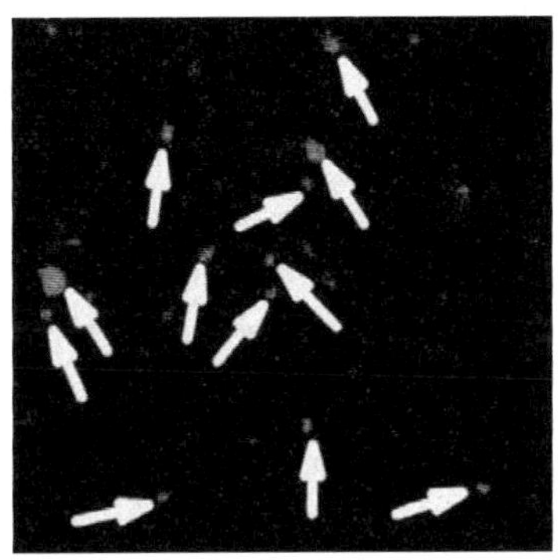

Figure 6. Denoised figure 5 using ImageJ Non-Local Mean Denoising routine. Detection events are identified on the base of size, shape and intensity.

that the dark noise may form such spots of intensities is excluded by putting a threshold in the spot size at 25 pixels and checking the number of events identified outside the localized beam position. Neutrons detected in blue scintillators appear similar in shape and size but their signal to noise ratio is smaller. Also the number of visible neutrons is reduced of at least one order of magnitude with respect to green scintillators. The direct detection of γ-rays (cross checked by exposing a camera sensor to a γ source) or charge particles in the silicon pixels is easily discriminated from a neutron as it has usually a well defined peak of intensity well above the dark noise of the camera, at least twice as much up to saturation. The characteristic intensities of those events are evaluated by the events showing a clear track like the one in figure 4. The number of neutron events is counted by the standard ImageJ routine Analyze Particles which search pictures for spots in a chosen range of intensities yielding for each identified object the total intensity, position and shape information. Direct detection in the sensor can be filtered on the base of this data. An analysis algorithm is under development and additional experiment campaigns are scheduled in order to test improvements in the setup in terms of optical coupling, focus and stability of the electronics.

Figure 7. Neutron detection in a blue scintillator as seen in the raw picture, 200 ms exposure.

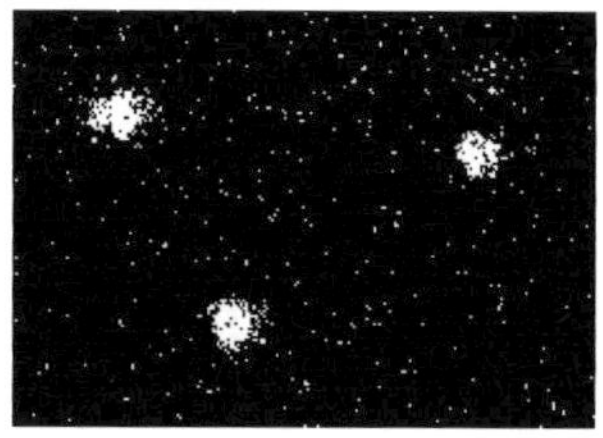

Figure 8. Denoised figure 7 using ImageJ Non-Local Mean Denoising routine.

4. Conclusions

Visualization of detections of single neutrons in ZnS based scintillators is possible by using a sensitive camera and an optic with a transmission achievable also in commercial objectives. The possibility to visualize neutron detections in ZnS(Ag), the scintillator employed in many neutron scattering instruments, makes this technique suitable for fast detections of neutrons by the use of a suitable electronics or a fast enough camera. Additional experiments will be done to assess the technique by further improving the optical coupling and establish the correct measurement procedure. The determination of the best picture analysis algorithm is ongoing.

Acknowledgments

The authors would like to thanks the ESS detector group for the initial support, in particular Dorothea Pfeiffer for her enthusiastic participation, Manuel Morgano and Eberhard Lehmann from the Neutron Imaging group of PSI for the illuminating discussions.

References

[1] Piscitelli F, Messi F, Anastasopoulos M et al. 2017 *JINST* **12** (3) P03013
[2] Stefanescu I, Christensen M, Fenske J et al. 2017 *JINST* **12** P01019
[3] Geiger H and Marsden E 1913 *Philosophical Magazine* **25**
[4] Crookes W 1902 *Proceedings of the Royal Society of London* **71** 405–408
[5] Regener E R A 1909 *Sitzungsberichte der Königlich Preussischen Akademie der Wissenschaften* 948–965
[6] Miller B W, Gregory S J, Fuller E S et al. 2014 *Nucl. Instr. and Meth. in Phys. Res. A* **767** 146–152
[7] Van Eijk C W 2001 *Nucl. Instr. and Meth. in Phys. Res. A* **460** 1–14
[8] Spowart A 1969 *Nuclear Instruments and Methods* **75** 35–42
[9] Van Esch P, Mutti P, Ruiz-Martinez, E et al. 2018 *EPJ Web of Conferences ANIMMA 2017* **170** 01018 URL `https://doi.org/10.1051/epjconf/201817001018`
[10] Dxo mark `www.dxomark.com` accessed: 2016-11-28
[11] Mutti P, Ruiz-Martinez E, Platz M et al. 2015 *Proceedings of ICALEPCS2015* (Melbourne, Australia) ISBN 978-3-95450-148-9 URL `https://accelconf.web.cern.ch/AccelConf/ICALEPCS2015/`
[12] Schneider C A, Rasband W S, Eliceiri K W 2012 *Nature Methods* **9** 671-675

IOP Conf. Series: Journal of Physics: Conf. Series **1021** (2018) 012023 doi:10.1088/1742-6596/1021/1/012023

Variable focusing system for neutrons

Christian Schanzer[1], Michael Schneider[1], Uwe Filges[2], and Peter Böni[1]

[1] SwissNeutronics AG, Brühlstrasse 28, CH-5313 Klingnau, Switzerland

[2] Laboratory for Scientific Developments and Novel Materials, Paul Scherrer Institut, CH-5232 Villigen PSI, Switzerland

christian.schanzer@swissneutronics.ch

Abstract. We have developed a variable focusing system for neutrons that allows varying the vertical and horizontal beam size at the sample position independently. It is based on a focusing parabolic neutron guide coated with supermirror m = 6 times the critical angle of reflection of Ni. The divergence of the incident beam is adjusted by solid-state collimators with horizontally and vertically arranged absorbing blades. The performance of the prototype set-up has been benchmarked at the beamline BOA at SINQ confirming the expected behavior. In particular we show that the beam size at the sample position can be varied between 0.6 and 2 mm without exchanging the focusing guide or by introducing a beam-defining aperture between the exit of the guide and the sample, i.e. the beam size can be adapted more than half a meter away from the sample.

1. Introduction

The investigation of novel materials using neutron scattering techniques is challenging because these materials are often only available in small quantities, e.g. encompassing a volume of a few mm^3. Moreover, special experimental conditions, e.g. high pressure, high/low temperatures and/or high magnetic fields can only be realized for small volumes. Confining the neutron beam to such small areas using slits reduces the number of neutrons delivered by the rather large neutron beams (e.g. 50-100 cm^2) significantly because for the minimization of the background, the penumbra must be suppressed too, to reduce scattering of neutrons from the sample environment. Moreover, it would require an aperture that is placed close to the sample which is usually not compatible with the sample environment. Therefore, it is desirable to focus the neutrons precisely to an area corresponding to the given size of the sample. Ideally, the useful neutrons would be selected already far away from the sample area using neutron optical devices.

As already shown in previous works, elliptic mirrors [1]-[4] are well suited to focus neutron and x-ray beams at the sample. With the rather dramatic increase of the reflection angles of supermirrors exceeding m = 8 times the critical angle of reflection of Ni [5], neutrons encompassing a wide range of wavelengths and large divergence can be directed towards the sample position. While adjusting the beam size at the sample with Montel mirrors [6] is straightforward using an aperture at the focal point at the entrance of the mirror [6], the selection of the beam size using elliptic or parabolic guides is more complicated and essentially given by the design of the optics [7]. In the following we propose a combination of a focusing parabolic guide with a variable collimation stage in front of the entrance of the guide that allows varying the size of the beam spot at the sample position in the horizontal and

IOP Conf. Series: Journal of Physics: Conf. Series **1021** (2018) 012023 doi:10.1088/1742-6596/1021/1/012023

vertical dimension independently thus avoiding the replacement of the focusing guide and its tedious alignment for different setups [7].

2. Experimental Set-Up

It is well known that parabolic mirrors focus beams following parallel trajectories (i.e. beams with a divergence $\psi = 0^0$) to a point at the focus of the parabola (see **Figure 1**). As soon as ψ is increased, however, the size a of the focal spot increases. According to Liouville's theorem the product $B = a\psi$ is a constant if the volume of the phase space is conserved [2]. Therefore, the beam size $A_f = a_{xf}a_{yf}$ is directly related to the solid angle $\Omega_i = \psi_{xi}\psi_{yi}$ of the incident beam for a parabola with a given geometry. Here, the indices i and f designate "incident" and "focal" position and the indices x and y indicate the directions transverse to the flight direction z of the neutrons.

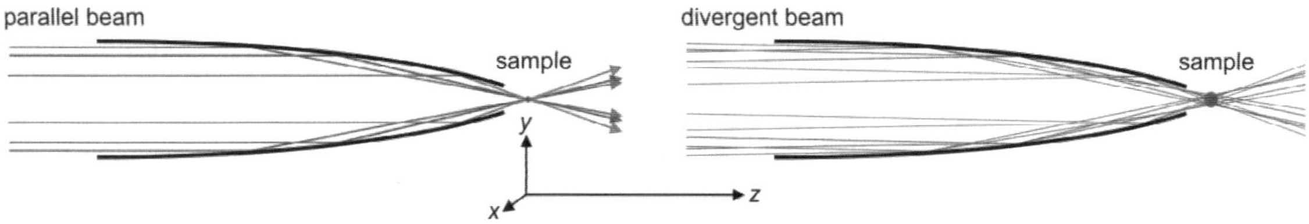

Figure 1. Focusing of neutrons using a parabolic guide segment. By varying the divergences ψ_{xi} and ψ_{yi} of the incident beam the beam size $A_f = a_{xf}a_{yf}$ at the focal position can be adjusted.

In order to prove the concept of focusing neutrons by adjusting the divergence of the incident neutrons, we installed at the beam line BOA [8] at SINQ a variable focusing set-up comprising two solid-state collimator stages in series (each consisting of 4 individual collimators) for defining the beam divergence independently in the x- and y-direction followed by a focusing guide. The collimations ψ_{xi}, ψ_{yi} = 10', 20', 40', 80' were defined by the length and the thickness of the Si-wafers, which were coated with absorbing TiGd. The focusing neutron guide with a length of 500 mm has truly curved parabolic profiles in both dimensions, i.e. the curved sides are made from glass substrates with a thickness of 2 mm which can be easily bent. The exact geometry is determined by precise forming parts glued to the backside of the glass substrates, thus the waviness (figure error) of the reflecting surface is only in the order of 10 µrad (RMS). Four parabolic mirrors are glued together to build the focusing guide. The apertures are 15.00 mm × 15.00 mm and 5.59 mm × 5.59 mm at the entrance and exit, respectively. The nominal distance of the focal point of the parabolas is 80 mm away from the exit of the guide. The substrates are coated with supermirror $m = 6$ having an excellent reflectivity of $R = 0.67$ at the critical edge of the supermirror [5].

A monochromatic beam with a wavelength λ = 3 Å was provided by a double-bounce HOPG monochromator. The neutrons were recorded by a CCD camera imaging detector system (resolution ≈30 µm). All components were aligned by means of a laser that was directed along the optical axis of the neutron beam. The alignment was refined with neutrons. In particular, the vertical and horizontal orientation of the collimators and the focusing guide were adjusted such that a symmetric beam profile was observed. We point out that the horizontal divergence of the beam was restricted by the geometry of the monochromator setup and the horizontal dimensions of the neutron beam to become less than 10'. Therefore, the horizontal collimators did not affect the beam divergence in the x-direction. In the vertical y-direction, however, the large opening of the neutron guide of 120 mm in combination with the supermirror coating of $m = 2$ at the top/bottom of the BOA guide provided a beam divergence of about 60' [9] allowing a variation of the divergence by the vertical collimators.

IOP Conf. Series: Journal of Physics: Conf. Series **1021** (2018) 012023 doi:10.1088/1742-6596/1021/1/012023

3. Results and discussion

Figure 2a) shows the intensity distribution versus the x-position on the detector and the z-position of the detector after integrating the data along the vertical y-direction. The horizontal and vertical collimations are 10'. With increasing z, the neutrons reflected from the sides of the converging guide evolve from a double peak structure to a single peak at the focal point. The further increase of z leads again to a double peak structure as expected [2]. The horizontal beam profile at the nominal focal point is shown in **Figure 2b)**. The highly symmetric shape confirms the precise geometry and alignment of the beam components. The profile is composed of a sharp component, the focused neutrons, with a FWHM of 0.8 mm sitting on a broad plateau having a width of about 5 mm, which is caused by neutrons that are not reflected by the sides of the focusing guide. These neutrons illuminate the sample environment and increase the background. Therefore, they should be blocked by a beam stop at the centre of the entrance of the guide that covers an area corresponding (or being slightly larger) than the cross section of the guide exit (5.59 mm × 5.59 mm). Clearly, the suppression of the plateau is most efficient for incident beams with low divergence, which yield the best focused beams at the sample position. As mentioned in section 2, the horizontal divergence of the incident beam from BOA is lower than the smallest collimation. Therefore, the horizontal beam profile is independent of the choice of the horizontal collimation and cannot be varied with the present set-up.

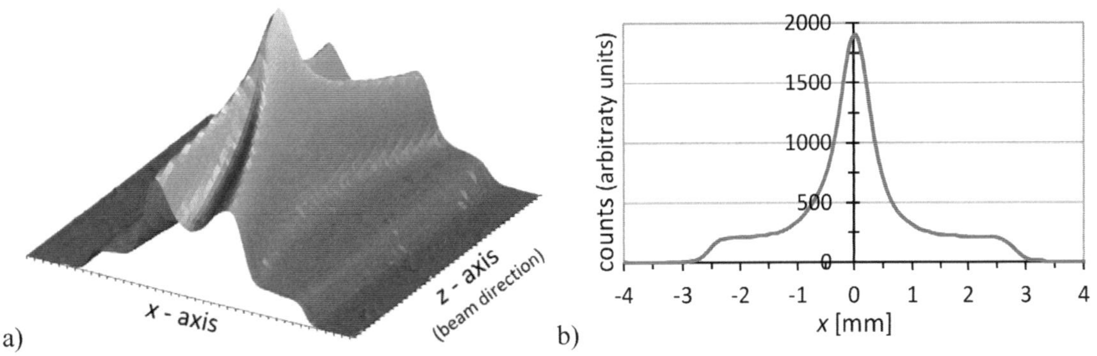

Figure 2. Horizontal beam profile for a horizontal and vertical collimation of 10'. The measured intensity is integrated along the vertical y-direction of the detector. a) Horizontal x-distribution of the neutrons at various distances z from the guide exit. b) Intensity profile at the nominal focal point (z = 80 mm).

The vertical beam profiles for the collimations 10' and 80' are depicted in **Figure 3** as intensity maps in the (yz)-plane. Again the double peak structure is observed before and after the focal point due to the reflection of the neutrons from the top and bottom sides of the guide. As expected, the focusing is very pronounced for the 10' collimation when compared with the 80' collimation. The broad 5 mm wide hump can be removed by an appropriate beam stop as discussed above.

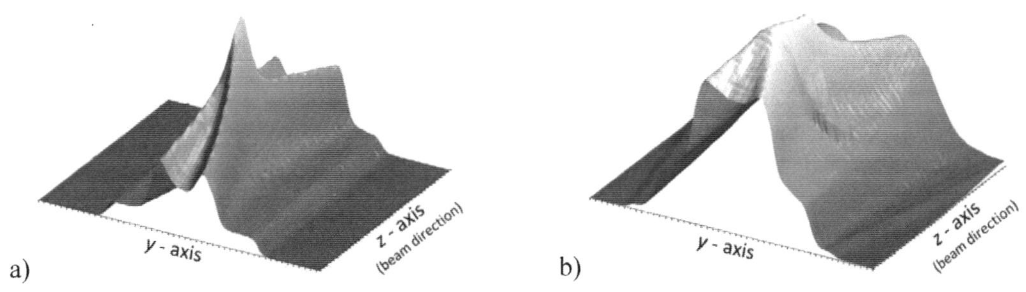

Figure 3. Vertical beam profiles, i.e. (yz)-intensity maps for a vertical collimation of a) 10' and b) 80'. The horizontal collimation was fixed at 10'. The measured intensity is integrated along the horizontal x-direction of the detector.

Figure 4 summarizes the experimental data for the four different vertical collimations used. The intensity distribution consists of a broad plateau with a width of 5 mm due to the direct beam and the focused beam, which has a FWHM that increases with decreasing collimation. The inset provides the fitted FWHM of the sharp component after the broad plateau is subtracted from the data. The level of the plateau is determined by estimating the intensity at the kink of the beam profile near the $y = \pm 2.5$ mm position as indicated by arrows in **Figure 4**. The width of the beam at the focal spot can be reduced by almost a factor of four when the collimation is decreased from 80' to 10'. Note that the amplitude of the sharp component is essentially independent of the selected collimation as expected and that the maximum divergence of the beam before the first collimator is only about 60'.

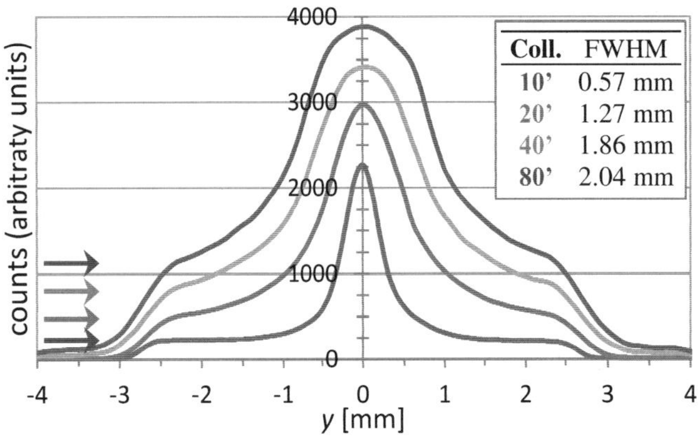

Figure 4. Vertical beam profile at the nominal focal point for various collimations. The measured intensity is integrated along the horizontal *x*-direction of the detector. The arrows at the left indicate the level of intensity of the neutrons, which pass the focusing guide without being reflected from the sides of the guide.

4. Conclusions

We have introduced a focusing system for neutrons that allows the selection of the vertical and horizontal beam size at the sample position over a wide range by means of a combination of two collimation stages and a 500 mm long focusing parabolic neutron guide. Because only neutrons being useful for the experiment enter the neutron optical system, the background is significantly reduced while the flux density at the sample position is improved. Longer guides improve the homogeneity of the focused beam [2]. Similar concepts can be applied to Montel optics using parabolic or elliptic mirrors yielding the additional benefit of a very homogeneous phase space [6] and allowing even more bulky sample environments. The presented focusing system installed on a mechanical support with provision for a highly reproducible positioning using kinematic mounts allows a very efficient, swift, and cost-effective conditioning of neutron beams at beamlines for diffraction, spectroscopy, imaging, and irradiation.

References

[1] Schanzer C, Böni P, Filges U and Hils T 2004 *Nucl. Instr. and Meth. A* **529** 63
[2] Böni P 2014, *J. Phys.: Conf. Series* **502** 012047
[3] Montel M 1957 *X-Ray Microscopy with Catamegonic Roof Mirrors, in X-ray Microscopy and Microradiography* (Academic Press, New York) pp 177-185
[4] http://www.x-ray-optics.de/index.php/en/types-of-optics/reflecting-optics/curved-mirrors
[5] Schanzer C, Schneider M and Böni P 2016 *J. Phys.: Conf. Series* **746** 1
[6] Stahn J, Panzner T, Filges U, Marcelot C and Böni P 2011 *Nucl. Instr. and Meth. A* **634** 12
[7] Brandl G, Georgii R, Dunsiger S R, Tsurkan V, Loidl A, Adams T, Pfleiderer C and Böni P 2015 *Appl. Phys. Lett.* **107** 253505
[8] Morgano M, Peetermans S, Lehmann E, Panzner T and Filges U, 2014 *Nucl. Instr. and Meth. A* **754** 46-56
[9] Bodek K, Böni P, Hilbes C, Lang J, Lasakov M, Lüthy M, Kistrin S, Markiewicz M, Medvedev E, Pusenkov V, Schebetov A, Serebrov A, Sromicki J and Vassiljev A 2000 *Neutron News* **11** 29

IOP Conf. Series: Journal of Physics: Conf. Series **1021** (2018) 012024 doi:10.1088/1742-6596/1021/1/012024

A Selene Guide for AMOR

Christine Klauser, Ryan Bergmann, Uwe Filges and Jochen Stahn
Paul Scherrer Institut, 5232 Villigen PSI, Switzerland
E-mail: christine.klauser@psi.ch

Abstract. We present the new guide design for the reflectometer AMOR at SINQ optimized for high intensity on the sample (gain factor 10) and reduced background. The design uses wavelengths longer than 3.5 Å and homogeneously illuminates horizontal samples with a 30 mm x 10 mm cross section. It allows for short counting times due to simultaneous measurement of $\Delta\theta = 1.5°$ in the specular regime. The beam transport relies on Selene guides using point-to-point focusing. We describe the application of this geometry to the conditions at SINQ including a focusing antitrumpet for transport to the virtual source. McStas simulations show a high performance of the new design. Shielding design has been checked with MCNP simulations.

1. Introduction

The demand for high performing neutron reflectometers is persevering. AMOR [1] is a TOF reflectometer at SINQ [2] with a polarized option. After the upgrade of the SINQ guide system, it is going to be the sole reflectometer available at SINQ. It is therefore designed as a vertical reflectometer accomodating all samples; including liquid interfaces. The new design relies on the Selene guide.

Neutron transport via Selene-type guides is a truly focusing technique. The footprint of the beam at the sample position, the final focal point, is defined at the (virtual) source, the initial focal point. The neutrons are transported by planar-elliptical guides in both dimensions normal to the beam. Coma aberration is corrected to first order by having two subsequent centrosymmetric guides sharing one common focal point. This type of guides has been proposed for neutron optics in 1963 [3], and a demonstrator for Selene-type guides has been successfully tested and is being used as an add-on at the current AMOR reflectometer [4].

With this type of transport system, measurements can be sped up significantly: The defined transported phase space makes it possible to measure simultaeously the specular reflectivity of all incident angles. Defining the footprint far away from the sample has the advantage that only neutrons that are going to touch the sample are being transported. This reduces background that would be created by beam-shaping elements close to the sample and detector environment. Furthermore, by focusing onto the sample, neutron contact with the sample environment is minimized, reducing neutronic background [5, 6, 4]. Neutrons that are being successfully transported by the geometry are being reflected exactly four times. This is advantageous over long distances where the number of reflection can be higher as a function of divergence.

2. The Selene Geometry at SINQ

AMOR is located at the RNR17 beamline supplied with neutrons by the cold D_2 source. The current AMOR guide shares the beamport and the first few meters of guide with the SANS-1

IOP Conf. Series: Journal of Physics: Conf. Series **1021** (2018) 012024 doi:10.1088/1742-6596/1021/1/012024

Table 1. Basic Selene parameters for implementation at AMOR

moderator to detector distance	40 m
distance virtual source to sample	30 m
minor half axis	130.8 mm
maximum size virtual source	30 mm x 10 mm
divergence	1.5° x 1.5°
coating	m=4
minimum wavelength	3.5Å

beamline. After that, the guides are separated into two individual (opposing) bent guides that share a vacuum vessel over a couple of meters. AMOR is bent over 24 m by a radius of 1234 m. A subsequent straight guide section transports the neutrons towards the chopper and polarizer.

The new transport system based on the Selene guide is shown in Fig. 1. The elliptic part

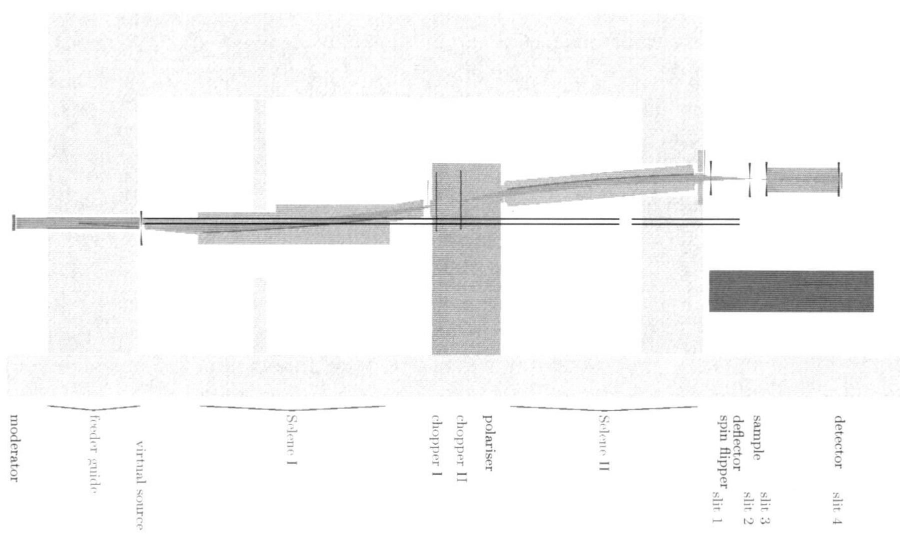

Figure 1. New design AMOR guide. Side view from cold source to detector including optical elements.

has a total length of 30 m where the individual guide elements Selene I and II are 9 m long. Tab. 1 summarizes the most important elliptic parameters. Opting for the transport of 1.5 °

of divergence, geometrical restrictions at SINQ make it necessary for the initial focal point to be at 6 m from the cold source surface. The maximum cross section of this rectangular virtual source is 30 mm x 10 mm and can be tailored to the sample size via diaphrams. The virtual source is fed by a rectangular linear anti-trumpet that assures the delivery of 1.5° divergence, pictured in Fig. 1: After sharing 1.5 m with SANS-1, the guides are split early on. The AMOR feeder at the lower beamport position decreases its cross section from 50 mm x 50 mm to 30 mm x 10 mm. The central axis is tilted downward and to the side by some tenth of a degree. This avoids collision with the SANS-1 guide further downstream and helps for irradiation of liquid samples. The major axis of the ellipse is tilted then upward such that the sample position is at a height of 1.93 m.

The performance of this set-up is compared to the current set-up via McStas simulation.

Table 2. Flux gain compared to current curved guide design

sample size [mm] x [mm]	gain
5 x 5	12.2
10 x 10	10.8
40 x 40	2.9

Results are presented in Tab. 2 comparing the current guide with the Selene guide, using the full divergence delievered. The Selene guide is clearly advantageous for small samples.

3. Optics

The main optical features of the current AMOR instrument are being maintained: The instrument offers wavelength resolution and polarization. The double disc chopper [7] is moved into the neutron guide bunker. It is placed symmetrically with respect to the shared focal point between the two guide segments. Polarization is achieved via logarithmic spiral polarizers [4], placed after the chopper. By placing major optical devices inside the neutron bunker, background can be further suppressed. A slit system close to the sample allows to tailor the incident divergence to the specific experimental needs, including off-specular measurements.

4. Shielding

MCNP simulations of the beamline have been conducted to study possible background sources in the energy range from thermal to 590 MeV neutrons except for background by the sample environement. The small opening of the feeder anti-trumpet limits the amount of background neutrons that enter the bunker region. In addition, it is inherent to the Selene guide to move out of direct line of sight efficiently. Optical beam manipulation can be done in the 6 m gap between the two guide segments inside the neutron bunker. This also reduces background at the sample position. Despite the good inherent shielding of the design, absorbing elements are added for further background suppression and to avoid contamination of neighbouring instruments: L-shaped absorbers are placed at the centre of each guide segment. A copper disc is placed at the shared focal point of the two guide segments to shield fast neutrons. A particular challenge for shielding is that cut-outs in bunker walls are relatively large compared to the beam size. This is due to the fact that the Selene guide needs an external vacuum vessel. At the last wall to the experimental hall a special shielding plug has been optimized: The guide and thus the vessel end inside the 3 m thick neutron guide bunker wall. A distance of 30 cm is available to insert a plug that is equivalent to 3 m of concrete. We opted for a combination of a first layer of steel (fast neutrons), followed by borated polyethylene (thermal neutrons). The thicknesses of the

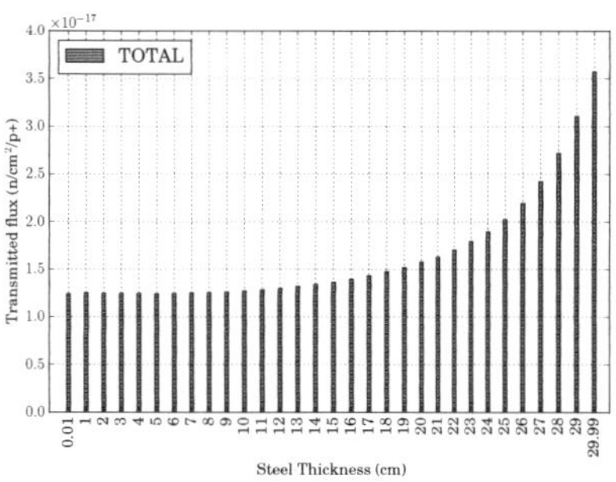

Figure 2. Neutron flux transmitted by the steel-polyethylene plug as a function of the steel thickness

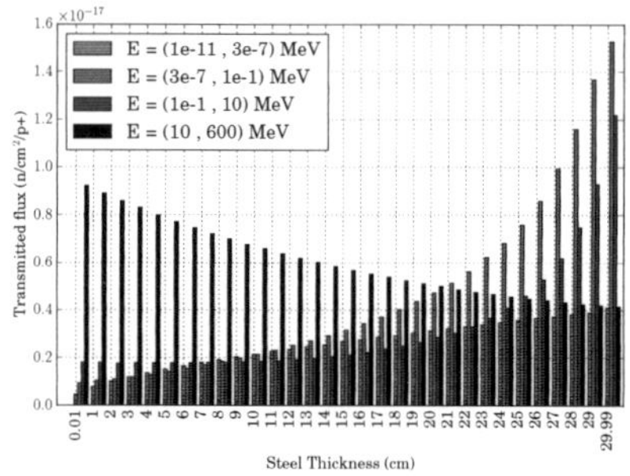

Figure 3. Neutron flux transmitted steel-polyethylene plug as a function of steel thickness for different energy regimes

two layers have been optimized for total neutron flux absorption and consist of 7 cm steel and 23 cm polyethylene, see Figs. 2 and 3. This configuration maximized the steel thickness while keeping the total number neutrons exiting the shielding near the minimum. The plug will end by a gamma shield to suppress the 2.2 MeV photons emitted by (n,γ) reactions in hydrogen (although this can be minimized by using a high concentration of boron in the polyethylene to ensure most of the absorption happens in ^{10}B).

5. Summary

We have shown the new design of AMOR implementing a 30 m Selene guide at the RNR17 beamport. Simulation show gain factors of 10 in flux for the standard 10 mm x 10 mm sample. Shielding elements were simulated and optimized to cater to the challenges risen by the guide geometry.

References

[1] Clemens D, Gross P, Keller P, Schlumpf N and Knnecke M 2000 *Physica B: Condensed Matter* **276278** 140 – 141 ISSN 0921-4526 URL http://www.sciencedirect.com/science/article/pii/S0921452699013861

[2] Blau B, Clausen K N, Gvasaliya S, Janoschek M, Janssen S, Keller L, Roessli B, Schefer J, Tregenna-Piggott P, Wagner W and Zaharko O 2009 *Neutron News* **20** 5–8 (*Preprint* http://dx.doi.org/10.1080/10448630903120387) URL http://dx.doi.org/10.1080/10448630903120387

[3] Maier-Leibnitz H and Springer T 1963 *J. Nucl. Energy, Pt. A B* **Vol: 17**

[4] Stahn J and Glavic A 2016 *Nuclear Instruments and Methods in Physics Research Section A: Accelerators, Spectrometers, Detectors and Associated Equipment* **821** 44 – 54 ISSN 0168-9002 URL //www.sciencedirect.com/science/article/pii/S0168900216300250

[5] Stahn J, Panzner T, Filges U, Marcelot C and Bni P 2011 *Nuclear Instruments and Methods in Physics Research Section A: Accelerators, Spectrometers, Detectors and Associated Equipment* **634** S12 – S16 ISSN 0168-9002 proceedings of the International Workshop on Neutron Optics {NOP2010} URL //www.sciencedirect.com/science/article/pii/S0168900210014117

[6] Stahn, J, Filges, U and Panzner, T 2012 *Eur. Phys. J. Appl. Phys.* **58** 11001 URL http://dx.doi.org/10.1051/epjap/2012110295

[7] van Well A 1992 *Physica B: Condensed Matter* **180** 959 – 961 ISSN 0921-4526 URL http://www.sciencedirect.com/science/article/pii/092145269290521S

IOP Conf. Series: Journal of Physics: Conf. Series **1021** (2018) 012025 doi:10.1088/1742-6596/1021/1/012025

Magnifying effect by neutron diffraction on cylindrically bent perfect crystal (BPC) of Si in fully asymmetric diffraction geometry

P Mikula, V Ryukhtin and J Šaroun

Nuclear Physics Institute ASCR, v.v.i., 250 68, Řež, Czech Republic

mikula@ujf.cas.cz

Abstract. Low absorption of neutrons in Si crystal enables using the fully asymmetric diffraction geometry (FAD) of the single crystal slab in which neutron beam can pass inside a rather long distance more than ten centimetres. A practical exploitation of this property can be namely in the case of bent perfect crystals when due to the homogeneous deformation a neutron beam can meet, on the path, the diffraction condition only once and its attenuation is mainly brought about by absorption or incoherent scattering. In this paper a large one dimensional magnifying property of the FAD geometry of the cylindrically bent perfect crystal slab of Si is demonstrated. This property can be used in high resolution small angle neutron scattering as well as in neutron imaging.

1. Inroduction

It is well known that depending on geometry of the single crystal and the diffractometer performance one can easy manipulate e.g. with resolution, cross-section, focusing properties of the diffracted beam in the scattering plane. In the case of manipulation of the cross-section of the beam, the simplest way is to use asymmetric diffraction geometry, which has been studied many times and described in many text books related to neutron or X-ray diffraction. Namely, it is often used for special monochromator systems on scattering instruments using synchrotron radiation. The extremum case of the asymmetric diffraction geometry is so called fully asymmetric diffraction geometry (FAD geometry). Such unique diffraction geometry employing the crystal of a large length is only practical in neutron diffraction. Neutrons have a high penetration ability and namely perfect Si crystals are very suitable for such experiments and studies. Two contrasts can be distinguished: the FAD geometry with the output beam compression and the opposite one - the FAD geometry with the output beam expansion. In the case of the former one e.g. a wide incident polychromatic beam is impinging the crystal and a narrow (compressed beam) is diffracted when passing the crystal perpendicularly to its thickness e.g. along its longest edge. This alternative was already tested in the eighties for exploitation in designing of new types of neutron monochromators [1-3]. In the latter case a narrow beam enters the bent-crystal slab through the end face and, after passing along the longest edge of the slab, the diffracted monochromated beam is expanded to a large cross section. Such FAD geometry has been then succesfully employed in the high resolution double bent crystal (1,-1) SANS diffractometer with the FAD analyzer as a second crystal [4-6] which provided a substantial enlargement of the range of analyzed angles and a possibility to use a one dimensional position sensitive detector for the analysis of the SANS signal instead of the step-by-step analysis with the conventional symmetric diffraction geometry of the crystal analyzer. Recently, double bent crystal dispersive (n,-m) arrangement with the second FAD crystal in the the output beam expansion diffraction geometry has been studied for imaging of refraction edge effects [7-9]. However, the FAD geometry can also be used for one dimensional magnifying as it is demonstrated in the following chapters.

2. Experimental performances

The experiments were carried out on the double crystal SANS diffractometer operating at the the neutron wavelength of λ=0.21 nm and the neutron optics diffractometer operating at λ=0.162 nm which are installed at the medium power reactor LVR-15 in Rez and figure 1 shows schematically the diffractometer arrangements and the detail of the FAD diffraction transforming the angular deviation $\Delta\theta$ from the mean position, e.g. in the middle of the FAD-crystal, to the linear spatial shift Δx. The arrangement related to figure 1b exploits a premonochromator - bent perfect Si(111) crystal slab - providing the monochromatic beam of a rather low collimation (no Soller collimators are used) and of a relatively large $\Delta\lambda$ spread.

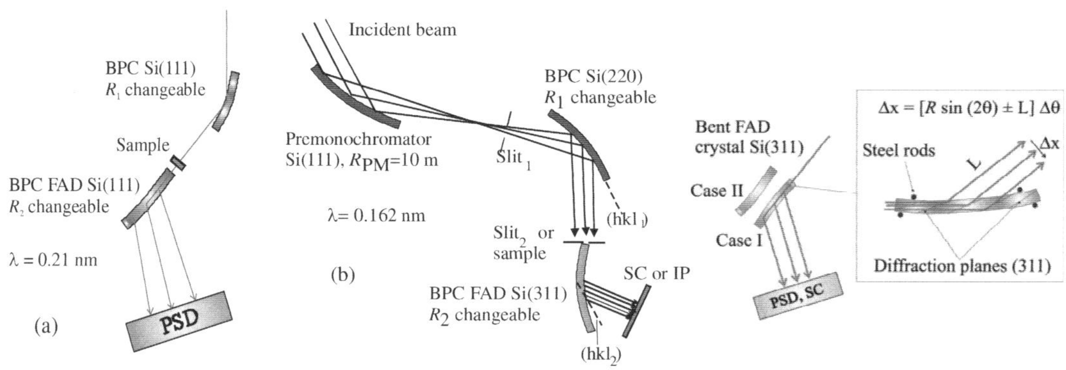

Figure 1. Schematic view of the basic diffractometer performances (a) and (b) used in the experiments and schematic sketch showing the detail of the FAD geometry of the Si crystal (c).

3. Experimental results

3.1. *Double bent crystal (1,-1) setting*

The double-crystal setting as schematically displayed on figure 1a is used for high resolution SANS measurements. In this case the curvature of both Si crystals is changeable and permits the choice of an optimum resolution required by a SANS experiment. Neutrons diffracted somewhere in the irradiated volume in the vicinity of the middle of the first bent crystal can be secondarily diffracted in the equivalent spatially limited volume in the second FAD Si-crystal. This diffraction volume, namely, its length (roughly expressed as $\Delta x/\sin 2\theta$) along the longest edge of the FAD crystal slab, depends on the relative values of the bending radii R_1 and R_2 of both crystals. For demonstration of the magnifying effect small holes in Cd sheet were used as a sample (see figure 2). Figure 3 displays the image of two verticaly situated holes of the diameter of 0.7 mm with the mutual distance of 19 mm situated just in front of the face of the FAD crystal for two different radii of curvature and for R_1=20 m. 2d-PSD used for imaging was situated at the distance of 0.9 m from the middle of the FAD crystal. It can be clearly seen that there is a strong one dimensional magnification in the horizontal scattering

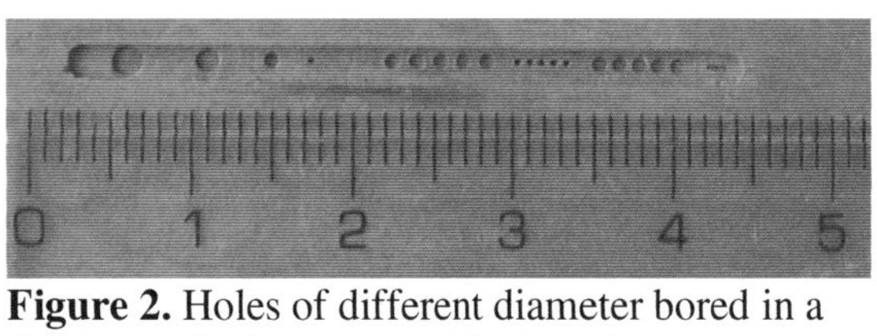

Figure 2. Holes of different diameter bored in a Cd sheet which were used for imaging.

IOP Conf. Series: Journal of Physics: Conf. Series **1021** (2018) 012025 doi:10.1088/1742-6596/1021/1/012025

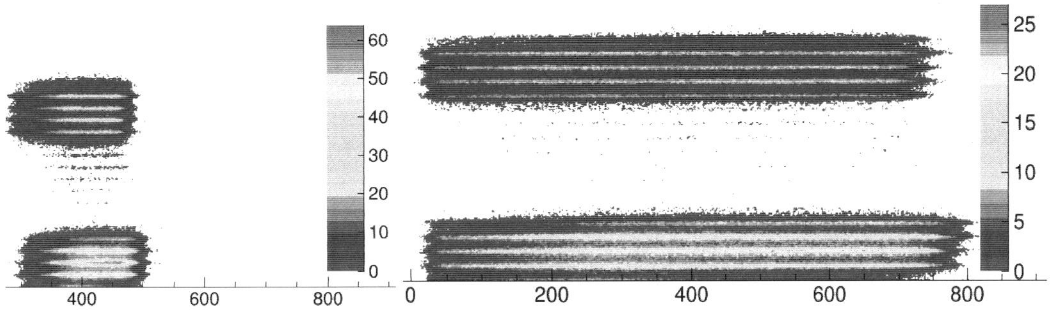

Figure 3. Image of two verticaly situated holes of the diameter of 0.7 mm with the mutual distance of 19 mm for R_{FAD}=56 m and R_{FAD}=131 m, respectively.

plane. Some vertical magnification is brought about by the vertical divergence of the monochromatic beam. Individual displayed fein lines in each hole image belong to the wires of the 2d-PSD detector having spatial vertikal resolution about 2.5 mm. 1 channel corresponds to 0.09 mm.

3.2. Three axis Si(111)+(n,-m) diffraction performance with the third FAD crystal slab

In this case, the bent Si(111) crystal, due to its large lattice constant, can be considered as premonochromator for the next quasidispersive double crystal (n,-m) setting when providing rather divergent monochromatic beam of a large $\Delta\lambda$ spread (see Fig. 1b). However, the used quasidispersive (n,-m) setting improves considerably $\Delta\lambda$ as well as $\Delta\theta$ resolution. The effect of cross section manipulation of the diffracted beam is demonstrated in figure 4 by imaging vertically arranged small holes in a Cd sheet situated just before the front face of the FAD crystal in the place of slit S_2. It can be seen from figure 4 that the FAD crystal works as a one dimensional magnifying glass. For larger radii of curvatures of the FAD crystal the spreading of the beam inside of it can be several centimetres. In this case the imaging plate (IP) was at a distance of 10 cm from the FAD crystal. The beam was relatively high, nearly 4 cm. Radius of curvature of the Si(220) slab was 36 m and the radii of FAD Si(311) slab was 9 m. After that we imaged only one hole from the middle of the sample with a nearly released bending of the FAD crystal ($R\approx$100 m) and the radius of Si(220) slab of 36 m (see figure 5). Similarly to the previous case described in paragraph 3.1, the magnifying factor depends on the relative combination of curvatures of the Si(220) and FAD crystal slabs. Generally, it can be said that the value of magnifying factor increases with the value of the radius of the FAD crystal slab. In the case of the larger hole (ϕ=0.7 mm) the influence of the vertical divergence making the image "thicker" is

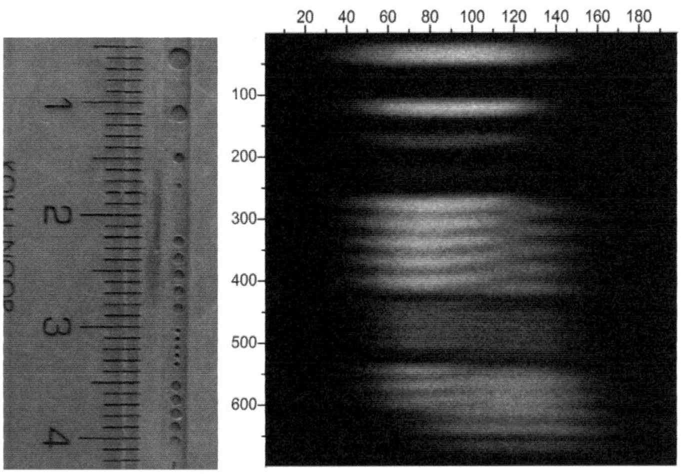

Figure 4. The photo of the sample used in the vertical position (left) and the magnified image of the holes taken by means of the FAD analyzer and IP.

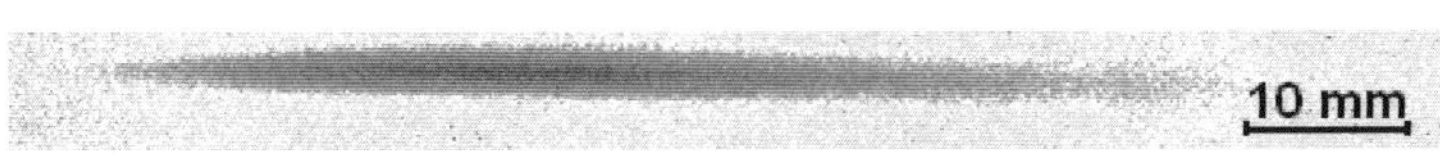

Figure 5. Magnified image of the 1 mm hole IP for a nearly released bending of the FAD crystal.

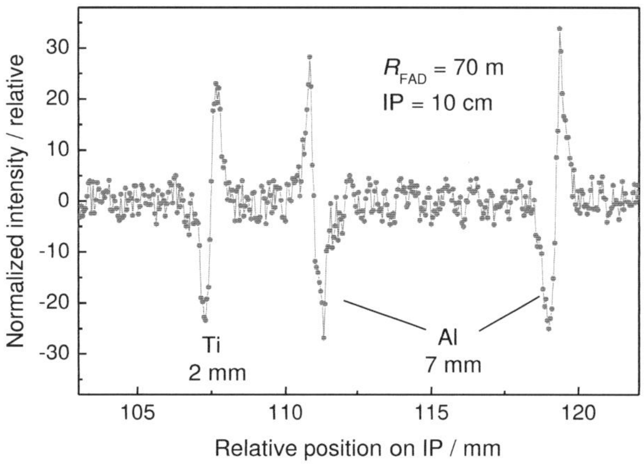

Figure 6. Refraction effects on the edges of Ti (2 mm) and Al (7 mm) for R_1=36 m, R_{FAD}=70 m as imaged on IP at the distance of 10 cm.

clearly visible. It should also be pointed out that the enlarged cross section of the monochromatic beam has locally much higher spatial resolution. High resolution properties of the double-diffracted beam in the scattering plane (in our case it is horizontal plane) can also be proved in the case of imaging some linear objects. In the present case we imaged the rectangular edges of Ti and Al plates (see figure 6). Inspection of figure 6 reveals that at the edges of the samples a typical refraction of neutrons takes place. It has been found that its observation depends on the curvature of the crystals and a proper distance of IP from the sample.

4. Conclusion

It has been demonstrated that the fully asymmetric diffraction geometry of the bent perfect crystal employed in the nondispersive (1,-1) and quasidispersive (n,-m) double-crystal settings permits a considerable one dimensional magnifying of small objects in the scattering plane. It is in fact a spatial enlargement of the analyzing angular scale. The observed slight increase of the height of the image is brought about simply by the vertical divergence of the neutron beam. As the neutron beam should pass inside the FAD crystal along its longest edge, thanks to low attenuation factor, perfect crystal of silicon appears as the best one for application. The choice of the crystal slabs was given mainly by some fixed experimental conditions. Of course, that some other settings employing the FAD second crystals could also be used for other neutron wavelengths. For design and optimization of such double bent crystal settings, Monte Carlo simulations are desirable.

Acknowledgements

Measurements were carried out at the CANAM infrastructure of the NPI CAS Řež supported through MŠMT project No. LM2015056. Bragg diffraction optics investigations are in the Czech Republic supported by GACR project no. 14-36566G and by the ESS project MEYS LM 2015048.

References

[1] Mikula P, Kulda J, Vrána M and Chalupa B 1984 *J. Appl. Cryst.* **17** 189.
[2] Mikula P, Chalupa B, Kulda J, Vrána M, Sedláková L and Michalec R 1985 *J. Appl. Cryst.* **18** 135.
[3] Mikula P, Kulda J, Horalík L, Chalupa B and Lukáš P 1986 *J. Appl. Cryst.* **91** 324.
[4] Lukáš P, Mikula P, Šaroun J and Strunz P 1994 *Nucl. Instrum. Methods in Phys. Research*, **A 338** 111.
[5] Šaroun J, Lukáš P, Mikula P and Alefeld P 1994 *J. Appl. Cryst.* **27** 80.
[6] Hempel A, Eichhorn F, Reichel P and Boede W 1995 *Nucl. Instrum. Methods in Phys. Res.* **A 381** 466.
[7] Mikula P, Vrána M, Seong B S, Woo W, Em V and Korytár D 2014 *IOP Publishing Journal of Physics: Conference Series* **528** 012004.
[8] Mikula P, Vrána M, Šaroun J, Pilch J, Seong B S, Woo W and Em V 2014 *J. Appl. Cryst.* **47** Part 2, 599.
[9] Mikula P, Vrána M and Korytár D 2015 *Physics Procedia* **69** 320.

IOP Conf. Series: Journal of Physics: Conf. Series **1021** (2018) 012026 doi:10.1088/1742-6596/1021/1/012026

VESUVIO+: The Current Testbed for a Next-generation Epithermal Neutron Spectrometer

Matthew Krzystyniak,[a,b] Giovanni Romanelli,[a] Margit Fabian,[c] Matthias Gutmann,[a] Giulia Festa,[d] Laura Arcidiacono,[d] Martyn Gigg,[e] Kacper Drużbicki,[f] Carla Andreani,[d,g] Roberto Senesi,[d,g,h] and Felix Fernandez-Alonso[a,i]

[a] ISIS Facility, Rutherford Appleton Laboratory,
Chilton, Didcot, Oxfordshire OX11 0QX, United Kingdom
[b] School of Science and Technology, Nottingham Trent University,
Clifton Campus, Nottingham, NG11 8NS, United Kingdom.
[c] MTA Centre for Energy Research,
H-1525 Konkoly Thege st. 29-33, H-1525 Budapest, Hungary
[d] CENTRO FERMI- Museo Storico della Fisica e Centro Studi e Ricerche "Enrico Fermi,"
Piazza del Viminale 1, Roma - 00184 - Italy
[e] Tessella Ltd., Abingdon, Oxfordshire, UK
[f] Faculty of Physics, Adam Mickiewicz University, Umultowska St. 85, 61-614 Poznan, Poland
[g] Università degli Studi di Roma Tor "Vergata", Dipartimento di Fisica,
via della Ricerca Scientifica 1, 00133 Roma, Italy
[h] Consiglio Nazionale delle Ricerche, CNR-IPCF, Sezione di Messina, 98122, Italy
[i] Department of Physics and Astronomy, University College London,
Gower Street, London WC1E 6BT, United Kingdom

E-mail: matthew.krzystyniak@stfc.ac.uk

Abstract. We present an overview of ongoing developments in epithermal neutron spectroscopy using the VESUVIO+ beam line at the ISIS Facility. In its current incarnation, VESUVIO+ provides a suitable platform for further and much-needed progress in the judicious exploitation of epithermal neutrons at a pulsed spallation source, as well as constitutes a necessary milestone towards a next-generation station for Epithermal and Thermal Neutron Analysis, hereafter ETNA. In particular, we discuss recent improvements in capability relative to its predecessor VESUVIO. These include the concurrent use of mass-resolved neutron spectroscopy, transmission, and diffraction to explore the properties of complex functional materials, as well as the implementation of techniques unique to pulsed neutron sources such as γ-ray dopplerimetry and energy-resolved prompt-γ activation analysis.

1. Introduction

Over the past decade, the VESUVIO spectrometer at the ISIS Pulsed Neutron & Muon Source [1–3] has supported an international community across many areas of contemporary science including chemistry, physics and cultural heritage [2, 4, 5]. Deep Inelastic Neutron Scattering (DINS) has been the primary technique for the measurement of Nuclear Momentum Distributions (NMDs). The past few years have witnessed a progressive shift of DINS beyond fundamental condensed-matter systems [3, 5–7]. This has been the primary driver behind ongoing improvements to the instrument, hereafter denoted as VESUVIO+. Figure 1 provides a visual summary of the recent transition from VESUVIO to VESUVIO+. In particular, a

IOP Conf. Series: Journal of Physics: Conf. Series **1021** (2018) 012026 doi:10.1088/1742-6596/1021/1/012026

series of concurrent techniques are now available, including transmission, diffraction, (n,γ)-resonance dopplerimetry, and Prompt-Gamma Activation Analysis (PGAA) [8]. Recent changes to the water moderator in the ISIS Target Station 1 have lead to a substantial increase in the thermal neutron flux without compromising the resolution of the instrument [8]. Methodological advances in neutron transmission have allowed the determination of total cross sections over a broad energy range, as well as the treatment of total-scattering data, especially for hydrogenous systems [9]. Moreover, (n,γ) resonance dopplerimetry for incident neutron energies between 1 and 100 eV creates the opportunity to explore nuclear quantum effects in heavy atomic species with a much-higher count-rate and spectral resolution than DINS [7]. PGAA has the potential of providing enhanced capabilities in the quantitative determination of chemical composition [10]. VESUVIO+ offers the possibility of expanding the remit of PGAA beyond its implementation in reactor-based neutron sources use by exploiting the inherent time resolution of pulsed spallation neutrons over a wide range of incident energies. These improvements in instrumentation have also been accompanied by advances in data analysis and computational materials modelling. The first task has now been largely achieved with a new data-analysis package developed within the MANTID framework [7, 11], including the use of Bayesian inference methods [12]. Ongoing efforts also aim to develop *ab initio* techniques to predict neutron-scattering observables across the epithermal and thermal regimes, as recently illustrated in Ref. [14]. Overall, these developments provide a suitable testbed for the subsequent development of a fully fledged ETNA, schematically shown in the right panel of Fig. 1 and explained in more detail in Ref. [2]. The order-of-magnitude increase in count rates and underlying information content possible with ETNA will enable the full exploitation of epithermal neutrons in materials research, particularly in studies requiring parametric scans as a function of an external variable or across a set of specimens within a given family of materials.

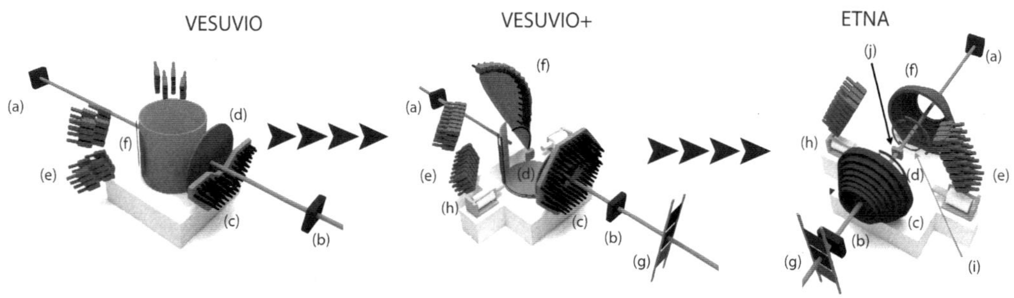

Figure 1. From VESUVIO to ETNA. Left: VESUVIO. Middle: VESUVIO+, currently under development. Right: ETNA. Beamline components: (a) transmitted and (b) incident beam monitors; (c) backscattering Li-glass neutron detectors; (d) backscattering Final Neutron Energy Analyser Mechanism (FNEAM); (e) forward-scattering YAP detectors; (f) forward scattering FNEAM; (g) incident foil-changing mechanism; (h) Ge detectors for PGAA; (i) FNEAM in Debye-Scherrer geometry; (j) sample position and concurrent *in-situ* techniques such as Raman spectroscopy [13]. See text for details.

2. A Glimpse at the Future

To illustrate the enhanced capabilities of VESUVIO+, we present a case study on $Mo_{4,5}Nd_{12,5}B_{20,8}O_{62,2}$ (MoNdBO), a rare-earth molybdate phase characterised by a complex stoichiometry involving both light and heavy nuclides [15]. Figure 2 shows the neutronic response of MoNdBO at Room Temperature (RT) measured in a simultaneous fashion using:
(a) MAss-selective Neutron SpEctroscopy (MANSE) in backscattering geometry;
(b) (n, γ) resonance dopplerimetry for Mo and Nd, using the forward-scattering YAP detectors;
(c) high-resolution powder diffraction.

The main panel in Fig. 2 demonstrates the use of MANSE to isolate the spectral response of heavy nuclei in a complex material well beyond previous attempts restricted to atomic masses below 25 amu [16–21]. As a first step, a high-precision instrument calibration was performed, extending previous protocols [22]. This procedure involved a Le-Bail fit to Pb-powder diffraction data [Fig. 2 (c) – inset], followed by a full Rietveld refinement, both using JANA2006 [23]. The result of the Le-Bail refinement is shown in Fig. 2 (c). A monoclinic unit cell was found (space group P21/c), with lattice parameters a = 10.0425±0.0011, b = 4.1630±0.0003, c = 11.9077±0.0009 Å, $\alpha=\beta=90$ degrees, and $\gamma=116.462\pm0.005^o$.

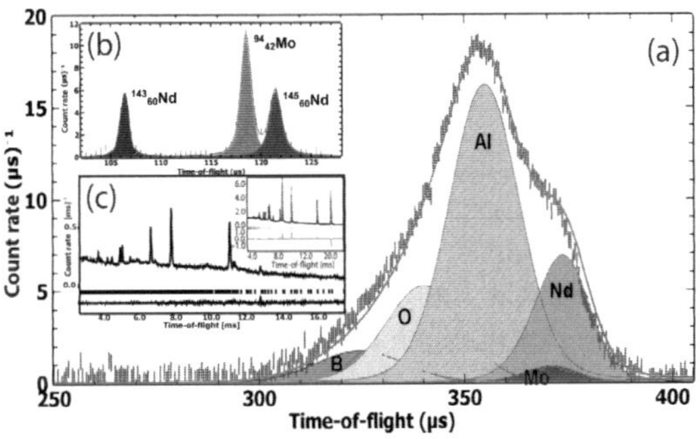

Figure 2. VESUVIO+ data for MoNdBO at RT, all measured simultaneously: (a) MANSE data; (b) (n,γ) resonances from Mo and Nd; and (c) powder-diffraction patterns. See text for details.

The (n, γ) resonances of $^{143}_{60}$Nd, $^{94}_{42}$Mo, and $^{145}_{60}$Nd were fitted with Voigt line shapes [22]. In fitting, the peak positions and Lorentzian line widths were fixed at 55.3±0.4, 44.7±0.4, and 42.5±0.3 eV, and 112, 150, and 411 meV, respectively [24], whereas Gaussian Doppler widths were fitted and yielded NMD widths of 47±4, 27±2, and 55±5 Å^{-1}. The Mo NMD width was then fixed to 27 Å^{-1} in fitting the MANSE data of largely overlapping Mo and Nd recoil peaks. MANSE data fitting yielded NMD widths of 8.6±0.5, 10.1±0.2, 14.0±0.1, and 47.5±1.1 Å^{-1} for B, O, Al, and Nd, respectively. The Nd NMD widths obtained from fitting (n,γ) resonances and MANSE data agree, thus validating the methodology adopted to model these complex data.

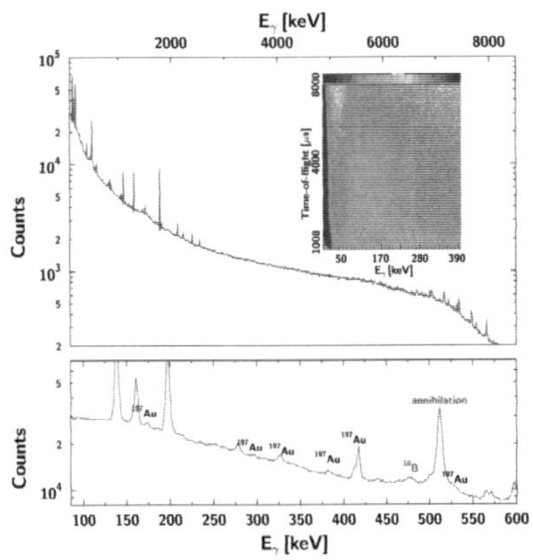

Figure 3. PGAA and multi-dimensional TPGAA spectra on an ancient golden coin. Upper panel: PGAA data over the γ-energy range 100 keV $\leq$ E$_\gamma$ $\leq$ 9 MeV. Lower panel: 100 keV $\leq$ E$_\gamma$ $\leq$ 600 keV, with labels indicating specific features associated with ^{197}Au, ^{10}B, and electron-positron annihilation. Inset: TPGAA spectrum in the time-of-flight range between 1000 and 8000 μs.

As a second example, we present PGAA and Time-of-flight-resolved multidimensional PGAA (TPGAA) [25] capabilities for simultaneous chemical and isotopic analysis, currently under development. In particular, the epithermal and thermal fluxes available on VESUVIO+ offer the potential for the further development of TPGAA, thus providing innovative ways to enhance the sensitivity and isotopic selectivity to intermediate- and heavy-mass isotopes using resonance scattering in the epithermal region. As an example, Fig. 3 reports PGAA and TPGAA spectra on an ancient golden coin. Spectra were taken with a high purity germanium detector using a coaxial close-ended N-type crystal of 8 cm diameter and 16 cm height. Detector resolution was tested using ^{60}Co source and assessed to be 3% in energy corresponding to 3 keV at the

IOP Conf. Series: Journal of Physics: Conf. Series **1021** (2018) 012026 doi:10.1088/1742-6596/1021/1/012026

^{60}Co peak in the 1.33 MeV γ energy spectrum. The TPGAA acquisition electronics chain was composed of a digitizer CAEN DT5724. Each channel of the digitizer was provided with a SRAM memory organized in a programmable number of circular buffers. When all buffers were filled, the board was considered full, that is, no trigger was accepted and the acquisition stopped. When the acquisition restarted, no trigger was accepted until at least the entire buffer was written. This means that the dead time was extended for a certain time (depending on the size of the acquisition window) after the board had exited the full condition. The dead time was estimated to fluctuate between the 10% and the 30%. There was a blind zone for the first 200 μs of each time-spectrum due to the high count rate and the difficulties for the digitizer to be as fast as needed to record all the signals. The digitizer could register 100 MS/s.

3. Outlook and Perspectives

VESUVIO+, an important stepping stone to develop ETNA, has been described. It already offers enhanced capabilities for the analysis of concurrent MANSE and diffraction data. In addition, the use of (n,γ) resonances and TPGAA enables a detailed characterisation of complex materials using both thermal and epithermal neutrons. The two examples presented in this work demonstrate the largely untapped synergies across a variety of experimental techniques, with a view to increasing the information content of a given experiment relative to what is currently possible using more conventional neutron-based methods.

Acknowledgements

UK Science & Technology Facilities Council is acknowledged for financial support and access to beam time at the ISIS Facility. This work was partially supported within the CNR-STFC Agreement (2014-2020) concerning collaboration in scientific research at ISIS. K.D. acknowledges the support from the Ministry of Science and Higher Education of Poland for mobility grants.

References

[1] WEB: `www.isis.stfc.ac.uk/instruments/vesuvio/`
[2] Andreani C *et al.* 2017 *Adv. Phys.* **66** 1.
[3] Seel A G *et al.* 2014 *J. Phys. Conf. Ser.* **571** 012014.
[4] Price D L and Fernandez-Alonso F (Eds.) 2013 *Neutron Scattering – Fundamentals* vol 44 (Academic Press)
[5] Price D L and Fernandez-Alonso F (Eds.) 2017 *Neutron Scattering – Fundamentals* vol 49 (Academic Press)
[6] WEB: `epubs.stfc.ac.uk/search/result?q=VESUVIOsortby=scoreorder=desc`
[7] WEB: `iopscience.iop.org/issue/1742-6596/571/1`
[8] Romanelli *et al.*, 2017 *Meas. Sci. Technol.* **28**, 9.
[9] Dawidowski J *et al.* 2016 *Ann. Nucl. Energy* **90** 247.
[10] Vértes, A., Nagy, S. *et al.* (Eds.) 2011 *Handbook of Nuclear Chemistry* vol 1 (Springer US)
[11] Arnold O *et al.* 2014 *Nucl. Instr. Meth. A* **764** 156.
[12] Romanelli G, Krzystyniak 2016 *Nucl. Instr. Meth. A* **819** 84.
[13] Adams M A *et al.* 2009 *Appl. Spectrosc.* **63** 727.
[14] Krzystyniak M, Druzbicki K and Fernandez-Alonso F 2015 *Phys. Chem. Chem. Phys.* **46** 31287.
[15] Lacorre P *et al.* 2000 *Nature* **404** 856.
[16] Seel A G *et al.* 2011 *J. Chem. Phys.* **134** 114511.
[17] Seel A G *et al.* 2012 *J. Phys.: Condens. Matter* **24** 365401.
[18] Krzystyniak M 2010 *J. Chem. Phys.* **133** 144505.
[19] Krzystyniak M and Abdul-Redah T 2010 *Phys. Rev. B* **82** 064301.
[20] Krzystyniak M and Fernandez-Alonso F 2011 *Phys. Rev. B* **83** 134305.
[21] Krzystyniak M *et al.* 2013 *Phys. Rev. B* **88** 184304.
[22] Mayers J and Adams M A 2011 *Nucl. Instrum. Meth. A* **625** 47.
[23] Petricek V, Dusek M and Palatinus L 2014 *Z. Kristallogr* **229** 345.
[24] Chadwick M B *et al.* 2011 *Nucl. Data Sheets* **112** 112.
[25] G Festa *et al.* 2016 *J. Instrum.* **11** C03060.

ToF-Backscattering spectroscopy at the ISIS Facility: Status and Perspectives

F. Demmel, D. McPhail, C. French, D. Maxwell, S. Harrison, J. Boxall, N. Rhodes, S. Mukhopadhyay, I. Silverwood, V. Garcîa Sakai, F. Fernandez-Alonso

ISIS Facility, Rutherford Appleton Laboratory, Didcot, OX11 0QX, UK

E-mail: `franz.demmel@stfc.ac.uk`

Abstract. The cold neutron spectrometers IRIS and OSIRIS at the ISIS Facility have evolved over the past decades continuously. Both spectrometers are inverted geometry instruments using time-of-flight to determine the incoming energy and Bragg reflection for final energy analysis. The scientific applications are diverse and range from quasielastic investigations of molecular diffusion in nanoporous materials to high-resolution spectroscopy of magnetic excitations. Here we will present upgrade opportunities toward improved energy resolution and increased flux which will keep the instruments competitive over the next decade and beyond.

1. Introduction

The backscattering technique with neutrons was developed about five decades ago in Garching, Germany [1]. This type of spectrometer promised the highest energy resolution of about 1 μeV at that time and hence access to 1 ns relaxation times. Most of the following developments focused on enhancing the flux and to this end the most modern instruments exploit phase space transformation [2, 3, 4].

At pulsed sources the idea of backscattering from an analyser crystal has also been applied and combined with an analysis of the incoming energy by measuring the time-of-flight (tof) to the detector. One of the first instruments of this type was IRIS at the ISIS Facility, UK [5, 6]. The chosen setup was different to the classical backscattering spectrometers as the cooled pyrolytic graphite analysers are placed in a near-backscattering geometry. The energy resolution was relaxed to about 15 μeV, but the dynamic range was significantly larger compared to classical reactor-based backscattering spectrometers. That combination of high resolution over a wide range in momentum transfers with a large dynamic range compared to classical backscattering spectrometers allowed novel applications in cold neutron spectroscopy. With the OSIRIS spectrometer a large gain in intensity was achieved by using a supermirror guide and an increased analyser unit [7, 8]. A decisive step toward the μeV resolution regime at spallation sources was undertaken at the spallation neutron source SNS at Oak Ridge, by combining a long flight path with a large silicon analyser (SA) [9]. As a natural further development tof backscattering spectrometers with a silicon analyser using a pulse-shaping chopper have been built or proposed [10, 11].

IRIS is a day-one instrument which opened in the 1980s and uses a Ni-coated guide terminated with 2.5 m of supermirror focusing guide. OSIRIS is the last of the first-generation of instruments at ISIS and has a curved $m = 2$ supermirror guide. Both spectrometers use the reflections of cooled pyrolytic graphite crystals to define the final energy. To remove higher order reflections, a cooled beryllium filter is installed between sample and analyser [12]. More details about the energy resolution and performance have been obtained through extensive Monte Carlo simulations, see e.g. Refs. [8, 13, 14]. IRIS is additionally equipped with a Mica analyser bank to achieve higher energy resolution of e.g. 11 μeV with the 006 reflection.

The strength of both spectrometers lies in the combination of relative high energy resolution due to the near backscattering geometry with a large dynamic range (up to several meV in energy transfer) due to the tof-technique. That virtue enabled unique experiments in the field of correlated electrons [15], research on Energy related materials and catalysis [16]. The need to cover a wide range of relaxation dynamics over several decades in time in complex materials and limitations in studying ionic conductivity in novel battery materials are exemplary drivers for new capabilities. The IRIS Mica option is often not viable due to the loss in intensity and the restricted momentum transfer range for lower reflection orders.

These performance limitations triggered a proposal to upgrade the OSIRIS spectrometer with a SA bench [17]. It will improve the spectral resolution by a factor of 2.5 to $\Delta E \approx 10\ \mu eV$. In the following the proposed SA upgrade will be discussed in more detail and at the end we will present further directions of development, which are under investigation.

2. New capabilities

From a previous study on the resolution function of the OSIRIS spectrometer it became clear that the dominating part for the resolution stems from the analyser geometry [14]. To improve the energy resolution, silicon is considered as analyser material which is the standard at classical backscattering spectrometers and nowadays also on high-resolution tof-backscattering instruments [18].

There are several advantages in using silicon compared to the present pyrolytic graphite. Because Si(111) has no second order reflection, there is no need for the installation of a cooled beryllium filter. Furthermore, silicon does not show strong thermal diffuse scattering as it was demonstrated recently by the DNA spectrometer with a very high signal to noise ratio, approaching 10^5 [10]. However, there is a disadvantage with silicon crystals concerning its neutron reflectivity: the crystals are too perfect for the required resolution, in particular, for the more relaxed resolution in near-backscattering geometry. For the Si(111) reflection, the energy resolution contribution from the Darwin width is 0.077 μeV [18] and hence about a factor 100 too small for OSIRIS. That mismatch with the resolution of the primary spectrometer would result in a substantial intensity loss. Since many decades standard practice is to deform elastically the crystals, which increases the reflected intensity. The elastic bending of the wafers creates a gradient in lattice spacing which depends on the curvature radius of the analysers R_A and the thickness of the wafers D [2]: $\frac{\Delta d}{d} = \mu_{eff} \frac{D}{R_A}$ with $\mu_{eff} = 0.44$ an average Poisson ratio and the negligible contribution from the Darwin width has been omitted. To calculate analytically the expected energy resolution for a silicon analyser setup on OSIRIS, we are using the same formalism and parameters as previously [14, 19]:

$$\frac{\Delta E}{E} = 2\{(\frac{\Delta t}{t})^2 + (\frac{\Delta L}{L})^2 + (\frac{\Delta d}{d})^2 + (cot\theta\Delta\theta)^2\}^{1/2} \quad (1)$$

For monochromatic focusing conditions of a Bragg-reflecting spherically bent crystal the focusing distance f is $f = \frac{R_A}{2} sin\theta$ with θ the Bragg angle and the object and image distances are equal to $2f$ [20]. The angular term of the analyser resolution is determined by the near-backscattering Bragg angle $\theta = 83^0$, which is imposed by engineering constraints, and the angular spread $\Delta\theta$. For a crystal with incoming $\alpha_i = 2^0$ and outgoing $\alpha_f = 0.5^0$ divergences and negligible mosaic spread $\Delta\theta$ can be written in the following form [21]: $\Delta\theta = \frac{\alpha_i \alpha_f}{\sqrt{\alpha_i^2 + \alpha_f^2}}$. For the thickness of the wafers, we assume $D = 0.8\ mm$ and a radius $R_A = 880\ mm$. With all these input parameters, we calculate the energy resolution around the final energy $E_f = 2.11\ meV$. Figure 1 shows the

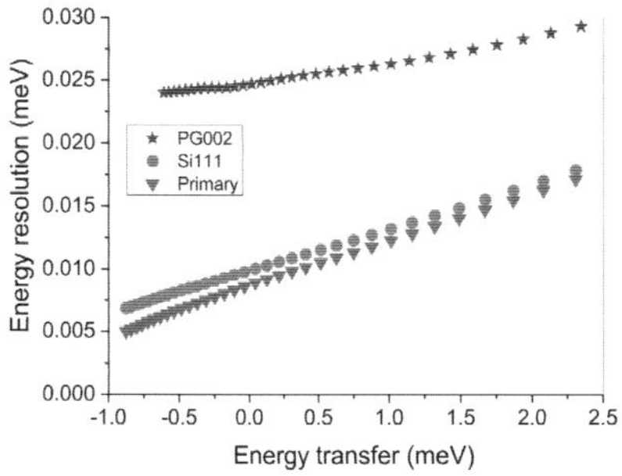

Figure 1. The analytically calculated energy resolution is shown for the PG analyser (stars) and the silicon analyser (circles). In addition, the contribution from the primary spectrometer is shown (triangles).

resulting energy resolution for energy transfers around the elastic line. The calculated resolution for the PG analyser agrees perfectly with the measured value from a vanadium measurement at the elastic line and with Monte Carlo simulations [14]. With the SA the energy resolution will approach 10 μeV at the elastic line, hence an improvement by a factor 2.5. That value is very similar to the resolution attainable with the Mica006 reflection on IRIS. Included is also the contribution solely from the primary spectrometer, which dominates now the resolution. The part of the secondary spectrometer is $\Delta E_f = 4.6\ \mu eV$. In the time domain, relaxation times up to around 400 ps will be accessible and in total about 3 decades in relaxation times will be covered over a wide range of wave vectors ($Q = 0.2 - 1.9\ \AA^{-1}$) in a single setting.

In Figure 2 an engineering view of the new setup is shown inside the OSIRIS tank. Clearly visible are two SA benches and their respective detector arrays. The detectors are arranged on a circle around the 400 mm diameter sample bin. The detectors for the new setup will be linear position sensitive He3 tubes with a half inch diameter, a half inch position sensitivity and an active length of 6 inches. The position sensitivity will provide two advantages compared to the detector setup on the PG side. At first, the position sensitivity will allow to measure momentum transfers in the vertical direction, a quality that will benefit mostly the single-crystal community with now access to full four-dimensional $S(Q, \omega)$ spectra. Secondly, the arrival times of the neutrons from different vertical sections can be measured separately. To increase the solid angle the height of the analysers will reach about 600 mm and hence it will be necessary to implement two separate detector benches. For the expected intensity of this new high-resolution

Figure 2. The engineering view of the new setup. Two analyser benches will be installed. The preamplifier electronics will be positioned near the tubes in air boxes. Between detectors and analysers sits a static radial collimator.

capability an estimate can be provided. The measured intensity from the Mica(006) reflection on IRIS is about an order of magnitude less than the intensity with the PG analyser. On OSIRIS, the height of the analyser will be about a factor 3 larger compared to the Mica installation on IRIS, the silicon crystals will be a factor 3-7 thicker and hence through elastic bending more neutrons are reflected. The flux on OSIRIS is about a factor $1.5-2$ larger at $\lambda = 6$ Å, therefore we expect a measured intensity on OSIRIS with the silicon analyser similar or even higher than IRIS provides with the PG analyser.

Beyond the increase of dynamic range, the need and opportunity for more intensity has been identified. The OSIRIS spectrometer utilizes a non-ballistic curved $m = 2$ supermirror guide with a straight focusing end and the older IRIS uses a Ni-coated guide. Just from the accepted phase space, one would expect a factor 3-4 larger intensity transported to the sample position on OSIRIS compared to IRIS. However, all measurements indicate only a gain factor of $1.5-2$ at $\lambda \approx 6$ Å. Since some time now it is known that superior guide geometries exist, which provide a high transmittance of cold neutrons over long distances [22, 23, 24]. Monte Carlo simulations are now planned to determine the exact guide geometry along a previous study [25]. A further direction of development is providing a platform for high-throughput QENS research. IRIS would serve as a first port of call for new research groups and this high-throughput approach to QENS research would pave the way for spatially and temporally resolved maps of the dynamical response of complex materials as a function of temperature or other external stimuli.

3. Conclusions

We have presented the status of the near-backscattering spectrometers IRIS and OSIRIS at the ISIS Facility. Both are highly productive instruments and provide cold neutron spectroscopic capabilities for quasielastic and low-energy inelastic problems. In a changing neutron scattering landscape the proposed new capabilities will ensure that the instruments remain competitive over the coming decade and beyond.

References

[1] Alefeld B 1967 *Bayerische Akademie der Wissenschaften* **11** 109
[2] Meyer A et al 2003 *Rev Sci Instrum* **74** 2759
[3] Wuttke J et al 2012 *Rev Sci Instrum* **83** 075109
[4] Hennig M, Frick B and Seydel T 2011 *J of Appl Cryst* **44** 467
[5] Carlile C J and Adams M A 1992 *Physica B* **182** 431
[6] Carlile C J et al 1994 *Nucl Instr Meth A* **338** 78
[7] Martin y Marero D et al J 1996 *J Phys Soc Jpn* **65** 245; Andersen K H et al 2002 *Appl Phys A* **74** Materials Science S237
[8] Telling M T F and Andersen K H 2005 *Phys Chem Chem Phys* **7** 1255; Telling M T F et al 2016 *Phys Chem Chem Phys* **18** 8243
[9] Mamontov E and Herwig K W 2011 *Rev Sci Instrum* **82** 085109
[10] Takahashi N et al 2009 *Nucl Instr Meth A* **600** 91; Shibata K et al 2015 *Jap Phys Soc Conf Proc* **8** 036022
[11] Tsapatsaris N et al 2016 *Rev Sci Instrum* **87** 085118
[12] Demmel F et al 2015 *EPJ Web of Conferences* **83** 03003
[13] Zsigmond G et al 2001 *Nucl Inst Meth A* **457** 299
[14] Demmel F and Pokhilchuk K 2014 *Nucl Instr Meth A* **767** 426
[15] Coldea R et al 2010 *Science* **327** 177; Stock C et al 2016 *Phys Rev Lett* **117** 017201
[16] Pili S et al 2016 *J Am Chem Soc* **138** 6352; Guilbert A A Y et al 2015 *Chem Mater* **27** 7652; Yang S et al 2015 *Nat Chem* **7** 121
[17] Demmel F et al 2014 *Technical Report RAL-TR-2014-014*
[18] Frick B, Mamontov E, vanEijck L and Seydel T 2010 *Zeit Phys Chem* **224** 33
[19] Demmel F and Andersen K H 2008 *Meas Sci and Tech* **19** 034021
[20] Scherm R H and Krüger E 1994 *Nucl Instr Meth A* **338** 1
[21] Kalus J and Dorner B 1973 *Acta Cryst A* **29** 526
[22] Böni P 2008 *Nucl Instr Meth A* **586** 1
[23] Schanzer C, Böni P, Filges U and Hils T 2004 *Nucl Instr Meth A* **529** 63
[24] Kleno K W et al 2012 *Nucl Instr Meth A* **696** 75
[25] Pelley C et al 2010 *J of Phys: Conf Ser* **251** 012063

IOP Conf. Series: Journal of Physics: Conf. Series **1021** (2018) 012028 doi:10.1088/1742-6596/1021/1/012028

Improvement for Neutron Brillouin Scattering Experiments on High Resolution Chopper Spectrometer HRC

Shinichi Itoh[1], Tetsuya Yokoo[1], Takatsugu Masuda[2], Hideki Yoshizawa[2], Minoru Soda[2], Masahiro Yoshida[2], Takafumi Hawai[1], Daichi Kawana[2], Ryosuke Sugiura[2], Toshio Asami[2], Yoshiaki Ihata[3]

[1]Neutron Science Division, Institute of Materials Structure Science, High Energy Accelerator Research Organization, Tsukba 305-0801, Japan
[2]Neutron Science Laboratory, The Institute for Solis State Physics, The University of Tokyo, Tokai 319-1106, Japan
[3]Materials and Life Science Division, J-PARC Center, Tokai 319-1195, Japan

shinichi.itoh@kek.jp

Abstract. Neutron Brillouin scattering (NBS) experiments can be performed on the High Resolution Chopper Spectrometer (HRC). Recently, an intensity gain of a factor 2.4 with an improved resolution for NBS has been obtained by improving the detector system, the Fermi chopper, and the collimator system. This improved performance provides opportunities for intense studies of dynamics in condensed matter by using non-single-crystal samples.

1. Introduction

Neutron Brillouin scattering (NBS), that is inelastic neutron scattering near to the forward direction, is effective for observing coherent excitations in non-single-crystal samples such as ferromagnetic spin waves in powder samples and acoustic phonons in liquids and polycrystalline samples. In order to detect such excitations, access to the energy-momentum space near to the (000) wave-vector is required. In fact, the scattering intensity from ferromagnetic spin waves remains near to (000) only and is rapidly decayed with increasing Q by the powder average. The energy-momentum space near to (000) can be accessed by using low-angle detectors, high-energy neutrons and high resolution.

The High Resolution Chopper Spectrometer (HRC) is being operated at the Materials and Life Science Experimental Facility (MLF), Japan Proton Accelerator Research Complex (J-PARC) to study dynamics in condensed matter with high resolutions using relatively high energy neutrons [1-5]. The HRC is a chopper spectrometer: a pulsed neutron beam is monochromatized by a Fermi chopper [6] and the neutrons scattered by the experimental sample are detected with a large area detector system. From the scattering angle (ϕ) and the time-of-flight of the detected neutron, the neutron counts are converted to the dynamical structure factor as a function of the momentum transfer Q and the energy transfer E. On the HRC, a range of incident neutron energies E_i = 5 - 2000 meV is available and an energy resolution of $\Delta E/E_i$ = 2% can be achieved by using $E_i \leq 300$ meV for NBS, where the energy resolution ΔE is defined as the energy width (full width at half maximum) of the elastic scattering.

The HRC has a detector array of ^{3}He position sensitive detectors of 2.8 m long and 3/4 inch diameter located at 4 m from the sample position covering $\phi = 3 - 62°$ for conventional experiments, and another detector array of ^{3}He position sensitive detectors of 0.8 m long and 1/2 inch diameter

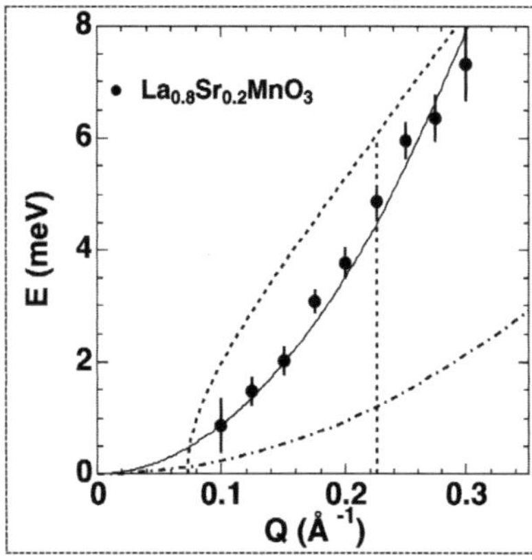

Figure 1. Energy-momentum space of NBS on HRC. The dashed curve is the scan locus for a detector at ϕ = 0.6° with E_i = 100 meV, and the chain line is the upper limit of the accessible energy-momentum space for $\phi \geq 5°$. The dispersion relation of spin waves in $La_{0.8}Sr_{0.2}MnO_3$ observed at 245 K with the HRC is indicated with a fitted curve $E(Q) = DQ^2$ with a constant D (solid line) [3]. The short vertical bars represent the statistical errors. The vertical dashed line at Q = 0.225 Å^{-1} is the scan for the data in Fig. 2(c).

located at 5.2 m at ϕ = 0.6 – 2.8° for NBS [1]. The partial pressure of ^{3}He gas is 1.8 MPa and 2 MPa for the detectors of lengths 2.8 m and 0.8 m, respectively. The collimator system has two collimators composed of vertical slits of sheets of Cd, and it is effective for background noise reduction at low angles [2,4]. Either of the two collimators with 1.5° or 0.3° collimations can be selected, reducing the background noise at low angles; these collimations can reach a minimum of ϕ = 3° for conventional experiments and ϕ = 0.6° for NBS experiments, respectively. The area of each collimator is 60 mm × 60 mm, that is larger than the beam cross section, and its length is 150 mm. Therefore, a setup for NBS has been provided on the HRC since its initial installation, and thus, results from NBS experiments were obtained [3].

The experimental conditions required for NBS are realized with the HRC, since it has detectors at low angles down to ϕ = 0.6°. In Fig. 1, the scan locus for a detector at ϕ = 0.6° with E_i = 100 meV and the upper limit of the accessible energy-momentum space for ϕ = 5° are indicated. The area between these two lines is the energy-momentum space expanded by using the NBS option of the HRC, and, at J-PARC, NBS experiments can be performed only on the HRC [3,4]. The principle of NBS is not new [7,8], and the energy-momentum space accessible by NBS has been extended by utilizing higher energy neutrons in spectrometers such as the BRISP spectrometer at ILL [9].

In this paper, recent progress on improving the performance for NBS on the HRC is described.

2. Improvement for NBS on HRC

Since the solid angle of detecting area for NBS experiments is little and limited to very low scattering angles, improvement for gaining the scattering intensity is essential. We made some efforts to gain the neutron flux in some ways.

The configuration of the detectors for NBS was improved. In the low angle detector bank, detectors are aligned in a double layered configuration to increase the counting rate: the detectors in the front layer face the sample and the detectors in the rear layer are placed just at the lower stream of the front layer. Neutrons transmitted through the front layer are detected with the rear layer. The counting rate for the double layer configuration is increased by a factor 1.4 to that for the single layer configuration for E_i = 100 meV, as shown in Fig. 2(a). The detectors are now mounted in the range of ϕ = 0.6 – 5.1°.

The Fermi chopper is composed of a slit package, which is a layered assembly of components of shielding plates and spacers sandwiched by curved walls. The slit package of a vertical slit, with cross section of 75 mm (width) × 64 mm (height) and length of 100 mm, is inserted into a cylindrical rotor with diameter 125 mm and it rotates around the vertical axis [6]. The rotor and the slit package are distorted by the rotation stress at the frequency up to f = 600 Hz, and its performance becomes worse than the designed one. In the Fermi chopper in the initial setup (Fermi-B), the slit package, where the sum of the widths of the components equals to 75 mm, was inserted. In this assembly, a large distortion of the slit package occurs by the rotation stress. In the present arrangement, the slit package

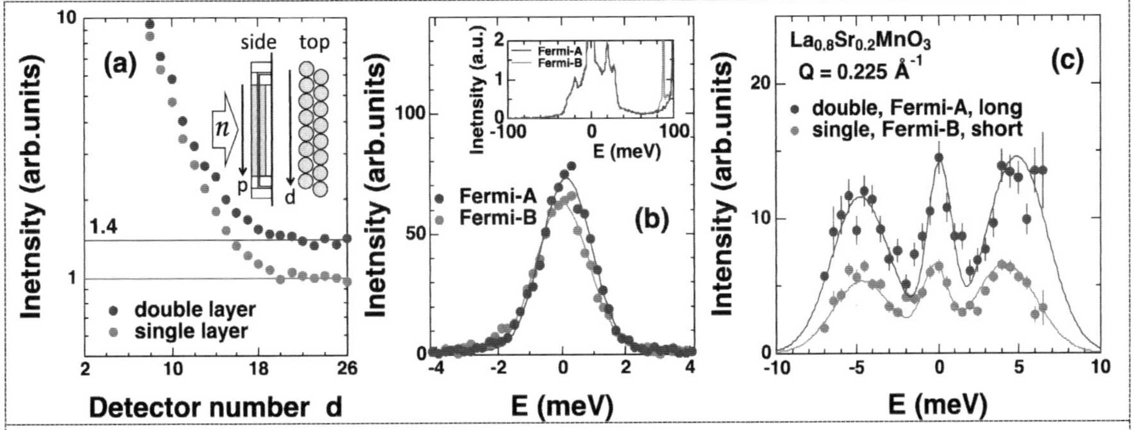

Figure 2. Improvement for NBS experiments on HRC. (a) Detector configuration. Observed scattering intensities from the vanadium sample, which are integrated over the positions p along each detector at E_i = 100 meV, are plotted as a function of the detector number d, where d = 2 and 26 correspond to ϕ = 0.6 and 2.8° for the front layer, respectively. Due to an incomplete mask for the sample, a direct beam spreads from low angles. The intensities at higher angles (larger d) are free from the spread and show the actual intensity gain. The inset shows the double-layered configuration in the side and top views, the colored areas indicate detectors. (b) Slit package of Fermi chopper: Fermi-A (improved) and Fermi-B (initial setup). Inelastic spectra from the vanadium sample observed at 5.2 m from the sample using the 0.3° collimator (short) are plotted with fitted curves with Gaussian functions. The inset shows inelastic spectra from the vanadium sample using the 1.5° collimator observed at 4 m from the sample. (c) Total intensity gain. Spin waves in $La_{0.8}Sr_{0.2}MnO_3$ were observed with E_i = 100 meV by using the double-layered detector system, the Fermi-A, and the long collimator, also by using the single-layered detector system, the Fermi-B, and the short collimator. The solid lines are curves fitted to the scattering function for resolution-limited spin waves.

(the sum of the widths of the components) was increased to 78.8 mm by increasing the number of the components, and this was pressed into the width of 75 mm and inserted into the rotor (Fermi-A). It was visually confirmed that the distortion after the rotation up to 600 Hz was greatly reduced. Figure 2(b) shows the inelastic spectra of the elastic scattering from a vanadium sample observed for the Fermi-A and Fermi-B with E_i = 100 meV, f = 600 Hz at the 5.2 m position from the sample, and with the 0.3° collimator. By this improvement, the energy width became sharper from ΔE = 2.0 meV to 1.9 meV, and the integrated intensity was increased by a factor 1.1. In addition, the transmitted intensity at the half rotation of the Fermi chopper, which was observed at around E = 90 meV for the Femi-B, disappeared for the Fermi-A arrangement, and therefore, the observable energy-transfer range was expanded.

The collimator system was also improved. The transmission of the 0.3° collimator is 0.2, that is only 40 % of the designed one. The transmission can be estimated by $T = w/(w+t)/[1+\{\Delta\theta/(w/L)\}^2]^{1/2}$, where t (= 0.1 mm), w, L (= 150 mm), and $\Delta\theta$ are the thickness of the Cd sheet, the spacing of the slit, the length of the collimator, and the beam divergence of a neutron beam incident on to the collimator. The collimation is defined by w/L. For E_i = 100 meV, $\Delta\theta$ = 8 mrad. The transmission can be estimated to be T = 0.48 for the 0.3° collimator with w = 0.8. This is because it is hard to maintain the flat plates of the thin Cd sheet with t = 0.1 mm. If the distortion of the Cd sheet is constant, T should be increased for longer L with the collimation w/L unchanged. A long collimator was introduced, where w = 1.55 mm, L = 295 mm, i.e., w/L = 0.3°, and the cross section was slightly reduced to 59 mm × 54 mm. Hereafter, the previous 0.3° collimator is called the short collimator. At present, we can select one of the four: the long collimator, the short one, the 1.5° one, and the blank.

In order to evaluate the performance of the presently-improved system, an NBS experiment for a polycrystalline sample was performed. The sample was a nearly-cubic perovskite ferromagnet $La_{0.8}Sr_{0.2}MnO_3$ with the ferromagnetic transition temperature T_C = 319 K [3], with weight 23 g, and the beam cross section was defined to be 32 mm × 34 mm by Cd plates. The measurement was performed at 245 K (< T_C) with E_i = 100 meV at f = 600 Hz, by using the double-layered detector system, the Fermi-A arrangement, and the long collimator. Figure 2(c) shows an observed spectrum, where the background measured with an empty can was subtracted. The spectrum is well fitted to the scattering function for resolution-limited spin waves [3]. Also, the observed spectrum in the same condition but by using the single-layered detector system, the Fermi-B arrangement, and the short collimator [3] is plotted in Fig. 2(c). This spectrum is also fitted to the scattering function for resolution-limited spin waves. From the intensity factors obtained by the fits, a factor 2.4 was obtained as the intensity gain. Since the gain for the double-layered detector system to the single one is 1.4 and the gain for the Femi-A to the Fermi-B is 1.1, the gain for the long collimator to the short one can be estimated to be 1.6.

3. Summary

From NBS experiments on the HRC, results have been obtained from studies on spin waves and phononic excitations mainly under the initially realized conditions [3-5]. By using the improved conditions, further results are expected. The NBS option differentiates the HRC from other chopper spectrometers, and opens opportunities for contributing to current science by measuring coherent excitations from non-single-crystal samples.

Acknowledgments

This study was approved by the Neutron Scattering Program Advisory Committee of the Institute of Materials Structure Science, High Energy Accelerator Research Organization (No. 2016S01).

References

[1] Itoh S, Yokoo T, Satoh S, Yano S, Kawana D, Suzuki J, Sato T J 2011 *Nucl. Instr. Meth. Phys. Res.* A **631** 90
[2] Itoh S et al. 2013 *J. Phys. Soc. Jpn.* **82** SA033
[3] Itoh S, Endoh Y, Yokoo T, Kawana D, Kaneko Y, Tokura Y, and Fujita M 2013 *J. Phys. Soc. Jpn.* **82** 043001
[4] Itoh S, Yokoo T, Kawana D and Endoh Y 2013 *J. Phys. Soc. Jpn.* **82** SA034
[5] Itoh S et al. 2017 *AIP Conf. Proc.* in press
[6] Itoh S, Ueno K, Yokoo T 2012 *Nucl. Instr. Meth. Phys. Res.* A **661** 58
[7] Passell L, Dietrich O W, Als-Nielsen J 1976 *Phys. Rev.* B **14** 4897
[8] Robinson R A , McQueeney R J, and Kelley T M 1998 *Physica* B **241-243** 161
[9] Aisa D et al. 2006 *J. Non-Cryst. Solids* 352 5130

IOP Conf. Series: Journal of Physics: Conf. Series **1021** (2018) 012029 doi:10.1088/1742-6596/1021/1/012029

The TOSCA Spectrometer at ISIS: the Guide Upgrade and Beyond

R.S. Pinna,[1,2] **M. Zanetti,**[1,2] **S. Rudić,**[1] **S.F. Parker,**[1] **J. Armstrong,**[1] **S.P. Waller,**[1] **D. Zacek,**[1] **C. Smith,**[1] **S.M. Harrison,**[1] **G. Gorini,**[2] **and F. Fernandez-Alonso**[1,3]

[1] ISIS Facility, Rutherford Appleton Laboratory, Chilton, Didcot, Oxfordshire OX11 0QX, United Kingdom

[2] CNISM, Università degli Studi di Milano-Bicocca, Piazza della Scienza 3, 20126 Milano, Italy

[3] Department of Physics and Astronomy, University College London, Gower Street, London WC1E 6BT, United Kingdom

E-mail: `felix.fernandez-alonso@stfc.ac.uk`

Abstract. We describe progress to date with the first major upgrade of the TOSCA spectrometer since it first became operational over fifteen years ago. This major project to boost the incident flux on the instrument by over an order of magnitude has been implemented in the past couple of years and has involved the complete redesign of the primary spectrometer to house a state-of-the-art, high-m neutron guide and associated chopper system. Engineering design and subsequent installation and commissioning efforts have been supported by extensive neutron-transport simulations and baseline studies of neutronic response in the context of the ISIS TS1 Project. Looking further ahead, we also outline ongoing feasibility studies to upgrade the secondary spectrometer, with a view to additional order-of-magnitude gains in neutronic and scientific performance.

1. Overview

The TOSCA instrument at the ISIS Pulsed Neutron and Muon Source is an indirect-geometry inelastic neutron spectrometer optimised for high-resolution vibrational spectroscopy in the energy-transfer range 0–4000 cm^{-1} [1]. The instrument has been operational since the turn of this century and has become the gold standard for broadband chemical spectroscopy with neutrons [2–4]. As such, both TOSCA as well as its vibrant and growing science programme have been the inspiration for recent and exciting developments in novel neutron instrumentation across the globe. These include VISION at the SNS in the USA [5], LAGRANGE at the ILL in France [6], and VESPA at the ESS in Sweden [7]. Key factors to the success of TOSCA have been its high spectral resolution over a wide spectral range, its ease of use, as well as the full integration of contemporary computational materials modelling into its science programme. At present, its *INS Database* [8] contains over 700 entries, and these data continue to be mined by a growing community worldwide, in addition to fresh and exciting science on the instrument across materials chemistry, energy research and catalysis, to name a few areas of strength.

In 2013, an international panel of experts recognised the pressing need to boost the incident flux on the instrument, in order to move beyond the state-of-the-art in key scientific areas such as gas and charge storage [2, 3]. Following this exercise, extensive neutronic calculations have

IOP Conf. Series: Journal of Physics: Conf. Series **1021** (2018) 012029 doi:10.1088/1742-6596/1021/1/012029

been performed to benchmark the existing instrument in the context of the ISIS TS1 Project, as well as to provide a well-defined scientific and engineering specification. For further details, see Refs. [9–12]. These efforts have been followed by the construction and installation of the new guide over the past couple of years as well as by feasibility studies to upgrade the secondary spectrometer, as reported below.

2. The Guide Upgrade

The original geometry of the TOSCA primary spectrometer relied on simple (passive) collimation of the incident beam over a distance of 17 m. This configuration has served the instrument well over the past two decades, yet at the same time it did not capitalise on advances in neutron-guide technology for the effective delivery of neutron flux at the sample position. The first step in this project involved extensive Monte-Carlo simulations using the McStas package [13] to explore a suitable geometry for optimal neutron transport over a wide wavelength range without compromising the spectroscopic capabilities of the instrument. From these simulations, a tapered geometry with an increasing m-number over distance from the spallation target was chosen. These simulations predicted an average gain in flux well above one order-of-magnitude over the wavelength bandwidth of the instrument [2, 4, 10–12]. At the longest wavelengths accessible on TOSCA, these calculations indicated that gain factors could approach two orders of magnitude.

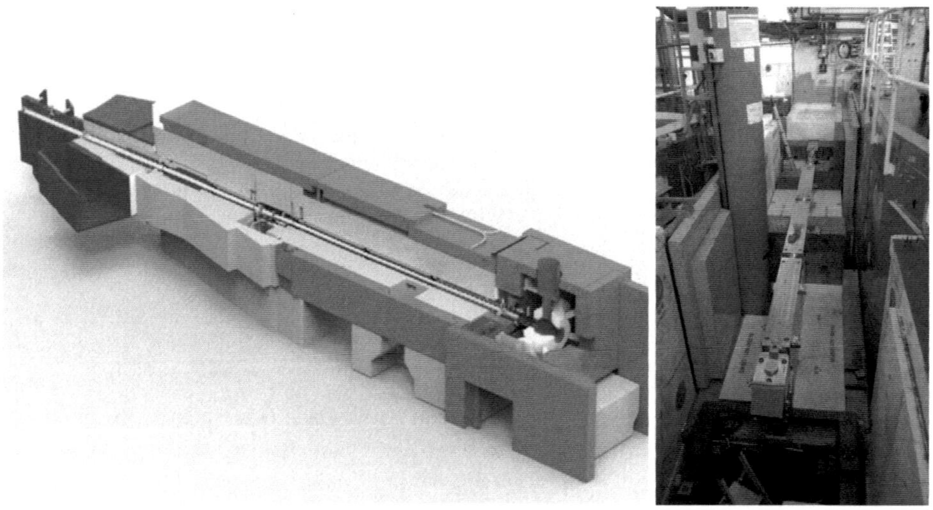

Figure 1. Left: Engineering design of the TOSCA guide following neutronics simulations. The ISIS TS1 target area is located on the left of the figure, whereas the TOSCA secondary spectrometer is depicted in yellow and green on the right-hand side. Right: Final stages of the first phase of the installation, completed in November 2016.

Figure 1 shows a schematic diagram of the engineering design of the guide according to the above specifications. Following manufacture by *Swiss Neutronics* over the period 2015-16, the guide was installed on TOSCA during the second half of 2016. In a first installation phase, all guide sections except the shutter insert inside the target monolith were installed, and this step was followed by a brief and preliminary commissioning period during December 2016. These two installation phases were required because of the radiation hazards associated with the handling of beamline components in close proximity to the spallation target. The preliminary commissioning tests that we have conducted during the first phase are very encouraging. The beam profile was found to be homogeneous across the sample area and the observed gains were somewhat higher than expected from the neutronic simulations of the *baseline* scientific specification, particularly at long wavelengths. The second and final phase of the installation schedule was completed in the second half of February 2017 and has been followed by extensive commissioning tests on the instrument, due for completion at the end of March 2017. A detailed analysis of gains and other performance criteria is currently underway.

3. A New Secondary Spectrometer

TOSCA uses the (002) Bragg reflection from Highly Oriented Pyrolytic Graphite (HOPG) to effect final-energy selection. At present, each analyser bank on TOSCA consists of a flat square of area $\sim$144 cm^2, sitting at an average distance of 31 cm from the sample position. Its surface normal is placed at an angle of 45° relative to the scattered beam, leading to a final energy of *ca.* 28 cm^{-1}. A beryllium filter equipped with cadmium foils is placed between the HOPG analyser and the detectors to suppress higher-order HOPG reflections. This filter is cooled to *ca.* 30 K to improve the rejection of neutrons with energies above *ca.* 40 cm^{-1}. In its current incarnation, the HOPG analyser and the beryllium filter are very close to each other. While being a compact and practical solution requiring a relatively small HOPG analyser area, this feature also poses severe constraints on the overall neutron-collection efficiency of the secondary spectrometer.

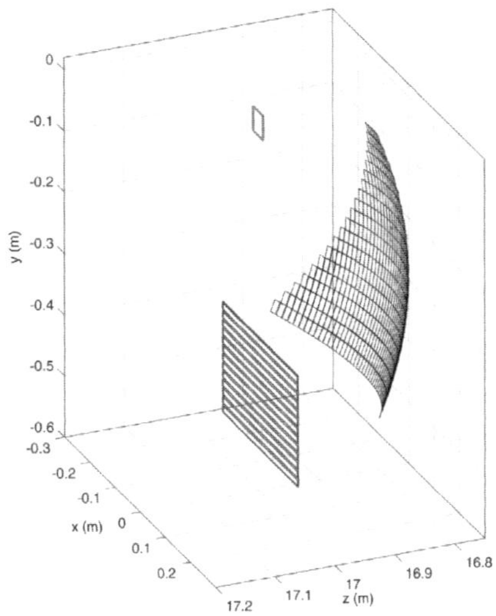

Figure 2. Curved HOPG analyser (black), sample (red), and detector array (blue).

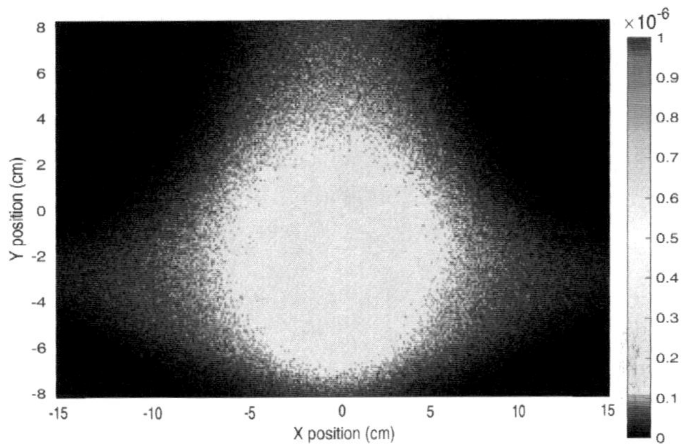

Figure 3. Spatial intensity distribution of detected neutrons from a parabolic HOPG analyser of total area 960 cm^2.

To circumvent these limitations, we have started assessing other potential geometries that also build upon recent experience with similar instruments worldwide [5, 6]. Using the McStas package [13], we have created a custom component where HOPG tiles are arranged on either a spherical or parabolic surface of variable area. Figure 2 shows a three-dimensional view of the resulting computer model of a curved HOPG analyser, and Fig. 3 reports a typical spatial intensity distribution of detected neutrons. Preliminary Monte Carlo simulations and associated geometric optimisations using McStas [13] and iFit [14] have been performed as a function of analyser area and radius of curvature. These calculations show that an order-of-magnitude increase in detected flux is well within reach, particularly via the use of a parabolic (as opposed to spherical) analyser geometry. We have also found that the detected flux can be increased by optimising parameters such as the mosaicity of the HOPG array without compromising spectroscopic performance. In ideal circumstances, the surface of the curved HOPG analyser for a given detector bank could be increased as much as to occupy an angular span of 60°, provided that engineering constraints can allow for it. Calculations have already been performed for three different sizes of doubly bent HOPG analysers of either spherical or parabolic shape. Our largest-area analysers amount to *ca.* 1500 cm^2, a similar size to the one used earlier by Smee *et al.* at NIST to improve the performance of cold-neutron spectrometers [15, 16].

In a wider context, we note that the original conceptual design and subsequent scientific specification of TOSCA (and its predecessor TFXA) decades ago were solely performed on the basis of analytical calculations, a task that necessarily relied on a number of hard-to-test assumptions at the time. This situation is particularly true for the trade-off between flux and spectral resolution, as well as for the conditions required to attain an optimal match of the many different contributions to the neutronic response that are associated with both the primary and secondary spectrometers. The present work using modern neutron-transport codes, therefore, offers the exciting prospect of assessing in quantitative and unprecedented detail the relative importance of these parameters, including analyser shape and size, as well as time- and energy-focusing conditions. We, therefore, anticipate that the lessons learnt from this work will also have important implications for the optimal neutronic design of similar instrumentation in the foreseeable future.

Acknowledgements

The authors gratefully acknowledge the *Science & Technology Facilities Council* (STFC) for financial support for this project, including computing resources from the STFC Scientific Computing Department's SCARF cluster. This work has been partially supported by the 2014-2020 CNR-STFC Agreement concerning collaboration in scientific research at the ISIS Facility. We are also immensely grateful to *Swiss Neutronics* for the timely delivery of the guide components within specification.

References

[1] WEB: www.isis.stfc.ac.uk/Pages/Tosca.aspx
[2] Rudić S, Ramirez-Cuesta AJ, Parker SF, Fernandez-Alonso F, Pinna RS, Gorini G, Salzmann GC, McLain SE, and Skipper NT 2013 *Rutherford Appleton Laboratory Technical Report* RAL-TR-2013-015, Chilton. WEB: epubs.stfc.ac.uk/work/11216706
[3] For a summary of the *TOSCA International Review* and its recommendations, see www.isis.stfc.ac.uk/Pages/TOSCA-Instrument-Review.aspx
[4] Parker SF, Fernandez-Alonso F, Ramirez-Cuesta AJ, Tomkinson J, Rudić S, Pinna RS, Gorini G, and Fernández Castañón J 2014 *J. Phys. Conf. Ser.* **554** 012003.
[5] Seeger PA, Daemen LL, and Larese JZ 2009 *Nucl. Instr. Meth. Phys. Res. A* **604** 719.
[6] Jimenez-Ruiz M, Ivanov A, and Fuard S 2014 *J. Phys. Conf. Ser.* **549** 012004.
[7] Fedrigo A, Colognesi D, Bertelsen M, Hartl M, Lefmann K, Deen PP, Strobl M, Grazzi F, and Zoppi M 2016 *Rev. Sci. Instrum.* **87** 0651014.
[8] WEB: www.isis.stfc.ac.uk/Pages/INS-database.aspx
[9] Pinna RS, Rudić S, Capstick MJ, McPhail DJ, Pooley DE, Howells GD, Gorini G, Fernandez-Alonso F 2017 *Nucl. Instr. Meth. Phys. Res. A* **870** 79.
[10] Pinna RS, Rudić S, Parker SF, Gorini G, and Fernandez-Alonso F 2015 *EPJ Web of Conferences* **83** 03013.
[11] Pinna RS, *MSc Thesis* (Università degli Studi di Milano-Bicocca, Italy, 2014). WEB: epubs.stfc.ac.uk/work/12257011
[12] Fernández Castañón J 2013 *Rutherford Appleton Laboratory Technical Report* RAL-TR-2013-002, Chilton. WEB: epubs.stfc.ac.uk/work/65395
[13] Willendrup PK, Udby L, Knudsen E, Farhi E, and Lefmann K 2011 *Nucl. Instr. Meth. Phys. Res. A* **634** S150. WEB: www.mcstas.org
[14] Farhi E, Debab Y, and Willendrup P 2014 *J. Neut. Res.* **17** 5.
[15] Smee SA, Orndorff JD, Scharfstein GA, Qiu Y, Brand PC, Broholm CL, and Anand DK 2002 *Appl. Phys. A* **74** S255.
[16] Smee SA, Brand PC, Barry DD, Broholm CL, and Anand DK 2001 *Nucl. Instr. Meth. Phys. Res. A* **466** 513.

IOP Conf. Series: Journal of Physics: Conf. Series **1021** (2018) 012030 doi:10.1088/1742-6596/1021/1/012030

Status report of the chopper spectrometer 4SEASONS

R Kajimoto[1], M Nakamura[1], Y Inamura[1], K Kamazawa[2], K Ikeuchi[2], K Iida[2], M Ishikado[2], N Murai[1], H Kira[2], T Nakatani[1], S Ohira-Kawamura[1], R Takahashi[1], N Kubo[1], W Kambara[1], K Nakajima[1] and K Aizawa[1]

[1] Materials and Life Science Division, J-PARC Center, JAEA, Tokai, Ibaraki 319-1195, Japan
[2] Neutron Science and Technology Center, CROSS, Tokai, Ibaraki 319-1106, Japan

E-mail: ryoichi.kajimoto@j-parc.jp

Abstract. 4SEASONS is a medium-resolution thermal neutron chopper spectrometer in the Materials and Life Science Experimental Facility (MLF) at J-PARC. Although 4SEASONS is routinely used for many experiments by internal and external users, upgrading and maintenance work is still underway. This paper reviews the recent improvements of the instrument.

1. Introduction

4SEASONS (or *SIKI*) is a time-of-flight direct geometry chopper spectrometer in the Materials and Life Science Experimental Facility (MLF) at the Japan Proton Accelerator Research Complex (J-PARC) [1]. It is installed at BL01 beam port viewing the coupled moderator which is 18 m upstream from the sample position. The incident neutrons are monochromatized by a fast-rotating Fermi chopper positioned at 1.7 m upstream from the sample position. In addition, the instrument has a T0 chopper for suppressing fast neutrons and two disk choppers for band definition. Neutrons scattered by samples are detected by the ^{3}He position sensitive detectors placed at 2.5 m from the sample position. 4SEASONS is designed for measurements of spin and lattice dynamics in the 10^0–10^2 meV energy range. The instrument has been used by researchers from a variety of countries for studies of magnetic and lattice dynamics in condensed matters such as superconductors, magnetic materials, dielectrics, catalysts, and thermoelectric materials. Upgrading and maintenance work is still ongoing, which will improve the performance, usability, and safety of the instrument. In this paper, we review the improvements of the instrument achieved in the recent two years for the Fermi chopper, detector system, sample environments, and software.

2. Newly developed Fermi choppers

The original Fermi chopper of 4SEASONS holds a slit package with 65 mm × 70 mm cross-section and 100 mm depth. Its designed maximum rotation frequency was 600 Hz [1]. However, in 2011, the rotation was found to be unstable at high frequencies because of a secular change in the slit package, and we limited the rotation frequency at 350 Hz and lower. This situation severely restricted the available energy resolution, especially for high-energy experiments. Accordingly, we developed a new Fermi chopper, which was manufactured by MIRROTRON Ltd. and SKF Magnetic Bearings. The new Fermi chopper has a more compact slit package with cross-section and depth of 58 mm × 53.4 mm and 20 mm, respectively. The spacing of the slits is designed so that it gives the same energy resolution as the original chopper. The new chopper can rotate stable up to 600 Hz because of its compact design. Figure 1(a) shows the energy spectra of $C_4H_2I_2S$ at 5 K measured with the incident energy $E_i = 220$ meV and the

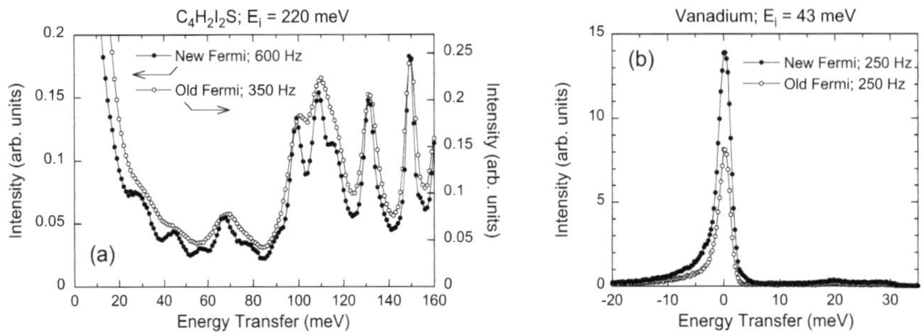

Figure 1. Comparison of energy spectra measured with the old and new Fermi choppers. (a) Energy spectra of $C_4H_2I_2S$ at 5 K measured with $E_i = 220$ meV. Scattering intensities were integrated over $Q = 6$–11 Å^{-1}. Solid and closed circles show data measured with the new Fermi chopper rotating at $f = 600$ Hz and old chopper rotating at $f = 350$ Hz, respectively. (b) Energy spectra of vanadium measured with $E_i = 43$ meV and $f = 250$ Hz. Scattering intensities were integrated over $Q = 3.4$–5.0 Å^{-1}. Solid and closed circles show data measured with the new and old Fermi choppers, respectively.

maximum achievable rotating speeds for both the choppers, i.e., $f = 350$ Hz and 600 Hz for the old and new choppers, respectively. The high-speed rotation of the new Fermi chopper considerably improves the data quality. It provides better peak separations and improved signal-to-noise ratio in the low-energy region, where inelastic scattering intensity often suffers from the tail of elastic scattering.

Another important benefit of the shallow depth slits is that the chopper provides much higher neutron transmission than the original chopper. Figure 1(b) shows a comparison of the energy spectrum of vanadium measured with the new and old Fermi choppers. In this example, the choppers rotate at $f = 250$ Hz to produce the incident energy $E_i = 43$ meV. The new Fermi chopper gives 1.7 times higher peak intensity than the old chopper, whereas the peak width is almost identical for both the choppers. The intensity gain by the new chopper becomes particularly high for low energy neutrons and high rotation speeds. (For example, it reaches 17 for 18 meV and 300 Hz.) This feature expands the practical energy range for the multi-E_i measurements [2]. Notably, Fig. 1(a) also proves the high transmission of the new chopper. The data for the new chopper have a comparable intensity with that for the old chopper in spite of the much higher speed of rotation in the former.

In parallel, we are developing another Fermi chopper, called MAGIC chopper, which utilizes supermirror-coated slits [3–5]. In late 2014, we tested the prototype of the MAGIC chopper on BL01. Unfortunately, we found that the obtained neutron flux was much lower than the expected value. Nevertheless, we could verify the basic performance of this type of chopper. We performed Monte Carlo simulations to find that the low neutron flux is mainly caused by the low reflectivity of the supermirrors. We are currently developing a new slit package with higher-reflectivity supermirrors.

3. Detectors

Partial recovery of the detector deficiency. The detector banks of 4SEASONS cover a scattering angle in the range of $-35.3°$ to $+130.5°$ in the horizontal plane [1]. However, the detector banks are not fully occupied due to the recent shortage and resulting high price of ^{3}He gas. The initial number of detectors was 191 pieces covering the range of $-35.3°$ to $+54.5°$ in the horizontal plane. In 2013–2015, we installed 75 pieces of detectors [Fig. 2(a)]. The maximum horizontal scattering angle is currently $+90.7°$. The newly installed detectors expanded the accessible momentum transfer (Q) as shown in Fig. 2(b). This value is still lower than the designed one, and a further increase in detectors is undoubtedly important. However, for studies on low-dimensional systems in which only some particular crystal axes are of interest, these axes are often set perpendicular to the incident beam. In such a case, the scattering angle of 90° can provide the maximum Q range.

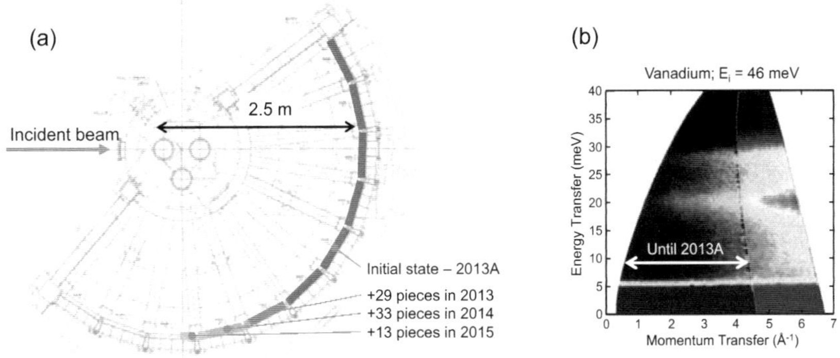

Figure 2. (a) Top view of the vacuum scattering chamber of 4SEASONS and the detector banks where detectors are installed. Detector banks indicated by gray thick lines show the initial detector arrangement. Detector banks colored in red, pink, and orange show the positions where detectors were installed in 2013, 2014, and 2015, respectively. (b) Momentum (Q)-energy (E) map of inelastic scattering spectra of vanadium measured with $E_i = 46$ meV. The areas displayed with a gray scale and with a color scale show the Q-E regions covered by the initially installed detectors and newly installed detectors, respectively.

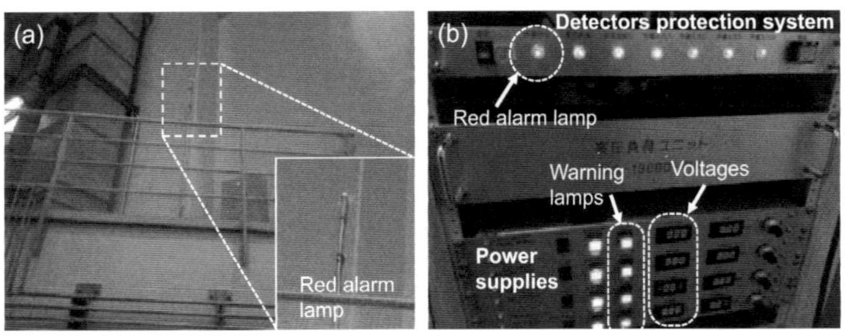

Figure 3. (a) Seismometer in MLF. (inset) The red alarm lamp is on by detecting an earthquake. (b) The detector protection system and the new high-voltage power supplies for detectors in the 4SEASONS cabin.

Detector protection. ^{3}He detector is vulnerable to vibration, especially when the electric power is supplied. Our experience with the big earthquake in 11 March 2011 reminded us of the importance of the protection of the detectors against earthquakes. Accordingly, in 2015, we introduced an interlock system to prevent detector damage caused by that type of disaster. This system is connected to seismometers placed in the experimental halls. Once it detects an earthquake whose intensity is bigger than some specified value (currently bigger than 4 on the Japanese seismic scale), it cuts the electric power supply to the detectors. The interlock system is expandable to adopt other input signals of incidents such as an unexpected vacuum break. In addition, we replaced the power supplies for the detectors with new ones which are more compatible with the interlock system. The new power supplies can cut off the power within 5 seconds. The practical shutdown time is longer depending on the capacitance of the detector electronics, and we have never experienced any damage on the detectors by the shutdown. Figure 3(a) shows the seismometer in the No. 1 Experimental Hall of MLF, which lit up its red warning lamp by detecting an earthquake with a seismic intensity of 4 on 22 November 2016. It successfully cut off the power supply to the detectors as shown in Fig. 3(b).

4. Sample environments

Preparing for magnet. To date, no magnets are available on 4SEASONS. However, the dimension of the sample environment attachment flange of the scattering chamber can accommodate the commonly-used 7-T magnet in MLF. One of the difficulties in operating magnets is that many iron components surround the sample area. Moreover, the turbomolecular pumps with magnetic bearings are attached just below the sample position, and we must be careful of the influence of the stray field on these pumps.

IOP Conf. Series: Journal of Physics: Conf. Series **1021** (2018) 012030 doi:10.1088/1742-6596/1021/1/012030

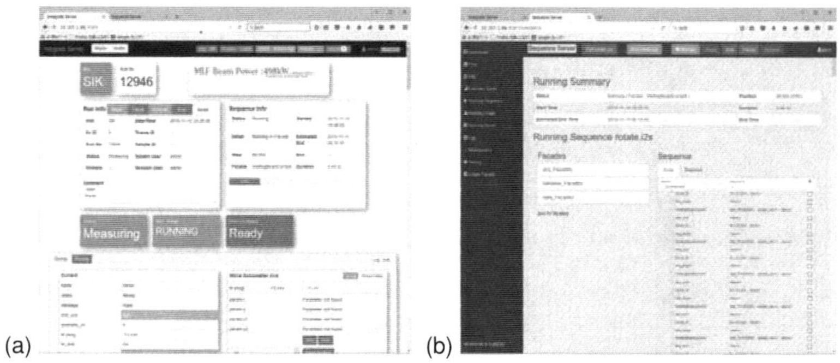

Figure 4. Web interfaces of the new instrument control system of 4SEASONS. (a) Integrate Server to control the devices and run measurements. (b) Sequence Server to manage measurement sequences.

Thus, we investigated stray fields using the computer aided engineering system Femtet®, considered the iron components, and evaluated the field intensity on the turbomolecular pumps. To validate our calculations, we operated the commonly-used magnet on 4SEASONS for the first time, and measured the intensity of the magnetic field at several points around the magnet. Although the maximum applied field in this test operation was only 1 T, the observed values were consistent with the calculations. Numerical calculations of the stray field are currently underway to find an appropriate measure to protect the turbomolecular pumps against the field and estimate the maximum available field on the instrument.

Improvement in the oscillating radial collimator. To reduce background scattering from the sample environment, 4SEASONS can be equipped with an oscillating radial collimator (ORC), which was developed by the Technology Development Section of MLF. It has a number of thin shielding blades made of cadmium-plated aluminum sheets, which surround the space for the sample environment device. The ORC is installed on the vacuum scattering chamber, and it is usually oscillated horizontally to avoid uneven intensity distribution on the detectors caused by shadows from the shielding blades [6]. However, this type of oscillation is not simple, because the unevenness in intensity depends on the oscillating speed and angle. Even after optimizing these parameters, we found that the shielding blades inevitably made shadows when they stopped at both ends of oscillating angle to change the oscillating direction, even though the dead time was small ($\sim$20 ms). To solve this problem, we recently developed a new oscillating mode, called "shift-mode", which can shift the oscillating angle range gradually. We confirmed that by changing the angle range by 0.1° per oscillation, we can obtain more even intensity distribution than the normal oscillating mode.

5. Softwares

New instrument control system. Many of the instruments in MLF, including 4SEASONS, utilize instrument control systems based on the control software framework IROHA [7]. A new control software framework, called IROHA2, was recently developed by the MLF Computing Environment Group to replace IROHA [8]. We developed a new instrument control system based on IROHA2, which has been available for user experiments since autumn in 2015. This control system has a web-based user interface, which allows users to control the instrument intuitively and remotely from any place connected to the server through LANs (Fig. 4). Registered users (currently they are limited to instrument group members) can view the instrument status even from their smart phones. The user interface is not the only benefit of the new control system. Another important feature of the new control system is that it enables us to control different devices uniformly and introduce new devices easily. This feature is particularly useful when a nonstandard sample environment is used on the instrument.

4D mapping by continuous rotation of single crystal samples. One of the recent trends in experiments on a single crystal using a chopper spectrometer is to grasp whole excitations in the 4D energy-momentum space by rotating the crystal over a wide angular range. The data analysis software suite

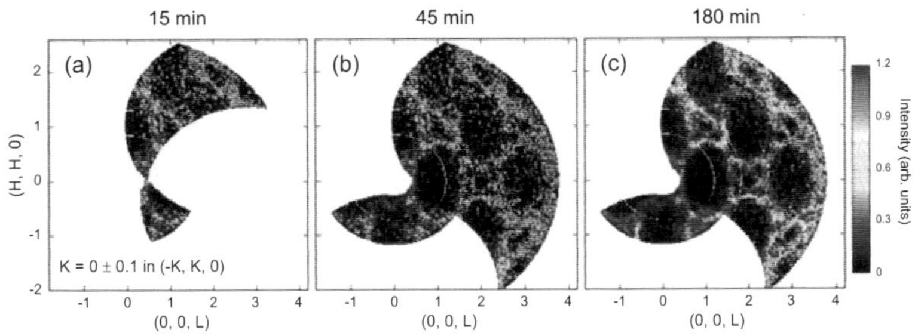

Figure 5. Phonon spectra of a single crystalline copper obtained by continuously rotating the crystal. The crystal was rotated horizontally at a rate of 160°/h. (a), (b), and (c) show the constant E slices at $E = 13 \pm 1$ meV, 15, 45, and 180 min after the start of the measurement, respectively.

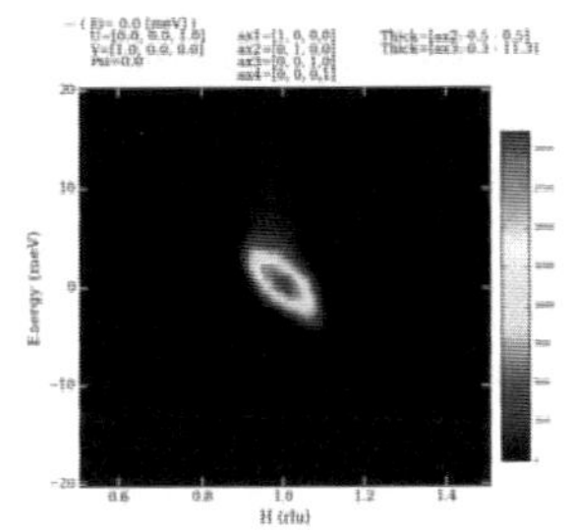

Figure 6. Example of the resolution function for 4SEASONS, simulated and displayed using McStas and Utsusemi, respectively.

in MLF, Utsusemi, is compatible for such measurements [9], which makes 4D mapping a routine at 4SEASONS. So far, implementation in Utsusemi has not been so flexible, because a crystal has been rotated stepwise with a small (0.5°–1°) step angle. For such a step-by-step rotation, the angular range and step size of the rotation have to be determined before the measurement. Real-time visualization of acquired data, which is often important to decide whether the current measurement is successful, is difficult. On the other hand, the data acquisition system of MLF based on event data recording can potentially provide flexibility in crystal-rotating measurements. The 4D mapping by continuous rotation of a single crystal was recently developed at the ARCS spectrometer at the Spallation Neutron Source utilizing the event data recording [10]. Recently, a new measurement and analysis system for crystal-rotating measurements, which maximizes the potential of the event data recording, was developed by the MLF Computing Environment Group. This system takes advantage of the data acquisition electric module, TrigNET [11], and a data processing function called online monitor [12]. The former can record any sample environment information as events synchronized with neutrons, whereas the latter enables pseudo real-time data visualization by incremental data analysis. With these features, we can perform 4D mapping by continuously rotating a single crystal without determining the step size in advance and then visualize the data in real time. The new system has been incorporated in Utsusemi, and we introduced it on 4SEASONS in late 2016. Figure 5 shows the phonon spectra of a single crystal of copper measured by continuously rotating the crystal using the new system. Given the online monitor functionality, we can observe the progress of the measurement. By rotating the crystal at a reasonably fast speed, we can obtain the rough picture at an early measurement stage [Figs. 5(a) and 5(b)]. The statistics improves with time by repeating the rotation [Fig. 5(c)], and we can stop the measurement anytime soon after the statistics reaches the required level. These features are useful to optimize the measurement condition and determine the experiment plan.

Instrument resolution. Estimation of instrumental resolution is important for neutron scattering instruments to derive physical quantities from the experimental data. We have already established the estimations of 1D energy and momentum resolutions for 4SEASONS, i.e., energy resolution for incoherent scattering and momentum resolution for powder samples [13, 14]. In general, however, the resolution function for inelastic neutron scattering instruments is expressed as an ellipsoid in the 4D momentum and energy space. Recently, numerical methods to evaluate the resolution by Monte Carlo simulation have been developed [15–18]. This type of method can be easily applied to complicated instruments, and detail shape of the resolution function such as tail due to the moderator time structure can be reproduced. A novel method to estimate the resolution function of 4SEASONS using the Monte Carlo simulation package McStas [15, 16] is currently under development. Another important issue for resolution analysis is how to visualize the 4D resolution function. Traditionally, an ellipsoid defining the half values of the resolution functions is displayed as functions of momentum axes parallel and perpendicular to the scattering vector. However, this general way of representation is not always convenient to compare the resolution with experimental data. For this purpose, we modified the *Q-E* visualizer in Utsusemi [9] to accommodate it to the resolution data generated by the McStas simulation [19]. Subsequently, the simulated resolution functions can be displayed in the same manner as a single crystal experiment. Figure 6 shows an example of a simulated resolution function displayed on a 2D map, defined by energy transfer and one of the principal axes in the reciprocal lattice. An overall feature of the inclined ellipsoid is clearly captured. The widths of the simulated resolution functions show reasonable agreements with the values obtained by simple analytical calculations, thereby validating the performed Monte Carlo simulation [19]. Moreover, a tail of the resolution function extending toward high energies is also visualized. Although the integration of the McStas simulation and Utsusemi visualization is still primitive, we continue developments to improve the integration.

6. Summary

We reviewed recent improvements in the 4SEASONS spectrometer at MLF in J-PARC. The performance of the Fermi chopper and the coverage of the detectors remarkably improved. The sample environment and software were also developed in their performances and usability. Protecting the instrument against hazards is important, and we have made progress in this aspect by developing detector protection.

Acknowledgments

We thank the Sample Environment Group, the Computing Environment Group, and other technical support staff members in MLF for their technical support. The neutron scattering experiments at MLF were performed under the user programs (nos. 2014P0801, 2014I0001, 2015I0001, and 2016I0001).

References

[1] Kajimoto R *et al.* 2011 *J. Phys. Soc. Jpn.* **80** SB025
[2] Nakamura M *et al.* 2009 *J. Phys. Soc. Jpn.* **78** 093002
[3] Nakamura M *et al.* 2008 *J. Neutron Res.* **16** 87
[4] Nakamura M *et al.* 2014 *Nucl. Instrum. Methods Phys. Res.* A **737** 142
[5] Ikeuchi K *et al.* 2013 *J. Phys. Soc. Jpn.* **82** SA038
[6] Nakamura M *et al.* 2015 *JPS Conf. Proc.* **8** 036011
[7] Nakatani T *et al.* 2009 *Proc. ICALEPCS 2009 (Kobe)* p 676
[8] Nakatani T *et al.* 2016 *Proc. NOBUGS 2016 (Copenhagen)* p 76
[9] Inamura Y *et al.* 2013 *J. Phys. Soc. Jpn.* **82** SA031
[10] Weber F *et al.* 2012 *Phys. Rev. Lett.* **109** 057001
[11] Seya T *et al.* 2011 *MLF Annual Report 2010* (J-PARC 11-03/KEK Progress Report 2011-4) p 102
[12] Inamura Y *et al.* 2017 *Proc. ICANS-XXII (Oxford)* these proceedings
[13] Iida K *et al.* 2014 *JPS Conf. Proc.* **1** 014016
[14] Kajimoto R *et al.* 2016 *Proc. ICANS-XXI (Mito)* (JAEA-Conf 2015-002) p 319
[15] Lefmann K and Nielsen K 1999 *Neutron News* **10** 20
[16] Willendrup P *et al.* 2004 *Physica B* **350** 735
[17] Vickery A *et al.* 2013 *J. Phys. Soc. Jpn.* **82** SA037
[18] Granroth G E and Hahn S E 2015 *EPJ Web Conf.* **83** 03006
[19] Kajimoto R *et al. AIP Conf. Proc.* to be published

IOP Conf. Series: Journal of Physics: Conf. Series **1021** (2018) 012031 doi:10.1088/1742-6596/1021/1/012031

Recent Issues encountered by AMATERAS: A Cold-Neutron Disk-Chopper Spectrometer

Kenji Nakajima[1], Seiko Ohira-Kawamura[1], Tatsuya Kikuchi[1], Maiko Kofu[1], Yukinobu Kawakita[1], Yasuhiro Inamura[1], Wataru Kambara[2], Kazuhiko Aoyama[1], Daisuke Wakai[1], Masahide Harada[3] and Motoki Ooi[3]

[1]Neutron Science Section, Materials & Life Science Division, J-PARC Center, Japan Atomic Energy Agency, Tokai, Naka, Ibaraki, 319-1195, Japan
[2]Technology Development Section, Materials & Life Science Division, J-PARC Center, Japan Atomic Energy Agency, Tokai, Naka, Ibaraki, 319-1195, Japan
[3]Neutron Source Section, Materials & Life Science Division, J-PARC Center, Japan Atomic Energy Agency, Tokai, Naka, Ibaraki, 319-1195, Japan

kenji.nakajima@j-parc.jp

Abstract. We report recent issues that have been encountered by AMATERAS, which is a cold-neutron disk-chopper spectrometer installed at the Materials & Life Science Experimental Facility at J-PARC. After the last ICANS meeting held in 2014, AMATERAS has continued its user program. On the other hand, the spectrometer has faced several problems, such as a weak shield at the beam dump, failure of one of the chopper disks and many such problems, which we are currently investigating. Further, several instrumental improvements have been performed, such as background reduction, fixing badly designed parts and other such improvements. Our extensive experience obtained in the period of more than 7 years on one of the world's first pulse-shaping chopper spectrometers may help other researchers who are operating or planning to construct similar spectrometers.

1. Introduction

AMATERAS is a cold-neutron disk-chopper spectrometer that is installed at the BL14 port of the neutron source at the Materials & Life Science Experimental Facility (MLF) at J-PARC [1, 2]. As one of the world's first direct geometry chopper spectrometers having pulse-shaping operation (other first generation pulse-shaping chopper spectrometers are CNCS at SNS [3] and LET at ISIS [4]), the spectrometer was designed to perform efficient measurements in inelastic and quasi elastic neutron scattering experiments on powders, single crystals and liquids to investigate versatile subjects that were obtained from corrective excitations in crystalline systems to the diffusion of atoms, molecules and ions. AMATERAS initiated its user program in December, 2009. By the end of 2016, AMATERAS conducted more than 90 proposals (both general user proposals and in-house proposals), which resulted in 30 papers, 8 theses and 3 press releases [5].

Along with performing a user program, we are continually maintaining and improving the performance of the spectrometer. After the last ICANS meeting held in 2014, we encountered some serious problems with our spectrometer. These problems were caused by the increase in the proton beam power at J-PARC or by the aging effects. Many efforts were devoted to newly developed devices that were employed by AMATERAS. In this situation, it was quite difficult to predict specific problems that the spectrometer may encounter at the beginning. Therefore, potential problems were often overlooked. More than seven years have elapsed since we started the operation of the instrument, and some hidden issues have been revealed now.

In this report, we describe recent issues of AMATERAS, which we encountered during the period from 2014 to 2016, to share our experiences with those who are engaged in the operation of similar spectrometers. This is one of the important tasks of the ICANS meeting, as demonstrated, for example, by J. R. D. Copley's report of his chopper spectrometer [6].

2. Spectrometer Issues

The causes of the problems that we have experienced while using AMATERAS in the period ranging from 2014 to 2016 can be categorized into two major causes. The first cause is an increase in the proton beam power of J-PARC. Until the end of 2014, J-PARC stably continued an operation of 300 kW for two years (including nine months of interruption due to the Hadoron accident). Further, a short time trial of 1 MW operation was performed at the beginning of 2015. The continuous 1 MW proton beam power operation came in sight at that time. Thereafter, considerations regarding the radiation dose rate became serious. The radiation dose rate at the site boundary of MLF was discussed severely, because the dose rate is proportional to the beam power and maintaining this condition becomes difficult when the operational beam power becomes higher. The second cause of the issues that have been encountered by AMATERAS were the aging problems. AMATERAS was thought to be able to operate for more than a decade when it was designed. Further, a careful maintenance scenario was considered when the design was prepared. However, the entire AMATERAS system is complicated, and several new devices, of which we had no experience, were employed. Therefore, some of the potential aging issues were overlooked in the original setup.

2.1. A dose rate-related issue

In March 2014, one of the area monitors that were located at the MLF radiation-controlled area boundary recorded a dose rate of 0.26 μSv/h. At that time, J-PARC was operated at a 300 kW proton beam power. According to our regulation, the dose rate at the boundary of the radiation-controlled areas should be less than 0.5 μSv/h. The recorded value implies that the dose rate will exceed this maximum value when the J-PARC beam power will reach 1 MW. Because AMATERAS is the closest instrument to that area monitor, investigation was immediately conducted at AMATERAS. It was soon discovered that when AMATERAS was operated in the white beam mode (all chopper disks are stopped at their open positions), an extremely high dose rate was observed on the surface of the beam dump of the spectrometer. The highest value that was recorded was observed to be 45 μSv/h at 300 kW of operation at J-PARC.

This high dose rate was unexpected when we designed the spectrometer. An investigation, including neutronics calculation studies, was performed. It was discovered that when the direct beam strikes the inside of the get-loss tube (Fig. 1), a high amount of γ-rays were emitted. The AMATERAS beam dump consists primarily of concrete with borax

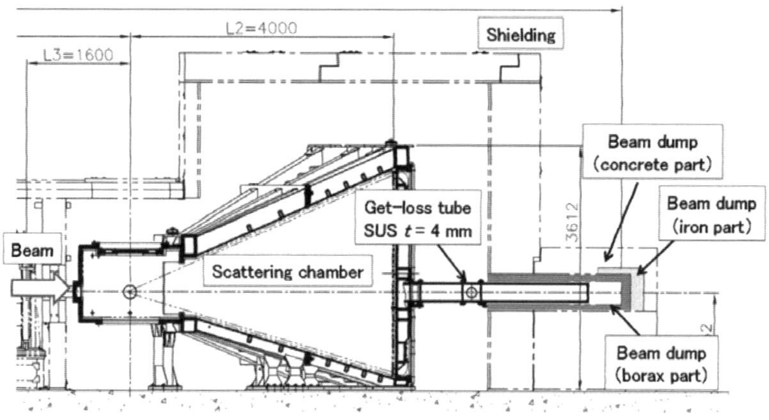

Figure 1. AMATERAS get-loss tube and beam dump.

around the get-loss tube and an iron plate at the end of the hole for the get-loss tube. When we designed the shielding for the AMATERAS, including the beam dump, a get-loss tube was not included in the design, because we did not expect a large contribution from it to the dose rate at that time.

To solve this issue with the help of neutronics calculations, we decided i) to add an additional shielding (100-mm-thick iron) to cover the entire outside of the beam dump and ii) to cover the inside of the get-loss tube with a 10-mm-thick B_4C resin plate. These were installed at the end of 2014, and the dose rate at the surface of the beam dump decreased to a maximum value of 0.4 μSv/h under similar conditions in which we had previously observed the high dose rate.

2.2. Aging-related issues

The major part of the AMATERAS construction was completed during the beginning of 2009. During its operation span of more than eight years, we have experienced a number of issues related to the breakdown of parts in the instrument. In the beginning, these were incipient failures that were caused by failures in manufacturing, design mistakes or misfittings. However, the situation has recently altered, and defects were more recently caused by aging.

In March, 2016, a vibration monitor at the housing of the fast disk chopper #3 (a RRM frame overlap chopper [7]) indicated 0.8 m/sec^2 at 300 Hz of operation, which is normally observed to be less than 0.2 m/sec^2. We suspected that the disk lost balance due to defects, deformation, or any other such reasons. During the developmental stage of the AMATERAS fast disk choppers, we experienced minor defects in the disks, which involved pealing between the prepreg layers of the carbon fiber-reinforced plastics (CFRP) in most situations, in addition to the major crashes. After initiating the operation of AMATERAS, we encountered two incidents related the disks in the fast disk choppers. We observed the deformation of one of the disks of the fast disk chopper #1 (a pulse-shaping chopper) in 2011. Deformation occurred at the slit. Therefore, we replaced the disk with a new one. Additionally, we realized that liquid was oozing off from the CFRP disks from the beginning [8]. Such a continuous leaking of liquid from the disks may hamper the balance of the disks over the long term.

The fast disk chopper was developed in collaboration with Kobe Steel Co. Ltd. However, the aforementioned company terminated their fast disk chopper business in 2013. Thereafter, we have not been able to find any replacement disks. Currently, we have limited the maximum operation revolution rate

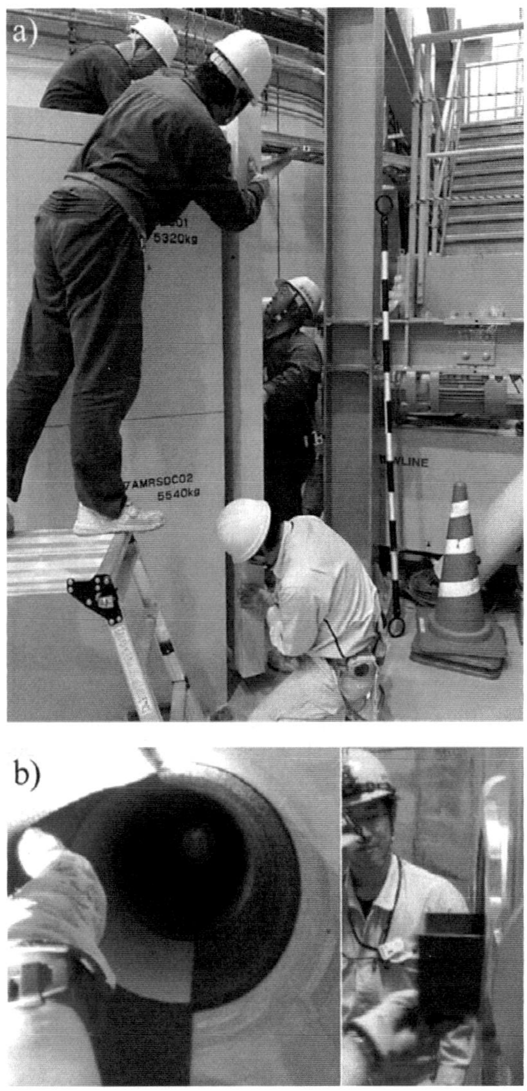

Figure 2. Solution to reduce the high dose rate. a) Covering the beam dump with a 100-mm-thick iron plate, and b) inserting a 10-mm-thick B_4C resin to cover the inside of the stainless steel get-loss tube.

IOP Conf. Series: Journal of Physics: Conf. Series **1021** (2018) 012031 doi:10.1088/1742-6596/1021/1/012031

to below 275 Hz, which affects the finest achievable energy resolution of AMATERAS. We have initiated a new development project for choppers with another company, and a new disk is expected to be delivered in 2017.

Regarding other major aging issues that were experienced from 2014 to 2016, we experienced i) the failure of a cryopump to evacuate the scattering chamber and ii) DAQ electronic issues, including high-voltage supplies to the main detectors. We have continuously repaired the cryopump, but have been unable to completely repair it. Therefore, we are planning to replace the entire system. Regarding the defective DAQ electronics, we will be replacing them with new systems until the end of Japanese fiscal year 2016. Computers for the instrument controls and data analysis, data stages, and pumps are regularly replaced and have encountered no serious problems until now.

3. Spectrometer Improvements

Major upgrades are difficult to plan because we cannot expect a huge budget. Regardless, we can plan and perform many small refinements that contribute to the users obtaining improved data. During the period ranging from 2014 to 2016, we primarily devoted our efforts to studying and reducing the background, modifying poorly designed parts of the instrument, and improving utilities to improve the usability of the instrument.

3.1. Background reduction

We have been continuously reducing the nose level of AMATERAS. As we have reported at the previous ICANS meeting [9], the background level in the measurements became approximately similar to the value that was obtained when the accelerator was terminated. The time-independent background mainly originates from electric noise and cosmic rays. Preliminary tests indicated that covering the floor near and underneath the scattering chamber with shielding materials is effective to reduce the cosmic-ray origin background. In 2015, we prepared B_4C tiles that were 10-mm-thick and placed them on part of the floor (Figs. 3 and 4(a)). Thus, we achieved more than 20% reduction of in the time-independent background. Therefore, we have to extend to the covered region to obtain a further reduction in the background, which has been planned for the future.

3.2. Refurbishment of vanes

AMATERAS has 14 detector banks that are accommodated in the scattering chamber. The detector banks are separated from each other by vanes to reduce the cross talk background from the detector surfaces. The vanes have a 1-m depth and are covered using 0.5-mm-thick Cd plates. These vanes were previously distorted and poorly aligned, and some pixels of the detectors were hidden by these vanes. These old vanes were stainless steel flames covered by the Cd plates, which hanged from the top of the inside of the scattering chamber. The bottom ends of the vanes were clipped on the bottom inside of the chamber, but contained no tension which caused the distortion. Thereafter, we refurbished these vanes. In the new vanes, each flame experienced an applied tension from both the top and bottom using spring coils and wires to avoid distortion. Careful alignment was performed using a laser marker. Therefore, the shadows of the vanes on the detectors have been completely eliminated, which is equivalent to installing new detectors.

Figure 3. The floor is covered with B_4C tiles (10 mm thick). The upper surface of the tiles are covered with flooring material.

IOP Conf. Series: Journal of Physics: Conf. Series **1021** (2018) 012031 doi:10.1088/1742-6596/1021/1/012031

3.3. Other efforts

We introduced automatic butterfly valves for two roughing pumping lines for the scattering chamber. So far, the initial process of vacuum evacuation has been manually performed. Further, we are improving the utility lines, such as gas, cooling water, electric power, He-recovery, and auxiliary vacuum lines for sample environments. Additionally, other minor improvements have been made to improve usability, reduce human burden, and make the experimental work more comfortable so that users and instrumental staff can concentrate easily during their experiments.

4. Summary

We reported the current status of our cold-neutron disk-chopper spectrometer, AMATERAS. AMATERAS is renowned as one of the most advanced neutron spectrometers. However, every spectrometer have their ugly sides [6], which we continuously aim to improve upon as part of the instrumentation business. We hope our experience will help others who operate or are planning to operate or build similar instruments.

Acknowledgments
The work on AMATERAS was supported by JSPS KAKENHI Grant Numbers 24510133. The on-beam tests were performed under the following user programs: Nos. 2014I0014, 2015I0014 and 2016I0014.

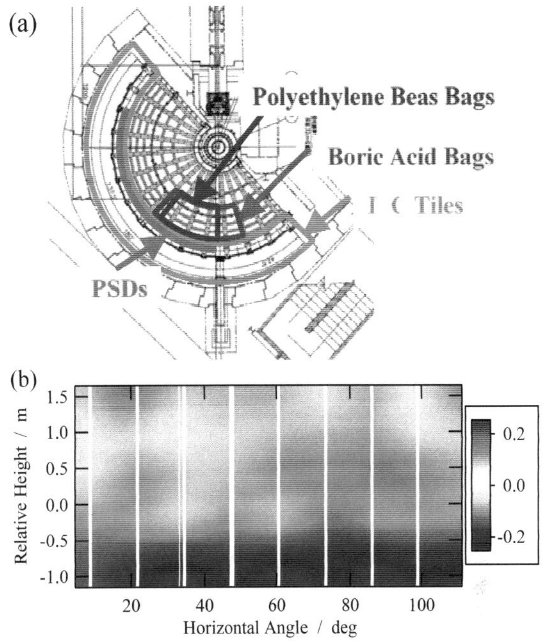

Figure 4. (a) The floor map of the AMATERAS shielding housing where the scattering chamber is accommodated. B_4C tiles (10-mm-thick) were placed in the corridor and just under the PSDs. Bags of borax and polyethylene beads are also placed under the scattering chamber. (b) The resulting reduction of time-independent background is described on the detector map image. A large reduction was observed, especially in the downward pixels.

References

[1] Nakajima K *et al.*, J. Neutron Res., **15**, 13 (2007).
[2] Nakajima K *et al.*, J. Phys. Soc. Jpn., **80**(Suppl. B), SB028 (2011).
[3] Ehlers G *et al.*, Rev. Sci. Instrum., **82**, 085108 (2011).
[4] Bewley R I *et al.*, Nucl. Instrum. Methods Phys. Res.: Sect. A, **637**, 128 (2011).
[5] Detail can be found in MLF web site, http://mlfuser.cross-tokai.jp/ja/.
[6] Copley J D R, *Proc. of the 17st Meeting of the International Collaboration on Advanced Neutron Sources (ICANS-XVII) (Santa Fe).*
[7] Mezei F, J. Phys. Soc. Jpn., **82**, SA025 (2013)
[8] The liquid is some kinds of ester, but the chemical analysis done by the disk vendor could not provide futher information. We are considering that the liquid comes from resin (CFRP binder) destroyed by the high-centrifugal force in the disk operation.
[9] Nakajima K *et al.*, *Proc. of the 21st Meeting of the International Collaboration on Advanced Neutron Sources (ICANS-XXI) (Ibaraki).*

IOP Conf. Series: Journal of Physics: Conf. Series **1021** (2018) 012032 doi:10.1088/1742-6596/1021/1/012032

Figure-of-Merit for a Cold Coupled Moderator at the SNS Second Target Station suited for Direct Geometry Inelastic Spectrometers

Georg Ehlers and Gabriele Sala

Quantum Condensed Matter Division, Oak Ridge National Laboratory, Oak Ridge 37830-6475, TN, USA

Franz Gallmeier and Kenneth W. Herwig

Instrument and Source Division, Oak Ridge National Laboratory, Oak Ridge 37830-6466, TN, USA

E-mail: ehlersg@ornl.gov

Abstract.
The suite of moderators for the Second Target Station (STS) at the Spallation Neutron Source (SNS) in Oak Ridge is currently undergoing the planning stage at which various options and configurations are being considered. The current plans for the instrument suite include two direct geometry inelastic spectrometers, CHESS and HERTZ, which will receive beam from cold moderators and work mostly in the 1-20 meV incident energy range. CHESS will be a relatively short (wide bandwidth) instrument that is dedicated to studying small samples ($\sim$ mm^3) with coarse energy resolution ($2.5 - 5\%$ of $E_{\rm i}$). HERTZ on the other hand will be designed to achieve very good energy resolution, $1 - 3\%$ of $E_{\rm i}$, using a long secondary flightpath with a $\sim 4 \times 4$ cm^2 beam on sample. In this contribution, we will discuss the implications of these design requirements for the optimal moderator pulse width, and for the repetition rate that would be best suited for the spectrometers.

1. Introduction

One of the most important limitations of the neutron scattering technique is that it requires, in comparison to x-ray scattering, relatively large samples, in particular for inelastic scattering. It is thus a priority to use the latest advances in source and instrumentation technology to design instruments that can overcome this limitation. Being able to study smaller samples will not only widen the range of an instrument for possible uses but will also enable inelastic neutron scattering to make important contributions earlier in the development process for new materials.

Direct geometry inelastic spectrometers are workhorse instruments at each major neutron scattering facility. They are very flexible and powerful instruments that can address a wide range of scientific problems in condensed matter physics, chemistry and biology. To address an area of critical importance, the first instrument of this kind, named CHESS, to be constructed at the Second Target Station (STS) will specifically aim to enable inelastic scattering experiments with $\sim$ mm^3 samples. It will focus a small beam of high (and tunable) divergence on the sample.

Naturally, the energy resolution has be rather coarse, and $\sim 2.5 - 5\%$ of E_{i} appears to be a reasonable target range for the energy resolution.

The high peak brightness of STS is achieved by compressing the linac output using the accumulator ring to produce a short proton pulse [1]. For CHESS this has the consequence that it needs to be a rather short instrument, ~ 30 meters from source to sample, because otherwise the source pulse will be sharper than optimal for the anticipated resolution range.

A second instrument that is being planned, called HERTZ, will use the high source brightness to achieve very good energy resolution. This is often needed at high scattering angles to study phonons. Better resolution means that this instrument is optimally somewhat longer (which has advantages for the optics design) and that it will operate with smaller bandwidth than CHESS.

This paper will discuss the consequences of these design requirements for the optimal moderator pulse width, and for the repetition rate that would be best suited for the spectrometers. We will establish a figure of merit that can be used to optimize the moderator performance for these instruments.

This paper is organized as follows: In section 2 the main terms governing the energy resolution are discussed and the conclusions one can draw from the resulting matching conditions. It will also be discussed how one can increase the flux on sample by relaxing the resolution at low incident energies. Section 3 discusses the main figure of merit for the moderator which will be expressed as the wavelength dependent product of pulse width and neutron speed at a particular neutron wavelength. Section 4 summarizes the paper.

2. Energy Resolution

A direct geometry inelastic spectrometer at a spallation source employs one or two high speed choppers, which define the working neutron wavelength (energy) through their phase relative to the source pulse timing, and the resolution via the burst times. These are often referred to as the P-chopper (pulse shaping chopper) and the M-chopper (monochromating chopper), respectively. The P-chopper is optional if the source pulse is short enough in the working energy range of the spectrometer so that it does not need to be cut further. This is typically the case for thermal neutrons but not for cold neutrons, because the moderator pulse width increases with neutron wavelength [2, 3, 4].

At the STS, the lengths (moderator to detector) of CHESS and HERTZ will be defined by the *coarsest* energy resolution setting to which the instruments will be optimized to operate in. This is $\Delta E/E_{\mathrm{i}} = 5\%$ for CHESS and 3% for HERTZ. These numbers are derived from operating experience: It is often found that data become much less informative if the resolution is relaxed beyond 5%. However, targeting small samples, it will be important for CHESS to be able to relax the resolution with adequate intensity gain, that is to say, one will want to increase the P- and M-chopper burst times until the full source pulse width is used at $\Delta E/E_{\mathrm{i}} = 5\%$. For the discussion that follows, it is therefore convenient to write down the energy resolution as if there was no P-chopper and the natural source pulse width was used. Assuming elastic scattering (incident and final neutron velocities are equal, $v_i = v_f = v$), one has three terms that add in quadrature [5, 6]

$$\Delta^2 E = \frac{m^2 v^6}{L_1^2} \cdot \left(1 + \frac{L_2}{L_3}\right)^2 \cdot \Delta^2 t_{\mathrm{S}} + \frac{m^2 v^6}{L_1^2} \cdot \left(1 + \frac{L_1 + L_2}{L_3}\right)^2 \cdot \Delta^2 t_{\mathrm{M}} + \frac{m^2 v^6}{L_3^2} \cdot \Delta^2 t_{\mathrm{D}} \,, \quad (1)$$

where L_1 is the distance from the source to the M-chopper, L_2 is the distance from the M-chopper to the sample, L_3 is the distance from the sample to the detector, Δt_{S} is the source pulse width, Δt_{M} is the burst time of the M-chopper, and Δt_{D} is the uncertainty in time-of-flight between the sample and the detector (which is related to the pixel and sample sizes). This approximation neglects line shape effects on the energy resolution but for the present purposes it is accurate enough. For best "value" (intensity vs resolution), the three contributions have to match. One of the matching conditions that one can derive is

$$\frac{\Delta t_{\mathrm{S}}}{\Delta t_{\mathrm{M}}} = \frac{L_1 + L_2 + L_3}{L_2 + L_3}\ .$$

Traditionally chopper systems have been designed to deliver burst times that are independent of wavelength (energy) although they can be changed between instrument settings such as "high flux" and "high resolution". In such a scenario one can read off of eq. (1) that $\Delta E \propto v \cdot E_{\mathrm{i}}$, which means that the relative energy resolution $\Delta E/E_{\mathrm{i}}$ becomes *better* (smaller) as E_{i} *decreases*, approximately like $\Delta E/E_{\mathrm{i}} \propto E_{\mathrm{i}}^{1/2}$ [6, 7, 8]. This is only an approximation because the last term $\Delta t_{\mathrm{D}} = \Delta L_3/v$ makes a constant contribution to $\Delta E/E_{\mathrm{i}}$.

At a modern spallation source that features a low frequency and emphasizes slow long-wavelength neutrons, direct geometry spectrometers are nowadays designed to employ repetition rate multiplication (RRM) [8, 9, 10]. This means that a number of incident neutron energies are used to measure separate spectra simultaneously. In such a scenario it is often more convenient to achieve an approximately constant resolution $\Delta E/E_{\mathrm{i}}$ that is independent of energy. Practical experience at high resolution instruments such as LET [11] and CNCS [12] also shows that at

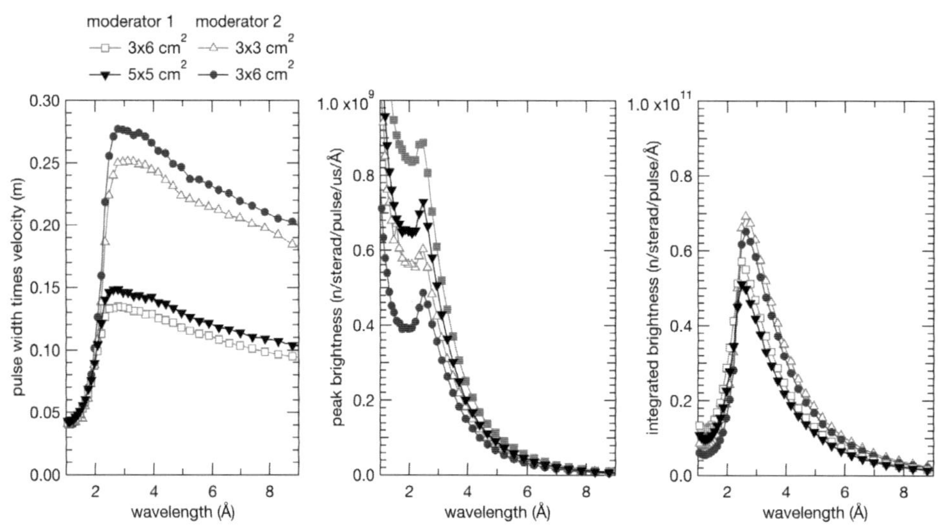

Figure 1. Performance comparison of two moderators for STS that have been optimized to different specifications. Moderator 1 was optimized for peak brightness which resulted in pulse widths of $v \cdot \Delta t_{\mathrm{S}} \sim 0.12$ m in the range of interest, $\lambda \sim 2 - 12$ Å. Moderator 2 was optimized for integral brightness, which resulted in a larger size and $\sim$ 2 times broader pulses. Left panel: pulse width multiplied with neutron speed. For each moderator two traces are given, which represent optima for different viewing areas. Note, that even moderator 2 still delivers less pulse width on average than what CHESS could use in its coarsest resolution setting. Center panel: peak brightness that one would pick off from a 2×2 cm^2 area (this is about what CHESS would need). The source brightness is normalized to 500 kW at 15 Hz with 1.3 GeV proton energy. One can see that a moderator optimized for a smaller viewing area gives a higher peak brightness, which is expected. Also, moderator 1 is better with this figure of merit because it was optimized for it. Right panel: Pulse brightness integrated over full width of the pulse, which is better approximating what a direct geometry inelastic instrument would accept as a beam. This shows that moderator 2 performs about 50% better. Again, a moderator optimized for a smaller viewing area yields a higher brightness.

low energy, $E_\mathrm{i} \sim 2$ meV and lower, the resolution is often *too good* and it would be preferable to be able to trade coarser resolution for higher intensity. In this energy range the brightness of a cold moderator goes rapidly down with increasing wavelength, approximately as $\propto \lambda^{-5}$ [1]. The issues of a variable $\Delta E/E_\mathrm{i}$ and too good resolution at low energies can be mitigated with a chopper system that is able to deliver *wavelength-dependent* pulse widths, which obey the condition

$$\Delta t_\mathrm{P},\ \Delta t_\mathrm{M} \propto v^{-1} \propto \lambda\ ,$$

where Δt_P is the burst time of the P-chopper. This would ensure constant $\Delta E/E_\mathrm{i}$ and a more balanced resolution at low energy. Recently it has been shown how this can be done with a "blind" chopper system [13].

3. Desirable Moderator Performance

It is instructive to plug in some numbers in the equation above, rearranging for the source term contribution to the energy resolution,

$$\frac{\Delta E_{(S)}}{E} = \frac{0.05}{\sqrt{3}} = \frac{2v}{L_1} \cdot \frac{L_2 + L_3}{L_3} \cdot \Delta t_\mathrm{S}\ . \tag{2}$$

Using the following reasonable length estimates for CHESS, $L_1 = 28$ m, $L_2 = 1.5$ m, $L_3 = 2.5$ m, one gets for matching conditions a moderator pulse width $v \cdot \Delta t_\mathrm{S} \sim 0.25$ m, which is expressed as the product of the pulse width and the neutron speed, v, at the corresponding neutron wavelength. HERTZ would view the same moderator, and the better energy resolution $\Delta E_{(S)}/E \sim 0.03/\sqrt{3}$ would be achieved by making the instrument longer, $L_1 = 37.5$ m, $L_2 = 1.5$ m, $L_3 = 5$ m.

The original design of a compact cold coupled moderator for STS, that was optimized for peak brightness over a viewing area of 3×6 cm^2, led to a figure $v \cdot \Delta t_\mathrm{S} \sim 0.12$ m averaged over the range of interest, $\lambda \sim 2 - 12$ Å [1]. This is shown in Fig. 1 (left panel). With a diameter of ~ 10 cm, and only few cm tall, this moderator is quite compact. On the other hand, a moderator optimized for integral brightness produces pulse widths much closer to the figure above, $v \cdot \Delta t_\mathrm{S} \sim 0.23$ m on average (also in Fig. 1). This moderator is larger, diameter ~ 20 cm, and delivers $\sim 25\%$ less brightness at the peak, but integrated over the useful peak width it performs nearly 50% better than the smaller moderator.

It would be difficult if not impossible to shorten the incident flight path of CHESS until the moderator figure $v \cdot \Delta t_\mathrm{S} \sim 0.12$ m is matched, and still optimize the neutron guide and leave adequate room for choppers, polarizer and other components. Furthermore, an instrument that is too short will also have to use an excessive bandwidth.

4. Conclusion

It can be concluded that direct geometry inelastic instruments at STS with energy resolution requirements as laid out above are best served with a moderator optimized for integral brightness. An appropriate figure of merit is $v \cdot \Delta t_\mathrm{S} \sim 0.25$ m for the pulse full width at half maximum multiplied by the neutron speed at the corresponding wavelength. In the case studied above, an optimization for peak brightness alone results in lower, by $\sim 1/3$, useful flux on sample.

The source frequency also enters the discussion via the bandwidth of incident wavelengths that such an instrument would use in a single pulse. No matter what the moderator will ultimately look like, CHESS is going to be a rather short instrument. This is necessary to match the source pulse width to the resolution needs. In fact, with $L_1 \sim 28$ m and $L_1 + L_2 + L_3 \sim 32$ m it is about as short as one can build it without seriously compromising neutron transport. At this length and $f = 15$ Hz, the bandwidth will be ~ 7 Å after factoring in that the edges of the spectrum need to be cut off in order to avoid frame overlap. By all measures, this is a *wide* bandwidth for this type of instrument, and all reasonable instrument configurations will always measure at least one pulse around $E_\mathrm{i} \sim 1$ meV or lower. The HERTZ instrument will be somewhat longer, $L_1 + L_2 + L_3 \sim 46$ m, but the bandwidth will still be rather wide, ~ 5 Å after accounting for frame overlap. For a direct geometry inelastic spectrometer, a wide bandwidth is a-priori not a disadvantage [14]. The science program at such an instrument will focus on exploration type experiments in which a wide energy range of dynamic responses in the sample is probed simultaneously. In the case of STS, where the source frequency is constrained to be a divider of 60 Hz, the choice between 10 Hz, 15 Hz, or 20 Hz results in incremental changes to the bandwidth, while increasing the repetition rate will give a corresponding increase in the useful flux across the range of incident energies used. Thus, both CHESS and HERTZ would benefit from a higher source frequency, and a broader moderator pulse width that more closely matches the resolution needs of the instruments.

Acknowledgments
We greatly appreciate discussions with P. Deen (ESS) and R. Bewley (ISIS). This research used resources of the Spallation Neutron Source, a DOE Office of Science User Facility operated by Oak Ridge National Laboratory.

References
[1] Gallmeier F X, Lu W, Riemer B W, Zhao J K, Herwig K W and Robertson J L 2016 *Review of Scientific Instruments* **87** 063304
[2] Stone M B, Niedziela J L, Abernathy D L, DeBeer-Schmitt L, Ehlers G, Garlea O, Granroth G E, Graves-Brook M, Kolesnikov A I, Podlesnyak A and Winn B 2014 *Review of Scientific Instruments* **85** 045113
[3] Kajimoto R, Yokoo T, Nakajima K, Nakamura M, Soyama K, Ino T, Shamoto S, Fujita M, Ohoyama K, Hiraka H, Yamada K and Arai M 2007 *Journal of Neutron Research* **15** 5–12
[4] Nakajima K, Ohira-Kawamura S, Kikuchi T, Nakamura M, Kajimoto R, Inamura Y, Takahashi N, Aizawa K, Suzuya K, Shibata K, Nakatani T, Soyama K, Maruyama R, Tanaka H, Kambara W, Iwahashi T, Itoh Y, Osakabe T, Wakimoto S, Kakurai K, Maekawa F, Harada M, Oikawa K, Lechner R E, Mezei F and Arai M 2011 *Journal of the Physical Society of Japan* **80** SB028
[5] Carpenter J M, Price D L and Swanson N J 1978 *Argonne National Laboratory, Report ANL-78-88*
[6] Ehlers G, Podlesnyak A A, Niedziela J L, Iverson E B and Sokol P E 2011 *Review of Scientific Instruments* **82** 085108
[7] Copley J R D and Cook J C 2003 *Chemical Physics* **292** 477 – 485 ISSN 0301-0104 URL `//www.sciencedirect.com/science/article/pii/S0301010403001241`
[8] Mezei F 2006 *Physica B: Condensed Matter* **385–386, Part 2** 995 – 999 ISSN 0921-4526 URL `//www.sciencedirect.com/science/article/pii/S0921452606012178`
[9] Nakamura M, Nakajima K, Kajimoto R and Arai M 2007 *Journal of Neutron Research* **15** 31–37 URL `http://www.tandfonline.com/doi/abs/10.1080/10238160601048767`
[10] Russina M and Mezei F 2009 *Nuclear Instruments and Methods in Physics Research Section A: Accelerators, Spectrometers, Detectors and Associated Equipment* **604** 624 – 631 ISSN 0168-9002 URL `//www.sciencedirect.com/science/article/pii/S0168900209004252`
[11] Bewley R I, Taylor J W and Bennington S M 2011 *Nuclear Instruments and Methods in Physics Research Section A: Accelerators, Spectrometers, Detectors and Associated Equipment* **637** 128 – 134 ISSN 0168-9002 URL `//www.sciencedirect.com/science/article/pii/S0168900211002750`
[12] Ehlers G, Podlesnyak A A and Kolesnikov A I 2016 *Review of Scientific Instruments* **87** 093902
[13] Vickery A and Deen P P 2014 *Review of Scientific Instruments* **85** 115103
[14] Deen, P P, Vickery, A, Andersen, K H and Hall-Wilton, R 2015 *EPJ Web of Conferences* **83** 03002 URL `http://dx.doi.org/10.1051/epjconf/20158303002`

Progress on The Time-of-Flight Ultra Small Angle Neutron Scattering Instrument at SNS

M. Agamalian[1], L. Heroux[1], K. C. Littrell[1], and J. M. Carpenter[2]

[1] Oak Ridge National Laboratory, Oak Ridge, TN 37831
[2] Argonne National Laboratory, Argonne, Il 60439

Abstract. We describe the Time-of-Flight (TOF) Ultra Small Angle Neutron Scattering (USANS) instrument at the Spallation Neutron Source, Oak Ridge National Laboratory (SNS, ORNL), which is now operating. The instrument, a double-crystal diffractometer (DCD), consists of a pair of triple-bounce symmetrically Bragg-reflecting Si(220) channel-cut crystals with a nominal Bragg angle of 70°, based on the Bonse-Hart multi-bounce concept. We briefly describe the instrument, present the results of measurements of double-crystal rocking curves, and exhibit a few early scientific results.

1. The TOF Ultra-Small-Angle Neutron Scattering Instrument at SNS

The Figure 1 schematically illustrates the SNS instrument [1] based on the Bonse-Hart principle [2]. The T_0 chopper, guide, and premonochromator are not fundamental components but serve to reduce the background and crowding against neighbors. The scattering angle is greatly exaggerated in the figure. Figure 2 is a photograph of the instrument.

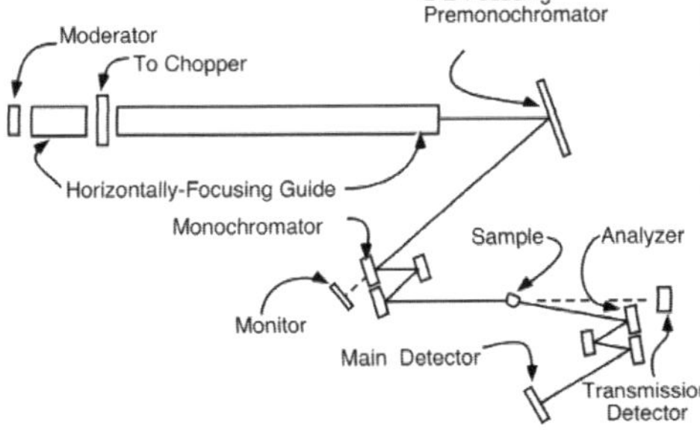

Figure. 1. The TOF USANS instrument

Figure. 2. A view of the SNS USANS instrument.

The beam at SNS beamline 1A from the 30-K hydrogen moderator, reflected by the 2-D focused premonochromator, enters from the lower left and strikes the triple-bounce crystal monochromator (small table at lower right). The reflected beam passes through the sample

IOP Conf. Series: Journal of Physics: Conf. Series **1021** (2018) 012033 doi:10.1088/1742-6596/1021/1/012033

position (round table with boron nitride apertures). The scattered neutrons pass to the triple-bounce analyzer crystal (upper left) and are reflected from there to the detector., the detector (SNS 8-pack, staggered 0.5-inch diameter ^{3}He proportional counters) rests in a massive shield (green, upper right). Stepping motors drive the analyzer angles with precision ~0.03 arc-sec. (This is about equal to the angular resolution of the best terrestrial optical telescopes.) The monochromator-analyzer assemblies rest on a vibration-isolated granite slab (dark platform). All are within a room with temperature held ±.2 C.

2. Single-crystal reflectivity

The single-crystal single-bounce reflectivity function, $R(\theta)$, is similar to the Darwin-Ewald function, with perfect reflection for small deviations of the reflection angle from the nominal Bragg angle and very sharp fall-off for angles greater than the half-width of the Darwin plateau, derived in terms of the (originally, x-ray) theories of diffraction [3,4],

$$\delta\theta_D = \frac{\lambda^2}{\pi V_{cell}\sin\left(2\theta_B\right)}\exp(-W)\left|F_{hkl}\right|. \quad (1)$$

Atomic motions are accounted for in (1) by including a "temperature factor," the Debye-Waller factor, $DWF(q)=\exp(-W(q))$, which is of Gaussian form in q.

It is convenient to express the reflectivity as a function of the dimensionless ratio y:

$$y \equiv \left(\theta-\theta_B\right)/\delta\theta_D(\lambda), \text{ where } \qquad \theta_B(\lambda)=\sin^{-1}\left(\lambda/2d\right). \quad (2)$$

For $|y| \leq 1$, the Darwin and Ewald reflectivities) are

$$R(y)=1.0, \quad (3)$$

and for |y| > 1,

$$R(y)=\begin{cases} R_D(y)=\left[|y|-\sqrt{y^2-1}\right]^2, \textbf{ Darwin, thick crystals, and} \\ R_E(y)=1-\sqrt{1-y^{-2}}, \textbf{ Ewald, finite crystals} \end{cases} . \quad (4)$$

The reflectivity function for the triple-bounce channel-cut crystals is

$$R_3(\theta)=R^3(\theta). \quad (5)$$

The rocking curve in the SNS USANS instrument is the convolution of the two triple-bounce reflectivity functions,

$$R_{meas}(\Delta)=R^3 * R^3 = \int R^3(\theta+\Delta)R^3(\theta)d\theta \quad (6)$$

which is approximately a triangular function with perfect reflectivity at the peak and a full width at half-maximum (FWHM) equal to the full width of the Darwin plateau. The difference between the Darwin and Ewald functions is not significant here. The more accurate functions (4,5,6) are used in data analysis.

With no sample, when the second multibounce crystal is offset by angle θ_s, the instrument samples the scattering at small angles near θ_s. Ignoring the wings on the reflectivity functions, one has a triangular weighting function, the resolution function, which is the convolution of the two rectangular functions (6). The FWHM $2\delta\theta_D$ of the triangular distribution, the rocking curve, as a function of $(\theta-\theta_B)$, is measured by scanning θ through

IOP Conf. Series: Journal of Physics: Conf. Series **1021** (2018) 012033 doi:10.1088/1742-6596/1021/1/012033

angle around $\theta = 0$. With a sample in place, the minimum value of q that is accessible without the analyzer resolution function overlapping the response to the unscattered beam is $q_{\min} = 4\pi\delta\theta_D / \lambda$, which is quoted as the instrument resolution of USANS instruments.

In operation for diffraction measurements, the measured scattered-neutron counting rate for each wavelength (orders separated by time-of-flight) as a function of the offset angle θ_S is a slit-smeared version averaged in the plane perpendicular to the scattering plane, a Kratky camera image. The nominal wave-vector change is $q_n = 4\pi \sin(\theta/2) / \lambda_n$ for each wavelength λ_n. The convolution-like form is not amenable to direct deconvolution methods.

Feigin and Svergun [5] discuss the data treatment in detail, based on known slit-smearing and wavelength distribution functions, which are specific to each instrument. For USANS this process is carried out by slit-smeared optimized cut-and-try modeling methods. A qualitative feature of the relationship between the point-image and slit-smeared measurements is that d-dimensional power-law q^{-d} pinhole-scattering functions, exhibit $q^{-(d-1)}$ dependence. That is, the slit-smeared function is one dimension lower than the pinhole function.

3. The Debye-Waller factor

For the nth-order wavelength, the DWF is

$$DWF_n = \exp(-q_n^2 \langle u^2 \rangle_T / 3) = \exp(-B \sin^2 \Theta_B / \lambda_n^2), \quad (7)$$

in which $\langle u^2 \rangle_T$ is the mean-squared atomic displacement due to thermal motions,

$$B = (4\pi)^2 \langle u^2 \rangle_T / 3 \quad (8)$$

and the factor 3 applies for single-crystal cubic materials. Butt, et al. [6], gives $B = 0.45 Å^2$.

4. Measured rocking curves

In rocking curve measurements, for each wavelength the monochromator is fixed in orientation, and the analyzer crystal is rotated in small steps through small angles around the straight-through position. summarizes the most recent measurements. Figure 3 shows the rocking curves, representative of recent measurements for the lowest-order reflections, $1 \le n \le 6$; Table 1 and figure 4 show the fitted FWHMs.

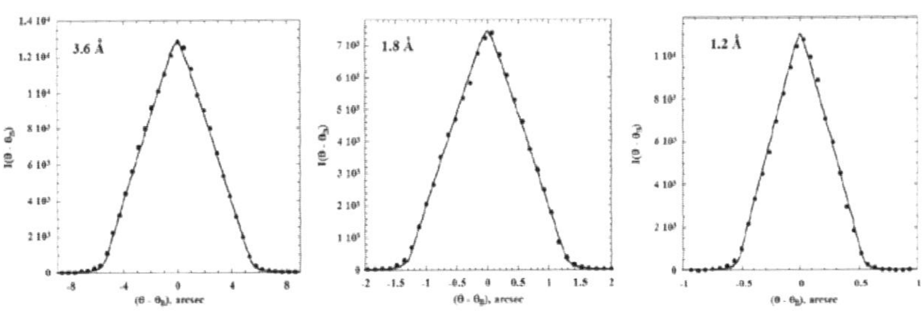

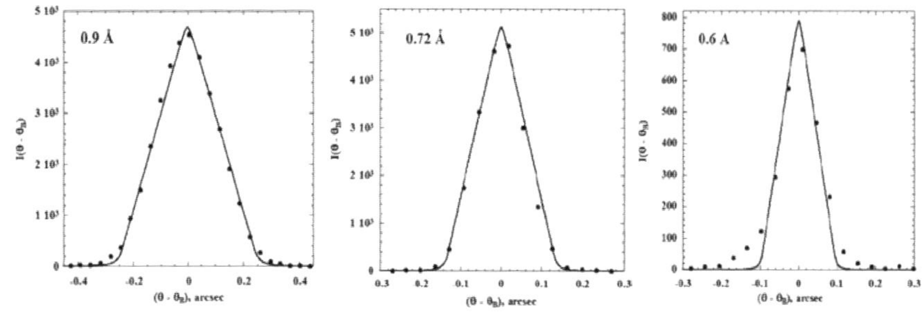

Figure 3. Rocking curves for wavelengths (orders n = 1,2,3,4,5,6), and fitted functions. Angle scales differ for the rocking curves for different orders of reflection.

Table 1. Measured Rocking Curve Widths
$FWHM_n = 2\delta\theta_{D_n}$, $\lambda_n = 3.6\AA/n$

Order, n	Wavelength, Å	FWHM, arc-sec
1	3.60	5.306±0.008
2	1.80	1.259±0.003
3	1.20	0.516±0.0009
4	0.90	0.244±0.0008
5	0.72	0.128±0.0006
6	0.60	0.097±0.001

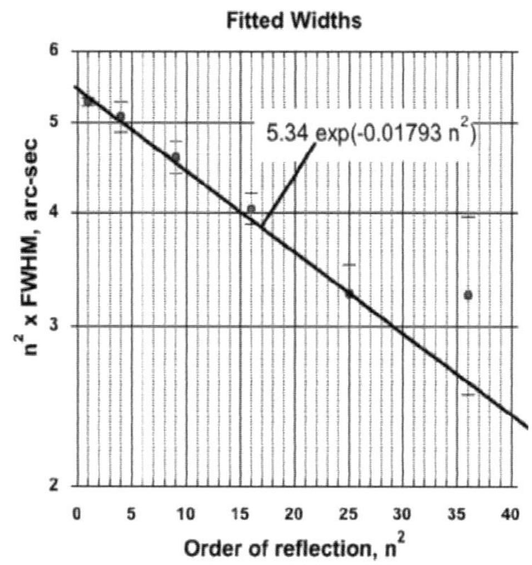

Figure 4. Rocking curve widths.

5. Conclusions

Resolution distorts measurements of the rocking curves, now under study. The effects on scientific measurements are small, but significant in the rocking curve measurements. We avoid detector dead time effects, but are developing improvements. Meanwhile, we introduce attenuating filters to reduce the counting rates when they are excessive. We are addressing the (thermal) stability of the instrument and the effects of vibrations on scattering measurements. When the instrument thermal environment is disturbed, angle calibration is lost, requiring time to recover. Vibrations driven by nearby rotating equipment (a neighbor's T0 chopper) sometimes disturb the instrument; this problem has been corrected. None of these preclude scientific use of the Time-of-Flight USANS instrument, which is ready for business.

IOP Conf. Series: Journal of Physics: Conf. Series **1021** (2018) 012033 doi:10.1088/1742-6596/1021/1/012033

6. Results of early measurements

Figure 5 displays SANS and TOF USANS measurements on tetragonal- and hexagonal-phase FeS powders, showing also the slopes of the Porod plots.

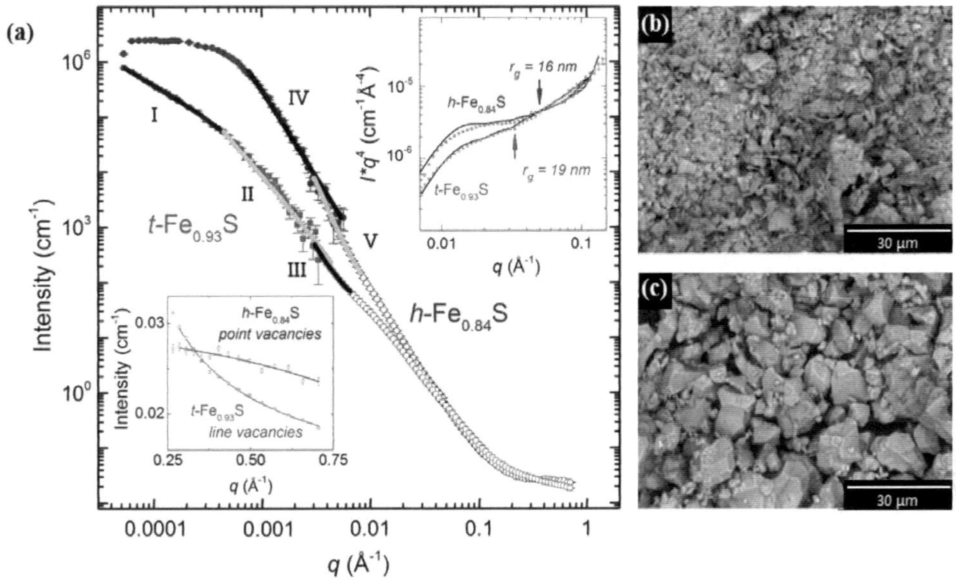

Figure 5 SANS and USANS diffraction of tetrahedral and hexagonal FeS.

USANS provides the opportunity to investigate particles across a wide range of 0.1-1000 nm length scales. USANS and SANS data (a) show that in both phases, there are two general length scales. Figures (b) and (c) are SEM images. (S. J. Kuhn, et al.,2016)

Figure 6 shows Shear cell measurements on complex fluids under shear.

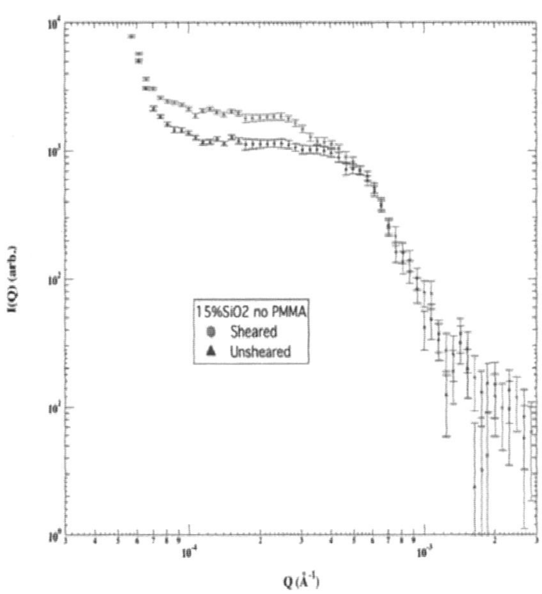

Figure 6. Shear cell measurements.

A new type of electrolytes for Li-ion batteries that exhibit a shear thickening response promises to prevent shorting between electrodes.

(unpublished, G.M Veith, B. Armstrong, and K. Browning, M. Doucet, J. Browning, R. Sacci, and B. Shen, 2017).

IOP Conf. Series: Journal of Physics: Conf. Series **1021** (2018) 012033 doi:10.1088/1742-6596/1021/1/012033

References

[1] Agamalian, M., J. M. Carpenter, K. C. Littrell, P. Thiyagarajan and Ch. Rehm. (2003), *J. Appl. Cryst.* 36, 763.
[2] Bonse, U., and M. Hart (1965). *Appl. Phys. Lett.* 7, 238-240.
[3] Darwin, C. G. (1914). The theory of x-ray reflection. Part II, *Philos. Mag.*, 27, 675.
[4] Ewald, P. P. (1917). Kristalloptik der Röntgenstrahlung *Annalen der Physik*, 359, 519.
[5] Feigin, L. A., and D. I. Svergun (1987). *Structure Analysis by Small-Angle X-ray and Neutron Scattering*. New York: Plenum Press., 277
[6] Butt, N. M., J. Bashir, B. T. M. Willis, and G. Heger (1988). *Acta Cryst.* A, 44, 396.

IOP Conf. Series: Journal of Physics: Conf. Series **1021** (2018) 012034 doi:10.1088/1742-6596/1021/1/012034

Activation of the ISIS muon beamline and corresponding gamma dose rates

Goran Škoro, Stuart Ansell[1], Stephen Jago, Kevin Jones, Sean Higgins and David Baker[2]

ISIS Neutron and Muon Source, Science and Technology Facilities Council, Rutherford Appleton Laboratory, Didcot OX11 OQX, United Kingdom

E-mail: goran.skoro@stfc.ac.uk

Abstract. The calculation of the activation of ISIS muon beamline components has been performed to predict the gamma dose rates and help in planning the replacement of the two quadrupole magnets during the first stage of the ISIS muon beamline upgrade project. The details about simulation procedure, obtained results and comparison with measurement results are presented in this paper.

1. Introduction

The ISIS muon beamline upgrade project has been recently successfully completed. The muon flux is now increased by about a factor of four with a significant improvement in reliability [1]. The first stage of this upgrade project was to replace the two quadrupole magnets near the production target with a new triplet. Those magnets were the parts of the original 1987 installation and over 25 years of operation they have been exposed to the particles produced in the interactions of proton beam with muon producing target.

2. Simulation procedure

The activation of the components of the muon beamline was calculated using the MCNPX Monte Carlo code [2] in association with the CINDER90 transmutation code [3]. The geometry of the model is based on the set of the engineering drawings [4] of the ISIS extracted proton beamline (EPB) section and the muon beamline. The CombLayer program [5] has been used then to build the corresponding MCNPX model. In addition to this, for the calculations described in this paper, the sub-set of C++ programs in the CombLayer was dedicated to prepare the neutron tallies and other files (and data) needed for subsequent run (using a set of in-house bash-shell scripts) of the transmutation code. In addition to the results on activation and inventory of radionuclides, CINDER90 is able to create gamma source definition to perform dose simulations. The set of in-house bash-shell scripts has been used to build gamma source terms for different cooling times using the results of the activation calculations. These, cell and energy dependent, gamma source terms have been used in a new MCNPX run to calculate the gamma flux (and

[1] Present address: Neutron Optics & Shielding Group, European Spallation Source, Lund, Sweden.
[2] Former member of Safety, Health and Environment Team at ISIS Neutron and Muon Source.

the spectrum) at particular distance from the objects in the muon beamline model. Then, the energy dependent dose per flux factors [6] have been used to obtain the gamma dose rate.

2.1. Model geometry

The cross-section of the geometry implemented in MCNPX is shown in Figure 1. To illustrate the dimensions of the model, the distance between the origin of the coordinate system (coincides with the centre of mass of the intermediate target) and point 10 in Figure 1 is 290 cm, between points 10 and 11 is 57 cm, and between points 11 and 12 is 195 cm. The proton beam orientation is from the bottom upwards. The effective thickness of the intermediate graphite target (IT) is 10 mm.

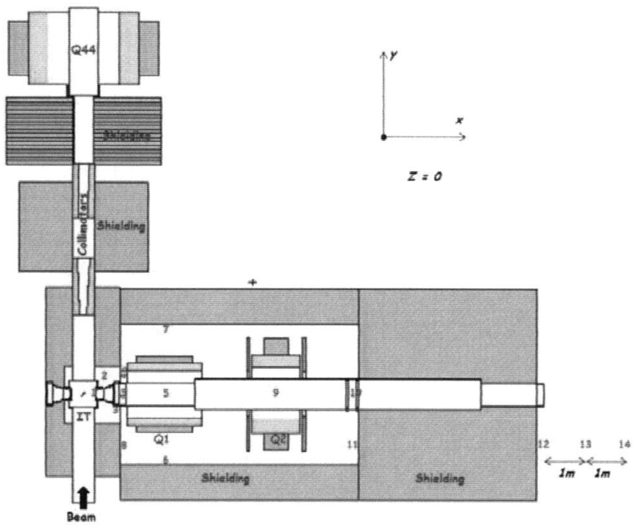

Figure 1. The MCNPX geometry of the EPB section and the muon beamline. Looking at the muon beamline branch of the geometry (located on the right hand side), the intermediate graphite target is followed by the beam window and two quadrupole magnets (Q1 and Q2). The orange colour is used for iron/steel (mild steel for shielding and the XC-06 steel as a part of the quadrupole magnets) and green colour is used for copper.

Because the aim of this analysis is to estimate the radiation field as a function of cooling time[3], the dose rates have been calculated (*case A*) at several positions denoted in Figure 1 with numbers 1 to 14. In addition, the gamma dose rates have been calculated for:

- the case where last part of shielding (between points 11 and 12) is removed. The results for this situation (*case B*) represent the dose rates expected at the beginning of the work in this area;
- the case where last part of shielding (between points 11 and 12) and the Q1 and Q2 magnets are removed. The results for this situation (*case C*) represent the dose rates that people would be exposed while installing the new quadrupoles;
- the case where only Q1 and Q2 quadrupole magnets contribute to the source term. The results for this situation (*case D*) represent the dose rates around Q1 and Q2 quadrupole magnets assuming they have been removed and are to be stored somewhere.

2.2. Beam and irradiation parameters

The incident beam is 800 MeV protons with Gaussian profile (both in x and y). The proton beam width at the IT position was assumed to be $\sigma_x = \sigma_y = 5$ mm [7]. Estimated and simplified irradiation time profile has been used: 180 days per year with beam **on** over 25 years and then final irradiation of 2 months, with assumed proton beam current of 180 μA. It has to be noted that the total number of protons on target have been intentionally overestimated in the initial calculations in order to have a safety margin on estimated radiation field in the area.

[3] To estimate the right time to begin with the process of removal of the magnets.

3. Results

As an illustration, Figures 2 and 3 show the calculated gamma dose rates for (*case A*) as a function of cooling time at the positions 1 – 11. The results for positions 12 – 14 are not shown because they are practically equal to zero (within statistical errors). Cooling time range of 7 months has been chosen expecting that work in this area will commence during such a time span after the last user cycle. As can be seen, the dose rates at these positions range from 10 to 5000 mSv/h depending on cooling time.

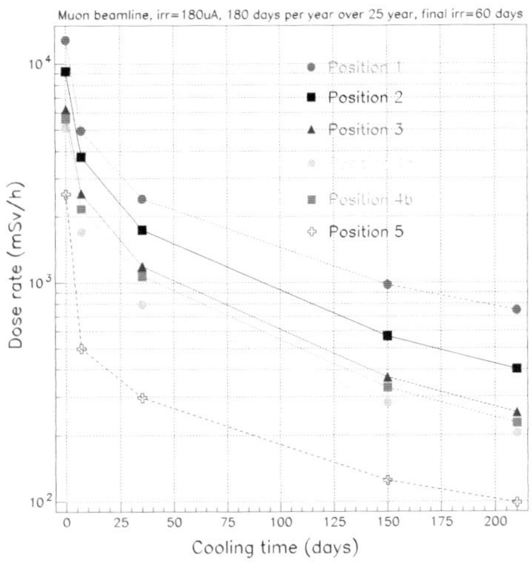

Figure 2. Calculated gamma dose rate (*case A*) as a function of cooling time at the positions 1 – 5.

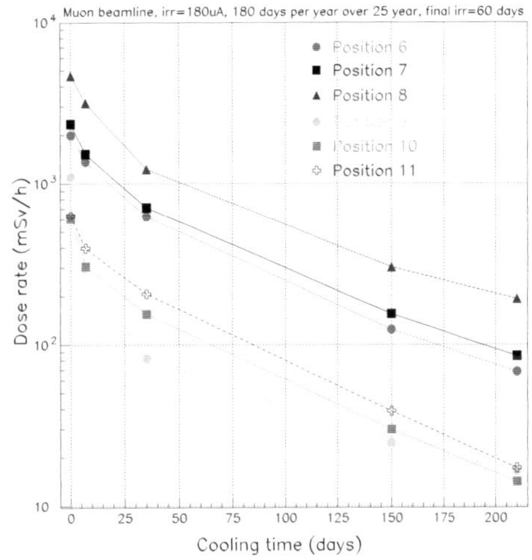

Figure 3. Calculated gamma dose rate (*case A*) as a function of cooling time at the positions 6 – 11.

The gamma dose rates at different positions have been measured (see Table 1) approximately 2 months after the end of the last cycle, and then again during the staged process of removal of the two quadrupole magnets (approximately 4 months after the end of the last cycle). It has to be noted that measurement results for *case C* are obtained under different conditions[4] than assumed in simulations (hence the symbol *C** in Table 1). Also, post festum analysis of the simulation input factors showed that real integrated proton current was 30% lower[5] so the scaled down dose rates[6] are shown in last column in Table 1.

Visual inspection of Table 1 shows a very satisfactory agreement between calculated and measured dose rates, especially if we keep in mind the simplifications used in the simulations. For example, the material composition of some of the muon beamline components are poorly known (such as 'mild' steel) so they have been estimated, all the changes and improvements in the area over many decades have been ignored (including target thickness changes[7]), simplified geometry of the components and irradiation time profile have been used, beam size/profile is known with uncertainty level of 20%. In addition, we have uncertainty of location where measurement is

[4] In addition to Q1 and Q2 magnets, the beam window assay (component between points 2 and 3 in Figure 1) has been removed as well.

[5] Based on detailed record between 1999 and 2014.

[6] Quick, first approximation of the corrected dose rate values.

[7] The target effective thickness was different during the different stages of muon beamline operation: between 2.5 and 7 mm in the period 1987 - 1997, between 3 and 10 mm in the period 1997 - 2007 and 10 mm from late 2007 onwards.

IOP Conf. Series: Journal of Physics: Conf. Series **1021** (2018) 012034 doi:10.1088/1742-6596/1021/1/012034

performed compared with calculation, statistical uncertainty of calculated values, uncertainty of measured values, influence of physics model used in simulations (CEM03 model used here) on final results, etc.

Table 1. The comparison between calculated and measured gamma dose rates during staged process of upgrading the south side muon beamline at ISIS.

Case	Position	Cooling time [days]	Measured dose rate [mSv/h]	Calculated dose rate [mSv/h]	Scaled calculated dose rate [mSv/h]
A	1	60	2200	2000	1460
A	4b	60	100	800	580
A	10	60	8.7	40	29
B	between 9 and 10	120	7	30	21
B	12	120	5	5	3.7
C*	between 12 and 13	120	1.5	10	7.2
C*	between 6 and 11	120	15	110	80
C*	3	120	200	1100	800
D	4a	120	50	55	40
D	4b	120	50	52	38

4. Conclusions

The calculation of the activation of ISIS muon beamline components has been performed to predict the gamma dose rates and help in planning the replacement of the two quadrupole magnets during the first stage of the ISIS muon beamline upgrade project. The obtained results have been benchmarked against the measured dose rate values (during the staged process of removal of the magnets) and very satisfactory agreement has been found.

References

[1] Lord J S et al. 2017 Upgrading the south side muon beamline at ISIS, *Nuclear Instruments and Methods A (to be published).*
[2] MCNPX 2.7.0 - Monte Carlo N-Particle Transport Code System for Multiparticle and High Energy Applications, https://mcnpx.lanl.gov/.
[3] Wilson W L et al. 1998 *Proceedings of the SARE4 Workshop, Knoxville, USA.*
[4] Jones K and Russell R 2013 *private communication.*
[5] Ansell S 2014 *https://github.com/SAnsell/CombLayer.*
[6] International Commission on Radiological Protection 1987 *ICRP Report No. 51, Pergamon Press, New York.*
[7] Jones B and Adams D J 2013 "Design and Operational Experience of Delivering Beam to ISIS TS1" *ISIS Internal Report.*

IOP Conf. Series: Journal of Physics: Conf. Series **1021** (2018) 012035 doi:10.1088/1742-6596/1021/1/012035

Neutron Beam Extraction and tailoring useful neutrons to instruments at ESS

M Arai[1], L Zanini[1], E Bryndt Klinkby[1], K Andersen[1], R Linander[1], F Mezei[1], K Niita[2], M Harada[3], F Maekawa[3]
[1] European Spallation Source ERIC, Lund Sweden
[2] Research Organization for Information Science and Technology, Tokai, Japahn
[3] J-PARC Center, Japan Atomic Energy Agency, Tokai, Japan

masatoshi.arai@esss.se

Abstract. The ESS has a characteristic design of the target and moderator system. The 2.5m-diameter target wheel needs a large horizontal vacant space around the moderator, and a thin moderator produces more high energy neutrons (HEN) in comparison with other facilities. Therefore, it is quite important to avoid HEN as much as possible before ejecting them into the beam line, although a curved guide or T_0 chopper can get rid of them. In this report we show the present situation of these issues and discuss the measures to be taken.

1. Introduction

The European Spallation Source (ESS) is the first long pulsed source under construction, at the outskirts of Lund, Sweden. The characteristic parameters of the accelerator are 5MW, 14Hz, 2GeV and long pulse of 2.7msec. This will be the most powerful accelerator-based neutron source in the world. Since it is a long pulsed source, therefore, it will be very suitable for production of cold neutrons. Moderators and instruments will be designed to extract the best performance of the source and would have a different character and unexpected issues compared with fairly well established short pulse sources such as ISIS, SNS and J-PARC. One possible choice of moderator system would be a thin moderator (3cm), which benefits from the low cross-section of para-hydrogen below 30meV [1] , and will enhance the brightness of the source. However, as we will discuss in this report, there will be a drawback which needs to be carefully dealt with.

The spallation process naturally produces high energy neutrons. Dose mal-effect drastically increases and shielding performance of any material decreases more than one order of magnitude in the 1MeV range compared with thermal neutrons. It is, therefore, quite important to get rid of high energy neutrons (HEN) within the target monolith shield as much as possible. Beam extraction ports to the instruments accepts both useful neutrons and unwanted HEN. However, by knowing the neutron distribution at the source, we can reduce HEN to the beam port by well designed optical components outside the monolith.

In this report, we briefly describe the character of the neutron distribution at the moderator, the structure of the beam course in the monolith and possible measures to get a clean beam at the sample in comparison with an example of J-PARC instruments.

2. Mechanical Structure around the Target

The shield/collimator design around the moderator is depicted in Fig. 1. Since ESS will take a 2.5m-diam Tungsten target wheel, there is horizontally a large void space around the moderators, and the beam port shield/collimator can only start at 2m out from the moderator due to the target structure, narrow beam separation and heat load on the guide system. We can contrast this situation with the beam port collimation in ISIS, SNS and J-PARC, where individual beam ports are separated horizontally and vertically. Therefore, it is not difficult to have a suitable collimation to avoid HEN to the beam port.

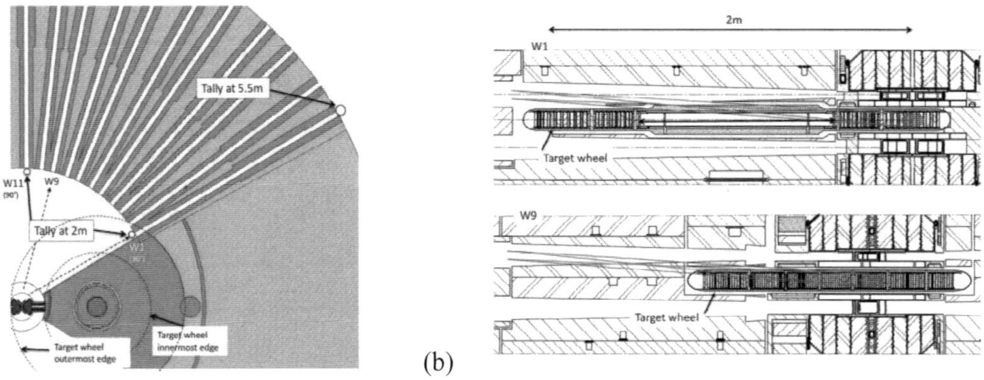

Fig. 1 (a) Horizontal cut-section of the target monolith shield for neutronics calculations. A tally was set at 2.0m at the entrance of the beam extraction system or at the beam port outside the monolith at 5.5m . (b) vertical cuts of the W1 and W9 beam axes, showing the steel plate shield above the target wheel highlighted in yellow [2].

Figure 1 (a) is the horizontal cut-section of the geometrical model and (b) is the vertical cut-section of the engineering model in the monolith, which shows how much shielding material, coloured in yellow, exists through a ray trajectory from the hot spot in the target (red) to the beam guide entrances of 5 cm in height at W1(30°), W9(78°) [2]. The W1 port (forward direction) seems not to have good enough shielding material, so that the high energy neutron flux is drastically enhanced as we will see in Fig. 2. However, neutrons from the hot spot can be shielded by the Tungsten target itself at W9.

3. Neutron Spectrum

Figure 2 shows the energy spectrum at the guide entrance at 2m from the moderator as shown in Fig.1(a). Here, we took two tally (measurement points in the simulation) methods of MCNPX, F4 and F5. F4 is the first principle calculation taking any nuclear reactions in the process and the tally counts particles which pass through it. F5 is an approximation method which assumes isotropic scattering processes in the physics model region above 150 MeV. Therefore the F4 tally gives a more accurate neutron flux above 150 MeV. Calculation was done for three different beam ports at the forward direction (30°), the perpendicular direction (90°), and the backward direction (150°). It is clearly seen that the high energy neutron intensity is very much enhanced in the forward direction at 30°. It is about one order of magnitude higher than the backward direction at 150° above 150MeV. We could not see a clear difference between F4 and F5 within the statistics, however,

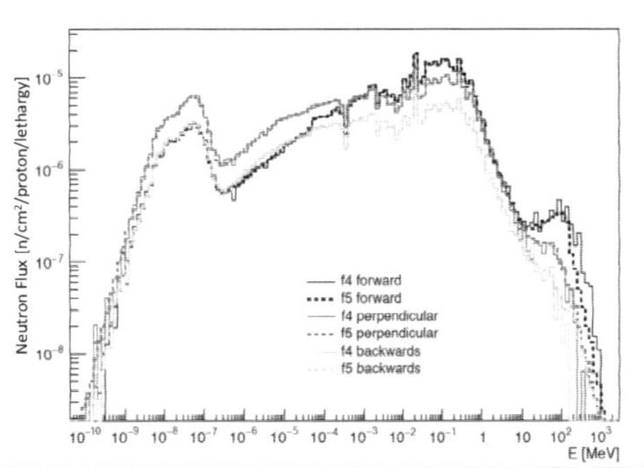

Fig. 2 Energy spectrum at the guide entrance at 2m. Comparison between F4 and F5 tallies (7(v) x 14(h) cm^2) for three direction at the forward direction (30°), the perpendicular direction (90°), and the backward direction (150°) are depicted.

IOP Conf. Series: Journal of Physics: Conf. Series **1021** (2018) 012035 doi:10.1088/1742-6596/1021/1/012035

4. Neutron Distribution at the moderator

The Neutron distribution around the moderator was estimated by tracing back the neutron trajectory from the 6 cm sphere tally at 2.0 m at 90° in Fig.1

Figure 3 shows the neutron distribution along the moderator height for 1.5, 3.0 and 10.0cm thick moderators. Fig.3 (a) shows the slow neutron distribution integrated between 0 and 5meV, where a brightness enhancement for the thinner moderator is clearly observed. Fig.3 (b) shows the distribution of high energy neutrons integrated above 100keV. Fig.4 (c) shows over-plot of 0-5meV and >100keV for 3cm thickness. There is a very sharp peak structure at the edge of moderators in Fig.3 (b), which comes from the 3mm-thick Al container. The intensity at the 13cm-position, which corresponds to the lower Al edge of the moderator, has a similar intensity for the different moderator thicknesses. But intensity at the other side of edge has a systematic decrease with thickness, and it seems to be almost proportional to $1/R^2$ (R is the distance from target). We should note that this edge effect can become more significant above 100MeV, where the cross section of Al largely exceeds that of H_2 but is just a half that of steel.

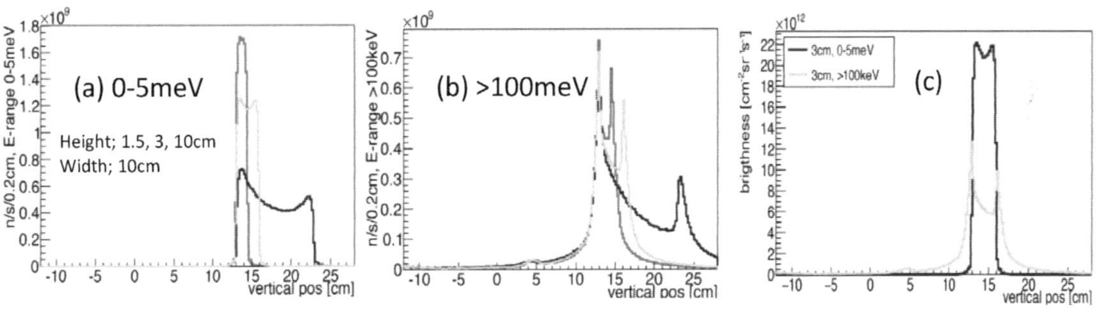

Fig.3 Neutron distribution on the moderator surface as a function of distance from the center of the target.; (a) 0-5meV, (b) >100keV for moderator size of 1.5, 3.0 and 10.0 cm, and (c) over-plots intensity of 0-5meV and >100keV for 3cm TDR-like moderator [3]. The calculation was truncated at the bottom of the bottom pre-moderator at 5cm in the figure, which produces an artificial hump in (b).

Figure 4 shows energy spectrum above 1MeV to show the contribution from the Al container. Here, the moderator thickness is 3cm and the aperture of the tally was changed from 11(w) x 3(h), 11(w) x 9(h) to 4.5(w) x 3(h) cm, with a different collimation condition, i.e. Here, "Art Coll" means the steel shield behaving as a perfect collimator, "no Al" means no Al moderator casing, "nothing" is the geometrical model shown in Fig.1. The figure clearly shows a larger enhancement for the larger aperture of tally above 100MeV. Probably the larger tally views a wider area around the moderator. The contribution from the Al container observed in no-Al is comparable nevertheless to the tally size. It is about factor 2.0. However, this contains the contribution from the front surface of the moderator, which cannot be avoided in any case, and the contribution for the Al edge can be 30% as seen from Fig. 3 (c). A message from Fig.4 is that good collimation is very important for beam extraction.

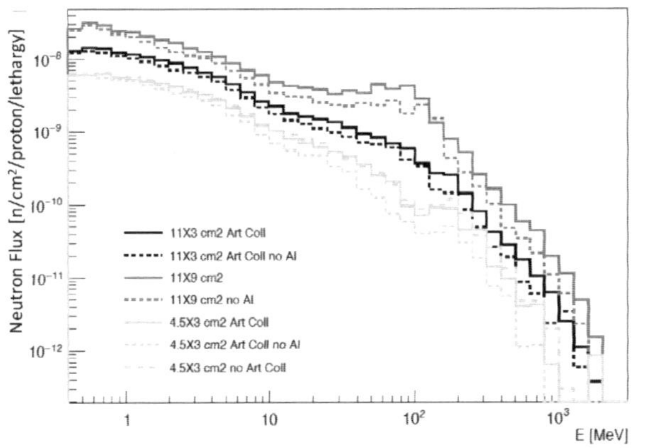

Fig. 4 Energy spectrum simulation with the F5 tally of MCNPX for various conditions as specified in the text.

IOP Conf. Series: Journal of Physics: Conf. Series **1021** (2018) 012035 doi:10.1088/1742-6596/1021/1/012035

5. Beam Extraction

J-PARC/MLF is a sharp pulse source, where a sharp neutron pulse is produced by the decoupling concept. A decoupled moderator has an intensity maximum at the center of moderator, Fig.5 [4]. Therefore, the moderator was designed to avoid high energy neutrons by having an off-set Al container from the beam course with a viewed surface of 10x10 cm^2. Also we chose to constrain the viewing area on the moderator surface with typically a 1cm-margin from the edge of the viewing surface. The good collimation system helped with rejection of high energy neutrons too. Therefore, essentially, the instrument does not directly see the Al casing of the moderator from the beam port except for viewing front surface.

6. General remark for high energy neutrons.

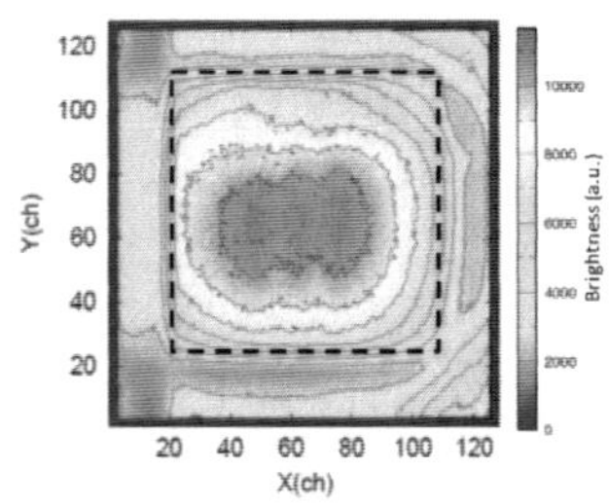

Fig. 5 observed neutron distribution, 5<10meV, at the decoupled moderator at BL10 [4]. Black dotted square: the moderator surface size 10 x 10 cm. X and Y values: detector channel.

There are mainly two remarks to keep in mind for high energy neutrons. One is the reduction of the cross-section of shielding materials in the high energy region. Roughly speaking the cross section decreases one order of magnitude for 1MeV range than that for thermal energies. Another factor is the biological dose effect of neutrons. The Dose Conversion Coefficient of neutrons is 100 times larger at 1MeV in comparison with that at thermal energies [5]. From the two factors, the amount of biological shielding material should be drastically increased to keep the required dose rate in the high energy range.

As we saw in Fig. 3 and 4 large HEN contribution to a beam port comes from the Al casing and surroundings around moderator. We cannot avoid to view the front surface of the moderator; however, better collimation can avoid HEN from the surroundings.

As we showed in Ref. [6] adoption of either a curved guide or T_0 chopper is quite effective to get rid of HENs to the sample. However, even after a line-of-sight of curved guide, HENs are transported because of elastic scattering by the guide substrate. The intensity difference between thermal neutron transported by the guide and HENs can be less than 1/million. This N/S ratio is good enough for most of instrument and can be improved further for long flight-path instruments at the ESS. Therefore, it is not unrealistic to achieve the world's best background performance at the ESS. Good collimation in the neutron beam extraction system (NBEX) can help the this aim; therefore, careful design of NBEX is highly required.

7. Conclusion and suggestion

We have reviewed the current neutronics calculations on the high energy neutron contribution to the instrument beam port at the ESS and compared with the performance at J-PARC. We found HEN produced from the Al edge and surroundings. Although we cannot avoid HEN from the front surface of the moderator, a good collimation system in NBEX and the downstream optical system can reject HEN from the Al collimator edge and surroundings, and may realize the best N/S ratio at the ESS.

References

[1] MacFarlane R E, Cold-Moderator Scattering Kernel Methods, LA-UR-98-655, Los Alamos National Laboratory, 1998.

[2] Andersen K, Jönsson B, Study of Vertical Support Structures in Neutron Beam Path of Outer Reflector, 12/1/2016

[3] Technical Design Report, edit Steve Peggs, ESS-2013-001 APR 2013

[4] Harada M et al., unpublished paper "Brightness measurement and its comparison with the simulation", 2017

[5] Conversion Coefficients for use in Radiological Protection against External Radiation, ICRP Publication 74, Ann. ICRP 26 (3-4), 1996

[6] Niita K et al., International Collaboration on Advanced Neutron Sources, ICANS, April 25-29, 2005, Santa Fe, New Mexico, p.640- 647

IOP Conf. Series: Journal of Physics: Conf. Series **1021** (2018) 012036 doi:10.1088/1742-6596/1021/1/012036

The SNS Moderator Demonstration Facility

E B Iverson[1], F X Gallmeier[1], R C Gillis[1], T Hügle[1] and T C McClanahan[2]

[1]Spallation Neutron Source, Oak Ridge National Laboratory, Oak Ridge, TN USA
[2]Department of Nuclear Engineering, University of Tennessee-Knoxville, Knoxville, TN USA

E-mail: iversoneb@ornl.gov

Abstract. We describe a moderator demonstration facility at the Spallation Neutron Source (SNS). We will use the Beam Test Facility (BTF) at SNS to provide a moderator neutronics test stand with which we will verify the performance gains expected from the innovative moderator concepts central to the SNS Second Target Station. These concepts include large-volume para-hydrogen moderators and high-brightness moderators, as well as new moderator materials. The BTF provides a test facility for the new Radio-Frequency Quadrupole accelerator (RFQ) to be installed at SNS and produce a pulsed H^- beam of 2.5 MeV energy. We will add to the BTF a beam chopper similar to that already used in the SNS at the RFQ exit, various beam transport components, a neutron-producing target, a cryogenic moderator test stand, a reflector-shielding assembly, and a performance assessment neutron beamline.

1. Introduction

The Spallation Neutron Source facility is currently developing the technical design for a Second Target Station (STS). One of the essential aspects of the STS concept is a high-brightness para-hydrogen moderator—this moderator will significantly outperform the coupled moderators on the first target station (FTS), with a significant fraction of that performance gain requiring complete conversion to equilibrium levels of para-hydrogen (approximately 99.8% at 20 K). [1, 2] In the design of the FTS, the moderators were designed to be relatively insensitive to the exact ortho/para-hydrogen fraction, with a consequent loss in performance. The success of the proposed STS concept relies on the a) reliable prediction of the neutronic performance advantages of a large-volume or high-brightness para-hydrogen moderator, b) the diagnostic capability to assess the ortho/para-hydrogen fraction in an operating moderator, and c) the ability to understand the interaction between natural conversion to para-hydrogen, radiolysis-induced back-conversion to ortho-hydrogen, and system catalysis. [3] These are long-standing challenges in the implementation of hydrogen moderators at high-power neutron sources. In short, successful execution of the Second Target Station concept requires solving these challenges.

This manuscript has been authored by UT-Battelle, LLC under Contract No. DE-AC05-00OR22725 with the U.S. Department of Energy. The United States Government retains and the publisher, by accepting the article for publication, acknowledges that the United States Government retains a non-exclusive, paid-up, irrevocable, worldwide license to publish or reproduce the published form of this manuscript, or allow others to do so, for United States Government purposes. The Department of Energy will provide public access to these results of federally sponsored research in accordance with the DOE Public Access Plan (http://energy.gov/downloads/doe-public-access-plan).

IOP Conf. Series: Journal of Physics: Conf. Series **1021** (2018) 012036 doi:10.1088/1742-6596/1021/1/012036

We will install a implement a moderator neutronics demonstration capability at the Beam Test Facility (BTF) at the Spallation Neutron Source with which we will verify these performance gains. These concepts include both high-brightness and large-volume para-hydrogen moderators. We will add to the BTF a proton beam chopper similar to that already used in the SNS at the RFQ exit, various proton beam transport components, a neutron-producing target, a cryogenic moderator test stand, a reflector-shielding assembly, and a performance assessment neutron beamline. The Moderator Demonstration Facility (MDF) will provide the ability to test a large-volume para-hydrogen moderator in a prototypic configuration, simultaneously measuring the neutronic performance of the moderator concept central to the anticipated STS gains and developing the instrumentation necessary to provide that performance in a production environment. Additional moderator concepts proposed for STS as well as FTS can also be tested at this facility, including the high-brightness moderators involving significantly restricted (transverse) moderator views. Testing these moderators requires a wing moderator configuration, not available at existing test facilities worldwide. [4, 5]

2. Project Overview

We envision four main components to the Moderator Demonstration Facility project:

(i) Provide a proton chopper and beam transport system to deliver 50 mA (peak) of 2.5 MeV protons from the BTF to a suitable neutron-producing target with pulse width between 0.5 and 10 μs.

(ii) Provide a neutron-producing target and reflector assembly optimized to represent prototypic moderator illumination (that is, moderator illumination from the target and reflector should replicate that predicted for the STS concept and accommodate large-volume and high-brightness moderator concepts.

(iii) Provide an instrumented moderator vessel, neutronically equivalent to a production moderator proposed for STS, but additionally capable of in situ Raman spectroscopy for assessing the spin-state distribution of the hydrogen in the moderator vessel, as well as providing ultraviolet irradiation of the moderating material to mimic the radiolysis rates in the production STS moderator independently from the neutron irradiation associated with measuring the neutronic performance.

(iv) Provide a neutron beamline for the assessment of neutronic performance of the prototypic moderator configuration as influenced by radiolysis and external catalysis.

Each of these four components is essential to the successful assessment of the STS volume moderator concept; each of these four components is also unique—no test facility currently exists that can adequately answer these questions in a prototypic manner.

3. Proton Chopper

The RFQ system itself is not capable of providing proton pulses of less than around twenty microseconds width, significantly larger than we prefer in order to assess the moderator slowing-down performance. SNS has developed a MEBT chopper system [6] to provide high-quality proton beam chopping prior to acceleration in the LINAC, in order to provide a beam-free extraction window in the accumulator ring proving sub-microsecond pulses to the SNS FTS (and eventually STS). The MEBT chopper system enables beam rejection of more than four orders of magnitude with rise/fall times of less than five nanoseconds. In SNS itself, this chopper imposes a beam-free period of approximately 300 ns in between beam mini-pulses lasting 700 ns; in the BTF-based moderator test facility we propose, we would instead use this chopper system to provide single pulses of variable duration as required for assessing different moderator performance characteristics; longer durations (up to ten microseconds) for more neutrons at very low energies where we can tolerate the resolution effects, shorter durations (around one microsecond) for prototypical studies going up to 0.5 eV, and even shorter durations (as low as 500 ns) for detailed investigation of the slowing-down capabilities of different moderator configurations. This ability to provide short-duration pulses is essential for the successful study of moderator performance for short-pulsed neutron sources. This capability is not currently available at low-powered test facilities such as LENS (which has a minimum proton pulse width around twelve microseconds). While sub-microsecond pulse durations are available at electron-bremsstrahlung facilities such as at Hokkaido University in Japan, those facilities necessarily provide copious fluxes of high-energy photons, complicating the study of moderator performance independently of radiolytic dissociation rates. The use of the SNS-developed MEBT chopper system is an essential and unique feature for a short-pulsed moderator test facility.

4. Target-Reflector Assembly

Most facilities have (for good reason) chosen to optimize the target apparatus to maximize neutron production. As a result, facilities such as those at LENS and Hokkaido University provide non-prototypic moderator illumination. We will optimize this facility to provide prototypic moderator illumination, with the recognition that the resulting characterization will require more advanced measurement techniques to offset the reduce intensity. At 2.5 MeV, protons produce far more neutrons off of lithium than off of beryllium (thick-target yields of 10^9 n/μC vs 10^8 n/μC). Additionally, the neutrons coming from the endothermic lithium reactions are of significantly lower energy (strictly less than 0.8 MeV) than the exothermic beryllium reactions (mean energy of around 5‘MeV). Accordingly, the shielding requirements are much reduced for a lithium target (on a per source neutron basis). Finally, with a primary neutron energy below 1 MeV, we will drastically reduce activation levels in the target-moderator-reflector assembly during operations; experience at LENS indicates that the primary source of residual dose is ^{24}Na from (n,α) reactions on aluminum, with a threshold of a few MeV. While a lithium target is somewhat challenging at these powers, the 50 mA peak current is strongly diminished by a duty factor of less than 10^{-5}. Additionally, our desire for a prototypic moderator illumination will push us toward a very extended target geometry—using a beam spot of greater than 300 cm^2, significantly reducing power deposition in the vulnerable lithium material.

5. Instrumented Moderator System

One nominal moderator vessel we would study will be neutronically similar in geometry to the volume moderator proposed [7] for the STS: a cylinder of diameter 15–20 cm, of height 10–15 cm. Such a moderator will, in some ways, represent a bounding case for accommodating large moderators. This moderator vessel will contain liquid or supercritical hydrogen at pressures less than 20 bar. Optical feed-throughs on the vessel will admit laser probe and collection optics for *in situ* Raman spectroscopy to study the spin-state distribution of the hydrogen. Additional ports will permit separately-controlled UV illumination to simulate the radiolysis effects of prototypic beam powers. The moderator vessel will be mounted on a standard closed-cycle refrigerator cold head similar (for maintenance convenience) to those used throughout the SNS facility. The moderator assembly will share vacuum space with the initial portion of the neutron beamline, as shown in Figure 1. This single penetration offers a number of functional advantages, including:

- permitting gas feed lines, thermal links, sensor access, and neutron beam egress without compromising moderator decoupling (when desired),
- horizontal moderator assembly extraction along a track, eliminating crane requirements and shield unstacking, while more reproducibly positioning the moderator system relative to both the target and the characterization beamline,

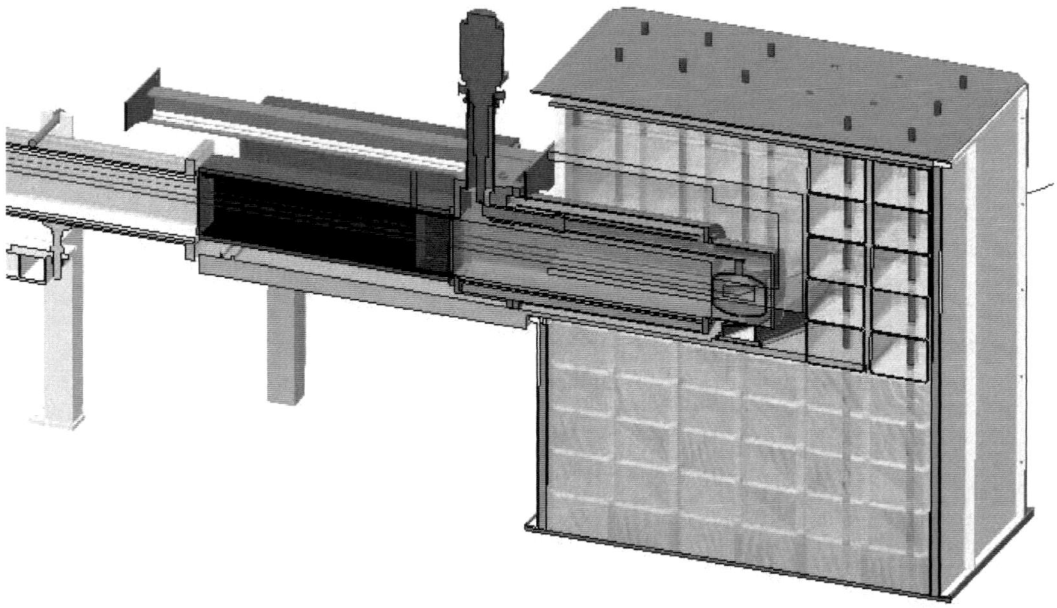

Figure 1. Elevation cross section of the target-moderator-reflector assembly.

- enabling much larger moderator systems than can be accommodated otherwise, or even large optical components to be placed in the first meter of beamline length.

6. Characterization Beamline(s)

A single, dedicated neutron beamline will enable detailed assessment of the neutronic performance of the moderator. In keeping with the prototypic nature of the facility concept, a relatively large-area beam will be permitted to leave the moderator and go down a flight-path of roughly 3–6 m to an array of analyzers and detectors which will measure the energy spectrum of the neutron beam from 0.1 meV to 10 eV, the energy-dependent emission time distribution of the neutron beam from 0.8 meV to 1 eV, and the energy-dependent spatial distribution of moderator emission intensity with position resolution (on the moderator surface) of order five millimeters. A conceptual layout for the emission time analyzers is shown in Figure 2. The single beamline hosts two arrays of analyzer crystals, both in fully time-focused geometry. [8] The first array (based around a fluorophlogopite analyzer array) will probe low energies (down to 0.3 meV), with good density of coverage at in the vital 2–10 meV range. The second array (based around an off-cut high-mosaic copper analyzer array) will provide significant coverage overlap at intermediate energies, but go significantly higher as well (up past 1 eV). Additionally, the independent analyzer arms will provide a cross check, improving our knowledge of instrument resolution.

7. Performance Estimates

We base performance estimates for this test facility on comparisons with the LENS facility [4] as used for moderator development. When used for short-pulsed moderator development measurements, LENS provides approximately 12 μA of 13 MeV protons on a beryllium target

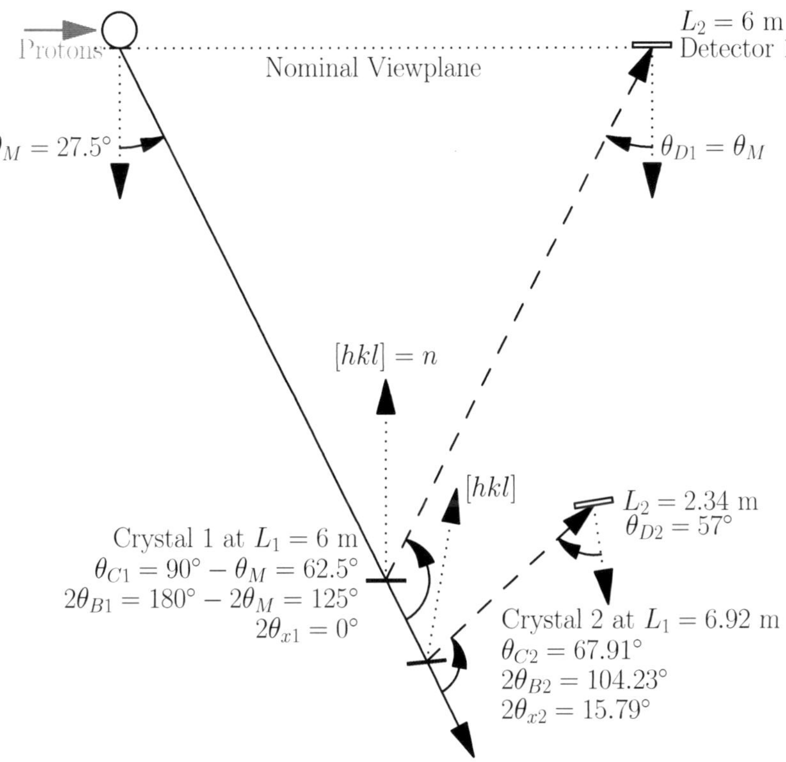

Figure 2. Layout of the emission time analyzer arrays neutron characterization beamline.

during a 12 μs pulse at up to 40 Hz, giving a primary source intensity of around 6×10^{11} n/s. The LENS facility hosts two neutron beam-lines used for moderator characterization experiments: a small-angle neutron scattering instrument (SANS) temporarily modified to provide spectral and emission time characterization and a dedicated Moderator-Imaging Station (MIS) providing energy-dependent brightness distribution on the moderator surface. Using 3 μA of 2.5 MeV protons (a 1 μs pulse width at 60 Hz) on a lithium target at BTF will reduce the primary source intensity by two orders of magnitude as compared to LENS. Additionally, the primary spectrum is much lower energy, with a 0.8 MeV maximum instead of a 5 MeV average. We will configure the BTF target-moderator-reflector system in a wing configuration as appropriate for prototypic SNS use, both FTS and STS, and essential for testing either volume-moderator or high-brightness concepts. Wing configurations are less tightly coupled to the primary neutron source than slab configurations; as a result, our base moderator brightness (for the same conventional moderator) at BTF (when operated with a 1 μs pulse width) will be around a factor two hundred lower than at LENS. While this is a significant disadvantage to overcome, LENS is configured primarily for neutron scattering applications, and is dramatically inefficient for moderator characterization. We will eliminate the ability to extract multiple distinct neutron beam-lines. We will employ analyzer arrays of approximately 300 cm^2 in fully time-focused geometry, where the SANS-beamline collimation at LENS limits the analyzer to around 20 cm^2. This analyzer array will

IOP Conf. Series: Journal of Physics: Conf. Series **1021** (2018) 012036 doi:10.1088/1742-6596/1021/1/012036

be placed closer to the moderator under test than the 8.5 m position at LENS (again dictated by the SANS-optimized geometry). As a result, the time-averaged beam incident upon the emission-time analyzer at the BTF facility will be reduced over that at LENS by only a factor three. At LENS, moderator characterization experiments require approximately ten hours of beam to provide adequate statistics for emission-time distribution measurements. At the BTF facility, accruing the same total counts will require some thirty hours of beam, but the sharper primary source pulse will provide the same ability to assess a given neutron pulse shape with far less statistics. Accordingly, we expect that the most time-consuming portion of a moderator characterization experiment, the measurement of emission-time distributions, will be comparable at the BTF facility to that at LENS, were the LENS target-moderator-reflector assembly to permit the study of large-volume and high-brightness moderators.

8. Future Outlook

This test facility is intended directly and primarily to demonstrate that the concepts proposed for the SNS Second Target Station can be successfully deployed. Once that demonstration has been accomplished, we will have a world-class test facility to continue to serve the needs of the SNS and HFIR facilities for the development, both conceptual and prototypic, of new advanced neutron sources. SNS already supports around four weeks worth of moderator development experiments at LENS each year, suggesting an internal demand on an ongoing basis of as much as six-to-ten weeks per year once we have accounted for offset travel costs and different capabilities. Additionally, the moderator test facilities already available world-wide are in considerable demand. We anticipate that a unique test facility as described here would become the premier location for the study of advanced moderators for slow neutron source applications, and would be able to function as a user facility in its own right, similar to those currently available at LENS, Hokkaido University, or the Lujan Center.

Acknowledgments

This material is based upon work supported by the U.S. Department of Energy, Office of Science, Office of Basic Energy Sciences, under contract number DE-AC05-00OR22725.

References

[1] Gallmeier F, Lu W, Riemer B, Zhao J, Herwig K and Robertson J 2016 *Review of Scientific Instruments* **87** 063304

[2] Remec I, Gallmeier F X, Rennich M J, McManamy T J and Lu W 2015 *In ANS MC2015 - the Proceedings of the Joint International Conference on Mathematics and Computation (M&C), Supercomputing in Nuclear Applications (SNA) and the Monte Carlo (MC) Method* vol 4 (American Nuclear Society) pp 2713–2726

[3] Iverson E B and Carpenter J M 2003 *ICANS-XVI, Proceedings of the International Collaboration on Advanced Neutron Sources, Neuss, Germany* ed Conrad H pp 707–718

[4] Baxter D V, Leung J, Kaiser H, Ansell S, Muhrer G, Iverson E B and Ferguson P D 2012 *Physics Procedia* **26** 117–123

[5] Kiyanagi Y, Ooi M, Ogawa H and Furusaka M 2003 *Journal of Neutron Research* **11** 3–11

[6] Aleksandrov A *et al.* 2008 *Proceedings of the European Particle Accelerator Conference* vol 8 URL `https://accelconf.web.cern.ch/accelconf/e08/papers/thpp074.pdf`

[7] Gallmeier F X 2013 Moderator studies for a SNS short-pulse second target station Tech. Rep. STS03-31-TR0004-R00 Oak Ridge National Laboratory

[8] Graham K F and Carpenter J M 1970 *Nuclear Instruments and Methods* **85** 163–171

IOP Conf. Series: Journal of Physics: Conf. Series **1021** (2018) 012037 doi:10.1088/1742-6596/1021/1/012037

Progress of the Scientific Commissioning of a fast neutron beamline for Chip Irradiation

Carlo Cazzaniga and Christopher D. Frost

ISIS Facility, STFC, Rutherford Appleton Laboratory, Didcot OX11 0QX, UK

carlo.cazzaniga@stfc.ac.uk

Abstract. ChipIr is a new beamline at the ISIS spallation source dedicated to accelerated testing of microelectronics. ChipIr features an atmospheric-like neutron spectrum to study the so-called Single Event Effects, a major concern for the reliability of modern electronics at ground level and flight altitudes. A bespoke design was needed to extract a fast neutron beam from a neutron source designed for thermal neutron applications.
The main objective of the Scientific Commissioning is to measure the flux, spectrum and profiles of the fast neutron beam in the different configurations of the beamline. In this work we present preliminary results of fast neutron measurements performed with a 70×70 mm^2 neutron beam and the accelerator at 700 MeV. The neutron flux above 10 MeV was found to be $4.9 \cdot 10^6$ cm^2s^{-1}. The beam profile was found to be uniform within the 70×70 mm^2 area.
ChipIr is now in an advanced stage of scientific commissioning and beta-users have started the first tests in late 2016. The first 'official' deliberate error was measured in a SRAM chip on the 29 June 2016; an 8 bit memory containing 1's had the 7th bit inverted to a 0 – just the kind of error that can occur in the atmosphere due to neutron interaction.
These results are promising for an imminent start of the ChipIr user programme.

1. Introduction

Atmospheric neutrons are a major cause of Single Event Effects (SEE), which disrupt the correct operation of microelectronics devices and systems. These neutrons are produced by high energy cosmic rays, mainly protons, interacting with the earth's atmosphere and generating cascading showers of secondary particles [1].

In recent years technology has seen a continuous drive towards miniaturization, low power, faster and ever more complex functionality. Greater vulnerability, combined with an ever greater prevalence of embedded micro-electronic devices, has meant that SEE problems have started to inevitably emerge as a serious issue for terrestrial ground based electronics [2]. The issue is even more important for avionics, because the atmospheric neutron flux is 300 times greater at typical aircraft altitudes with respect to the ground level.

The increasing need for accelerator based facilities to test SEE, particularly in Europe, has motivated a project to design and build a new fast-neutron test facility on the ISIS neutron source at the Rutherford Appleton Laboratory in the UK [3]. ChipIr, for Chip Irradiation, is an instrument designed to mimic the atmospheric neutron spectrum with a flux many orders of magnitude greater than the natural one to enable accelerated testing of electronic devices.

2. The beamline in a nutshell

The ISIS second target station is primarily used to simultaneously deliver thermal and cold neutrons (in the meV energy range) by means of hydrogenous moderators to a variety of instruments for the purpose of condensed matter science. ChipIr is the first instrument on this, or any other similar target station, to deliberately extract an atmospheric-like fast neutron spectrum (E_n up to hundreds of MeV) from such a source.

For this reason, a new section with a channel had to be inserted in the beryllium moderator for the line of view of ChipIr. In fact, the beryllium reflector is a key component of a low-energy neutron target station, being used by all the other beamlines to maximize neutron flux in the moderators.

Neutrons from this channel illuminate a secondary scatterer made with layers of different materials (the details of the scatterer design are confidential at the time of writing). The secondary scatterer is used for two reasons: 1) to optimize the hard atmospheric-like neutron spectrum and 2) to minimize the γ-ray flux by closing the line of view to the primary target.

ChipIr incorporates a collimator and jaw set to enable the beamline to operate in both 'flood' and 'pencil' beam configurations. A circular drum and a set of jaws, that can move independently, can define a beam from cm dimension (pencil) up to 0.4 m aperture (flood).

3. Neutron Measurements

The measurements of the beam profile and spectrum within the blockhouse are key for the commissioning of ChipIr. The following neutron measurements have been performed with the current configuration of the beamline, which will be probably changed in some minor details, and the accelerator running at 700 MeV. Therefore, the results have a preliminary connotation; more results will be presented in the coming year. The measurements were made at the forward position, which represent the maximum fluence rate available under the operating conditions, that is, with minimum beam attenuation due to divergence.

Fast neutron detectors included a Single Crystal Diamond Detector 2.0 × 2.0 × 0.3 mm^3, with a contact area of 1.5 × 1.5 mm^2 operated in pulse mode [4, 5], a fission chamber operated in current mode [6], and the Imaging Single-Effect Monitor (ISEEM), a neutron detector based on a CCD sensor [7]. ISEEM is used to benchmark neutron beams used for SEE testing [8].

ChipIr, as with every beamline at the ISIS source, has an independent shutter, so that its neutron beam can be closed and opened while the accelerator is running. The ChipIr shutter is also an active neutronics component, since it hosts the secondary scatterer. The shutter position therefore needs to be optimized. In Figure 1 we show the counting rate on the diamond detector, used at the centre of the beam, as a function of the shutter position. A maximum flux was found at the open position 650 mm. This result is compared in the figure to a measurement performed with the fission chamber and a good agreement is found.

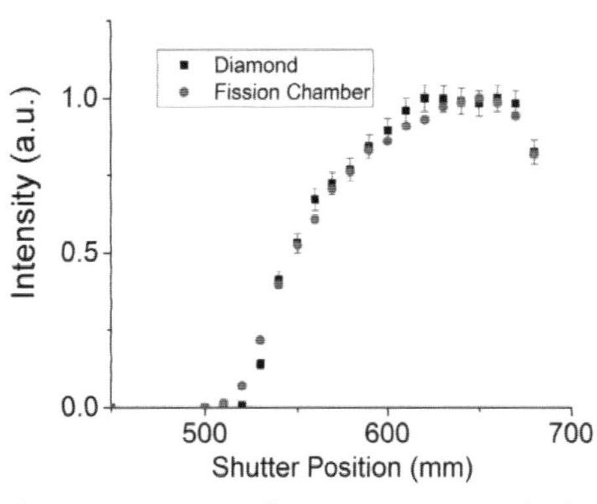

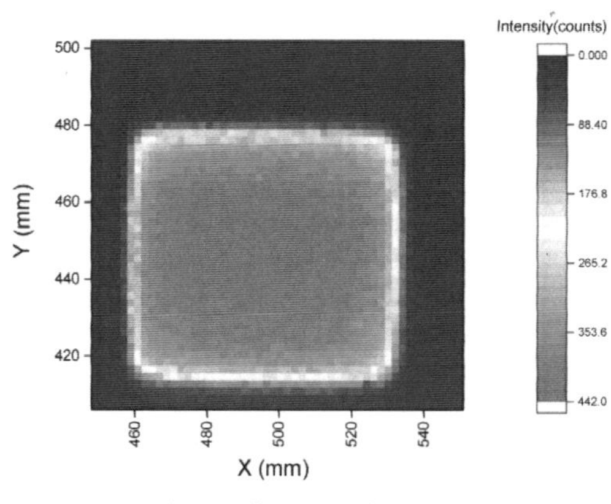

Figure 1. Neutron flux measurement during a shutter scan using a diamond detector and a fission chamber (Left). A map of the neutron beam measured with a diamond detector (Right).

One can notice a sharp increase of the neutron flux from 500 mm to 600 mm opening and a plateau from 600 to 700 mm. Differences can be due to the different size of the detectors and to different energy sensitivity as a function of neutron energy.

During this phase of the commissioning, we tested a collimator configuration that defines a beam of 70 × 70 mm^2. This dimension was chosen because it provides a typical configuration for electronic device irradiation. A map of the beam was measured using the diamond detector mounted on the XY stage. We demonstrate that this 70 × 70 mm^2 is uniform and the footprint of the beam is square. This is shown by the map in Figure 1 on the right. There is a good ratio, almost three orders of magnitude, between the centre of the beam and the penumbra.

The Imaging Single-Event effect monitor (ISEEM) was exposed to the ChipIr neutron beam at the centre of the 70 × 70 mm^2 collimated beam, and used to benchmark the beam against previous exposure at the LANSCE ICE House. This is a relative measurement with respect to LANSCE, which is a reference irradiation facility for electronics cited in JEDEC [9] and IEC [10, 11] standards. Independent measurements of the ChipIr neutron spectrum are being performed using a combination of activation foils technique and proton recoil telescope and results will be presented later in 2017.

The LANSCE-equivalent fluence rate at ChipIr was measured to be $4.9 \cdot 10^6$ cm^{-2} s^{-1} at 40 μA proton current to Target 2, reproducible to within 3%. This flux of ChipIr is about 10 times that at LANSCE operating at 1.8 μA (as in 2005). Figure 2 shows the spectrum of ChipIr compared to LANSCE and the atmospheric one increased by a factor of 10^9. Here the spectral shape comes from Monte Carlo simulations, but the intensity comes from the value measured with the ISEEM and activation foils.

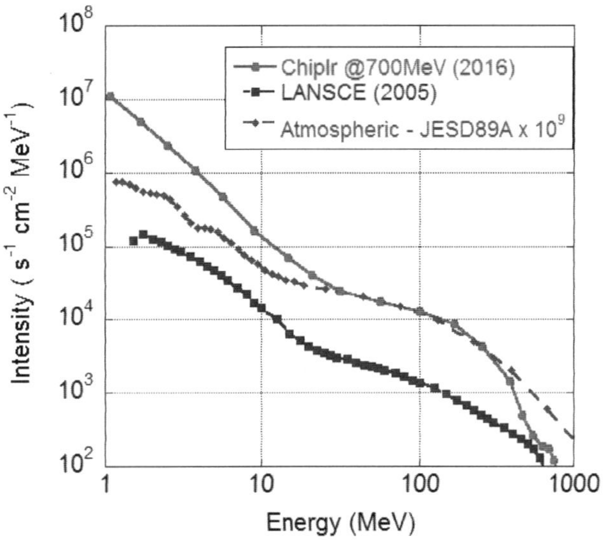

Figure 2. Neutron spectrum of ChipIr compared with the atmospheric [9] and LANSCE [12] spectra.

4. Conclusions and Outlook

In this paper we have presented first neutron measurements for the characterization of the ChipIr beamline, which is in an advanced stage of scientific commissioning. ChipIr is an instrument of the second target station of the ISIS spallation source and its neutronics has been designed to extract an atmospheric-like fast neutron beam.

IOP Conf. Series: Journal of Physics: Conf. Series **1021** (2018) 012037 doi:10.1088/1742-6596/1021/1/012037

A selected configuration was used for a preliminary characterization with the ISIS accelerator running at 700 MeV. Neutron measurements show a uniform 70 × 70 mm^2 neutron beam that can be used to irradiate microelectronics devices. An experiment was performed to compare the flux of this beam to LANSCE. The LANSCE-equivalent fluence rate was measured to be $4.9 \cdot 10^6$ cm^{-2} s^{-1} at 40 μA proton current to target 2.

This result is relevant for the forward position, which represent the maximum fluence rate available under the current operating conditions, that is, with minimum beam attenuation due to divergence. The characterization of a rearward position, designed for the flood beam, is in progress.

Moreover, other neutron measurements are needed to understand how the beamline operates in different configurations and how the neutron flux is affected. Spectroscopic measurements are being performed with the accelerator running at both 700 and 800 MeV and the results will be presented later in 2017.

References

[1] M. S. Gordon et al. "Measurement of the Flux and Energy Spectrum of Cosmic-Ray Induced Neutrons on the Ground", IEEE Trans Nucl Sci 51 (6) 3427 (2004).

[2] R. C. Baumann, "Radiation-induced soft errors in advanced semiconductor technologies" IEEE Trans. Dev. Mat. Rel 5(3) p.305 Sep. 2005

[3] C. Frost, et al. "A new dedicated neutron facility for accelerated SEE testing at the ISIS facility." *Reliability Physics Symposium, 2009 IEEE International*. IEEE, 2009.

[4] C. Cazzaniga, et al. "Characterization of the high-energy neutron beam of the PRISMA beamline using a diamond detector". *Journal of Instrumentation, 11*(07), P07012 (2016).

[5] M. Rebai, et al. "Response of a single-crystal diamond detector to fast neutrons", *Journ. of Instrum.* 8 P10007 (2013)

[6] *http://www.lndinc.com/products/category/33/*

[7] X. X. Cai and S. P. Platt, "Modeling Neutron Interactions and Charge Collection in the Imaging Single-Event Effects Monitor," in *IEEE Transactions on Nuclear Science*, vol. 58, no. 3, pp. 910-915, June 2011.

[8] S. Platt et al. "Charge-Collection and Single-Event Upset Measurements at the ISIS Neutron Source," in *IEEE Transactions on Nuclear Science*, vol. 55, no. 4, pp. 2126-2132, Aug. 2008.

[9] *Measurement and Reporting of Alpha Particle and Terrestrial Cosmic Ray-Induced Soft Errors in Semiconductor Devices*, Std. JESD89A, Oct. 2006.

[10] *Process management for avionics - Atmospheric radiation effects - Part 2: Guidelines for single event effects testing for avionics systems,* IEC 62396-2:2012, 2012.

[11] *Semiconductor devices - Mechanical and climatic test methods - Part 44: Neutron beam irradiated single event effect (SEE) test method for semiconductor devices*, IEC 60749-44:2016, 2016.

[12] B. Takala, Personal communication, 2005.

IOP Conf. Series: Journal of Physics: Conf. Series **1021** (2018) 012038 doi:10.1088/1742-6596/1021/1/012038

14 MeV neutrons for medical application: a scientific case for ^{99}Mo/^{99}Tc^{m} production

M. Capogni[1,2], A. Pietropaolo[1,*], L. Quintieri[1,2], A. Fazio[1,2], M. Pillon[1], P. De Felice[1,2] and A. Pizzuto[1]

1 ENEA Department of Fusion and Technologies for Nuclear Safety and Security, Via Enrico Fermi 45, I-00044 Frascati (Roma) Italy

2 National Institute for Meytrology of ionizing Radiation-ENEA, Via Anguillarese 301, Roma, Italy.

*__Corresponding author__: antonino.pietropaolo@enea.it

Abstract. ^{99}Tc^{m} is a widely used radionuclide for SPECT (Single Photon Emission Computed Tomography) diagnostics thanks to its short half-life (about 6h) and the low-energy gamma ray emission (140 keV) well suited for diagnostic devices. The specific method for producing ^{99}Tc^{m} via ^{100}Mo(n,2n)^{99}Mo reaction, as it was investigated in ENEA, is discussed in detail in this paper. The ^{99}Mo activity achieved by means of 14 MeV neutron irradiation on natural Molybdenum sample irradiated at the Frascati Neutron Generator (FNG) facility, at the Research Centre of ENEA-Frascati, was accurately assessed, by tracing it to the activity standards provided by the Italian National Institute of Ionizing Radiation Metrology (INMRI), located at the Research Centre of ENEA-Casaccia. The whole experiment carried out in ENEA was supported by simulations performed with the Fluka Monte Carlo code, whose predictions have been benchmarked against the experimental data collected at ENEA-FNG, relying on the traceability to the activity standards developed and maintained at the ENEA-INMRI Radioactivity laboratories.

1. Introduction

Metastable Technetium-99 (^{99m}Tc) is the principal radionuclide used worldwide in medical diagnostics with more than 30 million procedures per year, accounting for more than 80% of all Nuclear Medicine diagnostics [1,2]. The short half-life of ^{99}Tc^{m} ($T_{1/2}$ = 6.0067 (10) h [3]) and the related 140 keV gamma-ray make it well suited to medical imaging, using conventional gamma cameras such as Single Photon Emission Computed Tomography (SPECT). ^{99}Tc^{m} can be obtained from the decay of Molybdenum-99 (^{99}Mo) ($T_{1/2}$ = 2.7479 (6) d [4]). The known different technologies for producing ^{99}Mo/^{99}Tc^{m} are based on: a) HEU (Highly Enriched Uranium) and LEU (Low Enriched Uranium) targets in fission reactors; b) ^{98}Mo thermal neutron activation in nuclear reactors; c) direct cyclotron production; d) photo-fission reactions on ^{238}U; e) ^{99}Mo photo-production and f) fast neutron-induced reactions. Presently the almost complete supply of ^{99}Mo is provided by method a).

In 2010, the global demand for ^{99}Mo was estimated to be about 1.2×10^{4} 6-day Curie (6-day Ci) per week (444 TBq per week) and it is continuously growing. Global shortage of ^{99}Tc^{m} emerged in the late 2000s because the two main reactors (NRU in Canada and the High Flux Reactor-HFR in The

Netherlands), capable to provide about two-thirds of the world's supply of ^{99}Mo, were shut down repeatedly for extended maintenance periods. These events highlighted vulnerabilities in the classical supply chain of medical radionuclides based on nuclear reactors. At present, just a few nuclear reactors around the world can provide ^{99}Mo. Many of these are old and soon will stop their medical isotopes production, consequently leaving the world vulnerable to the shortage of this and several other radiopharmaceuticals. A complete landscape of the worldwide ^{99}Mo production from reactors is provided in reference [5]. Table 1 shows a list of the main production sites and their main features.

Table I: *Main characteristics of worldwide reactors used for ^{99}Mo production.*

Site	Location	operation days	Weekly Nominal production (6-day Ci)	Weekly % of world demand	Fuel/Target	First Commissioning
BR-2	Belgium	140	5200	25-65	HEU/HEU	1961
HFR	The Netherlands	300	4880	35-70	LEU/HEU	1961
MARIA	Poland	-	700-1500	-	HEU/HEU	1974
LVR-15	Czech Republic	-	> 600	-	HEU^e/HEU	1957
NRU	Canada	300	4680	35-70	LEU/HEU	1957
OPAL	Australia	290	1000-1500	-	LEU/LEU	2006
OSIRIS	France	180	1200	10-20	LEU/HEU	1966
SAFARI-1	South Africa	305	2500	10-30	LEU/HEU	1965
RA-3	Argentina	230	240	<2	LEU/LEU	1967

1.1. Feasibility study of ^{99}Mo production at the ENEA-FNG and perspectives on the New Sorgentina Fusion Source (NSFS).

^{100}Mo(n,2n)^{99}Mo reaction with 14 MeV neutrons [6] has been identified as a feasible and effective alternative method to the nuclear fission reactors for the ^{99}Tc^{m} production. Indeed, the ^{100}Mo(n,2n)^{99}Mo cross section is appreciable and exhibits a maximum around 14 MeV. This approach was tested in 2015 from a Japanese group, which provided a first complete assessment of ^{99}Tc^{m} production and the characterization of the radiochemical chain to produce the radiopharmaceuticals by 14 MeV neutron beam. The Frascati Neutron Generator (FNG) in ENEA is an accelerator driven continuous neutron source [7]. It relies on Deuteron-Tritium (D-T) fusion reactions, producing almost monochromatic 14 MeV neutrons with a maximum neutron emission rate of 10^{11} s^{-1}. A natural Molybdenum powder sample (density 1.94 g/cm^3) was irradiated at FNG for about 15 minutes with a FNG neutron emission rate of 2.89×10^{10} s^{-1}. The obtained results show that a ^{99}Mo specific activity of about (2.32 ± 0.05) kBq g^{-1} [8] at the reference time was produced, in good agreement (within 7%) with the Monte Carlo (MC) predictions, based on the Fluka code (version 2011.2c.3). In this context, the measurements performed at ENEA have to be considered as a mandatory step to build and assess a valuable procedure that allows computational and experimental verification of the ^{99}Mo production using 14 MeV neutron beams, in a controlled and reliable way. Indeed, it is important to highlight the peculiarities of the applied procedure: (1) use of an well-defined energy neutron source, determined by the D+T reaction, with a well-defined emission rate, determined by the absolute technique of the associated alpha particle; (2) MC calculations to optimize the experimental set-up and irradiation; (3) accuracy and traceability at high metrological level of the ^{99}Mo and ^{99}Tc^{m} activity measurements and MC benchmarking. Indeed, the activity measurements of ^{99}Mo and ^{99m}Tc were carried out by using a high-energy resolution HPGe spectrometer owned by the Italian National Institute of Ionizing Radiation Metrology (ENEA-INMRI), which was previously calibrated with long-lived multi-gamma-emitting radionuclides (such as ^{152}Eu) and with the ^{99m}Tc primary standard. In this way all the performed activity measurements are traceable to the activity standards maintained in ENEA-INMRI.

The results obtained at ENEA-FNG, after a proper normalization in order to take into account the different experimental conditions, are coherently comparable to those published by the Japanese group in reference [4], providing a consistent and reliable estimation of the measured specific activities. Indeed, properly considering the experimental conditions of the two experiments the specific activity obtained in Ref. [4] and at FNG are the same within experimental errors.

New Sorgentina Fusion Source (NSFS) [9,10] is a project of an intense double rotating targets 14 MeV neutron source, based on D-T fission reactions driven by deuteron accelerators and featuring a yield about 5×10^4 times higher than FNG, reaching in such a way an emission rate of about $4\text{-}5\times10^{15}$ s^1. NSFS main features make this facility a potential and interesting infrastructure for an extensive radiopharmaceutical production. In this framework, FNG can be considered as a useful laboratory in the European context to test and validate the NSFS's capabilities and performances. Indeed, a preliminary methodological approach has been already assessed involving both experimental measurements and Monte Carlo (MC) computations, as described in detail in the following.

1.2. Monte Carlo predictions for NSFS

Accurate and realistic predictions of the ^{99}Mo activity producible at NSFS by irradiating ^{100}Mo, have been obtained by MC simulations, taking properly into account the geometry of possible and "optimized" targets (w.r.t thermal issues) and identifying reasonable irradiation profiles (i.e. neutron rates and a realistic duty cycle'). A conservative approach has been adopted to estimate the maximum obtainable ^{99}Mo activity at NSFS. Supposing an operative neutron emission rate of 10^{15} s^{-1} and identifying an optimum irradiation time of 22 hours (coming mainly from thermo-physical considerations on the melting Molybdenum temperature but matching also the decay time curve of the transient equilibrium between ^{99}Mo, precursor, and ^{99}Tc^{m}, daughter radionuclide), two experimental scenarios can be established as delimiting the predictable NSFS ^{99}Mo production capabilities:

(a) An "optimistic" scenario that fixes the upper limit of ^{99}Mo weekly activity that could be provided by NSFS (about 75 TBq). This has been obtained by supposing to fill the irradiation volume between the two NSFS target wheels (0.5 liters) with a solid ^{100}Mo target (corresponding to almost 6 kg) and irradiating it continuously for 22 h with a neutron emission rate of 10^{15} s^{-1}. Anyway, the evaluated total power deposited in the target in this configuration would cause the maximum temperature of the target to exceed the Molybdenum melting point temperature (2900 K). This result was obtained assuming conservatively adiabatic heat transfer boundary conditions on the bulk target external surface.

(b) A "pessimistic" scenario that fixes the lower limit of ^{99}Mo weekly activity that could be provided by NSFS (about 40 TBq). This assumes 3 kg of ^{100}Mo (i.e. half of the volume of the previous case) irradiated for 22 h with a neutron emission rate of 10^{15} s^{-1}. With this constraint, it has been found the maximum admissible volume of a ^{100}Mo target that could be irradiated continuously for 22 h, without melting, that is a plate with the same planar surfaces of the previous case but thickness one half. Thus, the range 40-75 TBq can be referred as a reliable range, that confidently includes the designed performances of NSFS.

In a more realistic target layout (made of multiple piled ^{100}Mo sheets, properly dimensioned and distributed over the irradiation volume, e.g. each having 1 mm thickness) the weekly activity predicted by MC simulations, as expected, lies within the envisaged NSFS working range.

1.3. ^{99}Mo production: FoM definition

Referring to Table 1 and considering the nominal thermal power of the different reactors, a Figure of Merit (*FoM*) can be defined as $FoM = \frac{A_W}{NP_{th}}$, A_W being the weekly activity of ^{99}Mo (measured in Bq), and NP_{th} (measured in MW) is the facility nominal power that corresponds, respectively, to the nominal thermal power for reactors and the delivered beam power for accelerator driven facilities. The nominal thermal power and the delivered beam power are the specific parameters representative of the facility performances with respect to the achievable ^{99}Mo activity. The *FoM*, as defined above, can be assumed as an efficiency parameter for the specific production facility. Figure 2 reports the *FoMs* for the main reactors in Table 1 as compared to NSFS for which NP_{th} is 16 MW and $25 < A_W < 75$ TBq.

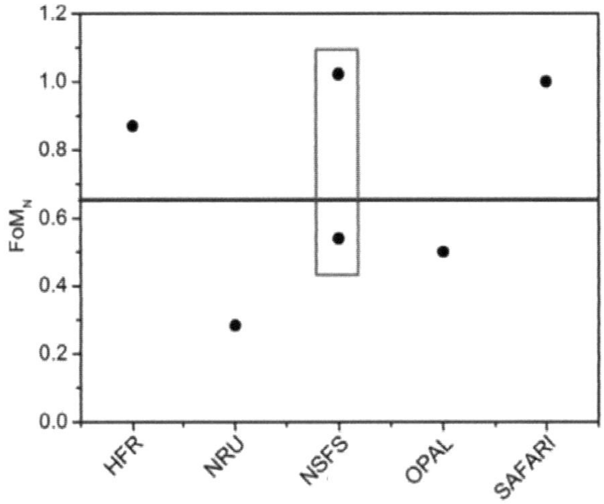

Figure 1: *Normalized FoM (FoM_N) for the main fission reactors and NSFS, the normalization value being the FoM of SAFARI fission reactor (see Table 1). The continuous blue line marks the average FoM over the whole set of fission reactors in the plot. The rectangle embeds the minimum and maximum value of the FoM for NSFS (see text for details).*

2. Conclusions and perspectives

The production of ^{99}Tc^{m} using methods alternative to nuclear fission reactors was identified as a social-driven issue to be addressed after the unexpected crisis in 2009. In this context, the use of 14 MeV neutrons to produce ^{99}Tc^{m} exploiting ^{100}Mo(n,2n)^{99}Mo reaction is clearly identified as a valuable solution, if really intense sources to feed radiopharmaceutical production were available. An experimental assessment of this approach was carried out in ENEA using a well-defined characterization procedure, based on accurately calibrated 14 MeV neutron beam delivered by FNG, MC simulations and high level traceability of the activity measurements to the national radioactivity primary standards at ENEAS-INMRI. The achieved results and the benchmarked predictive models allow for making a reliable extrapolation of the performances of New Sorgentina Fusion Source (NSFS) for the production of ^{99}Mo, providing a comparison with the radiopharmaceutical production classical chain, based on nuclear fission reactors, in terms of a proper defined figure of merit (*FoM*).

It is found that the FoM of NSFS is appreciably close to that one of the main (but aged) nuclear fission reactors in the global landscape which cover a quite large fraction of the present ^{99}Tc^{m} market.

Acknowledgements

Authors warmly acknowledge Prof. A. Duatti (University of Ferrara) for stimulating and precious discussion on the production and use of ^{99}Tc^{m} radiopharmaceutical.

References

[1] European Observatory on the supply of medical radioisotopes, Working Group 4 (WG4), Capacity and Infrastructure Development, July 2014
[2] Future Supply of Medical Radioisotopes for the UK, Report prepared by British Nuclear Medicine Society and Science & Technologies Facilities Council, December 2014.

[3] http://www.nucleide.org/DDEP_WG/Nuclides/Tc-99m_tables.pdf
[4] http://www.nucleide.org/DDEP_WG/Nuclides/Mo-99_tables.pdf
[5] NEA-OECD. Medical isotope supply in the future: Production capacity and demand forecast for the 99Mo/99mTc market, 2015-2020. NEA Report NEA/SEN/HLGMR(2014)2, Nuclear Energy Agency, Issy-les-Moulineaux, France (2014)
[6] Y. Nagai, EPJ Web of Conferences **66**, 10007 (2014).
[7] http://www.fusione.enea.it/LABORATORIES/Tec/FNG.html.en
[8] M. Capogni, P. De Felice, A. Fazio, L. Quintieri, A. Pietropaolo, M. Pillon and A. Pizzuto, ^{99m}Tc and ^{99}Mo produced at ENEA-FNG Facility of 14 MeV neutrons, submitted to the Scientific Secretariat of ICRM 2017 International Conference hosted by the National Energy Atomic Commission (CNEA) in Buenos Aires-Argentina on May 15th-18th 2017 and accepted as poster n. 110.
[9] M. Pillon, M. Angelone, A. Pietropaolo, A. Pizzuto, Fus. Eng. Des. **89**, 2141 (2014).
[10] P. Console Camprini et al. Fus. Eng. Des. **96-97**, 236 (2015).

IOP Conf. Series: Journal of Physics: Conf. Series **1021** (2018) 012039 doi:10.1088/1742-6596/1021/1/012039

A tale of two foils: ISIS TS-1 water moderators

Goran Škoro, Robert Bewley, Steven Lilley, Russell Ewings, Giovanni Romanelli, Matthias Gutmann, Ron Smith, Svemir Rudić and Stuart Ansell[1]

ISIS Neutron and Muon Source, Science and Technology Facilities Council, Rutherford Appleton Laboratory, Didcot OX11 0QX, United Kingdom

E-mail: goran.skoro@stfc.ac.uk

Abstract. Currently there are two, decoupled and poisoned, water moderators at the ISIS Target Station 1 (TS-1), located above the target, operating at room temperature and providing neutrons for the experiments in the fields of spectroscopy, diffraction, etc. In this paper, the details about neutronics model of ISIS Target Station 1, simulation results related to TS-1 water moderators and comparison with corresponding measurement results are presented. The importance of the quality neutronics calculations for planning the upgrade/changes in the target station (as in removal of the poisoning foil in upstream water moderator) and locating the operational issues (disappearance of poisoning foil in downstream water moderator) is discussed.

1. Introduction

The cross-sections of MCNPX [1] model of current ISIS [2] Target Station 1 (TS-1) target, reflector and moderators (TRAM) assembly are shown in Figure 1. The target consists of tantalum cladded tungsten rectangular plates. The target pressure vessel is made of stainless steel and the target coolant is a mixture of heavy water (82%) and light water (18%) [3]. Proton (800 MeV) beam orientation is from the left to right (vertical cut plot), from the bottom upwards (horizontal cut plots). The reflector is made of water-cooled beryllium rods (packing fraction by volume = 90% beryllium : 10% coolant). The reflector coolant is a mixture of heavy water (82%) and light water (18%) [3].

Currently there are four moderators at the TS-1. Below the target there are: coupled liquid hydrogen moderator and poisoned, decoupled liquid methane moderator. Two, identical in size, decoupled and poisoned water moderators operating at room temperature are located above the target (see Figure 1). The upstream water moderator has, asymmetrically positioned, poison layer (foil). The 'active' poison (gadolinium) layer, with thickness of 0.005 cm, is sandwiched between two aluminium layers (total thickness of aluminium layers is 1 mm). The downstream moderator (so-called Merlin) is centrally poisoned with a single gadolinium foil (layer). These two moderators (the main actors in the following story) serve 8 instruments at ISIS Target Station 1 (see Figure 2) and provide neutrons for the experiments in the fields of spectroscopy (MAPS, MERLIN, TOSCA, VESUVIO), diffraction (SXD, POLARIS), etc.

[1] Present address: Neutron Optics & Shielding Group, European Spallation Source, Lund, Sweden.

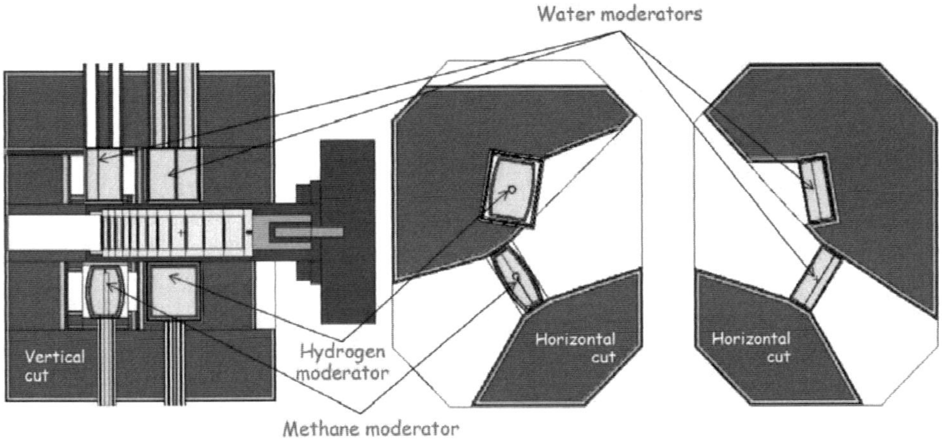

Figure 1. MCNPX model of current ISIS Target Station 1 TRAM assembly.

2. The upstream water moderator

At the beginning of 2016, the design of the upstream water moderator has been changed to what we have now (see Figure 1). The previous ("old") upstream water moderator had two gadolinium poisoning foils, which partitioned the moderator into three approximately equal volumes. Then, the decision has been made to install a new moderator where one (the South side) gadolinium foil is removed to give the corresponding instruments an increase in neutrons flux.

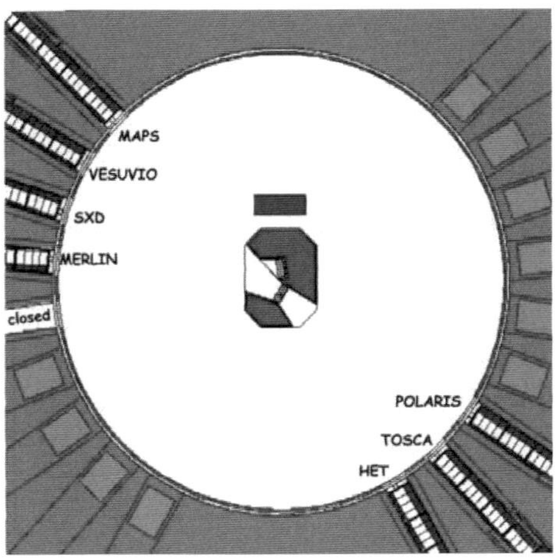

Figure 2. MCNPX model of current ISIS Target Station 1. The upstream water moderator serves 7 beamlines: MAPS, VESUVIO and SXD on the South side; POLARIS, TOSCA, INES (downstream of TOSCA) and HET on the North side. The downstream water moderator serves MERLIN instrument.

The planned change of design of the TS-1 upstream water moderator was successfully completed in February 2016. The neutronics simulations performed beforehand predicted a factor of 2 increase in neutron intensity for the South side instruments (see Figure 3) and no changes for the North side instruments. The measurements performed in November/December 2015 with old moderator and in April/May 2016 with new moderator confirmed that there are no changes in intensity and pulse time profiles for the North side instruments (POLARIS, TOSCA). All South side instruments, however, measured the gain which is significantly $(10-20\%)$ higher than expected (see Figure 3).

The reason for such an unusually big difference between experimental and simulation results was not hard to find. Calculations of expected gain factors have been performed assuming two

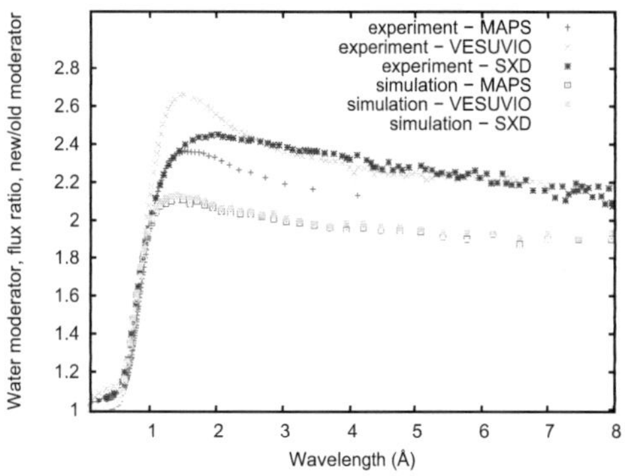

Figure 3. Predicted and measured flux gain on South side instruments after old water moderator was replaced with a new one.

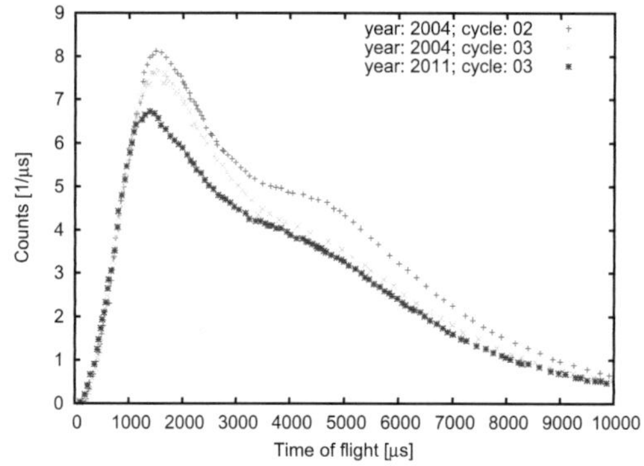

Figure 4. Neutron flux coming from old water moderator measured at MAPS instrument sample position (12 m distance) for different cycles.

"perfect" moderators. However, old moderator was not "perfect". It was installed in 2004 and after a first cycle its performance started to deteriorate. Figure 4 shows the sample of the history of neutron flux measurements at MAPS instrument. The loss of neutrons, at the level of $10-20\%$, coming from the (old) water moderator in period between 2004 and 2011 is clearly visible. If we take this into account, the agreement between predicted and measured flux gain is very good. However, the question about the reason(s) for such a change of old water moderator performance is still without the answer. Several possible scenarios have been proposed: from simple one (displacement of the south side poison foil) to very unusual (extraordinarily high production rate of ^{22}Na in moderator can). None of them has been convincing enough so the "mystery" of old water moderator flux loss is still under investigation. In the meantime, it is important to regularly monitor current (new) upstream moderator to see if there is any sign of such a strange behaviour.

3. The downstream water moderator (Merlin)

According to engineering drawings, downstream water moderator (so-called Merlin) is centrally poisoned with a single gadolinium layer (its thickness is 0.005 cm). The 'active' poison layer is sandwiched between two aluminium layers. The moderator is surrounded (except the viewing side and top side) with, 0.65 cm thick, boral layers. The original (old) Merlin moderator has been installed in, approximately, same period as the old upstream moderator (in 2004). Before installation, different tests were performed that confirmed that there is a strong neutron absorbing material (gadolinium) inside. Merlin moderator was in the TS-1 TRAM for almost 2 years before the MERLIN instrument [4] was ready for operation. Once the measurements started it was found that time resolution of MERLIN instrument was much broader than originally planned.

Current, very detailed, neutronics model [5, 6] of ISIS Target Station 1 provided the opportunity to investigate this issue. If the poison layer is removed from the model there is an excellent agreement (see Figure 5) between experimental and calculated time resolution[2] values [7]. This result was a strong indication that an effective poison layer is not in the moderator.

[2] In Figure 5 (and 6) apparently the tail of the pulse of 29 meV neutrons arrives before the 'main' pulse. This, at the first sight, 'unusual' behaviour can be explained if we keep in mind that the last chopper within the MERLIN instrument acts like a pinhole camera giving 'inverted' time-of-flight image.

This assumption has been confirmed in September 2016. New Merlin moderator (where presence of poison foil was experimentally confirmed beforehand) has been installed and measured flux and time profiles clearly show excellent agreement with simulation scenario with poison foil in (see Figure 6). This means that, for whatever reason, there was no poisoning in the old Merlin moderator [8]. On the positive side the MERLIN instrument will now have better line shapes and resolution though the disappearance of the gadolinium needs further investigation. One of possible scenarios is that gadolinium dissolved in water (because of potential issues with welding of the aluminium sandwich around) in the years before MERLIN instrument started up. If this is true it could have serious implications for the design of the water moderators within ISIS TS-1 project.

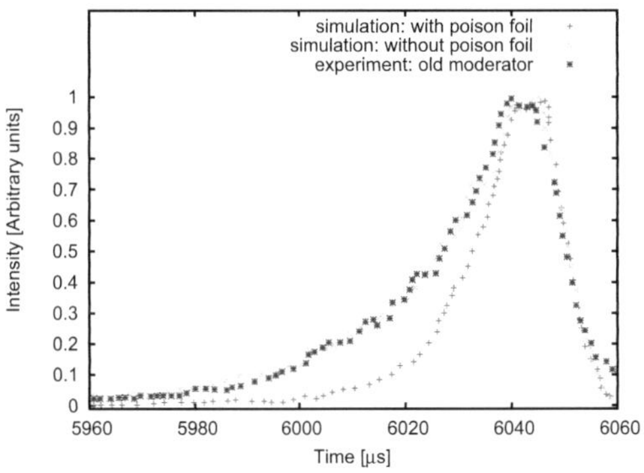

Figure 5. Simulated and measured time profile of 29 meV neutrons at MERLIN instrument with old (original) moderator.

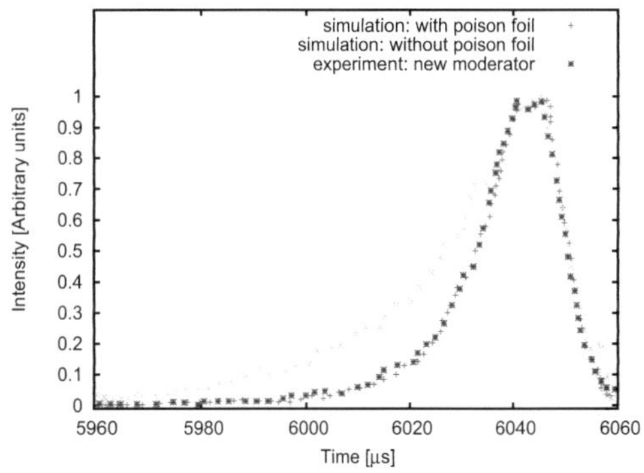

Figure 6. Simulated and measured time profile of 29 meV neutrons at MERLIN instrument with new moderator.

4. Conclusions

This tale of strange behaviour of ISIS TS-1 water moderators clearly shows the importance of the quality neutronics calculations for planning the upgrade/changes in the target station (as in removal of the poisoning foil in upstream water moderator) and locating the operational issues (disappearance of poisoning foil in downstream water moderator).

References

[1] MCNPX 2.7.0 - Monte Carlo N-Particle Transport Code System for Multiparticle and High Energy Applications, https://mcnpx.lanl.gov/.
[2] http://www.isis.stfc.ac.uk/.
[3] Jura M 2015 Determination of the D2O purity from the TS1 reflector cooling circuit *ISIS Internal Report.*
[4] https://www.isis.stfc.ac.uk/Pages/Merlin.aspx.
[5] Škoro G and Ansell S 2016 Proceedings of ICANS XXI Meeting, Mito, Japan, 2014; *JAEA-Conf 2015-002* p. 141-147.
[6] Škoro G et al. 2016 *Internal Report ISIS-TS1-UPG-NEU-Rp-0001.*
[7] Bewley R 2016 TS-1 baseline: experiment vs. simulatons *Talk at 4th ISIS TRAMSNEG Meeting.*
[8] Bewley R 2016 New MERLIN moderator: results *Talk at 5th ISIS TRAMSNEG Meeting.*

IOP Conf. Series: Journal of Physics: Conf. Series **1021** (2018) 012040 doi:10.1088/1742-6596/1021/1/012040

RF neutron spin flippers in time of flight Spin-Echo Resolved Grazing Incidence Scattering (SERGIS)

S.R.Parnell

Faculty of Applied Sciences, Delft University of Technology, Mekelweg 15, 2629 JB Delft, The Netherlands

R.M.Dalgliesh, N.J.Steinke

ISIS Neutron Source, Rutherford Appleton Laboratory, Chilton, Didcot, Oxfordshire, OX11 0QX, United Kingdom

J.Plomp, A.A. Van Well

Faculty of Applied Sciences, Delft University of Technology, Mekelweg 15, 2629 JB Delft, The Netherlands

Abstract. RF coils have been routinely used for Larmor labelling on the Offspec instrument at ISIS. These coils encode directional information via the neutron polarization using a series of parallelogram-shaped pole shoes which can be tuned to different angles with an RF gradient flipper in the centre of each magnet. We report on measurements of the magnetic field integral through the coils in reflection geometry for a range of scattering angles and different pole shoe angles. Such information is mapped out by measuring the phase of the neutron polarisation and the measurements are discussed in light of data on patterned silicon gratings with a dewetted polymer and the visibility of in-plane structures to SERGIS.

1. Introduction

In typical neutron scattering experiments information on the sample under investigation is obtained via measurement of the momentum transfer (Q) and/or the energy change of the neutron. Efforts to increase the resolution both in scattering angle and energy often result in tight beam collimation which is wasteful in terms of the usable neutron flux. Alternative techniques have been around for several decades now [1, 2, 3] which utilise magnetic fields to manipulate the neutron spin by encoding either the scattering angle or the energy transfer in the neutron polarisation. Scattering angle encoding involves shaped magnetic fields such that the accumulated Larmor precession is dependent upon the spatial trajectory of the neutron. A number of different technologies have been developed for such spatial encoding such as magnetised thin films [4], resistive shaped magnets [5, 6], superconducting coils [7, 8, 9] and RF coils which can be either tilted [2] or utilise shaped pole pieces [10].

For scattering angle encoding applications in SANS the Spin-Echo Small Angle Neutron Scattering (SESANS) [3] technique has been applied to measure 100's of nm's to micron size structures in for example soft matter [11], granular media [12] and food [13]. The high resolution

IOP Conf. Series: Journal of Physics: Conf. Series **1021** (2018) 012040 doi:10.1088/1742-6596/1021/1/012040

obtained in Larmor diffraction [14] allows changes in lattice spacings $\left(\frac{\Delta d}{d}\right)$ of the order 10^{-6} to be measured [15] .

However the Larmor labelling analogue in reflectometry, Spin-Echo Resolved Grazing Incidence Scattering (SERGIS) [16, 17] has only, with two exceptions, been successful in measuring well-defined periodic structures. For example, silicon gratings with ridges parallel to the neutron beam [18, 19]. In the two cases of non-periodic structures, Parnell *et al.* [20] looked at the formation of crystallites in a solar cell material on a silicon layer, showing that these were detectable after annealing and verified this with microscopy. This was performed on the instrument Offspec at ISIS [10]. At ILL on the Eva instrument Vorobiev *et al.* [21] looked at a dewetted polymer on a silicon substrate. They observed features in the SERGIS signal, which were attributed to in-plane correlations between the dewetted polymer islands and agreed well with microscopy measurements. They were able to obtain good signal to noise by measuring in the Yoneda region which enhances the scattering by a factor of four [22]. However, attempts to perform such measurements at ISIS on the Offspec instrument have thus far been unsuccessful.

One of the intrinsic problems of the SERGIS technique is the following. Analogously to spin-echo SANS (SESANS), where the spin-echo (SE) signal from the in-plane structures is hidden in the direct beam intensity, the SE-signal in SERGIS is superimposed on the specular intensity and the measured polarisation (P_{se}) is dominated by the reflected intensity from the specular component (I_r) and the diffuse component (I_d), given by $P_{se} = (I_d P_d + I_r P_0)/(I_d + I_r)$ with P_0 the polarisation of the reference beams and P_d the diffuse beams.

In a SESANS experiment the SE-signal can be enhanced by using thicker samples. Unfortunately, in SERGIS this enhancement is limited since the scattering volume in films is by definition small. However, the relative SE-signal in the off-specular scattering might be larger. But here we run into another problem. Whereas the SERGIS technique yields information of in-plane structures on the tens or hundreds of nm scale, detectable off-specular intensity occurs if there is in-plane structure on the micron scale, limiting the technique to samples exhibiting inhomogeneities on a broad range of length scales. A remedy to cope with these limitations could be by making a substrate having artificial in-plane structure on the micron scale, i.e. a grating and deposit the system under study onto this substrate. This creates scattering, however being on the micron range this should not add confinement effects or any perturbation upon structures on the order of the 10 nm range.

Spin-echo techniques rely on the exact cancelling of the field integral in the first arm by that of the second arm. Aberrations as a result of the different trajectories present in the experiment will lead to precession phase errors. The structure of the remainder of these proceedings is as follows. We first define a procedure to map out the phase as a function of scattering angle. Secondly we use this known phase in the determination of the polarisation (P_{se}) from a bare structured grating and also one onto which a dewetted polymer has been placed.

2. RF Coil layout and configuration for SERGIS

The schematic of figure 1 (a) is a generalised version of scattering angle encoding. The shaped field regions perform the angle encoding and are arranged with opposite magnetic field, in the case of an unscattered neutron it traverses both magnetic fields and the Larmor phase added in the first magnet is cancelled by the second magnet for all possible angles ϕ_i. Whereas if it is scattered the resulting path difference results in a different Larmor phase which varies with the strength of the magnet and the degree of scattering is measured in the change in P_{se}. As discussed in [10] the same Lamor phase encoding can be achieved using a series parallelogram-shaped precession regions with flippers in the centre of each magnet. Such a setup is shown in 1 (c) both top and side view. The scattering geometry (defined in figure 1 (b)) is such that the SERGIS encoding is along the Q_y direction, as reflectometry use a letter box shaped beam then there is normally no resolution along this direction and it only via SERGIS that this resolution

IOP Conf. Series: Journal of Physics: Conf. Series **1021** (2018) 012040 doi:10.1088/1742-6596/1021/1/012040

is obtained.

As illustrated in the side view in figure 1 (c) the path through the magnetic fields along the scattered arm vary with angle (θ_f). Hence the field integral changes. In order to correct for these changes we have developed a procedure which is detailed in the following section. As the length scales δ_{se} (termed the spin echo length) probed vary with the neutron wavelength (λ) and the magnet angle (θ_0) as $\delta_{se} = c\lambda^2 BLcot\theta_0/(2\pi)$ where B is the magnetic field strength, L is the separation between the magnets (see figure 1(c)) and $c = \gamma m/h = 4.632 \times 10^{14} T^{-1} m^{-2}$). Then it is necessary to correct for these integral changes as a function of λ, θ_0 and θ_f.

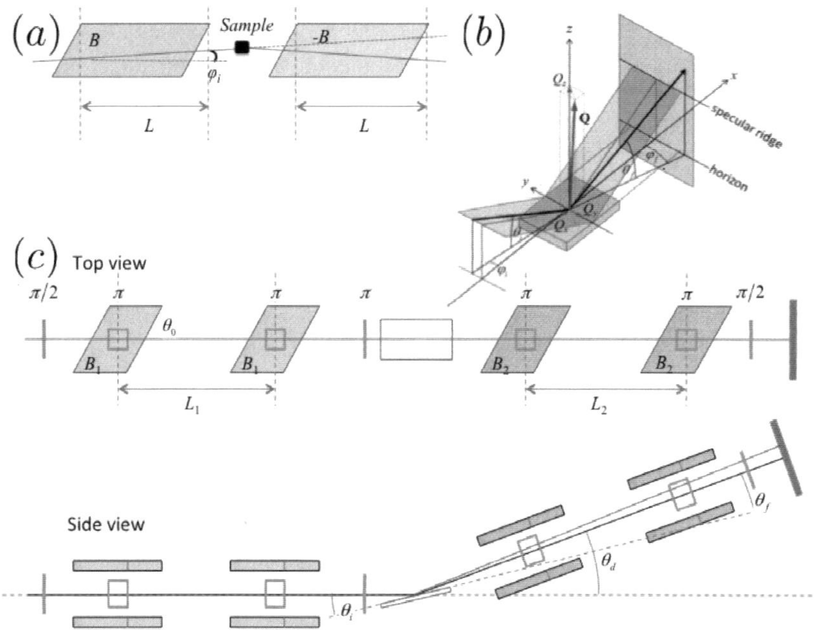

Figure 1. Schematic drawing showing an idealised case of scattering angle encoding in (a). In (b) the reflection geometry, illustrating the reference Cartesian co-ordinate system, the scattering vector Q and its coordinates as well as the angles θ and ϕ and (c) the layout of the spin encoding coils. In all cases the shaded regions indicate the magnetic field regions with changes from blue to red reflecting opposite magnetic field directions.

3. Measurement of the field integral as a function of scattering angle

To correct for such effects it is necessary to measure a series of polarisation echos on the detector and then to measure these over a range of θ_f values. This is the same procedure that was first used by Farago [23] for inelastic time of flight spin echo. Note that in order to avoid ambiguity we refer to the polarisation $P(\lambda)$ as the polarisation measured at a particular point whilst the echo polarisation, $P_0(\lambda)$is the maximum achievable polarisation when the echo is balanced.

The experiments were performed on the Offspec instrument at the ISIS Pulsed Neutron Source, UK. In these measurements we used a non-polarising supermirror (m=5) and measured the polarisation at different glancing angles ϕ for a fixed detector angle θ_d. The spin-echo polarisation for specular reflection occurs at an angle $\theta_f = 2\phi - \theta_i$, where θ_i is the incident angle and ϕ the glancing angle given by;

$$P(\lambda) = P_0(\lambda)cos\left(\frac{\gamma m}{h}\lambda\left(2B_1L_1 - 2B_2L_2\right)\right) \quad (1)$$

with $2BL$ being the effective magnetic field integral from the arms before and after the sample (as denoted by the subscripts), with the factor of 2 occurring due to the use of RF. This can be parameterised in terms of a current applied to the correction coil (I_{scan}) and we denote the current at which the echo is balanced in the straight through beam as I_0 and an offset current which varies as a function of reflection angle (θ_f), hence this becomes;

$$P\left(\lambda, I_{Scan}, \theta\right) = P_0(\lambda) cos\left(c\lambda\left(I_{scan} - I_0 - \Delta_I(\theta_f)\right)\right) \tag{2}$$

From the contour plots in figure 2 one period for $\lambda = 0.68nm$ corresponds to a change in current (ΔI_{scan}) of 0.85 A. Hence as $\gamma m/h\lambda Bl = 2\pi$ (i.e. one rotation) then the change in field integral of 0.85A is $\Delta(BL) = 2 \times 10^{-5}Tm$, hence a change in the scan current of 1 A corresponds to $\Delta(BL) = 2.4 \times 10^{-5}Tm$.

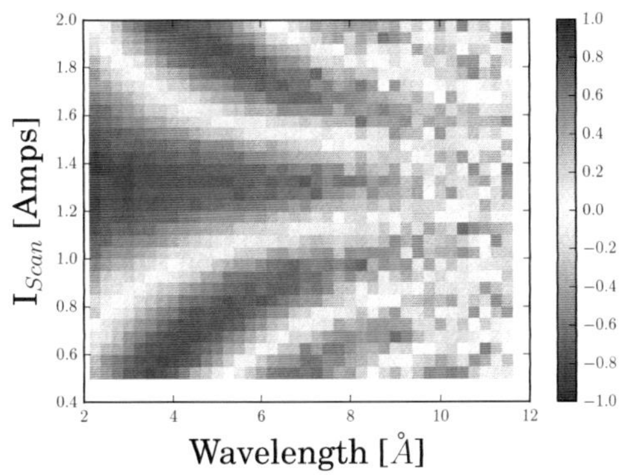

Figure 2. False colour plot of the polarisation as a function of wavelength for different values of the correction coil applied current I_{scan} for the m=5 supermirror at a glancing angle of 0.35° and a pole shoe angle of 50° with the side bar indicating the value of polarisation.

In order to extract the echo polarisation we fit for each wavelength (0.15 Åwide bins) to equation 2. This allows the determination of the echo polarisation as a function of reflection angle. From figure 2 we determine that the echo current is independent of wavelength at 1.3A, which is to be expected and hence we can parameterise this for all wavelengths, we can also quantify this offset for various pole shoe angles (see figure 3). In summary these measurements show that the phase varies in a predictable manner and can be parameterised.

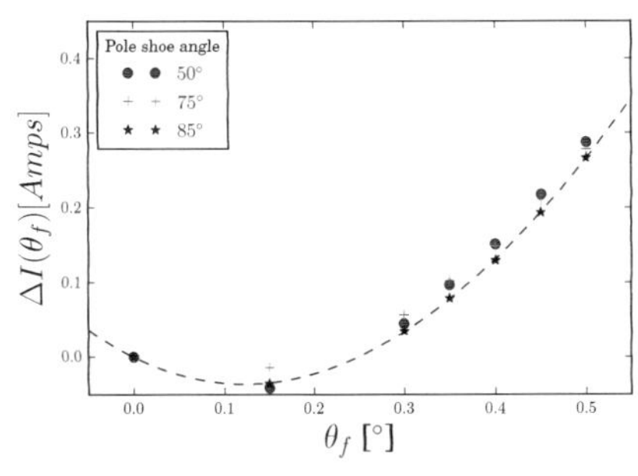

Figure 3. Field integral offset current ($\Delta_I(\theta_f)$) for measurements with the m=5 supermirror as a function of reflection angle.

4. Field integral correction in SERGIS example

The sample under investigation was a block co-polymer made of deuterated polystyrene and paramethylstyrene [24]. Grazing Incidence Small Angle Neutron Scattering (GISANS) data showed a peak corresponding between the distance between the droplets. In the GISANS there was also an additional peak which is attributed to the internal structure of the droplets. These samples were deposited onto silicon gratings with grooves $40\mu m$ wide (50 μm period and 26nm

IOP Conf. Series: Journal of Physics: Conf. Series **1021** (2018) 012040 doi:10.1088/1742-6596/1021/1/012040

deep). After annealing of the grating dewetting and island formation was observed which was characterised by AFM.

Both the specular and off-specular scattering was measured for a bare grating and one containing the polymer. For the grating structure a series of features are observed, these are simply due to the grating in-plane period (d) and the modulation with the grooves placed perpendicular to the neutron beam in $Q_x = \frac{2\pi n}{d} = \frac{\pi n}{\lambda}(\theta_i^2 - \theta_f^2)$ of the Bragg ridges. As θ_i is fixed these trace out curved trajectories in $\theta_f - \lambda$ space (with the exception of the specular beam n=0) and as we know this relationship of the phase along these curves we can constrain the fits to equation 2 and extract the polarisation P_0.

Shown in figure 4 is an example plot from such analysis showing the echo polarisation as a function of wavelength for a pole shoe angle (θ_0) of 50° (hence the spin echo length $z = c_{se}\lambda^2$ with $c_{se}/cot(\theta_0) = 2551nm^{-1}$). In all cases we compare with the specular beam from the bare grating. As expected with the grating grooves parallel to the spin-echo encoding direction there is no modification in the polarisation from that of the bare patterned substrate. The results are rather unpromising as they show no additional features from the dewetted polymer islands. In all cases the echo polarisation agrees with that determined from the specular ridge.

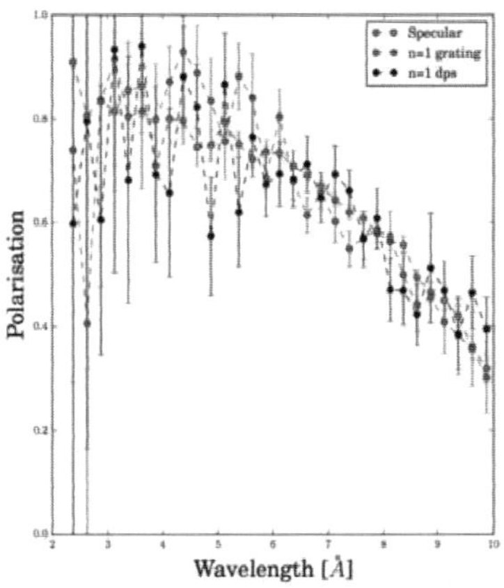

Figure 4. Spin Echo polarisation (P_{se}) as a function of wavelength for the n=1 reflection as labelled and the specular signal and a coated and uncoated grating with the dewetted polymer.

5. Discussion and conclusions

The work presented here shows that the RF gradient coils used in this work have small magnetic field inhomogeneities, however these can be corrected for and simultaneous measurement over a large angular coverage achieved, if the echo polarisation is determined. However despite this approach no additional scattering was observed from the dewetted polymers. The AFM measurements suggest that there are similar structures to those observed by Vorobiev *et al.* [21]. Also the question arises as to why SERGIS worked in the Parnell *et al* [20] investigation. Vorobiev at al used a 5.5 Åbeam of 30 x 2mm^2 around $0.5\alpha_c$ to $3\alpha_c$, where α_c is the silicon critical edge.

One key difference between the work of Parnell *et al* and Vorobiev *et al* is the amount of material in the beam. In Vorobiev's work the islands are $\approx 10nm$ tall whereas in Parnell's they are $\approx 1\mu m$, a factor 100 increase in material. This may explain the success of the Parnell study. However the overriding question is whether the technique can be applied in general to the investigation of buried in-plane structures. The quoted flux for Eva [25] is $\approx 9.8 \times 10^5 ns^{-1}cm^{-2}$, however no statement is made in the reference as to whether this is polarised and unpolarised. However just simply increasing the flux may not help as the visibility of the SERGIS signal is dependent upon measuring a small signal on a larger signal as explained in the introduction.

Hence it may be that only within the Yoneda region such a signal is visible as illustrated by Vorobiev *et al.* [21] in figure 4. If that is so then such experiments should be possible at ESS, if the integrated flux is of the same order of magnitude as that of the ILL. For one glancing angle, the monochromatic setup at ILL measures the off-specular intensity of one point of the Yoneda wing whilst the tof method measures the whole Yoneda wing simultaneously.

To conclude we remark that using the polarisation echo recovery method here we are able to correct for spatial aberrations in the magnetic field integral, such an approach is necessary when simultaneously looking over a range of scattering angles.

6. Acknowledgements

We would like to acknowledge discussions with Roger Pynn (Indiana), Andrew Parnell (Sheffield) A.Vorobiev. We acknowledge the STFC proposals RB1010439 and RB1510189 and thank Peter Muller-Buschbaum (TU Munich) for the polymer samples and Henk van Zeijl (TU Delft).

7. References

[1] Mezei F 1972 *Zeitscrihft Fur Physik* **255** 146
[2] Keller T, Gähler R, Kunze H and Golub R 1995 *Neutron News* **6** 16–17
[3] Rekveldt M T 1996 *Nuclear Instruments and Methods in Physics Research Section B: Beam Interactions with Materials and Atoms* **114** 366–370
[4] van Oossanen M, Kraan W H, Bouwman W G and Rekveldt M T 2000 *Physica B: Condensed Matter* **276–278** 134–135
[5] Pynn R, Lee W T, Stonaha P, Shah V R, Washington A L, Kirby B J, Majkrzak C F and Maranville B B 2008 *Rev Sci Instrum* **79** 063901
[6] Parnell S R, Washington A L, Li K, Yan H, Stonaha P, Li F, Wang T, Walsh A, Chen W C, Parnell A J, Fairclough J P A, Baxter D V, Snow W M and Pynn R 2015 *Review of Scientific Instruments* **86** 023902
[7] Li F, Parnell S R, Hamilton W A, Maranville B B, Wang T, Semerad R, Baxter D V, Cremer J T and Pynn R 2014 *Review of Scientific Instruments* **85** –
[8] Li F, Parnell S R, Bai H, Yang W, Hamilton W A, Maranville B B, Ashkar R, Baxter D V, Cremer J T and Pynn R 2016 *Journal of Applied Crystallography* **49** 55–63
[9] Li F, Parnell S R, Wang T, Baxter D V and Pynn R 2016 *Journal of Physics: Conference Series* **711** 012015
[10] Plomp J, de Haan V O, Dalgliesh R M, Langridge S and van Well A A 2007 *Thin Solid Films* **515** 5732–5735
[11] Parnell S R, Washington A L, Parnell A J, Walsh A, Dalgliesh R M, Li F, Hamilton W A, Prevost S, Fairclough J P A and Pynn R 2016 *Soft Matter* **12**(21) 4709–4714
[12] Andersson R, Bouwman W G, Luding S and de Schepper I M 2008 *Granular Matter* **10** 407–414
[13] Tromp R H and Bouwman W G 2007 *Food Hydrocolloids* **21** 154–158
[14] Rekveldt M T, Keller T and Golub R 2001 *EPL (Europhysics Letters)* **54** 342
[15] Repper J, Keller T, Hofmann M, Krempaszky C, Petry W and Werner E 2010 *Acta Materialia* **58** 3459–3467
[16] Major J, Dosch H, Felcher G, Habicht K, Keller T, te Velthuis S, Vorobiev A and Wahl M 2003 *Physica B: Condensed Matter* **336** 8 – 15
[17] Pynn R, Fitzsimmons M R, Rekveldt M T, Major J, Fritzsche H, Weller D and Johns E C 2002 *Review of Scientific Instruments* **73** 2948–2957
[18] De Haan V O, Plomp J, Bouwman W G, Trinker M, Rekveldt M T, Duif C P, Jericha E, Rauch H and Van Well A A 2007 *Journal of Applied Crystallography* **40** 151–157
[19] Ashkar R, Stonaha P, Washington A L, Shah V R, Fitzsimmons M R, Maranville B, Majkrzak C F, Lee W T, Schaich W L and Pynn R 2010 *Journal of Applied Crystallography* **43** 455–465
[20] Parnell A J, Dalgliesh R M, Jones R A L and Dunbar A D F 2013 *Applied Physics Letters* **102**
[21] Vorobiev A, Major J, Dosch H, Müller-Buschbaum P, Falus P, Felcher G P and te Velthuis S G E 2011 *The Journal of Physical Chemistry B* **115** 5754–5765
[22] Yoneda Y 1963 *Physical Review* **131** 2010–2013
[23] Farago B 2003 *Neutron Spin Echo Spectroscopy: Basics, Trends and Applications* Springer Lecture Notes in Physics (Berlin, Heidelberg: Springer)
[24] Müller-Buschbaum P, Bauer E, Wunnicke O and Stamm M 2005 *Journal of Physics: Condensed Matter* **17** S363
[25] *Ill Yellow Book* (Ill)

IOP Conf. Series: Journal of Physics: Conf. Series **1021** (2018) 012041 doi:10.1088/1742-6596/1021/1/012041

Current experiments at the irradiation facility of the IBR-2 reactor

M. Bulavin and S. Kulikov[1]

[1] Joint Institute for Nuclear Research

E-mail: bulavin85@inbox.ru

Abstract. Technical details, the principle of work and the distribution of a generated neutron spectrum are presented in the paper. Current experiments at the irradiation facility, as well as investigations at the radiation resistance of various materials, the radiation coloring of minerals, neutron activation analysis, etc. are also described. Conclusions about the high efficiency and the practical relevance of the facility as an instrument for the irradiation of samples with neutrons of various energy (0.25 meV – 10 MeV) with fluence from $5\cdot10^{11}$ n/cm^2 to 10^{18} n/cm^2 are presented as well.

1. Introduction

The research pulse reactor IBR-2 (FLNP JINR) is a powerful instrument for the research of radiation resistance and properties after irradiation for a wide range of materials in the most diverse fields of science and technology. Such research allows to solve various science problems such as the upgrade of LHC (HI-LHC), R&D of magnetic field sensors for tokamaks (ITER, DEMO) or colliders (LHC, NICA), neutron activation analysis, the radiation coloring of different minerals, the production of isotopes and others. Currently, an overview of the most important science experiments, carried out at the irradiation facility of the IBR-2 reactor, is represented.

2. The irradiation facility of the IBR-2 reactor and its neutron spectrum

The irradiation facility consists of two parts: an aluminum (Al) section and an elbowed (in horizontal plane) stainless steel pipe (Fig. 1). The Al section can be the closest to the surface of water moderator from the side of IBR-2 beamline №3 (Fig. 2). It has a length of 3 m. The elbowed pipe has an internal diameter of 160 mm, a length of 5 m and an elbow's radius of 25 m. The elbowed pipe is located into the cylinder with a diameter of 800 mm, filled with water as a biological shield. Another side of the cylinder is equipped with a dumper. It has a thickness of 1 m and closes off the elbowed pipe for protection from the direct neutron beamline and γ-rays during the irradiation of samples. When the reactor is stopped, dumper is opened and the personnel can install samples into the elbowed pipe. Samples are installed along the length of aluminum section and the elbowed pipe. One of the main advantages of the facility is the possibility of irradiation of large-size samples (180x180 mm^2). When samples are already installed, the facility begins to move to the surface of the water moderator. Distance from the moderator is chosen depending on the necessary neutron fluences at the surface of the sample. Irradiation lasts for 11 days on the 2 MW of the reactor's power. The differential neutron flux density of the beamline №3 was calculated

IOP Conf. Series: Journal of Physics: Conf. Series **1021** (2018) 012041 doi:10.1088/1742-6596/1021/1/012041

taking into account the experimental data in [1-2]. In addition, it is shown in Fig. 3. The neutron spectrum consists of three main parts: thermal, resonance and fast neutrons. Their proportions are 27% of thermal, 51% of resonance and 22% of fast neutrons.

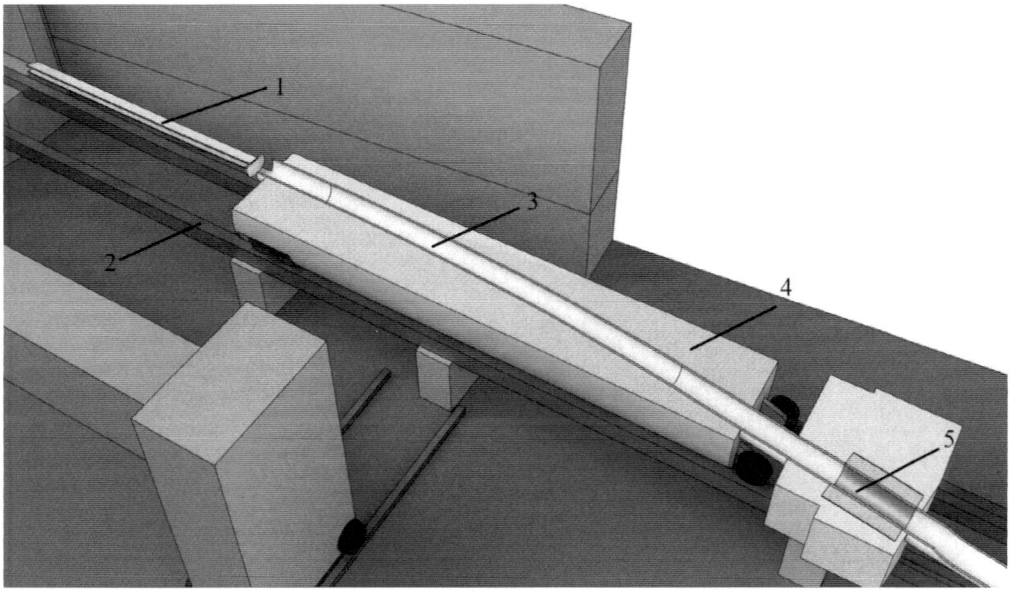

Fig.1. Irradiation facility of the IBR-2 reactor: 1 – Al section, 2 – railway, 3 – elbowed pipe, 4 –cylinder with water 4, 5 – the dumper

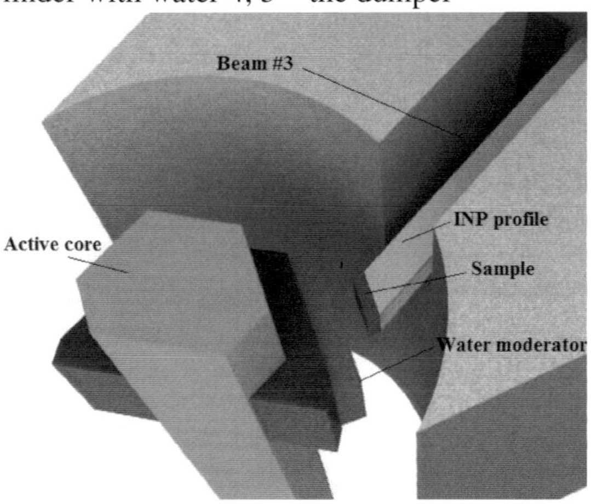

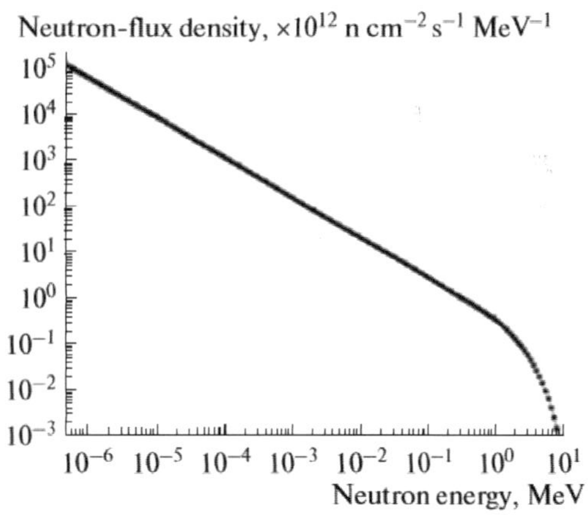

Fig.2. Al section (INP profile) of the irradiation the facility near the water moderator of the IBR-2

Fig.3. Differential neutron flux density of beamline №3

3. Current experiments at the irradiation facility

The study of changes in the properties of materials after the irradiation

The study of changes in the properties of materials after the irradiation is a basic research field at the irradiation facility of the experimental beamline №3. Experiments for research at the radiation resistance of different materials (poly- and single crystals of

IOP Conf. Series: Journal of Physics: Conf. Series **1021** (2018) 012041 doi:10.1088/1742-6596/1021/1/012041

diamonds, GaAs, printed circuit board (PCB) of G10, FR4, Arlon 85N, Rogers 4450B, polyimide and the rest) are carried out in the framework of the program to upgrade ATLAS and CMS detectors at LHC [3, 5]. Fast neutrons' fluence (E_n>0.4 MeV) due to irradiation was near $4 \cdot 10^{17}$ n/cm^2. In Fig. 4a,b the photos of PCB after irradiation are shown. The measurements of mechanical and electrical properties (resistant of interlayer isolation and capacitance) and measuring leakage current had been made before, during and after the irradiation.

Experiments in the radiation resistance of semiconductor sensors in radiation conditions close to ITER conditions are carried out in the framework of project for the design of ITER fusion reactor [4], Fig. 4c. During experiments, long-term and stability performance of sensors with neutron fluence of 10^{20} n/cm^2 was demonstrated.

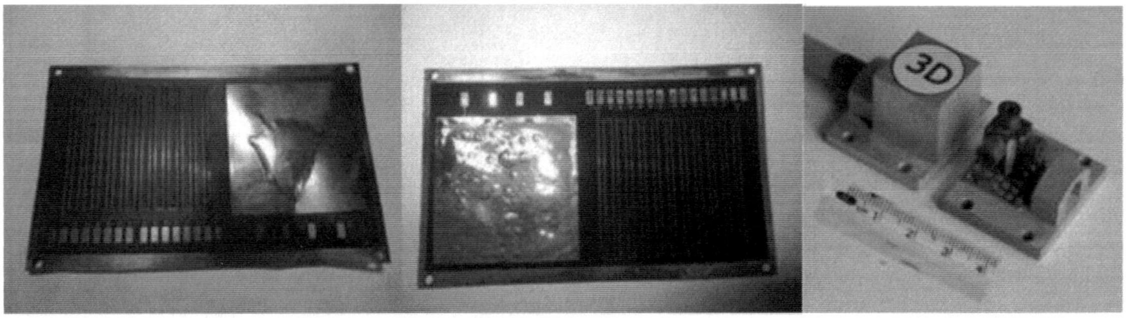

a b c

Fig.4. Materials for irradiation: a, b – PCB after irradiation, c – 3D Hall sensors for ITER

In addition, the research of material properties of scintillation detectors (Fig. 6) after the irradiation with small neutron fluence from $5 \cdot 10^{11}$ n/cm^2 to 10^{15} n/cm^2 were carried out [6].

Currently, experiments assessing the possibility of period extension in sustained operations of the pelletized cold moderator [6-7] based on the mixture of mesitylene and m-xylene are also carried out on the irradiation facility. The radiation resistance of the aromatic hydrocarbons of mesitylene and m-xylene mixture and the new substance – triphenylmethane as different aggregate state was studied in these experiments. Substances with the same mass were put into the cylindrical containers, which were installed at the minimal distance from the thermal moderator of the reactor and were irradiated to the neutron fluence of 10^{18} n/cm^2 (160 MGy). Containers joined with manometers are used as copper tubes for the determination of pressure, which is increased because of the formation of radiolytical hydrogen during the irradiation. The formation of the radiolytical hydrogen from triphenylmethane is 10 times less than from the mixture of mesitylene and m-xylene. This can indicate that the radiation resistance of triphenylmethane is better.

Radiation coloring of topazes

Blue topaz is very popular in the world as jewelry. Yet, naturally occurring blue topaz is quite rare. For that reason, on the facility for the radiation research colorless topazes are irradiated up to fast neutron fluence of 10^{18} n/cm^2 to produce a more desired darker blue. The photo of topaz after the irradiation is shown in Fig.5.

Fig.5. Topazes: 1 – colorless, 2 and 3 – after the irradiation with different doses

Neutron activation analysis

The neutron-activation analysis on the irradiation facility IBR-2 is used for various tasks: for the most part, it's the evaluation of the fast neutron flux density with an energy of E>1MeV (by capture reaction of $Ni^{57}+n^0\rightarrow Co^{58}+\gamma$ (811 keV)) and the elementary composition of irradiated samples, such as various kinds of isotopes, archeological or biological materials.

4. Conclusion

Facility for the radiation research at the IBR-2 reactor is efficiently used for carrying out experiments in various fields of science and technology. The irradiation of samples is generally needed for the investigation of radiation resistance of materials and their mechanical, physical, electrical and other properties after the irradiation. Simultaneously, the radiation coloring of topazes, the production of medical isotopes, the irradiation of biological samples and other challenging scientific tasks become getting actual. In addition, the irradiation facility can be used for experiments carried out by external users.

References

[1] Irradiation facility at the IBR-2 reactor for the investigation of material radiation hardness / A. Verkhogliadov [et al.] // Nuclear instruments and methods in physics research B. 2015. Vol. **343**. – pp. 26-29.
[2] The spectrum and density of neutron flux in the irradiation beamline №3 of the IBR-2 reactor / E. Shabalin [et al.] // Physics of particles and nuclei letters. 2015. Vol. **12**, №2. pp. 336-343.
[3] ATLAS Collaboration, Letter of Intent. Phase II Upgrade. CERN-2012-022, LHCC-1-023, December, 2012
[4] Experimental evaluation of the stable long term operation of semiconductor magnetic sensors at ITER relevant environment / I. Bolshakova [et al.] // Nuclear Fusion. – 2015. – Vol. **55**. – №8. – P. 083006-083016.
[5] Investigation of the SCSN-81 scintillator irradiated by neutrons / S.V. Afanasiev [et al.]// CMS internal note, CMS IN-2013/003. - 2013. - Vol. **2**. - p. 1-4
[6] The world's first pelletized cold neutron moderator at a neutron scattering facility / S. A. Kulikov [et al.] // Nuclear instruments and methods in physics – 2014. – Vol. **320**. – P. 70-74.
[7] The pelletized cold moderator of the IBR-2 reactor: current status and future development / M. Bulavin [et al.] // Journal of physics: conference series (ECNS 2015). – 2016. - №**746**, 012031, p. 6.

For further information or questions, please use the following contact information: ksa@nf.jinr.ru (Dr. S.A. Kulikov) and bulavin85@inbox.ru (Dr. M.V. Bulavin).

IOP Conf. Series: Journal of Physics: Conf. Series **1021** (2018) 012042 doi:10.1088/1742-6596/1021/1/012042

Off-gas processing system operations for mercury target vessel replacement at J-PARC

Tetsuya Kai, Toshitsugu Uchida, Hidetaka Kinoshita, Masakazu Seki, Motoki Ooi, Takashi Wakui, Katsuhiro Haga, Yoshimi Kasugai and Hiroshi Takada

Japan Atomic Energy Agency, Tokai-mura Naka-gun Ibaraki 319-1195 Japan

E-mail: tetsuya.kai@j-parc.jp

Abstract. An off-gas processing system was installed in the J-PARC spallation neutron source to reduce radioactivity of xenon-127 and tritium contained in a helium cover gas in a surge tank of a mercury circulation system to obey the regulation by law. In addition to this role it has been utilized to a purging process before the target vessel replacement and an air-flow control procedure to minimize uncontrollable radioactivity release during the replacement. An standard and urgent model plans of the off-gas processing system operation were established indicating that 31 days were required at least to replace the target vessel.

1. Introduction

The Japanese spallation neutron source (JSNS) [1] in the Materials and Life science experimental Facility (MLF) of the Japan Proton Accelerator Research Complex (J-PARC) started operation in 2008. The intense proton beam of 3 GeV with a power of 1 MW is injected into mercury in a mercury target vessel made by stainless steel. Mercury is supposed to be used until the end of the lifetime of JSNS while the target vessel is replaced periodically. Gaseous radioactivity are accumulated in a cover gas of a surge tank of a mercury circulation system. An off-gas processing system was introduced so that the cover gas including gaseous radioactivity was transferred before mercury draining and a target replacement. The transferred cover gas is kept for 6–12 months for the decay of ^{127}Xe having half-life of 36.4 days, where tritium in the gas is absorbed by molecular sieve (synthetic zeolite), and then is released to the environment after checking the radioactivity.

The first replacement of the target vessel was carried out in November 2011. At the beginning the radioactivity in the mercury circulation system was reduced by repeating three-step purging process: 1) filling the mercury circulation system with clean helium gas, 2) keeping for a while, and 3) transferring the helium gas to the off-gas processing system. The purging process reduced ^{127}Xe radioactivity in the mercury circulation system down to 1/100 [2], and then circular specimens were cut out from the head of the target vessel for post irradiation experiments (PIE). The replacement of the target vessel was carried out after the specimen cut-out. During the replacement gas was continuously transferred to the off-gas processing system to form inflow air aiming at reducing the tritium release. The amount of tritium released from the start of cutting process to the end of the replacement was 150 GBq [3], which was 25% larger than 120 GBq, the design value of annual tritium release from MLF. The proton beam irradiation of

450 MWh before the first target vessel replacement was only 9% of 5000 MWh, being assumed in the design. The tritium release in the nominal operation of 1-MW was estimated to be 1.7 TBq based on these values. This higher value could give an impact to other scenarios of radioactivity treatment in MLF, efforts to reduce tritium release have been started.

After the first replacement the mercury target vessel was replaced in 2014, 2015 and 2016. A standard model plan of the replacement has been developed to be scheduled during maintenance period. An urgent model plan has been also discussed for the case of the target failure during beam operation. In this paper improvements of processes to reduce radioactivity release in the model plans during the target vessel replacement are described based on real experiences.

2. Improvements of processes to reduce radioactivity release

2.1. Purging process

Figure 1 shows the radioactivity of ^{127}Xe in cover gas and purging gas to comparethe effects for different frequency of purging. In July 2015, the purging was performed in a scheduled summer maintenance period. The minimum intervals of the purging was one day. Even in the purging after longer interval, the ^{127}Xe radioactivity was in the range of extrapolation of daily purging. It was concluded that longer interval than one day brought no benefit.

In November 2015, the purging was performed due to the unexpected target failure. A quick replacement was desired since the failure occurred within one-month operation after a long beam absence from the end of April. Thus, the purging with shorter intervals of 0.5-day was performed expecting to early readiness for the target replacement. Although more purging were carried out in a short period, the duration required to reduce ^{127}Xe radioactivity was not drastically improved. The goal of ^{127}Xe radioactivity in the purging gas was determined to be 0.2 GBq not to initiate an alarm of an exhaust gas monitor of MLF assuming gas release of 200 L in 1 minute during the target vessel cutting of replacement processes. The goal was achieved on 9th and 6th day in the July and November 2015 purgings, respectively. The proton beam power at the target failure on November was 500 kW, and the ^{127}Xe radioactivity was only 28% of the saturated one after a long beam operation. Thus 7.2 times higher radioactivity needs to be taken into account for the nominal operation. Considering these results ten times of purging in 10 days were allocated for the purging process in both standard and urgent model plans of the target vessel replacement.

As a result of transferring the cover gas and the purging gas of ten times, the volume of gas to be stored in the gas holders reached to 3,000–5,400 NL, resulting that 2–3 out of 4 gas holders were occupied. During the target vessel cutting and replacement process, gas in the mercury circulation system needs to be transferred to the off-gas processing system as an air-flow control process, to be mentioned in the next section. At least two, preferably three, gas holders need to be empty to perform the air-flow control during the target vessel replacement. A liquid-nitrogen cooled charcoal trap (CCT) was introduced to separate noble gases from helium

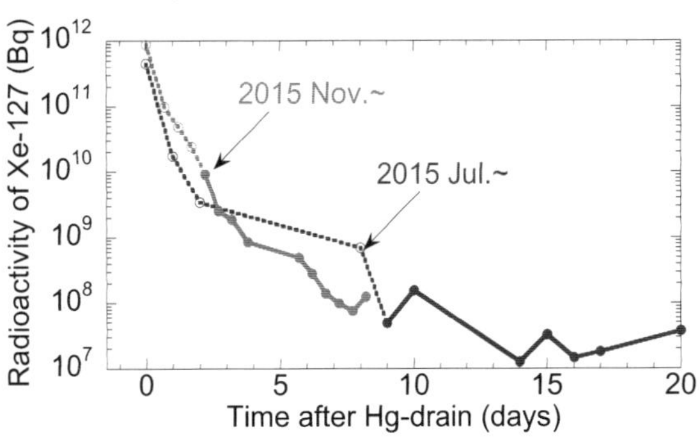

Figure 1. Radioactivity of ^{127}Xe in purging gas, which was reduced by repeated purging. Solid lines and filled circles were obtained by measuring sampled gas by a high-purity germanium (HPGe) detector, dashed lines and open circles were estimated by dose trends of the radiation monitor at the mercury absorber normalized by the results with the HPGe detector.

IOP Conf. Series: Journal of Physics: Conf. Series **1021** (2018) 012042 doi:10.1088/1742-6596/1021/1/012042

gas. The gases in the gas holders are circulated through the CCT, and other gases except for helium are trapped in it. The helium gas is released in the environment after the CCT process, and the trapped gas are returned to a gas holder by heating CCT. The volume of the trapped gas were about 600 NL so far. Six days were allocated for the standard model plan including a tritium removal process to avoid frost forming, the CCT and the release processes. Four days were allocated for the urgent plan assuming a shift work.

2.2. Air-flow cotrol

The mercury circulation system was vacuum at the PIE specimen cutting out in 2011. One of the reason why large amount of radioactivity was released assumed that a turbulent inflow might enhance desorption of tritium being trapped in stainless steel of the system. Thereafter we decided to fill the mercury circulation system with dry gas before the cutting for PIE.

The holes after cutting out PIE specimens were sealed with curing tape, but radioactivity was continuously released as shown in Fig. 2. Once the tape was removed, the radioactivity release rapidly increased. Then gas was transferred to the off-gas processing system in order to form an inflow from outside to inside the target vessel. As a result of this 'air-flow control', the radioactivity release was reduced down to almost background level. The air-flow control has been carried out frequently hereafter. However the volume of gas holders in the off-gas processing system is limited, silicon-rubber plugs were employed to seal such open holes and mercury flow pipes to minimize radioactivity release and gas volume to be stored in the gas holders.

Based on these efforts, three days were allocated in the standard model plan for the cutting for PIE, processing the gas used by the air-flow control, and release of the processed gas. In the urgent case, the cutting would be skipped. Five days were allocated in the both model plans for the target vessel replacement, and processing the air-flow control gas, and release of it assuming that three gas holders were occupied by the air-flow control gas.

3. Model plans of target vessel replacement and further upgrade

The improvements and estimations of gas processing procedures have been carried out, and the standard and the urgent model plans for target vessel replacement were established as shown

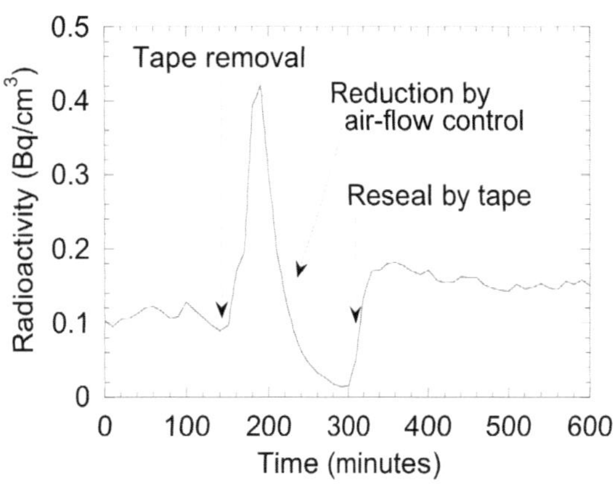

Figure 2. Effects of air-flow control to reduce radioactivity release from the hole of the target vessel head made by cutting out of PIE specimens. Continuous radioactivity release before and after this test indicated that the tape could not seal the hole well.

Figure 3. Photograph of newly installed gas holders on the shield of existing gas holders.

Table 1. Required time of model plans of target vessel replacement

Process	Standard (days)	Urgent (days)
Cooling/Cover gas transfer/Purging/Process & release	7/1/10/6	0/1/10/4
Cutting for PIE/Process & release	1/2	0/0
Target vessel replacement/Process & release	1/4	1/4
Leak-check/Hg-loading (both incl. Process & release)	6/6	6/5
Total	44	31

in Table 1 including processing of gas used in the leak check after an installation of new target vessel and that used to pressurize a drain tank for mercury loading. The urgent model plan was verified through the target replacement after the target failure occurred in November 2015, which took 32 days except for delays due to other non-technical reasons. There were some delays in the cooling and the process of purging gas. It was not easy to decide to replace the target vessel just after the failure. Four days for discussions and checking the situation were counted as the cooling time. Extra 2 days were used for a tritium removal process in the process of purging gas in order to convert HT gas to HTO by using a catalyst loop. The extra days were not necessary if the first tritium removal process was carried out with the catalyst loop. There were also some advances in the purging, the process of gases used in the air-flow control during the target vessel replacement and that used in the mercury loading due to the lower proton beam power and shorter operation time before the failure than those assumed for the model plan.

Upgrades of the off-gas processing system has started in order to shorten the time of the target vessel replacement. New gas holders (2.4 m^3 for each) were installed in March 2017 as shown in Fig. 3, and to be used after the next target vessel replacement. With the new holders, the whole volume of gases of the mercury loading can be transferred to the off-gas processing system. It will reduce 4 days from the model plan by making it possible to start beam operation without finishing the process and release of the gas used for the mercury loading. Parallel processing modes, in which carrying out two processes at the same time, will be fully added until March 2018 aiming at further reduction of 5 days from the model plans.

4. Summary

The off-gas processing system in JSNS was given the role of reducing the radioactivity release at the target vessel replacement in addition to that defined in the initial design. The model plans in the standard and the urgent cases of the target vessel replacement utilizing the off-gas processing system were established. The urgent model plan of the target vessel replacement needs 31 days, and will be reduced by 9 day after the undergoing upgrades of the off-gas processing system.

Acknowledgments

The authors also appreciate the staff of radiation safety section of the J-PARC center for their cooperation to measure tritium during the target replacement.

References

[1] Takada H, Haga K, Teshigawara M, Aso T, Meigo S I, Kogawa H, Naoe T, Wakui T, Ooi M, Harada M and Futakawa M 2017 *Quantum Beam Science* **1**,**8**

[2] Kai T, Kasugai Y, Ooi M, Kogawa H, Haga K, Kinoshita H, Seki M and Harada M 2014 *Progress in Nuclear Science and Technology* **4** 380–383 URL `http://www.aesj.net/pnst004/380_383.pdf`

[3] Safety Division, J-PARC 2016 Annual report on the activities of safety in J-PARC, FY2015 [in japanese] Tech. Rep. JAEA-Review 2016-032 URL `http://jolissrch-inter.tokai-sc.jaea.go.jp/pdfdata/JAEA-Review-2016-032.pdf`

IOP Conf. Series: Journal of Physics: Conf. Series **1021** (2018) 012043 doi:10.1088/1742-6596/1021/1/012043

ESS Target Station Ventilation - Managing Radiation Hazards

A Polato, A Bieliaieva, L Spanier, M Göhran, F Javier

European Spallation Source ESS ERIC, Lund, Sweden

andrea.polato@esss.se

Abstract. The ESS spallation target undergoes to a scheduled irradiation period of 5 years. After this time, the Target assembly (wheel and shaft) is transported to ESS Active Cell where a series of mechanical operations are foreseen in order to dismantle the target and sort the wastes before their disposal. The nature of the dismantling operations – mainly performed with sawing and shear cutting tools – generates activated aerodispersed particles (aerosols and gas) contributing to high contamination levels inside the Active Cell. The function of the HVAC is to ensure the dynamic confinement of the airborne contamination and to collect it, through filtration system, to a specific point before the release.

In this paper, the study of the HVAC system is focused on the fire event case. This situation, among others involving the Active Cell, can jeopardize the dynamic confinement hence creating the potential for a radiation hazard for workers and public. With reference data coming from existing literature, the study shows that a controlled operational mode of the HVAC ensures, at the same time, the protection against the fire spread and the dynamic confinement of the airborne contamination for the entire duration of the postulated fire.

1. Introduction

The current lifetime of the ESS target is set today at 5 years of full power beam time. After this time, a new target has to be installed in the target Monolith while the old target has to be removed from here and disassembled. The Active Cell constitutes a process enclosure (total volume of nearly 4000 m^3) devoted to the dismantling operations of irradiated components coming from the target Monolith. It hosts, in its premises, a process cell where the effective dismantling is performed, a maintenance cell and a series of storage pits to temporary host the waste before their final disposal.

Airborne contamination inside the Active Cell is the result of the dismantling operations occurring on a complete target assembly and, in particular, from activated dusts and gases from stainless steel shaft and shroud plus eroded tungsten from the target (Figure 1).

The levels of contamination were evaluated on the base of the estimated debris produced during dismantling (cf. [1]) combined with radioactive inventory present in the components (cf. [2] and [3]). In addition to this, a series of coefficients were considered to put in relation the total produced activity during dismantling with the quota of it that results airborne by effect of local phenomena (presence of a local extraction, suspension of dusts due to the air stream, suspension of particles with sufficiently small aerodynamic diameter). These coefficients are derived from experience on the same kind of operations in existing nuclear facilities. Figure 2 and Table 1 give an overview of the coefficients used.

IOP Conf. Series: Journal of Physics: Conf. Series **1021** (2018) 012043 doi:10.1088/1742-6596/1021/1/012043

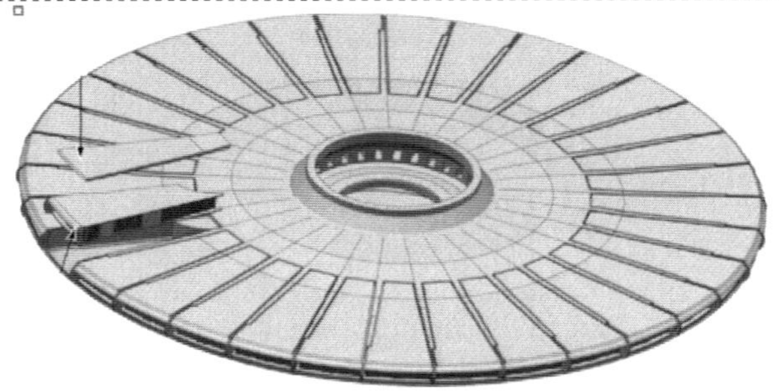

Figure 1. Detail of the target shroud with shown a tungsten sector

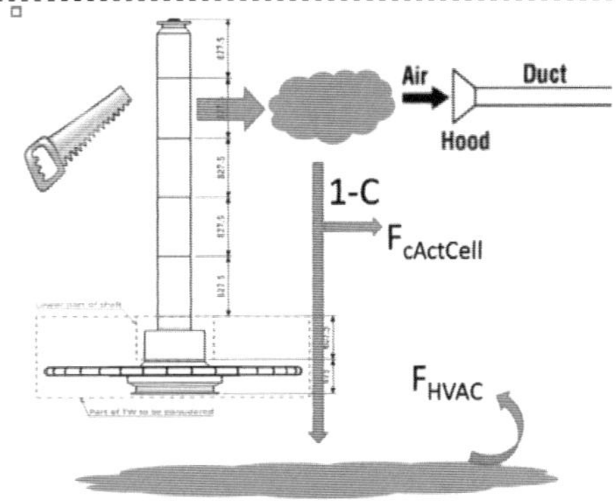

Figure 2. Airborne release coefficients

Table 1. Airborne release factors of the total produced activity during dismantling operation

Airborne fraction of the cutting debris $F_{cActCell}$	Efficiency of local extraction device C	Resuspension of floor dusts due to ventilation F_{HVAC}	Resuspension of floor dusts due to fire F_{fire}
1E-2	9E-1	1E-6	1E-2

The contamination levels were finally evaluated for different kind of situations [4] and the main results are recalled in Table 2 where the different ratios between total and specific activities are due to the different source terms considered for the different situation (normal, accidental, fire).

Table 2. Active Cell airborne contamination figures

	Total Activity	Specific activity	Derived Air Concentration (DAC)
Normal conditions	7.92E+10 Bq	1.47E+07 Bq/m^3	1164.0 DAC
Accidental conditions[a]	7.41E+11 Bq	1.47E+08 Bq/m^3	6558.5 DAC
Fire conditions	2.18E+11 Bq	4.04E+07 Bq/m^3	7376.3 DAC

[a] The dismantling performed in absence of local extraction device (C=0 in Table 1) was considered as the reference accidental contdition.

If not confined, leaks through the Active Cell structure would cause dose received by public and personnel working in the adjacent areas. As a consequence, a ventilation system is foreseen to ensure the dynamic confinement of the enclosure.

2. Active Cell HVAC system configuration

The contamination levels of Table 2 classify the Active Cell as a C4** containment area according to [5], which suggests also the implementation of a ventilation system architecture of the type IIIB (Figure 3). Also, the Active Cell being a fire cell, fire dampers are foreseen on the supply and extraction ducts.

In normal conditions the HVAC system supplies a total airflow of 6500 m^3/h (1.8 m^3/s) to the Active Cell and ensures, via the extraction flow (about the same flow as the supply), a negative pressure with respect to adjacent areas hence ensuring the dynamic confinement.

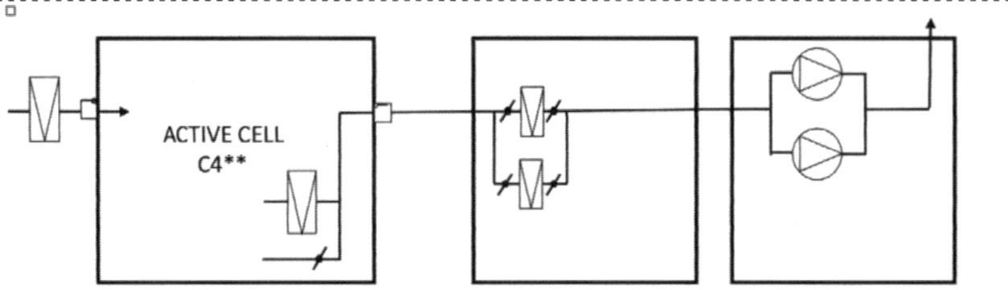

Figure 3. Active Cell HVAC configuration with, indicated, fire dampers on the Active Cell perimeter, filters, extraction fans stack. It is possible to notice a bypass on the first level of filtration and the redundancy on the last level of filtration as well as on the extraction fan

3. HVAC system operations during a fire event

The particular operational mode of the HVAC, during a fire event, is in line with what is described in [5] for areas presenting a high inventory of dispersible radioactive material. It is obtained via closing the supply fire damper upon fire detection, while the extraction fire damper remains open. In this configuration, the extraction can continue at a regime that is sufficient to keep the Active Cell at a pressure value that is lower than the adjacent rooms (ensuring the dynamic confinement). Additionally, the stop of the supply flow reduces the amount of oxygen delivered to the fire.

The goal of the analysis is to demonstrate that, during a fire event, it is possible to keep the ventilation extraction running for the entire duration of the postulated fire without compromising the fire sectorization of the building, ensuring the dynamic confinement of the volume and keeping the filtration at the same time.

The postulated fire is the one described in [6] with a maximum heat release rate of 320 kW and a total duration of 800 seconds. Reference [6] describes an additional fire (sensitivity analysis fire) with a maximum heat release rate of 1000 kW and a total fire duration of 1200 s. In Figure 4 and Figure 5 temperature and flow profiles connected to the fire scenarios are shown.

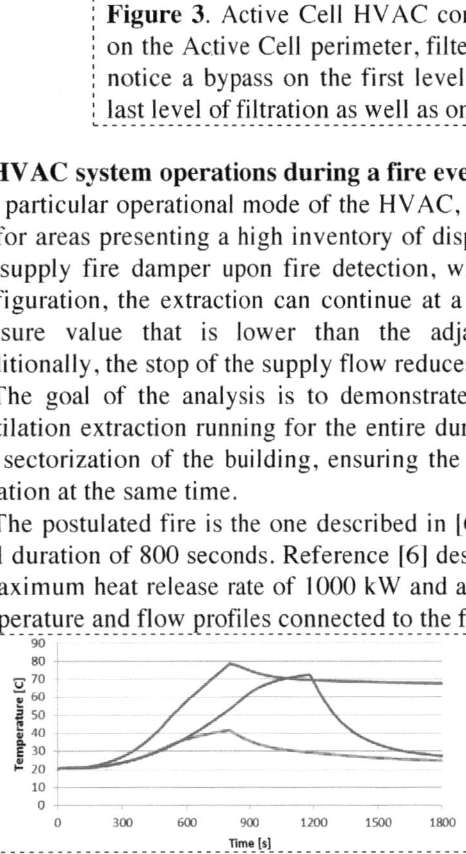

Figure 4. Temperature profile in the Active Cell during postulated (green) and sensitivity analysis (purple) fires. The red curve describes the adiabatic fire

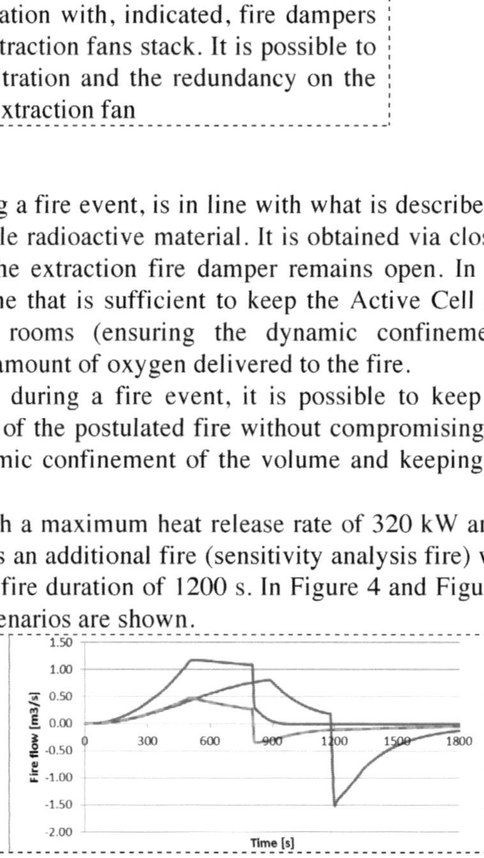

Figure 5. Flow increase due to air density variations during postulated (green) and sensitivity analysis (purple) fires. The red curve describes the adiabatic fire

4. HVAC system behaviour during the fire event

4.1. Effect of increased temperature

The continuous extraction during fire implies that air at a higher temperature circulates through the ducts with the potential of damaging them and in conclusion jeopardize the fire sectorization and possibly spread contamination within the facility (i.e. dynamic confinement function lost).

Temperature profiles of Figure 4 show that the maximum temperature in the Active Cell reaches 40°C for the postulated fire and around 75°C for the sensitivity analysis fire (similar values for the adiabatic fire). These temperatures are compared with the resistance of the HVAC components (for which a temperature resistance of 200°C for 2 hours was required) that in conclusion are not affected by the higher temperatures for the entire duration of the event.

4.2. Effect of the fire flow

The continuous extraction configuration has to take into account the increased flow result of reduced air density during a fire. If the HVAC is not capable to handle the increased flow, pressure inside the Active Cell can increase compromising the dynamic confinement of the airborne radioactivity.

The HVAC configuration described in Section 2. has, as a first action at the detection of a fire, the closing of the supply fire damper. This has an impact on the extraction flow that is reduced from the initial value of 1.8 m^3/s to the sum of the leaks through the Active Cell structure (0.014 m^3/s or class 3 according to [7]) plus the flow increase, shown in Figure 5. In this situation, the flow reaches approximately peak values of 0.5 m^3/s and 0.8 m^3/s for the postulated and sensitivity analysis fires respectively (similar values for the adiabatic fire). Given the maximum flow capacity of the system described in Section 2. the HVAC system can therefore handle the fire flow for the entire duration of the fire without an effect on the confinement function.

4.3. Effect of soot and dust on filters

One of the main concerns of the HVAC configuration of Section 2. is the clogging of filters due to deposition of soot on the filtration media. In this case, an elevated resistance can considerably reduce the extraction flow with repercussions on the dynamic confinement.

Assuming a heat of combustion of 34 kJ/kg with a soot production rate of 0.1 g/g (defined in [6]), the estimated quantity of soot is of 427 g and 1754 g for the postulated and the sensitivity analysis fires respectively. A dust amount of 100 g was added to these quantities to take into account the airborne dust already present in the Active Cell (cf. [4]) for a total of 527 g and 1854 g.

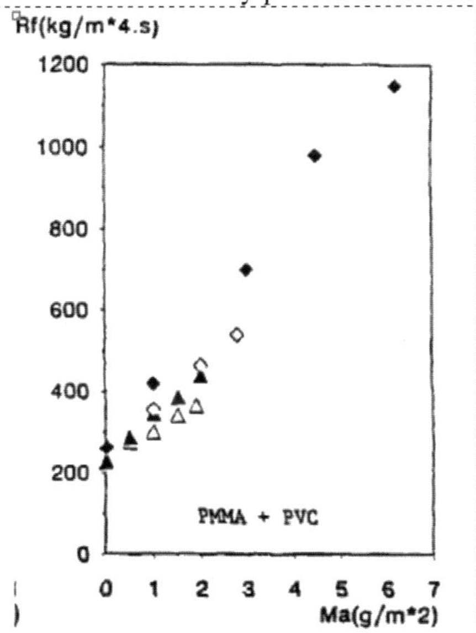

Figure 6. Variation of filters resistance as a function of the mass of filtered aerosols [8].

The filters selected for the Active Cell (HEPA filters specifically designed for nuclear applications) present filtrating media surfaces of 80 m^2 and 120 m^2 for the first and the second levels of filtration respectively (Figure 3) based on the airflow.

The variation of resistance as a function of the mass of filtered aerosols described in [8] sets at 10g/m^2 the maximum limit after which the resistance of a filter increases in a non-linear way. The same value was considered in the study as the extreme operating value for the Active Cell filters. After, filters were considered not functional anymore.

It is possible to see that in the case of the postulated fire soot and dusts can be deposited on the first level of filtration without this reaching its functional limits (500 g on 80 m^2 with a final value of 6.6 g/m^2).

For the sensitivity analysis fire, a two-step clogging takes place. In a first stage, the first level of filtration gets progressively clogged until the 10 g/m^2 is reached (total of 800 g ashes deposited). At this point, the bypass present on the system of Figure 3 opens, the extraction continues passing on the second level of filtration that progressively gets clogged. Of the initial amount of dust, 1054 g deposit on the 120m^2 for a mass of filtered aerosol of 8.8 g/m^2 confirming that the extraction can continue for the entire duration of the fire without the soot and dust compromising the resistance of the filters and in conclusion allowing the dynamic confinement function.

5. Conclusions

The study shows that the radiation hazard connected with a fire event in the Active Cell can be managed via a controlled operational mode of the HVAC system. This ensures, for the entire duration of the postulated - and the sensitivity analysis - fire, the respect of the fire sectorization and the dynamic confinement of the Active Cell.

References

[1] Erterius P, *Dismantling study of Target Wheel,* ESS-0020529 ESS internal document (Sweden)
[2] Muhrer G, *Target shroud and shaft inventory*, ESS-0058374 ESS internal document (Sweden)
[3] Muhrer G, , *Target inventory for normal operations*, ESS-0048608 ESS internal document (Sweden)
[4] Polato A, *Airborne contamination inside the Active Cell*, ESS-0042017 ESS internal document (Sweden)
[5] SS-ISO 17873 2011, *Criteria for the design and operation of ventilation systems for nuclear installations other than nuclear reactors* (Sweden)
[6] Åström L Winberg D, *Fire analysis – Confinement,* ESS-0047616 ESS internal document (Sweden)
[7] SS-ISO 10648-2 1994, *Containment enclosures – part 2: classification according to leak tightness and associated checking methods* (Sweden)
[8] IAEA International Nuclear Information System (INIS), *Impact of the filter clogging on the behaviour of the ventilation network in the event of fire*, CEA-DAS—771 (Austria)

IOP Conf. Series: Journal of Physics: Conf. Series **1021** (2018) 012044 doi:10.1088/1742-6596/1021/1/012044

Tritium permeation from the primary helium cooling loop for the ESS target

Y Lee and J Harborn

European Spallation Source ERIC, Tunavägen 24, 223 63 Lund, Sweden

E-mail: yongjoong.lee@esss.se

Abstract. The tritium permeation from the target cooling primary helium loop to the intermediate water loop through heat exchangers is studied, for the 5 MW operation of the target station at European Spallation Source (ESS). Both bi-metallic and mono-metallic heat exchanger concepts are studied, for the co-current and counter flow cases. The quasi two-dimensional temperature profiles in the helium, in the water and in the metallic barrier separating the two flow regions are calculated. The surface temperature of the metallic barrier facing to the helium side determines the tritium concentration dissolved in this region. The diffusion flux of the surface tritium through the wall thickness is calculated. The permeation rates via these different heat exchanger configurations are compared and a recommendation is made. It is shown that the tritium contamination level in the intermediate cooling loop allows controlled discharge of water to the city sewage system. The methodology presented in this paper can be applied to general engineering analyses of the tritium permeation problems in spallation facilities.

1. Introduction

A 5 MW proton beam will deposit up to 3 MW heat in the rotating tungsten target at European Spallation Source (ESS). The target is primarily cooled by gaseous helium flow and the heat is further transferred to the intermediate water loop via heat exchangers. The impinging proton beam on the target generates hydrogen isotopes in tungsten, which subsequently diffuses to the helium coolant flow, forming hydrogen molecules. The helium is constantly purified during normal operation, and the steady state tritium impurity in the helium loop at ESS is calculated to be 0.074 ppm. The permeation of tritium to the intermediate cooling loop water is a concern for the safe operation of the target cooling systems. In this paper, we calculate the tritium permeation rate from the target cooling helium loop to the intermediate water loop through heat exchangers, for the 5 MW operation of the target station at ESS.

A conventional mono-metallic heat exchanger made of 316L stainless steel is studied for a reference case. The structural material SS316L is chosen for its corrosion resistance in radiation environment. A bi-metallic heat exchanger concept is also studied, to evaluate the additional tritium barrier effect that additional aluminium alloy layer might provide.

Figure 1 shows a schematic diagram of the target cooling circuits. The heat exchanger W001 has the highest flow temperatures, as it transfers heat from the hot helium gas directly from the target to the intermediate cooling system. The W001 accounts for most of the tritium permeation between the two loops. Table 1 lists typical flow parameters for W001, which are available on the market.

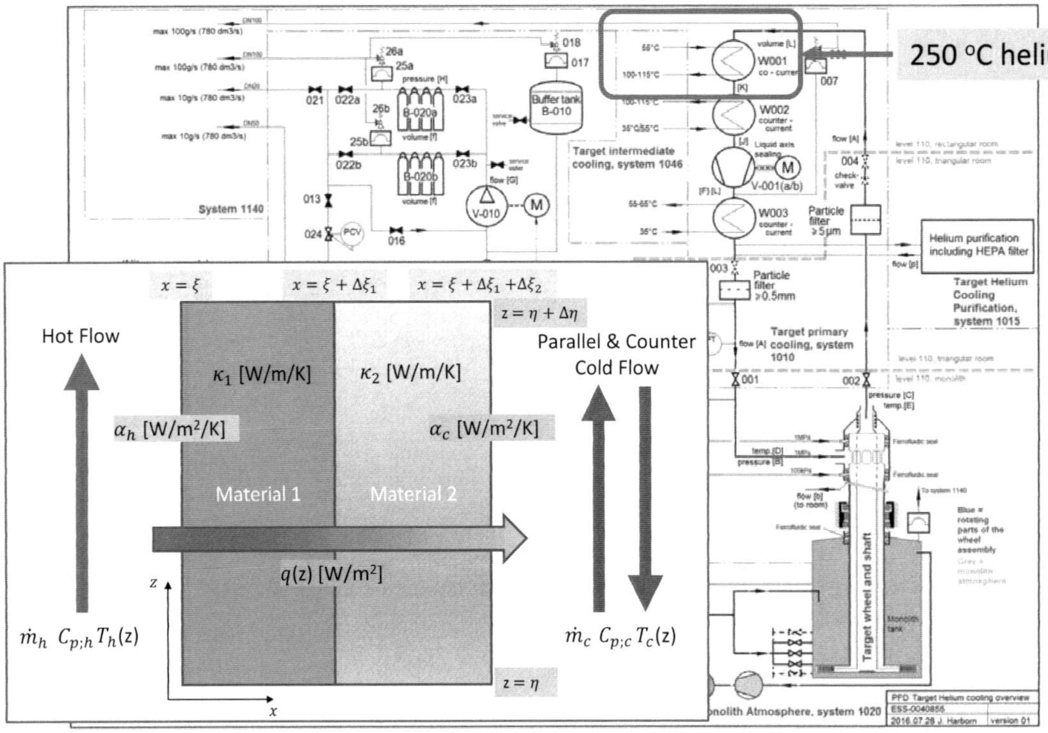

Figure 1. A schematic diagram of the target cooling circuit with the heat exchanger W001. Also shown is the quasi-2d differential volume for the thermal analysis of heat exchanger.

Table 1. The typical flow parameters for W001, which are available on the market.

Parameter	Symbol	Value	Unit
Mass flux: helium side	$\dot{m}_h$	2.85	kg·s^{-1}
Mass flux: water side	$\dot{m}_c$	7.0	kg·s^{-1}
HTC: helium side	α_h	700	W·m^{-2}·K^{-1}
HTC: water side	α_c	900	W·m^{-2}·K^{-1}
Total width of flow contact	$\hat{L}$	0.3	m
Total length of flow contact	L	2.0	m

2. Quasi two-dimensional tritium permeation

The tritium diffusion at heat exchangers can be simplified to a quasi two dimensional problem, one for the perpendicular direction to the helium flow and the other for the flow direction of the helium flow. We refer the direction perpendicular to the helium direction be "x" and the one along the flow direction be "z."

The tritium permeation rate is highly correlated to the properties of the wall material and temperature field. Therefore, we first calculate the temperature configuration in the heat exchanger. The heat exchanger in general can be described by a quasi two-dimensional geometry model, which consists of thin flat metal plate separating two flow regions. The caption figure in

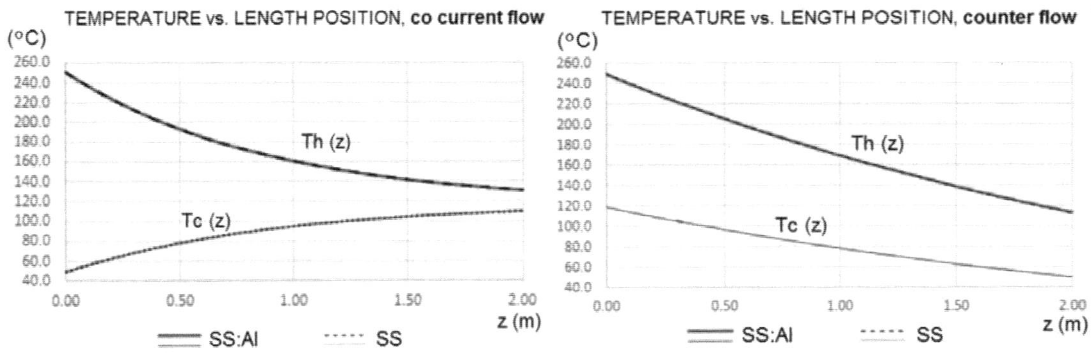

Figure 2. Temperature profiles for different heat exchanger wall types and flow configurations. The bi-metallic and mono-metallic wall types are denoted by SS:Al and Fe respectively.

Fig. 1 describes the geometry setup for the differential flow section on zx-plane. The bi-metallic wall made of austenitic steel 316L ("Material 1") and aluminium ("Material 2") separates the two flow regions. The wall thicknesses of Material 1 and Material 2 are respectively given by $\Delta\xi_1$ and $\Delta\xi_2$. For this study, the wall thicknesses of $\Delta\xi_1 = 3$ mm and $\Delta\xi_2 = 1.5$ mm are taken. The mono-metallic heat exchanger case is a special case of the bi-metallic case with $\Delta\xi_2 = 0$. The thermal conductivities of these two wall materials are given by κ_1 and κ_2 respectively. The specific heat capacities of the helium and water flows are denoted by $C_{p;h}$ and $C_{p;c}$ respectively.

The quasi two-dimensional temperature profile on zx-plane can be analytically calculated by solving the set of differential equations,

$$-\kappa\frac{\partial}{\partial x}T(z,x) = q(z) \quad \text{for} \quad 0 \le x-\xi \le \Delta\xi_1 + \Delta\xi_2\,, \tag{1}$$

$$\frac{d}{dz}T_h(z) = -\frac{q(z)\hat{L}}{\dot{m}_h C_{p;h}}, \quad \frac{d}{dz}T_c(z) = -\frac{q(z)\hat{L}}{\dot{m}_c C_{p;c}}\,. \tag{2}$$

Here, $T(z,x)$ is quasi-2D temperature in the bi-metallic wall. $T_h(z)$ and $T_c(z)$ are respectively the bulk flow temperatures of helium and water along the length of the heat exchanger. The convective heat transfer on the heat exchanger surfaces are represented by $q(z)$. The co-current flow and cross flow configurations are distinguished by setting two different boundary conditions for Eqs. (1) and (2),

$$\text{Co-current Flow Option} \;:\; T_h(z=0) = T_{h;inlet}, \quad T_c(z=0) = T_{c;inlet}\,, \tag{3}$$

$$\text{Counter Flow Option} \;:\; T_h(z=0) = T_{h;inlet}, \quad T_c(z=L) = T_{c;inlet}\,. \tag{4}$$

For this study, the inlet temperatures of $T_{h;inlet} = 250$ °C (523 K) and $T_{c;inlet} = 50$ °C (323 K) for the helium and water flows are taken. Figure 2 shows the one-dimensional temperature profiles for different heat exchanger wall types and flow configurations.

Given the temperature field, the tritium permeation at a metallic wall can be calculated. The local concentration of tritium that is dissolved in the metallic lattice and in equilibrium with the tritium gas at the wall surface facing the helium flow is given by

$$\Phi(z, x=\xi) = K\sqrt{P_{^3H}}, \; K = K_0 \exp\left[-\frac{\Delta H_S}{RT}\right]. \tag{5}$$

Here, K is the solubility of tritium and R is the universal gas constant, 8.314 J·mol^{-1}K^{-1}. The material specific values are the constant K_0 and the standard enthalpy of dissolution of tritium

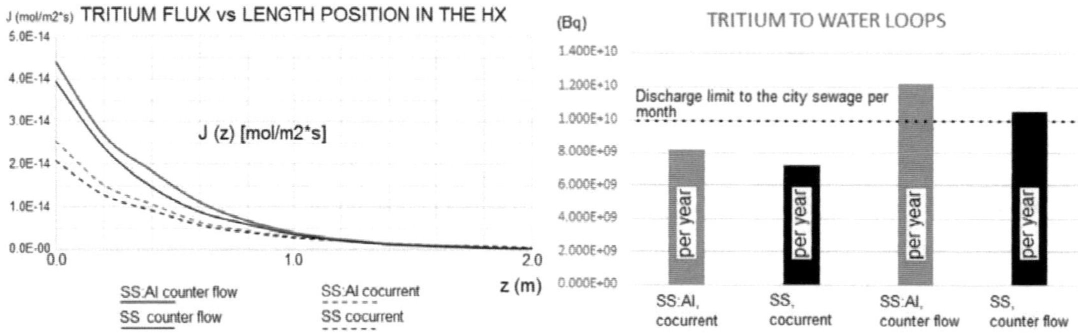

Figure 3. Diffusion flux profiles for different heat exchanger wall types and flow configurations, for the steady tritium impurity level of 0.074 ppm in the helium loop. The annual tritium contamination rate of the water in the intermediate cooling loop is shown on the right.

ΔH_S. In this study, we use the empirical values of K_0 and ΔH_S for the austenitic steel and aluminium, which are compiled in Ref. [1].

The tritium permeation rate can be anaytically calculated by solving the Fick's first law,

$$J(z) = -D(z,x)\frac{\partial \Phi(z,x)}{\partial x}, \quad D(z,x) = D_0 \exp\left[-\frac{\Delta H_D}{RT(z,x)}\right], \tag{6}$$

for the surface concentration of tritium atoms $\Phi(z, x = \xi)$, given by Eq. 5. Here, J(z) is the diffusion flux, $D(z,x)$ is the diffusivity and ΔH_D is the diffusion activation energy of tritium, which are barrier material specific. In this study, we use the empirical values of D_0 and ΔH_D for the austenitic steel and aluminium, which are compiled in Ref. [1]. We assume the tritium concentration on the wall surface facing the water flow is zero.

Figure 3 shows the one-dimensional diffusion flux profiles for different heat exchanger wall types and flow configurations. Also shown is the annual tritiation rate of the water in the intermediate cooling loop at ESS. The mono-metallic heat exchanger W001 with the co-current flow option results in the least annual tritium contamination of 7 GBq/year, among the different heat exchanger wall types and flow configurations compared. All the four cases satisfy the regulatory requirement of 10 GBq/month and 1 GBq/release. There are also tritium permeation via the heat exchangers W002 and W003. But, the amount of permeated tritium there is pretty much suppressed to a negligible level compared to W001 by orders of magnitudes, due to its much lower flow temperatures.

3. Conclusions and outlook

A co-current flow type mono-metallic heat exchanger made of austenitic steel provides a better tritium barrier between the target primary and intermediate cooling loops. The tritium amounts in the target intermediate water cooling loop is allowed for controlled discharge into the city sewage system, during normal operation.

In the future, sensitivity analyses will be performed for a optimal combination of the heat exchanger parameters, using the analytical model developed for the presented study and advanced CFD simulations.

References

[1] Causey R A, Karnesky R A, and San Marchi C 2012 Tritium barriers and tritium diffusion in fusion reactors *Comprehensive Nuclear Materials* Vol. 4 (Amsterdam, Netherlands) Elsevier 511

IOP Conf. Series: Journal of Physics: Conf. Series **1021** (2018) 012045 doi:10.1088/1742-6596/1021/1/012045

ESS Personnel Safety Systems Test Stand

D Paulic[1], S L Birch[2], M Mansouri[3] and Y K Sin[4]

[1] Deputy Group Leader, Protection and Safety Systems, European Spallation Source ERIC, Lund, Sweden
[2] Senior Engineer, Protection and Safety Systems, European Spallation Source ERIC, Lund, Sweden
[3] Engineer for Safety Critical Systems, Protection and Safety Systems, European Spallation Source ERIC, Lund, Sweden
[4] IEC61508 Electrical Controls Engineer, Protection and Safety Systems, European Spallation Source ERIC, Lund, Sweden

E-mail: denis.paulic@esss.se

Abstract. Providing and assuring safe conditions for personnel is a key parameter required to operate the European Spallation Source (ESS). The Personnel Safety System (PSS) primarily prevent workers from the facility's ionising radiation hazards, but also identify as well as mitigate against all other hazards such as high voltage and oxygen depletion. The PSS test stand is being developed in the ESS Protection Systems (PS) lab. It will be used for simulation, testing and verification purposes of any system, equipment, function and procedure developed by the PSS team. At the later stage of the project it will facilitate preparation for operation, maintenance and periodical testing of PSS, and hence it should be as similar as possible to real system. This paper describes the ESS Personnel safety systems' test bench design, devices, and implemented functions.

1 Introduction

The European Spallation Source ERIC (ESS), currently under construction, consists of a 600m long high power proton linear accelerator (Linac), accelerating proton pulses of 2.86ms length to the energy of 2GeV with a repetition rate of 14Hz. The proton pulses then interact with a rotating tungsten target wheel. Neutrons are created due to the spallation process and further guided through 22 different neutron beam lines towards the experimental stations. It is expected to have the first protons on target mid 2019 [1]

The Integrated Control Systems (ICS) division is responsible for the design, procurement, installation, commissioning and validation of all personnel safety systems at ESS.

2 Personnel safety systems

Like in many other facilities within the accelerator research field, the personnel safety systems (PSS) at ESS will be designed, manufactured, tested, commissioned, validated, operated and decommissioned in accordance with IEC 61508 international standard [2], which proved to be the best practice for developing such systems. On top of this, as part of the license application, the Swedish Radiation Safety Authority (SSM) has requested that ESS PSS meet SSM2014-127-1 [3] and SSMFS 2008-27 [4] requirements that include following:

- Radiation risk analysis shall be carried out before the facility is taken into operation.
- A formalised search of each PSS controlled area shall be carried out before the facility is operated.
- Two independent technical design solutions should be used in each system.
- External events, single failure, common cause failures, redundancy, diversity and separation shall be taken into account when developing the system.

The PSS primarily prevent workers from the facility's ionising radiation hazards, but also identify as well as mitigate against all other hazards such as oxygen depletion hazards (ODH), high voltage hazards, magnetic field hazards, laser hazards and motion control hazards.

The PSS controlled areas are hazardous areas where personnel might be affected if the beam is on or if any of hazardous equipment within the area is powered. They include: on-site cryogenic module test stand, Linac, target utility rooms, bunker and 22 neutron instruments (each instrument will have individual PSS). Some of these areas will be divided in smaller zones to facilitate various procedures. For example, the Linac has been divided into several zones that are separated by fenced gates, which facilitates the search procedure (i.e. reduces the time needed to search accessed areas).

There are four main personnel safety systems:

- Safety interlock system – ensures that if the beam is permitted there is no access to PSS controlled areas and vice versa, if access is allowed beam is not permitted. In case of emergency (e.g. intrusion to controlled area) it hard switches-off all hazardous equipment under control.
- Access control system - monitors the conditions for enabling the entry stations and controls door opening and access subsystems.
- ODH monitoring system – prevents the entry to controlled areas in case of ODH and reports the alarm by activating signalization lights.
- Radiation monitoring system – stops the proton beam by switching-off the designated equipment under control (i.e. ion source plasma generation and high voltage extraction systems) if radiation level goes beyond specific thresholds in some supervised areas.

3 PSS test stand

For simulation, testing and verifying different PSS systems, equipment, functions and procedures PSS team decided to build a test stand. Its development started in February 2015 and is still on-going in the Protection Systems (PS) lab at ESS. At the later stage of the project it will facilitate preparation for operation, maintenance and periodical testing of PSS, and hence it should be as similar as possible to final install system.

3.1 Layout

The PSS test stand consists of a caged area, which represents the PSS controlled area (see Figure 1), with double-door entry station, an emergency-exit door next to it, a section door which represents the gate between two PSS zones within the PSS controlled area, and two PSS cabinets painted in RAL 2000 (orange, ESS PSS specific color). In addition to above mentioned entry station, there are several access control subsystems (e.g. message display system, card readers, etc.) that will be used for simulating and testing different entry and exit procedures. The position of each door is monitored constantly in PSS test stand control system through pairs of mechanical and magnetic safety switches with self-diagnostic functions and in-built redundancy.

IOP Conf. Series: Journal of Physics: Conf. Series **1021** (2018) 012045 doi:10.1088/1742-6596/1021/1/012045

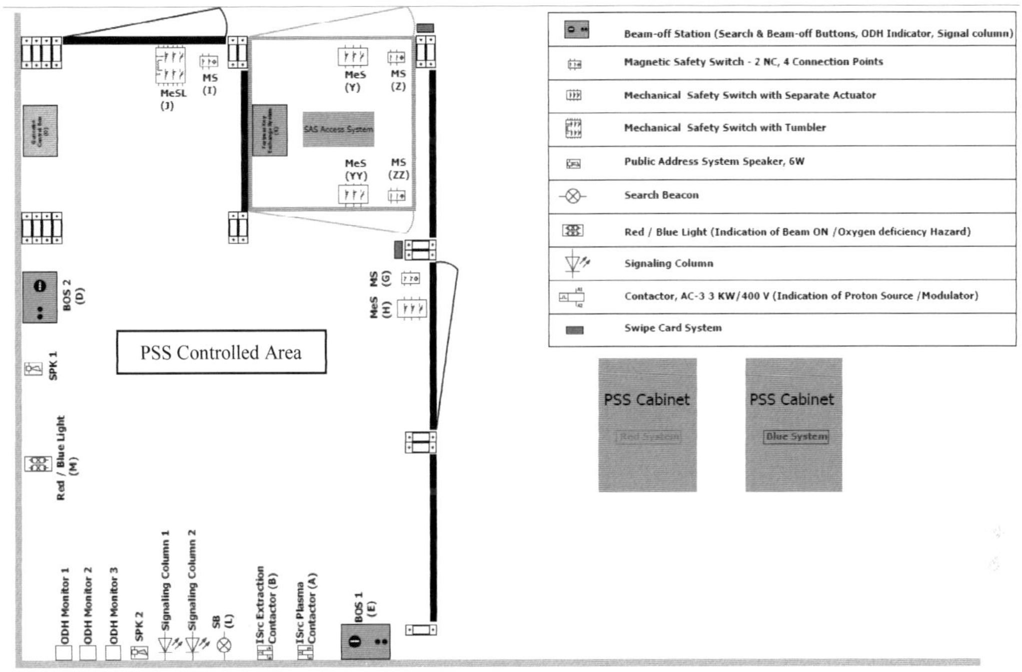

Figure 1 PSS test stand devices layout

3.2 *Test stand devices*

Figure 1 shows the layout of PSS test stand devices and equipment under control (EUC) inside the controlled area. The main functions of these devices are signalization and protecting people in case of hazard and preventing them from entering the hazardous areas in such cases. Below are listed some of the commercial-off-the-shelf (COTS) devices and equipment that are currently installed in PSS test stand:

- Physically separated two-train safety PLC system
- Rack HMI-s as PSS supervision stations
- Two-door security airlock system (SAS) entry station
- Beam-off stations
- Swipe card door access system
- Key exchange
- Public address system
- Message display system
- PSS blue and red area lights
- PSS contactors

3.2.1 Beam-off station. The beam-off station is an in-house designed PSS device which will be installed in all PSS controlled areas and will be used for several functions and procedures. The final number of beam-off stations in all controlled areas is not defined yet. The main purpose is to enable workers to switch-off beam (remove the hazard) from inside the PSS controlled areas in case of emergency. Each beam-off station will also have a search button, two light towers for signalisation and a buzzer for search procedure. Its preliminary design is shown in Figure 2. Two of them are currently installed at PSS test stand.

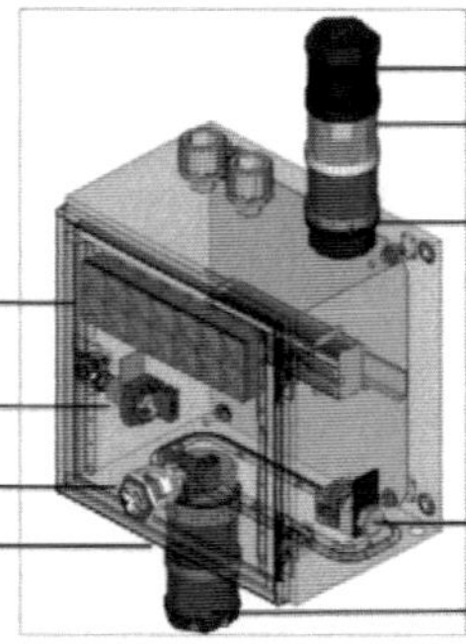

Figure 2 Preliminary design of the ESS PSS beam-off station

3.2.2 Entry station. A commercial-off-the-shelf (COTS) 2-door security airlock system manufactured by Gunnebo is used as an entry station at PSS test stand. It is a robust station which allows single-person entry, has two hours of battery back-up in case of power failure and glass doors so that personnel can see if the station is occupied, as well as the status of signalisation lights inside the controlled area before or during the entry process. A pair of Siemens safety switches is installed on each door for position monitoring, and doors are internally interlocked, which means that both cannot be open at the same time. There is also possibility for a real-time visual feedback to supervision system through the camera installed inside the station.

3.2.3 Safety PLCs. Two physically-separated, independent Siemens S7 1500 fail-safe programmable logic controllers (PLC-s) are used for functional safety implementation, principally through safety functions in the software, in order to bring the system to a safe state or maintain it in a safe state in case of a dangerous event [5]. Both PLCs run exactly the same safety program in parallel, but all standard components and non-safety functions are implemented only on one of them, which is therefore considered to be a main PSS PLC. The ODH monitoring system will be independent of this 2-train system, implemented in a separated rack and PSS team is still evaluating if a safety PLC will be needed for this. All PSS sensors and actuators are connected to these PLCs through local Siemens ET200SP distributed I/O stations with fail-safe I/O modules.

3.2.4 ODH monitors. Two Oxigraf O2iM oxygen deficiency monitors are installed and currently being tested at PSS test stand. They measure and display the oxygen concentration in gas sample drawn through the instrument. Gas is pumped sequentially from up to four input ports, through the oxygen sample cell, and out the exhaust port. There are several relays inside the monitor that can be used in ODH monitoring PLC control system.

4 Test stand functions

4.1 PSS test stand key exchange system

The key exchange system installed in PSS test stand has two different set of keys: one set in main PLC cabinet which represents the key exchange that will be placed in the Main Control Room (MCR) and will be used by the operator for selecting access modes and issuing beam permit; and other one inside the entry station for the type of access which requires having a key as a safety token (see the description of Restricted access in a following subsection).

This system is trapped key interlock type, which means that a safe sequence of operations is enabled through transfer of keys that are either trapped or released in a predetermined order. For example, a main key is used to remove the power for all hazardous EUC; this main key is then released and can be used to choose one of the access modes by inserting it into appropriate lock, which will then, in case of access with key, enable releasing appropriate access key. The action of taking this access key will lock (trap) the main key in place and it will remain trapped until the access key is returned to key exchange. Two access modes have already been developed and tested at PSS test stand and will be described in following sections. Table 1 shows all the modes that are currently implemented in PSS test stand control system.

Table 1 PSS test stand modes

Mode	Access status	Hazardous EUC permits
Beam ON	No access	All ON
Area closed	No access	According to requirements
Open access	Allowed – no key required	According to requirements
Search	Allowed – limited number of people, key required	According to requirements
Restricted access	Allowed – limited number of people, key required	According to requirements
Alarm	No access	All OFF

4.2 Search procedure
The purpose of formalised search is to make sure nobody's left in the controlled areas before issuing the beam permit. It will be done by authorised search team (or teams) of two people who will have a dedicated search key for entering the controlled areas. A safety program ensures that search is broken when required. Each beam-off station includes a search button, a search light and a buzzer that are used for search procedure. Search lights inform the search team which search button shall be pressed next, and buzzers are used to inform people inside the controlled area that search process is ongoing. If any door opening is detected during search or pre-defined search time is exceeded, the search process has to be re-done form beginning.

4.3 Entry to PSS controlled area in Open access mode
In Open access mode, a person can enter the PSS controlled area if authorized, but search is broken upon entering the area. Entries are counted by PSS software and can be limited inside the software if required.

4.4 Entry to PSS controlled area in Restricted access mode
In Restricted access mode search is not broken upon entry, but person must to take a key to enter the controlled area. It's very practical and time-saving to have such mode, especially in case of quick interventions, because the search process doesn't have to be done after and mechanically beam permit cannot be issued if all keys are not returned to key exchange. It's limited by number of keys in entry station's key exchange. The PSS test stand currently allows the entry of maximum 3 people in this mode.

4.5 Beam imminent warning and ODH signalization
For risk reduction purposes and to increase the awareness of some hazardous events, PSS will change all the lights inside controlled areas to blue when the beam permit is issued and to red if ODH alarm goes off. At PSS test stand, one blue and one red light are installed to test procedures in such cases. The outcome is showed in Figure 3. Besides these lights, each beam-off station will have its own red and blue lights for informing workers in case of hazard (see Figure 2). All mentioned lights are controlled by PSS safety program and their functionality testing will be part of PSS annual maintenance procedures. In parallel to above mentioned lights, a PSS public address system will inform personnel if hazard occurs.

Figure 3 Left: beam imminent warning blue area lights; right: ODH alarm red area lights

5 Conclusion

The ESS is close to an installation of first systems that will require functional personnel safety systems. Thus, it is important to test and verify the concept, design, devices and all procedures in PSS test stand to ensure the PSS safety functions, but also to provide proper system integrity. Since the requirements for PSS are still being defined, the current PSS design is not mature enough. Therefore, PSS test stand will play a crucial role in coming years to improve PSS design and deliver required systems on time.

References

[1] Yeck J 2015 Neutron facility: European Spallation Source is on track *Nature* **519** 7543 291
[2] IEC61508-2010 Functional safety of Electrical/electronic/programmable electronic safety related systems
[3] SSM2014-127-1 Review of application for licence for activity involving ionizing radiation chapter 10.
[4] SSMFS 2008-27 The Swedish Radiation Authority's regulations concerning operations at accelerators and with sealed radiation sources
[5] Birch S L *et al* 2015 Personnel safety systems for the European Spallation Source *Proceedings of IPAC2015* MOPTY046

IOP Conf. Series: Journal of Physics: Conf. Series **1021** (2018) 012046 doi:10.1088/1742-6596/1021/1/012046

The gaseous discharges at ISIS and the activated air composition effect

Goran Škoro

ISIS Neutron and Muon Source, Science and Technology Facilities Council, Rutherford Appleton Laboratory, Didcot OX11 0QX, United Kingdom

E-mail: `goran.skoro@stfc.ac.uk`

Abstract. Activated air at ISIS is discharged through five ventilation stacks. In each stack, a portion of activated air is diverted through a drum where the beta and gamma activity is measured using two Geiger–Muller detectors (one of them is surrounded by a beta particles shield). The procedure to get conservative estimate of the activity discharge rate from the stack(s) is presented in this paper. It is shown that without any knowledge about the relative proportions of radionuclides, gas mixing and details of ventilation system it is possible to have most conservative estimate (maximal value) of the specific activity of the activated air. This most conservative estimate of activity discharge rate is what is usually needed to check the status of compliance with applicable regulations.

1. Introduction

Activated air at ISIS Neutron and Muon Source is discharged through the ventilation stacks. Each stack has a simple, but effective, monitoring system that takes filtered air from downstream of the stack, feeds it into a drum (200 litre volume) and returns the air from the drum to the stack [1]. The $\beta\gamma$ activity of the air is measured using two Geiger–Muller (GM) detectors located in the middle of the drum [1, 2, 3, 4]. Air from the discharge stack is continuously passed through the drum so, one can say, the air in the drum has the same specific activity as the air being discharged up the stack. By measuring the count rate of the detectors and with knowledge of the gas flow rate up main stack, gas flow rate through a drum, detection efficiency and, what is most important, the relative proportions of radionuclides in activated air, the value of air specific activity can be obtained. In reference [4], equations for the rate of discharge of $\beta\gamma$ activity from a stack were given. In the complete mixing approximation (assuming that gas entering the drum is immediately distributed uniformly throughout the volume of the drum), the activity discharge rate (given in Bq/s, for example) is:

$$A'_{com-mix} = \frac{FC'}{f\epsilon} \frac{\sum_{i=1}^{N} \lambda_i r_i}{\sum_{i=1}^{N} \frac{r_i}{1+\frac{1}{\lambda_i T}}}. \qquad (1)$$

Here, F is the gas flow rate up main stack, $C^{'}$ is the total count rate from the Geiger–Muller detector[1] and f is the gas flow rate through a drum ($f << F$). λ_i is the decay constant[2] for the radionuclide i and $r_i = n_i/n_{total}$ is the relative proportion of the radionuclide i in the activated gas (n_i and n_{total} are given in number of atoms per unit volume). The detection efficiency ϵ is assumed to be equal for all i ($\epsilon_i = \epsilon$). T is the transit time through the drum ($T = V/f$, where V is the volume of drum).

In the no-mixing approximation (assuming that every molecule of gas entering the drum takes the same, transit, time to pass through the drum), the activity discharge rate is:

$$A^{'}_{no-mix} = \frac{FC^{'}}{f\epsilon} \frac{\sum_{i=1}^{N} \lambda_i r_i}{\sum_{i=1}^{N} r_i(1 - e^{-\lambda_i T})}. \tag{2}$$

Both equations have identical form when the half-lives of all radionuclides are much longer (or much shorter) than transit time through the drum:

$$A^{'} = A^{'}_{com-mix} = A^{'}_{no-mix} = \frac{FC^{'}}{f\epsilon T} = \frac{FC^{'}}{\epsilon V}. \tag{3}$$

The 'beauty' of equation (3) lies in the fact that all the variables can be directly measured so it can be used as a first estimate of $\beta\gamma$ activity discharges from the stacks.

2. Analysis

The main condition for using equation (3) for stack discharge calculations is that the half-lives of all radionuclides in activated air are significantly different than transit time T through the drum. The most frequently seen ($\beta\gamma$ emitting) radionuclides in activated air at proton accelerators are[3]: ^{11}C ($T_{1/2} = 20.4$ minutes), ^{13}N ($T_{1/2} = 9.96$ minutes), ^{15}O ($T_{1/2} = 2.04$ minutes) and ^{41}Ar ($T_{1/2} = 109.6$ minutes). The measurement of transit time T through the one of the ISIS drums using a ^{85}Kr source gave the following value: $T = 6$ minutes [5]. As can be seen, the half-lives of the radionuclides listed above are similar to the transit time. The consequence is that equation (3) underestimates stack(s) discharges.

This means that equations (1) or (2) must be used to obtain the 'correct' value of the activity discharge rate, or in other words, the relative proportions of radionuclides in air (at measuring/discharging point) must be known. Generally speaking, the relative proportions of radionuclides can be estimated via Monte Carlo simulations of production of airborne radionuclides. However, in some cases there are many different production spots of airborne radionuclides and/or complicated ventilation systems so performing such a simulation is not an easy task. ISIS Neutron and Muon Source with its accelerator, two target stations and very complicated ventilation system is a typical illustration of such a case. This means that in the ISIS case the relative proportions of radionuclides at the discharging point can be significantly different than corresponding values at the production spots. So the interesting question here is: Is it possible to have any sort of estimate of the air specific activity if the relative proportions of radionuclides in activated air are practically unknown?

Figure 1 shows how strongly the activity discharge rate depends on the relative proportions of four radionuclides listed above. The procedure for making this plot was to: vary r_i values

[1] As the radionuclides of interest all emit β^- or β^+ particles with 100% branching ratio on each decay, it is assumed that the total activity discharged will be proportional to the difference in counts of the unshielded and shielded GM detectors.

[2] $\lambda = (\ln 2)/T_{1/2}$, where $T_{1/2}$ is a half-life.

[3] Beta emissions from ^{3}H are below the minimum energy which can be detected by the GM detectors, and it is monitored using a bubbler system. ^{7}Be is particulate and it is trapped by the HEPA filters.

IOP Conf. Series: Journal of Physics: Conf. Series **1021** (2018) 012046 doi:10.1088/1742-6596/1021/1/012046

between 0 and 1 (taking into account that: $\sum_{i=1}^{4} r_i = 1$), use equations (1), (2) and (3) to calculate the following ratios:

$$M_{com-mix} = \frac{A'_{com-mix}}{A'}, \tag{4}$$

and

$$M_{no-mix} = \frac{A'_{no-mix}}{A'}, \tag{5}$$

and plot these ratios against all the possible combinations of the relative proportions of four radionuclides. Visual inspection of Figure 1 reveals that, in the worst-case scenario, 'real' activity discharge rate can be more than 3 times higher than the value calculated using equation (3). As expected, the correction factors for complete-mixing case are always higher than corresponding values in no-mixing case.

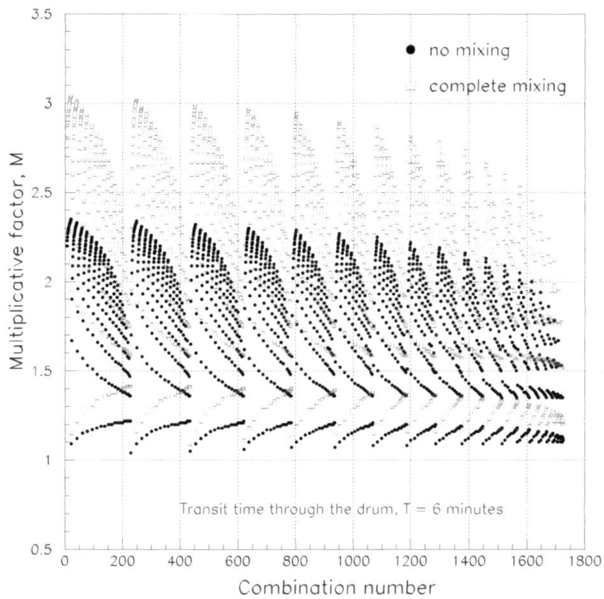

Figure 1. The multiplicative factor M (see text for definition) versus all the possible combinations of the relative proportions of four characteristic radionuclides (^{11}C, ^{13}N, ^{15}O and ^{41}Ar) in activated air. The step in varying the relative proportions was 0.05 so there is 1725 combinations in total.

It has to be noted, however, that worst-case scenario presented above is not realistic. Very high values of $M_{com-mix}$ and M_{no-mix} correspond to very big proportion of ^{15}O in activated air and this is highly unlikely. Available literature data shows that maximal expected fraction of ^{15}O in activated air at proton accelerators is not higher than 20%. The multiplicative factor values in this case are shown in Figure 2.

On the basis of the above mentioned, one could say that even without any knowledge about the relative proportions of radionuclides, gas mixing and details of ventilation system it is possible to have conservative estimate on the activity discharge rate from ventilation stack(s). The procedure to get this, most conservative, estimate (if a monitoring system described in this paper has been used) is to:

- measure count rate from the Geiger–Muller detector(s), detection efficiency, gas flow rate up main stack and gas flow rate through a drum;
- calibrate the drum (measure the transit time);
- use the equation (3) to calculate first estimate of the activity discharge rate;
- calculate factors M using the procedure described above;

- and multiply obtained result for the activity discharge rate by the maximal value of M (in this example $M = 2.7$ for transit time of 6 minutes).

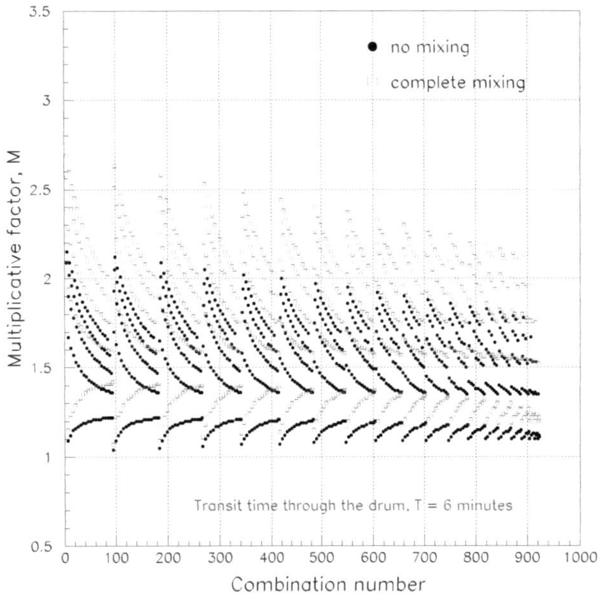

Figure 2. The multiplicative factor M (see text for definition) versus possible combinations of the relative proportions of four characteristic radionuclides (^{11}C, ^{13}N, ^{15}O and ^{41}Ar) in activated air. The relative proportion of ^{15}O is varied in the range $[0.0, 0.2]$.

3. Conclusions

In this paper, the procedure to get conservative estimate of the air activity discharge rate from the ventilation stacks is presented. It is shown that without any knowledge about the relative proportions of radionuclides, gas mixing and details of ventilation system it is possible to have most conservative estimate (maximal value) of the specific activity of the activated air. This most conservative estimate of activity discharge rate is what is usually needed to check the status of compliance with applicable regulations.

References

[1] Wright P 2011 Monitoring and calculation of beta-gamma discharges from ISIS using GM systems *ISIS Internal Report.*
[2] Findlay D J S 2011 ISIS gaseous discharge stack monitors equations - 1 *ISIS Internal Report.*
[3] Findlay D J S 2014 Gaseous discharges and environmental dosimeters *ISIS Report ISIS-DJSF-14-04-A.*
[4] Findlay D J S 2011 ISIS gaseous discharge stack monitors equations - 2 *ISIS Internal Report.*
[5] Burge F et al 2012 Calculation and measurement of gaseous radioactive discharges from ISIS stacks *ISIS Internal Report.*

IOP Conf. Series: Journal of Physics: Conf. Series **1021** (2018) 012047 doi:10.1088/1742-6596/1021/1/012047

Target Safety System design for the ESS target station

Atefeh Sadeghzadeh[1], Mikael Olsson[1], Linda Coney[1]

[1]European Spallation Source Eric, Tunavagen 24, 223 63, Lund, Sweden

Atefeh.sadeghzadeh@esss.se, mikael.olsson@esss.se, linda.coney@esss.se

Abstract. The European Spallation Source (ESS) will be a 5 MW neutron spallation research facility where an energetic proton beam incident upon a helium-cooled tungsten target is converted to neutron beams. The liquid hydrogen moderator, water moderator, and reflector systems within the target monolith slow the high-energy spallation neutrons down to cold and thermal neutrons suitable for use by experiments. Several layers of control, protection, and safety systems will be implemented in order to ensure safe operation of this advanced facility.
The Target Safety System (TSS) is part of the overall radiation safety plan for the Target Station. As such, in case of an abnormal event, the TSS will protect the public from exposure to unsafe levels of radiation, prevent the release of radioactive material beyond permissible limits, and bring the neutron spallation function into a safe state. This safe state is defined as no beam incident upon the target. Since the target is the source of the release of radiation in the presence of the beam, putting the target station into the safe state (no beam), brings the spallation process to the safe state.
The TSS safety functions are based on the results from hazard and accident analyses done to evaluate radiation hazards within the target station. The most critical hazards are related to increased temperature of the target wheel tungsten, which can be the result of cooling system malfunctions or stopping of the rotating target. Therefore, the TSS will monitor parameters related to the cooling conditions and rotation of the target.
As a safety system, the TSS is subject to requirements from the Swedish Nuclear Safety Authority (SSM), including single failure and common cause failure criteria. These requirements impact the system architecture in terms of redundancy, separation and diversity. Dedicated redundant and diverse TSS sensors are connected to two diverse and separated trains performing the same safety functions. Each train acts on diverse and separated mechanisms to stop the beam from reaching the target.
This paper will describe the design and architecture of the TSS, which implements the safety functions identified in the radiation safety analyses and fulfils the SSM conditions.

1. TSS safety functions

The ESS target radiation safety functions were derived from the hazard and accidents analyses of target station systems and areas. A qualitative hazard analysis was performed to identify and evaluate potential radiological accidents, from which a collection of bounding events was selected for further analysis. The accident analysis detailed the identified accidents to determine the related level of risk. This involves quantification of severity, as measured by the dose consequences to both workers and the public, and definition of appropriate control actions to enable an acceptable level of risk.

To mitigate consequences of the accidents, different functions were identified in different levels of defence in depth (DiD) for systems in the target station. Functions to be fulfilled by the TSS were identified in DiD level three. Depending on the level, different constraints shall be applied on the design in terms of conditions from SSM and design guidance from applicable standards.

Since the target contains a high inventory, many of the accident analyses address scenarios that could affect the target material or the helium cooling system. The most critical hazards tend to be related to increased temperature of the target wheel tungsten, which, if accompanied by oxidation of the tungsten and a loss of confinement, might have consequences of radiological releases. The following accident scenarios require TSS functions in order to prevent or mitigate the unacceptable consequences:

- AA1: Target wheel rotation stop during beam on target
- AA2: Proton beam events on target and proton beam window (non-rastered & focused beam)
- AA3: Loss of target wheel cooling during beam on target wheel

In these accident scenarios, the increase of temperature in the target material leads to unacceptable radioactive material releases. Since the target is designed so that the decay heat can be dissipated by passive means, removing the beam removes the source of heat and puts the spallation process into the safe-state.

The following safety functions are dedicated to the TSS. The TSS shall monitor process variables in the wheel, helium cooling, and monolith systems to identify if the:

- Target helium cooling outlet velocity is below a certain limit
- Target helium cooling outlet pressure is below a certain limit
- Target helium cooling inlet temperature is above a certain limit
- Target wheel rotational speed is below a certain limit
- Monolith atmosphere pressure is above a certain limit

If any of the above conditions occur, the TSS shall bring the ESS spallation process to a safe-state (in terms of radioactive releases) by turning off the proton beam to prevent escalation of the situation.

Figure 1 illustrates the accident analyses (AAs) and location of the process variable monitored for each radiation safety function.

Process variable / Accident	Target rotational speed (1)	Outlet pressure (2)	Monolith pressure (3)	Inlet temperature (4)	Outlet velocity (5)
AA1 Target Wheel stop during beam on target	X	X	X		
AA2 Beam event		X	X		
AA3 Loss of He cooling (blower fail)		X	X		X
AA3 Loss of He cooling (heat exchanger fail)		X	X	X	
AA3 Loss of He cooling (leakage)		X	X		

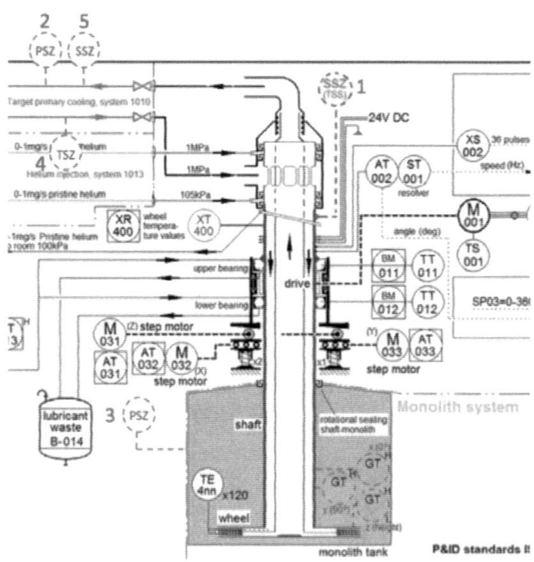

Figure 1 TSS process variables for safety functions

2. TSS Requirements

TSS is identified as a safety SSC (Structure, System, Component) in [1] which requires TSS to include redundancy, diversity, independence, and physical separation to the extent that it can meet the SSM

IOP Conf. Series: Journal of Physics: Conf. Series **1021** (2018) 012047 doi:10.1088/1742-6596/1021/1/012047

conditions for single failures and common cause failures.

2.1. Redundancy
In engineering, redundancy is the duplication of critical components or functions of a system with the intention of increasing reliability of the system [2]. Redundancy in TSS will be realized as:

- Two-train architecture: TSS is designed as a two-train interlock system. Either of the TSS trains can execute the expected safety functions.
- Redundant trip devices (contactors) to shut down proton beam production. The contactors will be placed in series on the relevant accelerator power circuits.
- Redundant sensors with a 2oo3 voting functionality in the target helium cooling system, the target wheel rotation system, and monolith system

2.2. Diversification
ESS general implementation rule states that diversity shall be implemented for a safety SSC to prevent common cause failures in redundant parts. For TSS specifically, diversity will be implemented to prevent internal failure of two independent components at the same time. Diversity will be realized as:

- Diverse technology for platforms (PLC vs. relays) and cable material (fibre vs. copper)
- Diverse trip mechanisms: Ion source vs. RFQ
- Diverse vendors for sensors and trip devices
- Functional diversity: diverse measurements to identify the same event

2.3. Independence
In order to fulfil functional separation, which is a condition from SSM to the safety systems, TSS shall be independent from other systems. Independence will be fulfilled by:

- TSS acts independently of other systems. There will only be one-way communication of monitoring data to other control systems.
- TSS trains act independently of each other with no inter-train safety-credited communication.

2.4. Physical separation
Physical separation shall be implemented for a safety SSC to minimize the risk that events give rise to failure in redundant parts. For TSS specifically, physical separation will be implemented to maintain functionality in case of an external event, such as fire, that could cause failures in both trains or in redundant components within one train. Realization of physical separation for TSS will be:

- TSS physically separated from other control systems
    - TSS dedicated sensors, logic and trip devices
    - TSS dedicated rooms for logic
    - TSS dedicated closed trunks/pipes/conduits for cables
- TSS trains physically separated from each other
    - Separate locations for logic devices
    - Separate locations of trip devices
    - Separate racks or cabinets in combined areas
    - Separate closed trunks/pipes/conduits for cables

3. TSS Architecture
The TSS architecture is fundamentally a two-train fail-safe interlock system. TSS will continuously monitor safety parameters in the target He cooling, wheel, and monolith atmosphere systems, evaluate their condition, and shut down the accelerator if necessary. Figure 2 shows the layout of TSS

IOP Conf. Series: Journal of Physics: Conf. Series **1021** (2018) 012047 doi:10.1088/1742-6596/1021/1/012047

equipment within the ESS facility, highlighting the locations relevant to each train. It also illustrates the physically separated cable routes of the two trains through different areas of the facility.

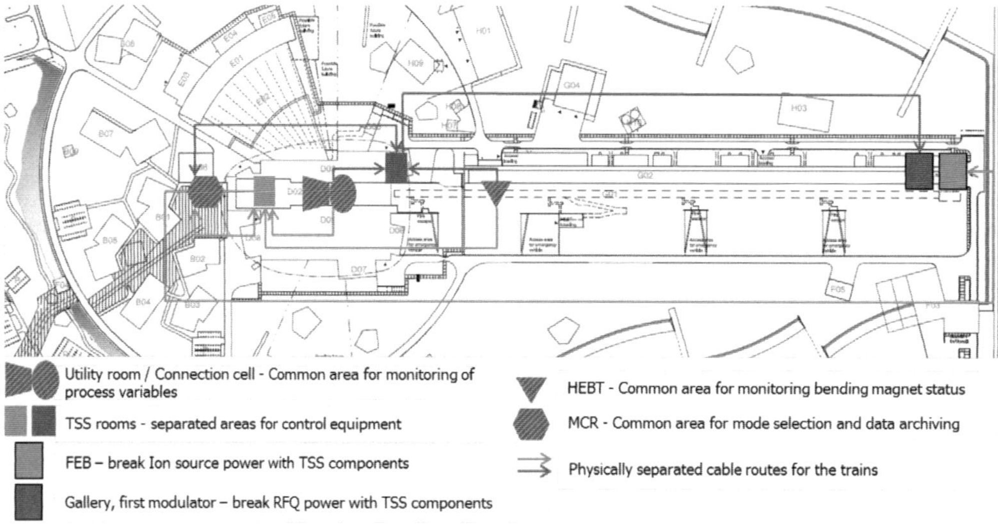

Figure 2 TSS layout

As shown in Figure 3, the two trains will measure the same process variables and perform the same logical evaluation of the variables, but will act on separate mechanisms to shut down the beam. The system will be independent of other systems and will rely solely on its own dedicated equipment.

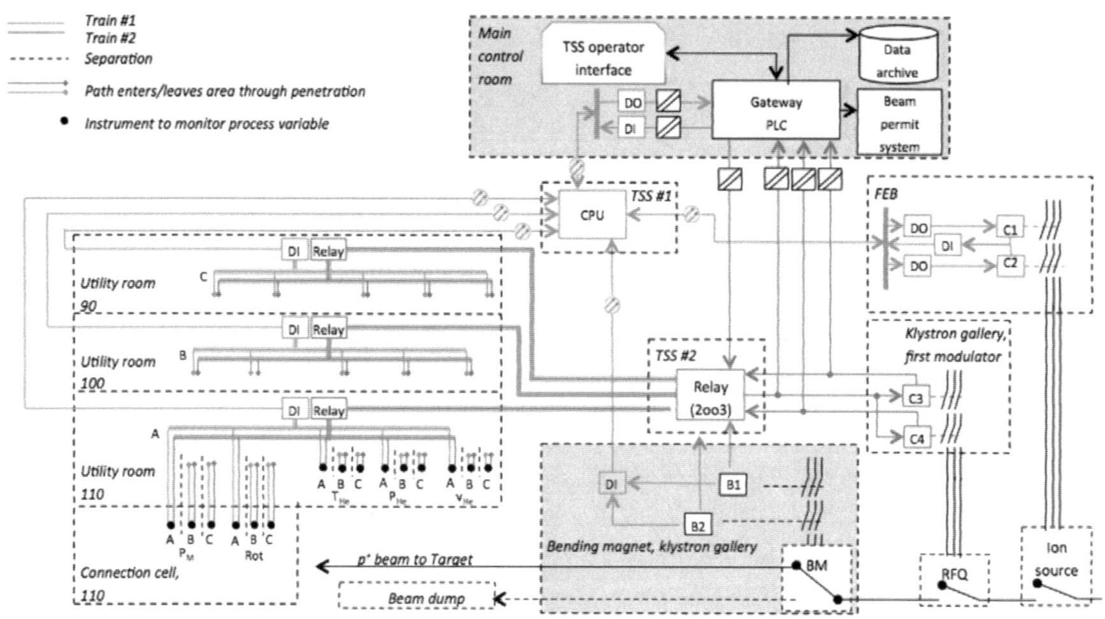

Figure 3 TSS architecture

Each train can prevent the beam from reaching the target by removing the active control signal to contactors on the power circuits to the ion source or to the RFQ. The control signal to the contactors must remain active in order to produce the proton beam, i.e. the system is fail-safe.

Diverse sensors will measure each process variable, and all sensor values will be sent to both trains. The logic handler of each train treats input from the redundant sensors with a 2oo3 voting function. The logic solvers are pieces of diverse control equipment to be located at physically separated areas within the Target Building. They will evaluate the sensor information and control the contactors.

The actuators are diverse and physically separated power contactors. They will be located at the front-end of the accelerator tunnel and act on the power circuits feeding the equipment that generates the proton beam at the ion source and the RFQ. If any of these power circuits is broken, no proton beam can be propagated through the accelerator to the target.

TSS is a fail-safe system. The PLC will continuously run diagnostic tests of status on its own HW, connected cables, and feedback from actuators. Any failure will force the system to the safe-state. Back-up power is not accredited for any of the TSS safety functions. TSS components have a pre-defined fail-safe state in case of loss of power, leading to the safe-state of the system. While not accredited, back-up power will be used for availability reasons.

4. TSS Interfaces

The TSS will interface with other ESS systems according to Figure 4. From the target station systems, critical process values are monitored and evaluated continuously. Any sign of failure will cause the action to go to the safe state. The TSS will extract information from accelerator systems related to operational modes. TSS has its own circuit breaker in the power line of the dogleg bending magnets to force the beam direction to the dump while continuously monitoring feedback from these breakers. In this way, the accelerator can send beam to the dump while the target is not ready to receive beam. The status of the TSS will be available for monitoring to operations personnel in the main control room.

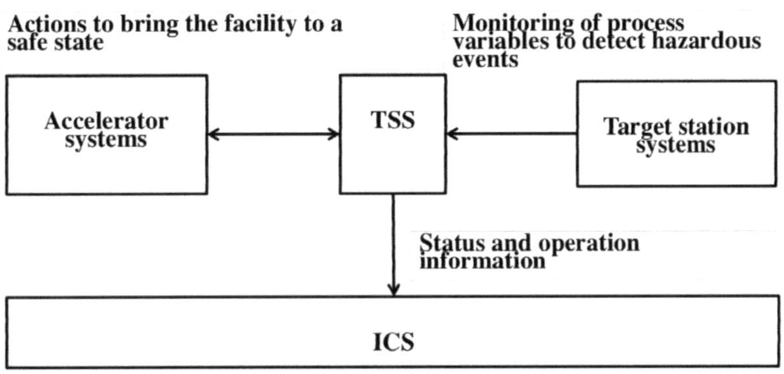

Figure 4 TSS interfaces

5. Conclusions

The TSS is a robust and independent safety interlock system. It has dedicated equipment for monitoring, evaluation, and actuation, and does not rely on any other system in order to perform its safety functions. The design is fail-safe and the actuation of the safety functions is performed by passive means. The TSS design includes redundancy and diversity, which fulfils SSM conditions for resistance to single failure and common cause failure.

References

[1] Francois Javier, ESS rule for identification and classification of safety important components, European Spallation Source Report ESS-0016468 (2016)
[2] https://en.wikipedia.org/wiki/Redundancy_%28engineering%29

Development of the sample environment system for the DN-12 diffractometer on the IBR-2M pulsed reactor (pressure – temperature – magnetic field). Project status.

A Chernikov[1], I Dobrin[2], N Kovalenko[1], S Kulikov[1], O Culicov[1,2] , I Popovici[2], D Enache[2] and A Dobrin[2]
[1] Joint Institute for Nuclear Research, 6 Joliot-Curie, 141980 Dubna, Moscow region, Russia
[2] National Institute for R&D in Electrical Engineering ICPE-CA, Splaiul Unirii, Nr. 313, Sector 3, 030138, Bucharest, Romania

E-mail: ksa@nf.jinr.ru

Abstact. The neutron diffraction experiments with microsamples use specific sample environment systems. One of such systems is under development in this project for the DN-12 diffractometer on the IBR-2M reactor. It is needed to keep of the investigated sample in the next simultaneous conditions: a high-uniformity magnetic field up to 4T, high pressure and low temperatures down to 4 K.
The arrangement of detectors of the diffractometer restricts the operations of the magnet construction and the magnetic field uniformity. The developing magnet is a superconducting split-coil type magnet. This provides the neutron scattering perpendicularly towards the vector of the magnetic field and to the beamline, as well as the scattering directed forward under the angle of 45 degrees.
The HTSC (high-temperature superconductor) magnet is made of YBCO tape with a critical current in the range of 300A@77K. The cryocooler ("cryogen-free") has been used for cooling the magnet. The cryostat also holds coaxial cylindrical channel from stainless steel provided by its own vacuuming system with an internal diameter of 79 mm for the access of the sample into the existing high pressure cell, which is cooled by the second cryocooler. Both of the cryostats are in a horizontal lay-out.
This report focuses on the description of the system and the parameters obtained at the first experiment.

1. Introduction

The investigation of structural mechanisms of the formation of magnetic states in various types of complex magnets is one of the vital problems in condensed matter physics. Important information on this subject can be obtained as a result of simultaneous research of crystalline and magnetic structures by varying structural parameters.

Neutron diffraction experiments under high pressures at various temperatures provide unique possibilities for studying mechanisms of the formation of magnetically ordered states in magnets via controlled variation of structural parameters (interatomic bond lengths and angles, atomic displacements from equilibrium positions).

On the pulsed high-flux IBR-2M reactor from JINR-Dubna, high-pressure investigations are performed on two neutron diffractometers – DN-12 and DN-6, which are equipped with horizontal cryostats for carrying out experiments in a temperature range of 4 - 300K.

2. System description

The preliminary calculations of the quantity of the required superconductor have shown that the generation of a magnetic field of 4 T requires 1480 meters of 12-mm-wide HTSC tape to form two double pancake coils. The manufacturer SuperOx [1] produces 200-m-long pieces of HTSC tape with guaranteed properties. SuperOx produces long tape by means of SnPb soldering separate pieces with 20 $*10^{-9}$ohms resistivity on junction at 77 K [2].

Changing sample temperature requires a cryostat-insert for the magnet. This cryostat is planned to be equipped with holders and containers of high-pressure cells. The high-pressure cells of non-magnetic materials are to be used.

It is proposed to employ RDK408S cryocooler to cool the magnet. In the current design, the expected temperature of the magnet is about 15 K. At this temperature a critical field for HTSC tape is significantly higher than the estimated field of the magnet.

Carrying out temperature measurements depending on the value of a magnetic field requires the application of RDK101D cryocooler with a final temperature of ~ 3 K. Figure 1 presents the layout of a cryogenic facility with a superconducting magnet.

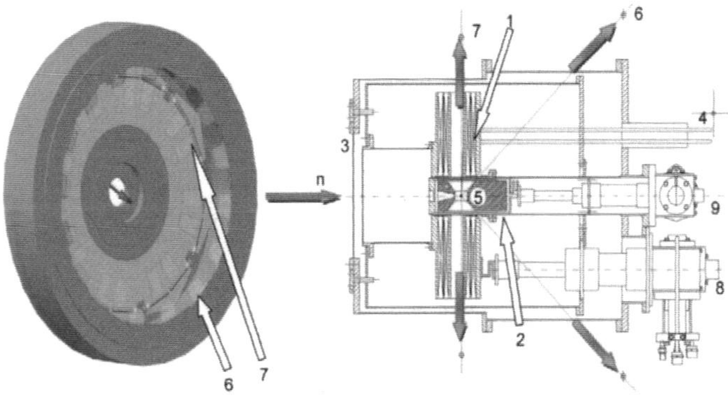

Figure 1 - The HTS magnet layout: 1 – superconducting magnet; 2 – cryostat-insert; 3 – entrance window for neutrons; 4 – current leads (up to 300 A); 5 – high-pressure cell; 6 and 7 – detectors of the scattered neutrons at the angles of 45° and 90°; 8 – cryorefrigerator RDK408S; 9 – cryorefrigerator RDK101D

2.1 Characteristics of the coils and the magnet

The required length of the 12mm wide tape YBCO type for each coil was achieved by joining the parts using SnPb solder. For each coil, SuperOx produced tape with two junctions. The quality of the junctions of tape pieces with SnPb solder was investigated in [2].

Next, two double pancake coils were wound, pre-coated with a layer of kapton 7 μm thick on one side. The winding was carried out using epoxy glue EPO-TEK 301-2. The inner diameter of the coils is 84mm, the outer diameter of the coils - 320 mm. The thickness of the double pancake coil is 27 mm. The electrical resistance of each coil is ~ 25 ohms at 300 K. The distance between the coils in the magnet frame is 35 mm.

The two HTS pancake coils were assembled together into a thermally conductive support made of aluminium alloy (Figure 2).

Figure 2 - Magnet assembly

2.2 Cryostat and magnet

The first tests of the magnet were carried out in the vertical orientation of the cryostat, which simplifies the assembly and adjustment work. The location of the magnet and HTSC current leads in the cryostat is shown in Figure 3. The magnet is mounted on the central pipe-shaft. The cryostat insert is injected into this shaft. The cryostat insert is a SRDK101D cryocooler on the second stage of which the sample chamber is mounted. Figure 6 shows the general view of the cryostat in a vertical position.

Figure 3 - The position of the magnet in the cryostat. HTSC current leads

3. Preliminary experimental results

3.1 Temperature decreasing during cooling

Since the magnet has a mass of about 20 kg, its cooling is a fairly long process. It takes three days to reach the working temperature. Figure 4 shows the dependence of temperatures of different parts of the cryostat on time in the cooling process. The DT-670A thermometers were located on the flange cooled by the first stage SRDK408S in the immediate vicinity of the HTSC current leads. The temperature of the hot ends of the HTSC current leads was also monitored. We note that the temperature of the upper part of the current leads is decreased below 60K. The upper part of the current lead between the main cryostat flange and the flange thermally coupled to the first stage of the cryocooler, is a copper tape of 1 mm thick, 30 mm wide and 400 mm long. An electrically insulated thermal contact of the copper current lead was made to the flange of the first stage of the cryocooler. We expect that for a current of 300 A, the upper end of the HTSC current lead to be heated to a temperature of less than 65 K. The temperatures of the upper and lower parts of the magnet were also monitored. The achieved temperature of the magnet was 16 K.The temperature of the second stage of the cryocooler was 8.2 K.

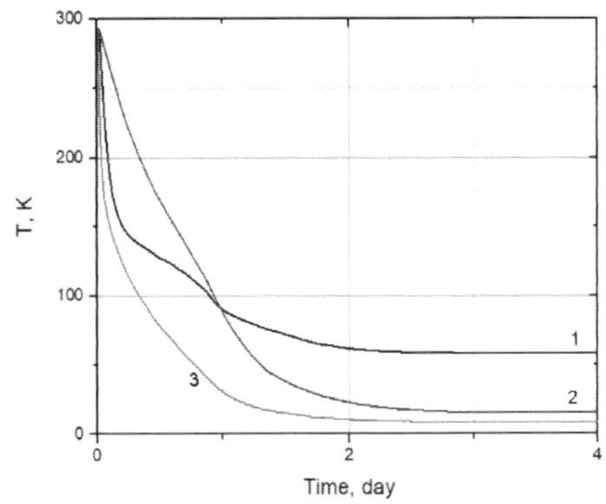

Figure 4 - The magnet cooling process: 1 – HTSC current leads at the first stage; 2 – the top and the bottom of the magnet, 3 – the second stage of SRDK408S

These final temperatures show a good agreement with computed ones by a Comsol Multiphusics[R] numerical simulation made on the CAD model of the system [3]. The results of the cooling process show also the adequacy of the system design for the purposes of the project.

3.2 Magnetic field generation, experimental results

Figure 5 shows the results obtained with the introduction of a current into the magnet. The permanent of the magnet was measured, which was 0.0154 T / A. A current of 90 A did not lead to a quench. A current of 105 A led to a quench. A, B, C are the values of the potentials at different points of the coils as a function of time; D, E are the potentials at the HTSC current leads including connection. The dashed line indicates the temperature of the top and the bottom of the magnet as a function of time. The time from 13:28:24 to 13:29:24 is a time period with a current of 105 A. Potential C is removed from the jumper between the coils, which is a copper plate. Potential C has linear increase, indicating the increase of resistance in this section, or heating. The increase in the value of the derivative of the temperature as a function of time at 13:29:24 indicates a quench, which was also monitored by measuring the total potential of the coils. As we assume, the heating in section C led to the quench because of the absence of thermal contact to the second stage of the cryocooler. Whereas HTSC current leads in the lower part had thermal contact with the second stage of the cryocooler, which gave sufficient cooling to the incoming and outgoing ends of the HTSC tape of the magnet.

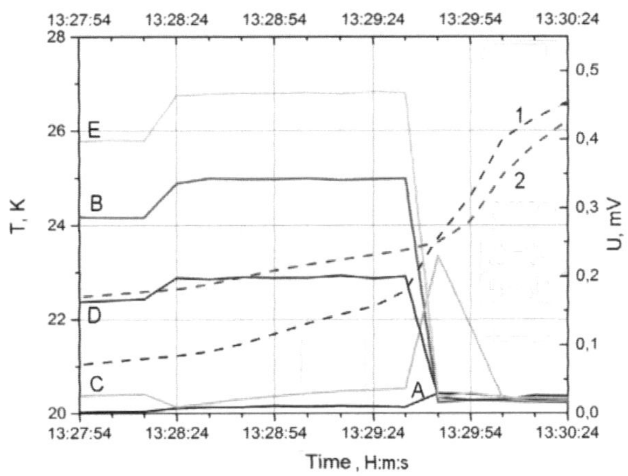

Figure 5 - Temperature and voltage evolution during charging/discharging the magnet: 1, 2 – temperature on the top and at the bottom of the magnet case; A, B, C – junctions potentials; D, E – HTSC current leads potentials. The magnet constant is 0.0154 T/A. The achieved field magnitude of 1.5 T in the first test. The reason – quench in the central connection of the coil windings. The connection will be upgraded

IOP Conf. Series: Journal of Physics: Conf. Series **1021** (2018) 012048 doi:10.1088/1742-6596/1021/1/012048

4. Conclusions and further study

Further work will include: improvement of the system of thermal and electrical contacts of the magnet and current leads; testing in the required temperature ranges and magnetic field values; test in the final, horizontal position and installation work on the DN-12 diffractometer of the IBR-2M reactor.

The authors acknowledge the project funded by Romanian contribution to the Joint Institute for Nuclear Research – Dubna, Russia under contract no. 08626319/156201211-74/2015. The authors are grateful to S. Kichanov and E. Lukin for formulating the problem.

References

[1] SuperOx Company http://www.superox.ru/en/
[2] Balashov N, Degtyarenko P and Ivanov S 2015 *Progress in Superconductivity and Cryogenics* **17** pp 25-27
[3] Dobrin I *et al* 2016 *IEEE Transactions on Applied Superconductivity* **26** 4500404.

IOP Conf. Series: Journal of Physics: Conf. Series **1021** (2018) 012049 doi:10.1088/1742-6596/1021/1/012049

Shielding analysis of Transmutation Experimental Facility

Hiroki Iwamoto[1], Hiroki Matsuda[1] and Shin-ichiro Meigo[1]

[1] J-PARC Center, Japan Atomic Energy Agency, 2-2, Shirakata, Tokai-mura, Naka-gun, Ibaraki 319-1195, Japan

iwamoto.hiroki@jaea.go.jp

Abstract. The Japan Atomic Energy Agency is planning to construct the Transmutation Experimental Facility (TEF) to promote research, development, and demonstration of elemental technologies for accelerator-driven systems (ADSs). At ADS Target Test Facility (TEF-T), 250-kW H^- beams accelerated by the 400-MeV LINAC will be derived to a lead-bismuth eutectic target. Toward the construction of TEF, a shielding analysis is carried out for the whole structure of the TEF-T in detail. The analysis for the beam tunnel shows that backward neutrons produced from the LBE target have large a impact on the radiation dose in the TEF beam tunnel. This paper presents the analytical results for components such as target station, hot cell and beam tunnel, and the shielding structure of the whole TEF.

1. Introduction

To promote research, development, and demonstration of elemental technologies for accelerator-driven systems (ADS), the Japan Atomic Energy Agency (JAEA) is planning to construct Transmutation Experimental Facility (TEF) at J-PARC. At ADS Target Test Facility (TEF-T), 250-kW H^- beams accelerated by the 400-MeV LINAC will be derived to a lead-bismuth eutectic (LBE) target for the material examinations aimed for the ADS. The TEF-T will operate as a spallation neutron source especially for research and development of semiconductor by using high-energy neutrons. For the license of this facility, the shielding design is important because unnecessary shield drastically increases the cost of construction.

In this study, shielding analysis was carried out for the whole structure of the TEF in detail toward the construction. Shielding design including the target station, beam tunnel, hot cell to maintain the LBE target and beam dumps for beam commissioning were also determined. This paper presents the results of the shielding analysis for these components and the shielding structure of the whole TEF.

2. Calculation conditions

The shielding analysis was conducted based on the design criteria shown in **Table 1**. In this table, for example, calculated dose values at a boundary between radiation shield and controlled area (CA) must be lower than 25 μSv/h. Here, the dose rate was calculated using the Monte-Carlo transport code system PHITS [1], where INCL4.6 [2] and GEM [3] were employed for simulation of intranuclear cascade and de-excitation processes, respectively. For neutron and photon transport below 20 MeV, the Japanese Evaluated Nuclear Data Library JENDL-4.0 [4] was adopted. For conservative estimation, the Tesch formula [5] was also used in determining the thickness of concrete shield.

Figure 1 illustrates a birds-eye view of the TEF-T facility. The 250-kW negative hydrogen (H^-) beams accelerated by the 400-MeV LINAC are transported through a joint section between LINAC and the TEF beam tunnel, and injected to the LBE target (15 cm in diameter ×100 cm in depth) installed at the center of the target station. In the beam tunnel, totally two beam dumps are installed for beam commissioning according to the intended use.

Table 1. Design criteria used in this analysis.

Areas	Laws and regulations	Effective dose rate
Boundary of JAEA-Tokai site	< 250 μSv/3 months (whole JAEA) < 50 μSv/3 months (whole J-PARC)*4	< 0.11 μSv/h*1 (whole JAEA)
Non-controlled area in JAEA-Tokai site	< 20 μSv/week	< 0.5 μSv/h*2
Controlled area where rad-workers can enter without special permission	< 1 mSv/week	< 25 μSv/h*2
Criterion of soil activation*3	< 5 mSv/h (< 11 mSv/h)	

*1 The 3 months correspond to 2184 hours (=91 days = 365/4 days) / According to a note of radiation safety division of Science and Technology Agency on Oct 23, 2000.

*2 The 3 months correspond to 5000 hours (1/4 of annual actual working hours, 2000 hours) and the 1 week corresponds to 40 hours.

*3 The criterion on the soil activation is set to an average of each facility but allows 11 mSv/h as a maximum local value.

*4 About 10 Sv/year for Transmutation Experimental Facility.

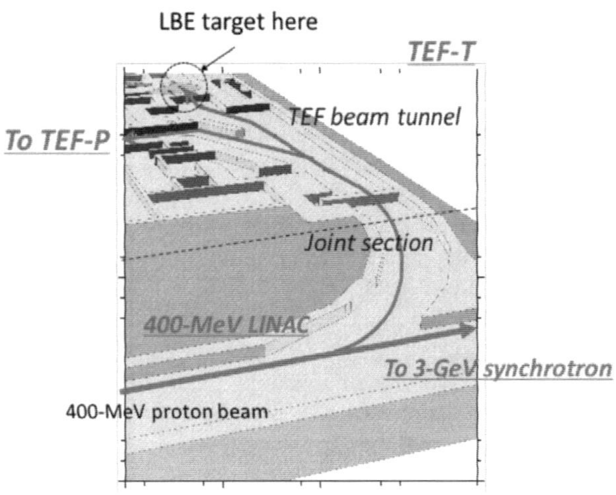

Figure 1. Birds-eye view of the TEF-T facility.

3. Results

3.1. Target station

To reduce radiation dose due to the intense spallation neutrons produced from the LBE target, the LBE target is surrounded with thick iron and concrete shields. As shown in **Figure 2**, at least a 3-m-thick iron and 2-m-thick concrete shield is required to satisfy the criterion of "< 25 μSv/h" at the boundary of CA. By this analytical result, the structure of target station was determined as illustrated in **Figure 3**.

3.2. Hot cell

The irradiated LBE target including its vessel is moved to a hot cell and utilized for various examinations. **Figure 4** shows dose rate distributions in the hot cell and radiation shield of concrete and shielding glass. From these results, it was found that a 1.5-m concrete shield or 1-m shielding glass is required to satisfy the criterion at the boundary of CA.

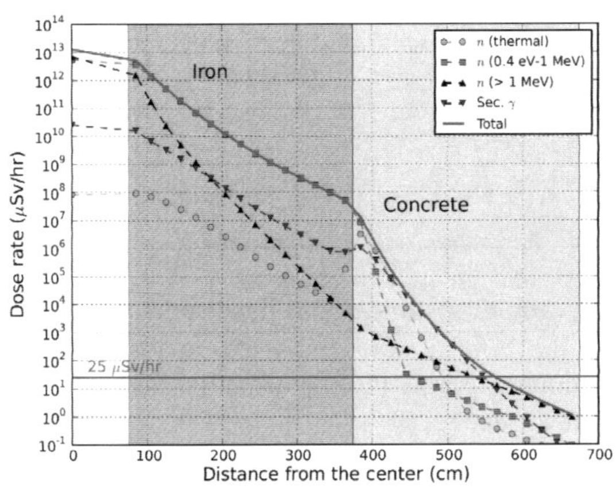

Figure 2. Dose rate distribution in radiation shield of the target station.

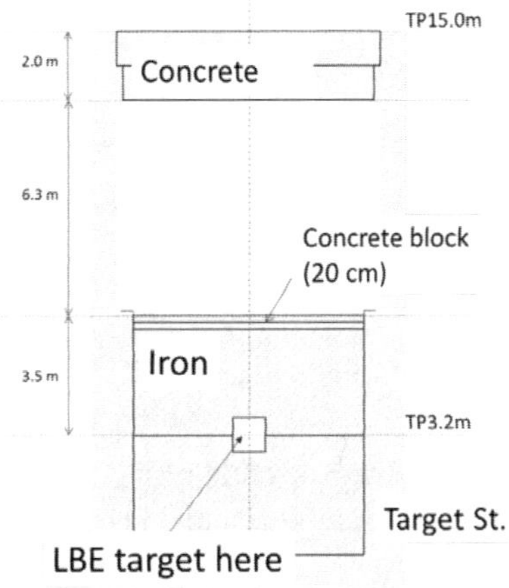

Figure 3. Vertical cross-section of the target station.

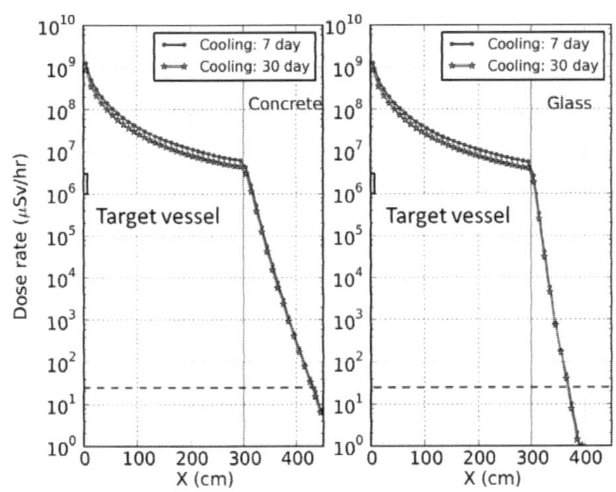

Figure 4. Dose rate distributions in the hot cell and radiation shield. (left panel: ordinary concrete, right panel: shielding glass).

3.3. Beam tunnel

The thickness of the concrete wall, floor, and ceiling in the beam tunnel is determined with the Tesch line-loss formula, where beam loss was assumed to be 0.1 W/m. In addition to the results of Tesch line-loss formula, contributions of backstreaming neutrons from the LBE target and spallation neutrons produced from the beam dump to the dose rate were taken into account. **Figure 5** shows a contour map of a contribution of backstreaming neutrons to the dose rate distribution in the beam tunnel. As indicated in this figure, it was revealed that backward neutrons dominate the radiation dose in the beam tunnel. To ensure the dose rate lower than the criterion at the boundary of CA, a maze structure was employed for the passage for carrying in/out equipment and a concrete wall dedicated for shielding the backstreaming neutrons was installed.

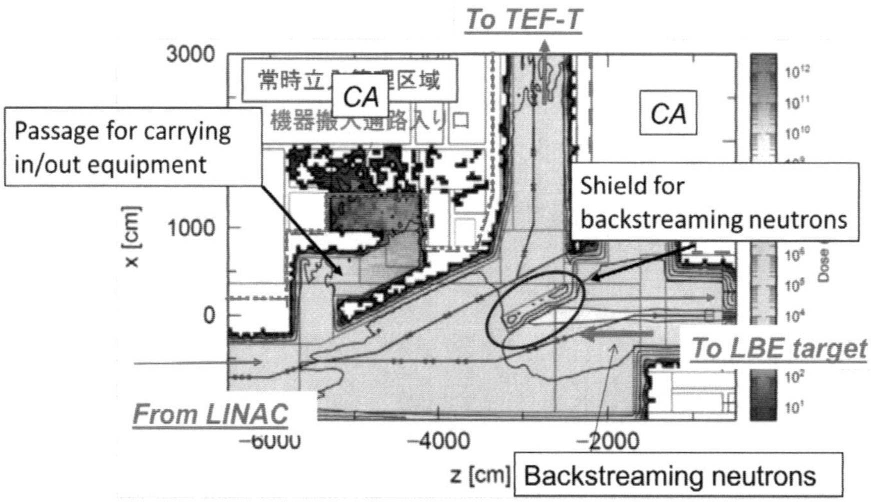

Figure 5. Contour map of dose rate distributions in the beam tunnel. Red dashed line represents a boundary of CA (< 25 μSv/h).

4. Conclusion

The detailed shielding structures including the target station, hot cell, beam tunnel and the beam dumps have been determined by the PHITS code based on the INCL4.6 and GEM models. The analysis for the beam tunnel has shown that the backward neutrons produced from the LBE target have a large impact on the radiation dose in the beam tunnel. This indicates that the prediction accuracy of the spallation models, i.e. INCL4.6/GEM, for the backward neutrons is important for the shielding design of the TEF-T beam tunnel. Future work, therefore, will pay attention to validation of spallation models especially for the prediction accuracy of backward neutrons.

References

[1] Sato T, Niita K, Hashimoto S, Iwamoto Y, Noda S, Ogawa T, Iwase H, Nakashima H, Fukahori T, Okumura K, Kai T, Chiba S, Furuta T and Sihver L 2013 Particle and Heavy Ion Transport Code System PHITS, Version 2.52 2013, *J. Nucl. Sci. Technol.* **50**:9 913-923.

[2] Boudard A, Cugnon J, David J.-C, Leray S and Mancusi D 2013 *Phys. Rev.* C **87** 014606.

[3] Furihata S, Niita K, Meigo S, Ikeda Y and Maekawa F The JEM code – a simulation program for the evaporation and the fission process of an excited nucleus – 2001, JAERI-Data/Code

[4] Shibata K, Iwamoto O, Nakagawa. T, Iwamoto N, Ichikawa A, Kunieda S, Chiba S, Furutaka K, Otuka N, Ohsawa T, Murata T, Matsunobu H, Zukeran A, Kamada S and Katakura J JENDL-4.0: A New Library for Nuclear Science and Engineering 2011, *J. Nucl. Sci. Technol.* **48**:1 1-30.

[5] Tesch K The attenuation of the neutron dose equivalent in a labyrinth through an accelerator shield, 1982, *Par. Accel.* **12** 169-175.

IOP Conf. Series: Journal of Physics: Conf. Series **1021** (2018) 012050 doi:10.1088/1742-6596/1021/1/012050

Preparation for activation measurements of concrete and PE-B4C-concrete to be applied for shielding at the European Spallation Source

E. Dian[1,2], E. Klinkby[2,3], C. P. Cooper-Jensen[2,4], D. Párkányi[1], D. Hajdú[1], J. Osán[1,5], G. Patriskov[1], U. Filges[1,6] and P. M. Bentley[2,4]

1) Hungarian Academy of Sciences, Hungary,
2) European Spallation Source ERIC, Sweden,
3) Technical University of Denmark, Denmark,
4) Uppsala University, Sweden,
5) International Atomic Energy Agency (IAEA), Nuclear Science and Instrumentation Laboratory, Austria,
6) PSI, Switzerland

E-mail: dian.eszter@energia.mta.hu

Abstract. To improve the effect of the concrete below 10 MeV where iron has resonances in the cross section, a new concrete have been developed. The PE-B4C-concrete utilizes hydrogen containing PE to thermalize the neutron and boron for in situ absorption. It is of utmost importance that the activation of the shielding material itself is well-understood, since it is planned to be used at the ESS. The first steps in this direction are shown the present study, in which concrete as well as reference aluminium samples are subject to XRF measurements to precisely determine the element content. This is compared to data sheets from the vender, and simulations are carried out to predict the sample activity.
The samples are planned for insertion into the the Budapest Research Reactor, followed be activity and spectral measurements.

1. Introduction

When the European Spallation Source (ESS) reaches its design configuration, protons of 2 GeV will impact a tungsten target at a rate of about $1.5 \times 10^{16} s^{-1}$, corresponding to 5 MW of proton beam power. Neutrons are generated by spallation processes in the target and will be released in substantial numbers with energies reaching up to the incident proton energy. The task of shielding instruments and personnel from high energy neutrons escaping the target-moderator-reflector system as well as from gamma photons will be undertaken by a combination of mainly steel and concrete. To improve the neutron absorption effect of concrete below 10 MeV where iron has resonances in the cross section, a new concrete has been developed. The PE-B4C-concrete utilizes hydrogen containing polyethylene to thermalize the neutron and boron for in situ absorption. It is of utmost importance that the activation of the shielding material itself is well-understood, both considering short-term effects on personnel during the operation phase and long-term effects on the decommissioning of the ESS facility.

IOP Conf. Series: Journal of Physics: Conf. Series **1021** (2018) 012050 doi:10.1088/1742-6596/1021/1/012050

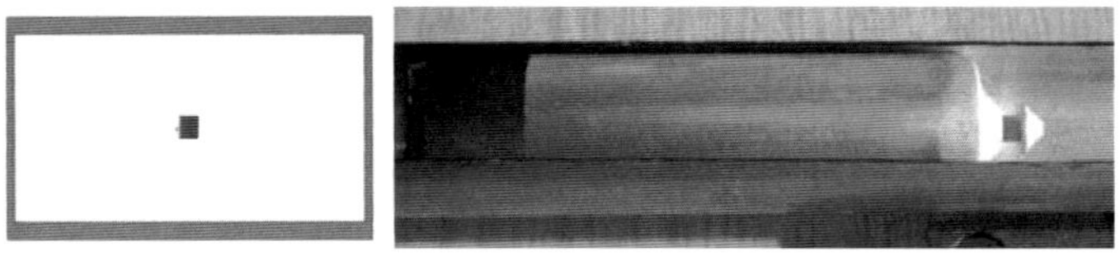

Figure 1: Simulated (a) and real (b) aluminum sample holder tubes.

2. Methodology of material testing

A study is carried out using samples of PE-B4C-concrete as well reference samples of regular concrete and aluminum, to establish the methodology. First, the elemental concentrations are determined using X-ray fluorescence techniques and the results are compared to the nominal composition of the material. Samples of the materials are planned for insertion in the Budapest Research Reactor [1] where they will be subject to irradiation in both thermal and fast neutron fluence for a duration ranging from a few minutes to hours. After irradiation the activities of the samples will be measured and compared to the corresponding measurements on standard concrete as a function of cooling time. In addition the decay gamma spectrum will be measured. Due to an unforeseen long-widened repair of the reactor, the samples have not yet been activated therefore only results from XRF measurements and Monte Carlo activation simulations are available at present.

3. MCNP activation simulation with measured and nominal sample compositions

Monte Carlo simulation [2] was performed under conditions identical to the planned irradiation experiment at the Budapest Research Reactor. The samples will be irradiated in a "fast" and a "thermal" vertical channel, in aluminium sample holders (see Fig. 1(b)). The neutron flux spectra of the two irradiation channels of the Budapest Research Reactor are well-characterized from Monte Carlo simulations of reactor core (see Fig. 2). Using these spectra as source distributions, a simplified model consisting of only the source (uniform volume source) surrounding an aluminum sample holder containing the sample is prepared as shown in Fig. 1(a). The resulting spectra incident on the sample are input into CINDER'90[3] activation calculation program, thus time-dependent radionuclide activities are mapped, and in the future compared to the measurements. The measured element compositions and predicted activity is shown in Table 1 and Table 2.

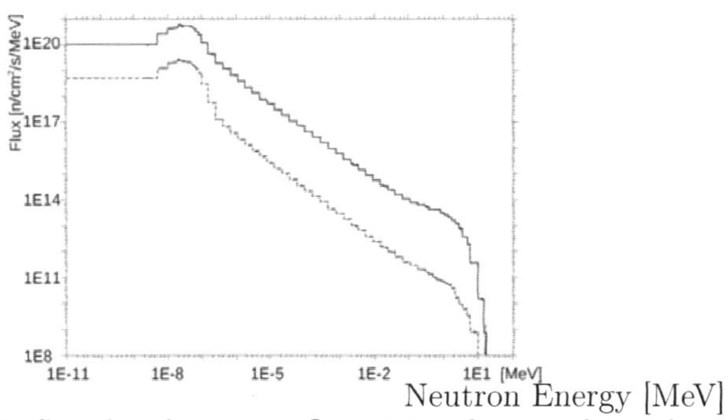

Figure 2: Simulated neutron flux at irradiation channels.

4. XRF-measurement on metal and concrete samples

Element-analysis was achieved with X-ray fluorescence (XRF) spectroscopy on metal and concrete samples. Three metal samples, $5 \times 5 \times 5$ mm^3 cubes of aluminum, copper and steel were measured with a handheld XRF device (Niton X3Lt GOLDD+, Thermo Scientific). In this paper we only report on the aluminium data while the steel and copper samples give similar results.

Two concrete samples, an average reference concrete and the PE-B4C-concrete, which contains 0.76 wt% B_4C and 10.2 wt% polyethylene were also studied with a polarizing XRF device (Epsilon5, PANalytical). The samples were received as grist. For sample preparation, both samples were ground again with ball grinder in tungsten carbide (WC) mortar, to increase the homogeneity (see Fig. 4(left)). Then pressed pellets of 2.5 g were prepared from both concrete samples, with the usage of 0.25 g wax (see Fig. 4(right)).

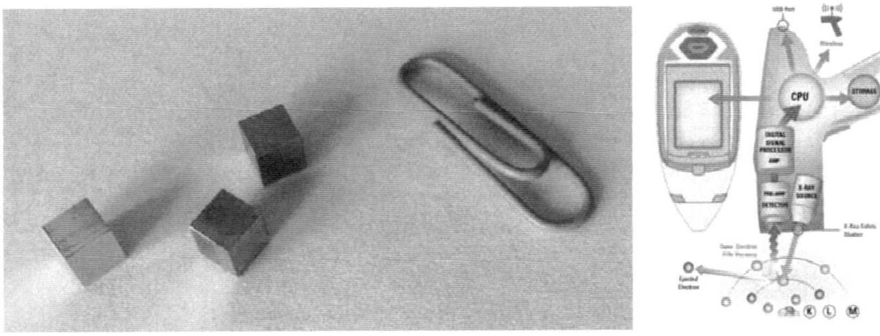

Figure 3: Metal samples (aluminium, stainless steel and copper) and handheld XRF device.

Figure 4: Grained concrete and pressed concrete pellets for XRF analysis.

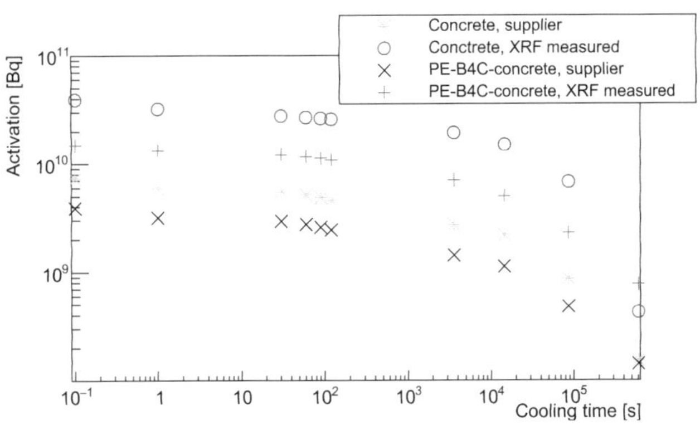

Figure 5: Simulated activity of PE-B4C-concrete and reference concrete with the XRF-measured and the nominal composition (1 week irradiation 1 week cooling).

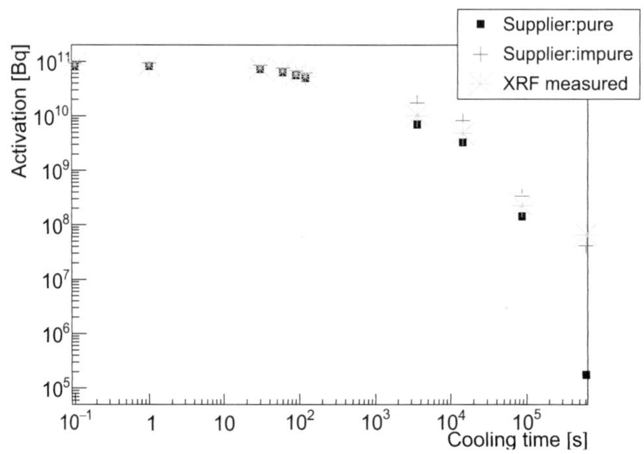

Figure 6: Simulated activity of aluminium sample with the XRF-measured and the nominal composition (1 week irradiation 1 week cooling).

5. Conclusions

The MCNP simulations show that PE-B4C-concrete have a much lower activation than normal concrete with both for the reference composition and for the XRF data. This is expected since the PE is 10.2 wt% which is about 20 vol% of the concrete and that is replacing the same volume of granite. Since PE does not activate significantly and we removed granite that activates. Also by adding PE and B increase the moderator power of the concrete and hence the neutron spectrum is driven towards lower energies, where the B absorption cross section increases (and B does not activate).

What is also clear that neither the reference values nor the XRF data give the full information of the activation. For this reason it is very important to perform the actual irradiation experiment to understand the long term effect of activation of the concrete and therefore a facility as ESS.

References

[1] www.bnc.hu

[2] Waters L. S. and others. The MCNPX Monte Carlo radiation transport code. *AIP Conf.Proc.* **896**, 81-90 (2007)

[3] W. B. Wilson, S. T. Cowell, T. R. England, A. C. Hayes & P. Moller, A Manual for CINDER'90 Version 07.4 Codes and Data, LA-UR-07-8412 (December 2007, Version 07.4.2 updated March 2008).

IOP Conf. Series: Journal of Physics: Conf. Series **1021** (2018) 012050 doi:10.1088/1742-6596/1021/1/012050

Table 1: Composition of aluminium sample. Reference composition that is the standard of the supplier and measured composition with handhelf XRF. (LOD: Lower Detection Limit)

Element	Reference material 6082-T6		Handheld XRF		
	Min	Max	Average	St Dev	LOD
			wt%		
C					
Mg	0.600	1.200	0.952	0.511	0.650
Al	95.200	98.300	97.032	0.570	0.518
Si	0.700	1.300	1.024	0.072	0.087
P			< LOD	-	0.003
S			< LOD	-	0.003
Ti		0.100	< LOD	-	0.007
V			0.010	0.004	0.008
Cr		0.250	0.438	0.290	0.014
Mn	0.400	1.000	0.591	0.100	0.030
Fe		0.500	0.220	0.036	0.016
Co			< LOD	-	0.010
Ni			< LOD	-	0.011
Cu		0.100	< LOD	-	0.005
Zn		0.200	< LOD	-	0.003
Se			< LOD	-	0.003
Zr			< LOD	-	0.003
Nb			< LOD	-	0.003
Mo			< LOD	-	0.003
Ru			< LOD	-	0.003
Pd			< LOD	-	0.003
Ag			< LOD	-	0.005
Cd			< LOD	-	0.004
Sn			< LOD	-	0.003
Sb			< LOD	-	0.004
W			< LOD	-	0.004
Au			< LOD	-	0.003
Pb			< LOD	-	0.003
Bi			< LOD	-	0.003

IOP Conf. Series: Journal of Physics: Conf. Series **1021** (2018) 012050 doi:10.1088/1742-6596/1021/1/012050

Table 2: Composition of concrete samples. Reference composition that is best theoretical estimation of the composition taking into account supplier data and average compositions, and measured composition with Polarized XRF Epsilon5. (LOD: Lower Detection Limit)
*Tungsten presence from grounding in tungsten-carbide mortar.

Element	Concrete			PE-P4C-Concrete			Lower
	Reference	Polarized XRF Epsilon5		Reference	Polarized XRF Epsilon5		Detection
	material	Average	St Dev	material	Average	St Dev	Limit
			Matrix wt%				
PE+B_4C				11.200	11.025	1.533	
			Major elements wt%				
Na	1.060	1.968	0.197	0.617	1.291	0.197	0.281
Mg	0.237	0.953	0.295	0.196	0.983	0.295	0.116
Al	3.700	6.656	0.036	2.350	5.512	0.036	0.250
Si	32.700	30.098	0.003	28.600	27.037	0.003	0.223
P	0.045	< LOD	-	0.026	0.000	0.000	0.050
S	0.236	0.169	0.015	0.278	0.237	0.015	0.009
K	2.120	2.190	0.001	1.260	1.947	0.001	0.002
Ca	7.120	6.634	0.000	8.100	8.543	0.000	0.002
Fe	1.160	1.343	0.001	1.160	1.467	0.001	0.001
			Trace elements ppm				
Cl	30.0	< LOD	-	35.5	130.0	10.0	30.0
Sc		14.3	2.1		11.8	3.4	6.8
Ti	910.0	1760.0	10.0	520.0	1590.0	10.0	10.0
V		55.5	0.2		57.6	7.1	7.1
Cr		44.3	0.3		82.2	1.3	2.3
Mn		230.0	0.0		230.0	10.0	10.0
Co		32.2	2.2		61.1	1.5	1.2
Ni		6.4	0.7		11.7	0.8	1.2
Cu		22.4	0.5		43.5	0.6	0.7
Zn		87.7	0.5		100.1	0.9	0.7
Ga		10.9	0.6		7.8	0.4	0.8
Ge		3.2	0.2		4.9	0.3	1.3
As		< LOD	-		2.9	0.1	1.3
Se		< LOD	-		0.0	0.0	1.3
Rb		76.0	0.7		59.4	0.4	0.4
Sr		380.9	1.3		311.7	1.2	0.4
Y		12.3	0.4		11.6	0.2	0.7
Zr		114.1	0.8		91.4	0.4	0.7
Nb		6.2	0.1		5.4	0.1	0.4
Mo		2.2	0.1		3.0	0.1	0.4
Ag		< LOD	-		< LOD	-	0.4
Cd		< LOD	-		< LOD	-	0.4
In		1.3	0.1		1.7	0.2	0.4
Sn		2.8	0.3		2.8	0.0	0.5
Sb		1.2	0.2		1.6	0.2	0.5
Cs		3.9	0.1		2.6	0.3	0.4
Ba		665.1	1.6		508.5	1.1	1.1
La		23.7	0.3		20.0	0.3	1.3
Ce		37.0	7.9		41.2	0.5	1.5
Pr		5.5	0.5		4.8	0.5	2.6
Nd		18.6	4.3		19.5	0.2	1.8
W*		111.4	0.4		164.8	0.5	0.5
Pb		19.4	0.4		18.3	0.9	0.9
Th		4.7	0.2		3.8	0.3	0.9
U		3.4	0.2		2.7	0.4	0.8

IOP Conf. Series: Journal of Physics: Conf. Series **1021** (2018) 012051 doi:10.1088/1742-6596/1021/1/012051

Radiation challenges of primary cooling return water at the ESS

Esben Klinkby[1,2], Günter Muhrer[1], H. Carlsson[1] and Björn Eriksson[1]

1) European Spallation Source ESS, Box 176, S-221 00 Lund, Sweden
2) DTU Nutech, Technical University of Denmark, Frederiksborgvej 399, DK-4000 Roskilde, Denmark

E-mail: esbe@dtu.dk

Abstract. At ESS water is used both as thermal moderators and to cool the beryllium- and steel reflectors, the shielding and plugs. This means that the water, in separate loops, will be subject to a significant proton and neutron irradiation which will activate the water. After irradiation, the water is led to delay tanks situated in one of the triangular rooms downstream from the target.This paper aims at determining the shielding required to ensure that the biological dose-rate requirements in the target building and instrument halls are met during operation of facility.

1. Introduction

The construction of the European Spallation Source (ESS) is ongoing in Lund, Sweden. At ESS it is necessary to operate water loops for the thermal moderators and for the cooling of the beryllium- and the steel reflectors as well shielding blocks and plugs. The water in each of these circuits is exposed to a very high level of radiation. This causes activation of the water and impurities therein. In this work the measures for safe handling of the return water are investigated. In particular the consequences in terms of shielding requirements of moving the delay tanks from their baseline location in the triangular room, to the adjacent Connections cell are investigated.

2. Methods and Modelling

The radiation shielding calculations are performed following the ESS Procedure for designing shielding for safety [1]. Monte Carlo radiation transport calculations are performed with the MCNPX-2.7 code [2]. Using a this tool, a model is prepared describing in detail the target, moderators and reflectors as shown in figure 1.

To determine the neutron flux in the water of the moderators and reflectors, the interactions of one million 2 GeV protons on target are simulated. The resulting neutron flux is recorded and handed to CINDER'90 (v. 1.05)[4, 5] for an activation calculation using as irradiation scenario the expected ESS operations schedule for two years of full power operations (5 MW).

As seen in figure 2 the piping leading the water through the various systems is complex, and is not modelled in detail in this study. Instead, a simplified MCNPX model is prepared (see figure 3 (left)), in which the approximate water volume of various fractions of the pipe circuits

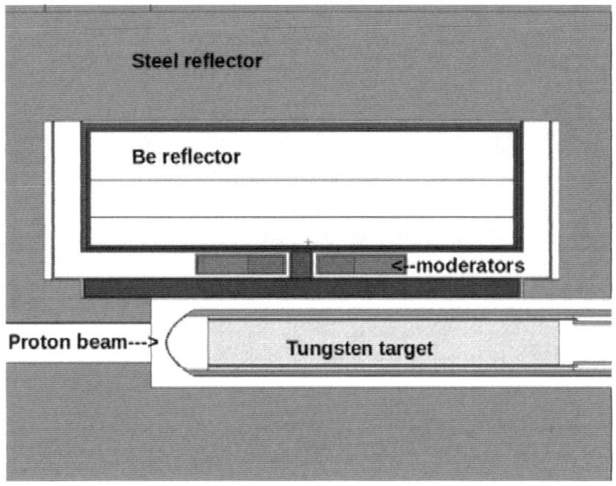

Figure 1: MCNPX geometry used in the proton on target simulations which are performed to evaluate the neutron flux at the location of the cooling water.

is used to define volumetric source terms (scaling the full source term), at various cooling times to account for the water flow.

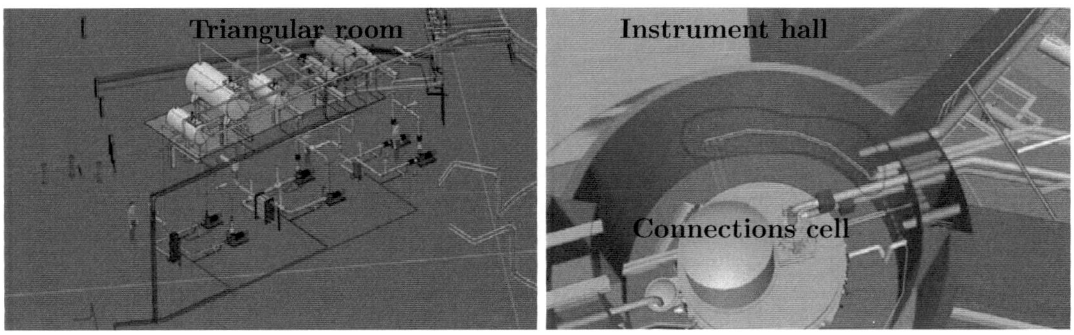

Figure 2: Left: CAD drawing of triangular rooms. The orange pipes and orange delay tanks contain the irradiated water under study. Right: The connections cell, located directly above the target-moderator-reflectors. The orange pipes, carrying the irradiated water the toward the Connections cell, are encircled in red.

3. Results

Table 1 lists the isotopes giving the largest contribution to the total activity at selected cooling times. As expected ^{16}N (^{16}O$(n,p) \rightarrow^{16}$N) is the main contributer at short cooling times whereas ^{7}Be plays this role at longer cooling times.

3.1. *Inlet pipe shielding*

To describe the irradiation due to the cooling water as it pass through the delay tank inlet pipe, a simplified MCNPX model is prepared describing only the Connections Cell, triangular rooms, delay tanks and inlet- and outlet pipes - (see figure 3). A volumetric source term is defined using that the 5 m long pipe results in a total water volume of about 1.5 L, or approximately 1/6 of the total water inventory. Thus, as a conservative approximation the source at zero cooling

IOP Conf. Series: Journal of Physics: Conf. Series **1021** (2018) 012051 doi:10.1088/1742-6596/1021/1/012051

Isotope	Half-life [s] [s]	Decay mode	Time step [0s]	Time step [90s]
^{3}H	3.89E8	β	621.3	621.3
^{7}Be	4.61E6	EC$\beta + \gamma[477keV]$	445.1	445.1
^{11}C	1223	β	1473	1400
^{14}O	70.6	$\beta + \gamma[2.3MeV]$	92.3	38.5
^{15}O	122	β	4511	2723
^{16}N	7.1	$\beta + \gamma[6.1MeV]$	4462	0.8
Total			12380	7161

Table 1: Total activity in Curies. Only isotopes contributing in excess of 10Ci at 60 s cooling are listed, but all are included in the total sum.

time is used with a weight reduced by a factor of 1/6. This source is placed in the delay tank inlet pipe and a gamma transport simulation is carried out using MCNPX. Using ICRP-116 gamma flux-to-dose conversion factors [6], the resulting biological dose-rate map is shown in figure 3(left). In this figure 12 cm lead shielding surrounds the pipe.

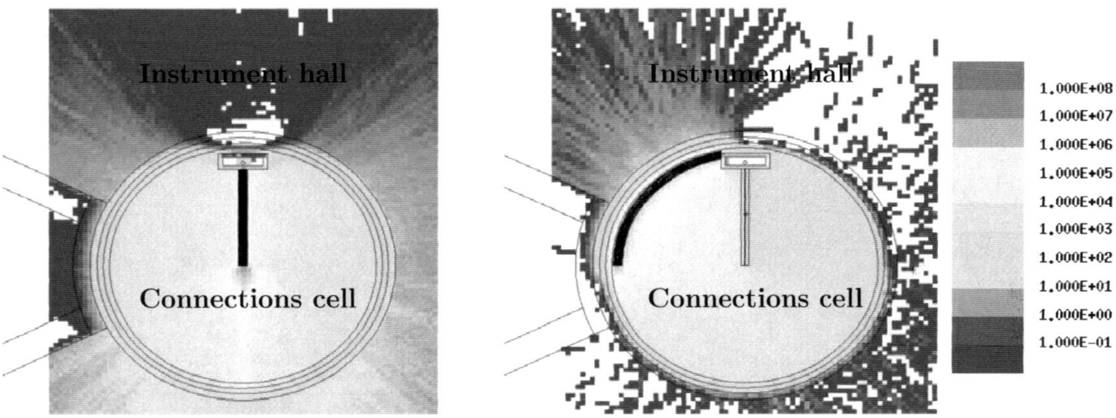

Figure 3: Left: Biological dose-rate in μSv/h due to the delay tank inlet pipe (central in the figure). Shielding thickness: 12 cm lead. Right: Biological dose-rate in μSv/h due to the delay tank outlet pipe (following the circumference of the wall). Shielding thickness: 2 cm lead.

3.2. Delay tank shielding

The water circuits, of course, are not exactly steady. Moreover the timescales relevant to the water flow are similar to the half-life of ^{16}N, which according to table 1 is main contributer to the activity on a few seconds timescale. To accurately model the source term to be used for delay tank shielding calculations, the following observations are made:

- Moderator water content: 9L
- Moderator exhaust speed: 0.6 L/s (or 2 m/s) so the average cooling time, at the time of exiting the moderator: 7.5 s.
- Pipes: 10 m. 5 m vertical and 5 m horizontal

IOP Conf. Series: Journal of Physics: Conf. Series **1021** (2018) 012051 doi:10.1088/1742-6596/1021/1/012051

Thus, source terms are prepared corresponding to cooling times between $7.5\text{s} + \frac{10\ \text{m}}{2\text{m/s}} \approx 13$ s and 37 s in 9 equidistant steps - see table 2.

Cooling time [s]	13	16	19	22	25	28	31	34	37
Source weight [%]	12	12	12	12	12	12	12	12	4

Table 2: Source definitions used to model the delay tank.

Several limits of biological dose-rate apply at various locations in the target building and the surrounding buildings. Focusing to the connections cell the most stringent limit driving the shielding is that of the neighbouring instrument hall: 3μSv/h - see figure 2.

The resulting dose-rate map due to the presence of activated primary coolant water in the delay tank is shown in figure 4. The dose-rate requirements in the instrument hall are met only after introducing 18 cm of lead shielding. It should be noted that the delay tank need shielding on all horizontal directions, though the shielding requirements on the side facing the center of the room, can be reduced by approximately two orders of magnitude due to the geometrical dilution between the delay tank and the instrument hall on the opposite side of the connection cell.

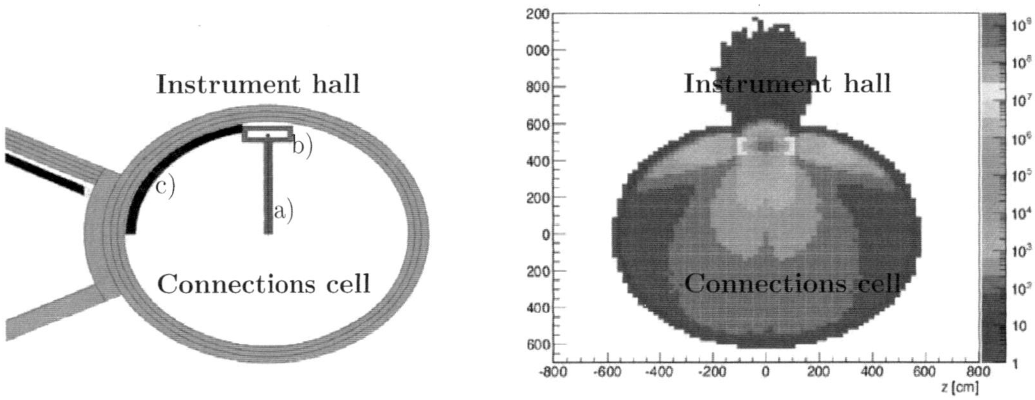

Figure 4: Left: simplified MCNPX geometry used throughout this study for calculations of biological dose rate. The source is placed either in the delay tank inlet pipe: *a*), in the delay tank: *b*) or in the delay outlet pipe *c*). The green structure is regular concrete (density 2.35 g/cm^3, thickness of circular wall: 80 cm). Right: Biological dose-rate in μSv/h due to the delay tank, which is located inside the Connections cell. Shielding thickness: 18 cm lead.

3.3. Outlet pipe shielding

A delay tank size of 90 s is assumed for calculations of the outlet pipe.

The outlet pipe is modelled as a 2 cm diameter pipe placed along the inner circumference of the Connections cell. Its volume correspond to about 30% of the moderator volume, which is applied to the source weight. The pipe shielding required to reach satisfactory dose-rate levels in the neighbouring instrument hall is 2 cm of lead - the resulting dose-rate maps are shown in figure 3(right).

IOP Conf. Series: Journal of Physics: Conf. Series **1021** (2018) 012051 doi:10.1088/1742-6596/1021/1/012051

4. Conclusions

The shielding requirements of the primary coolant return water are studied using MCNPX neutron and gamma transport calculations in combination with CINDER'90. Immediately after returning from the moderators/reflectors ^{16}N is the most challenging isotope, while after the cooling for 90 s in the delay tanks, ^{7}Be claims this role. For irradiated water present in the connections cell, the most challenging biological dose-rate limit to be met is that of the instrument hall. To meet the 3μSv/h limit, the following shielding must be installed:

- 12cm lead around delay tank inlet pipe
- 18cm lead between delay tanks and instrument hall
- 2cm lead around 90s delay tank outlet pipe

The above assumes that the delay are moved from the present baseline location in the triangular room to the Connections cell. In the opposite case, where the delay tanks remain in the triangular rooms, the 12 cm shielding requirement of the delay tank inlet pipe must be maintained throughout the length of the inlet.

References

[1] ESS Procedure for designing shielding for safety, ESS-0019931
[2] Waters L. S. and others . The MCNPX Monte Carlo radiation transport code. *AIP Conf.Proc.* **896**, 81-90 (2007)
[3] ESS-0045043. Sheet: 2.Input and output data
[4] W. B. Wilson, S. T. Cowell, T. R. England, A. C. Hayes & P. Moller, A Manual for CINDER'90 Version 07.4 Codes and Data, LA-UR-07-8412 (December 2007, Version 07.4.2 updated March 2008).
[5] Franz X. Gallmeier and Michael Wohlmuther, ACTIVATION SCRIPT VERSION 1.0 USER GUIDE, ORNL/TM-2008/031 (August 18, 2008).
[6] Conversion Coefficients for Radiological Protection Quantities for External Radiation Exposures. ICRP Publication 116. Ann. ICRP 40(2-5), 2010

IOP Conf. Series: Journal of Physics: Conf. Series **1021** (2018) 012052 doi:10.1088/1742-6596/1021/1/012052

SNS mercury target design optimization

S Kaminskas

Oak Ridge National Laboratory, Oak Ridge, TN, USA

kaminskass@ornl.gov

Abstract. The mercury target at the Spallation Neutron Source (SNS) in Oak Ridge National Laboratory, Tennessee, works under a number of severe conditions: high temperature, radiation, mercury flow, and proton beam pulse pressures. Insufficient strength and durability of the target stainless steel vessel from beam pulse and thermal cyclic loads can lead to mercury leaks and premature target life termination. The parametric and non-parametric shape optimization tools (Isight, Tosca) from Dassault Systemes Extended Abaqus Package have been adopted at SNS to develop robust and reliable target designs. The durability evaluation with recognized leading Fe-Safe (structural fatigue) and Verity (weld fatigue) software is also included in optimization process. Design of Experiment (DOE) was performed to create the response surface for steel vessel structure. The response surfaces are used for design exploration on a basis of Abaqus and Fe-Safe different analyses results. The shape optimization of steel vessel filled with mercury was performed first time using Tosca optimization tool. Obtained complex results enable the consideration of possible trade-offs for the most feasible design.

1. Introduction

The Spallation Neutron Source facility at Oak Ridge National Laboratory will be expanded by developing the Second Target Station. In relation to it, the maximum power on the targets is planned to increase from 1.4 MW to 2.0 MW. The target steel vessel high cycle fatigue with higher power was analysed using Fe-Safe and is presented in the form of lifetime curves for different locations in figure 1. The total number of pulse cycles through the entire regular target service time is expected to be at least 600 million or higher.

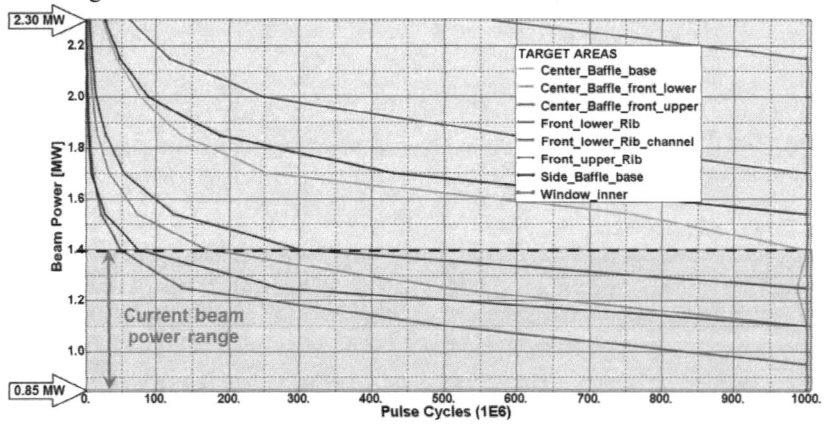

Figure 1. Target pulse fatigue life vs power at different locations.

IOP Conf. Series: Journal of Physics: Conf. Series **1021** (2018) 012052 doi:10.1088/1742-6596/1021/1/012052

Curves in figure 1 indicate the fatigue life reduction above 1.0 MW and significant reduction above 1.4 MW power in comparison with required target service time. That leads to the necessity of vessel optimization to address the demands of reliable operation. Two tools from Abaqus Extended Package are being used for target steel vessel optimization – Isight for parametric optimization and TOSCA for non-parametric optimization.

2. Parametric optimization of the target

The one case of parametric optimization is the search for optimal speed of mercury pump. Nominally pump provides the flow of mercury at 2 m/s but also can be run at different speeds. Higher speed provides better cooling of steel vessel and reduces the thermal stress. However, at the same time it leads to higher pressure from mercury what increases the stresses at vessel walls. Those factors are in contradiction with respect to fatigue life and actually represent the multi-objective optimization problem. The Isight capabilities were used to perform Design of Experiment (DOE) and generate the response surface for fatigue life on a basis of Abaqus and Fe-Safe simulation results. Obtained response surface is presented in figure 2 and clearly indicates the optimal point for maximal target life before the appearance of damages from fatigue.

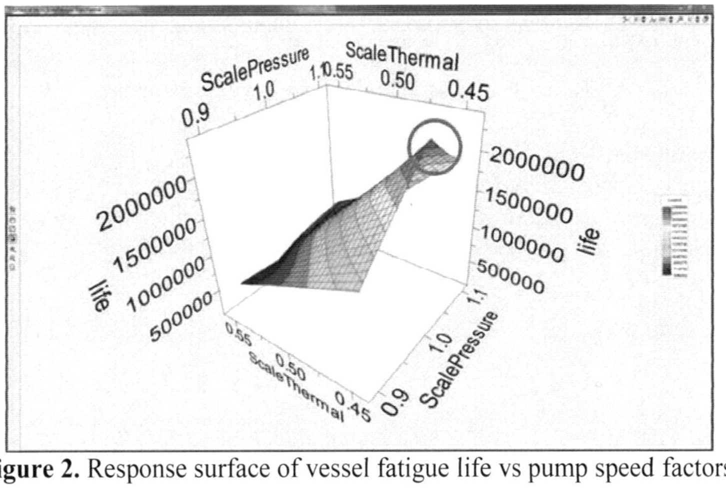

Figure 2. Response surface of vessel fatigue life vs pump speed factors.

3. Overall optimization of the target

Target overall optimization is an interactive process and is illustrated as general flowchart in figure 3. As an optimization problem it can be formulated in the following way:

Minimize: $$\mathbf{f_g}\left(\sigma_1^{\max}, \sigma_2^{\max}, \ldots, \sigma_n^{\max} \right) \quad (1)$$

Subject to: $$P_j^{min} \le P_j \le P_j^{max} \quad (2)$$

Where: $\mathbf{f_g}$ – aggregate function of max stresses $\sigma_i^{\max}$

$\sigma_i^{\max}$ – max stress at $\mathbf{i}^{th}$ stress concentration

P_j – parametrized design features (geometric, etc.)

No any efficient method exists to solve the problem (1), (2) for the mercury target with its highly complicated operational loadings and constraints. Only interactive, task by task approach there is possible as shown in figure 3.

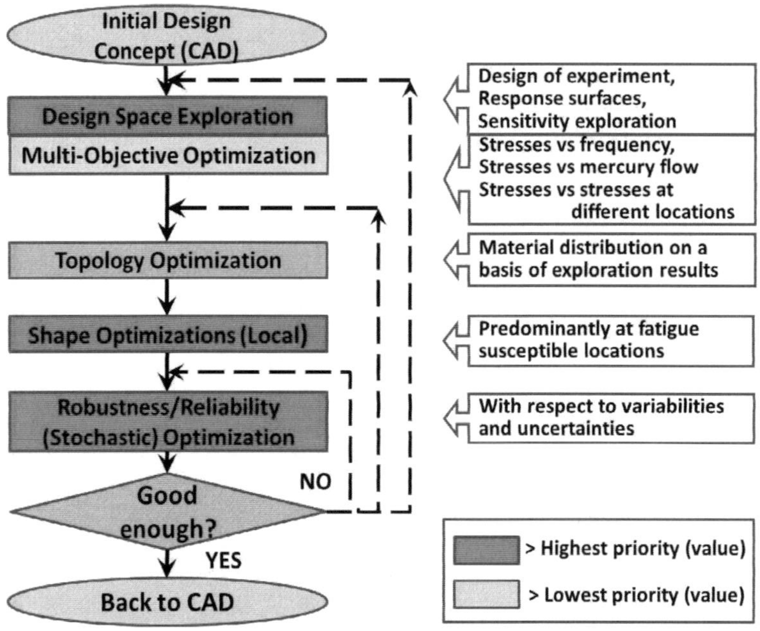

Figure 3. Target overall optimization flowchart.

4. Target steel vessel shape optimization

To minimize fatigue damage and extend the target life there is very important to reduce the stresses in local concentration areas. The TOSCA shape optimization capabilities had been efficiently employed for that purpose. Optimization problem here can be formulated as follows:

Minimize: σ_i^{max} (3)

Subject to: $p_j^{min} \leq p_j \leq p_j^{max}$ (4)

Where: σ_i^{max} – max stress at i^{th} stress concentration

p_j – local geometry features (nodal coordinates)

Direct application of problem (3), (4) for target steel vessel inner baffle shape optimization is illustrated in figure 4. The TOSCA optimizer successfully reduced local stress by 45% in 5 iterations. However, it appeared that the connection between mercury and steel finite element surfaces had been lost when the steel surface shape was undergoing modification in optimization process. Consequently, the pulse pressure induced in mercury had not been entirely transferred to steel resulting in unjustified reduction of stress.

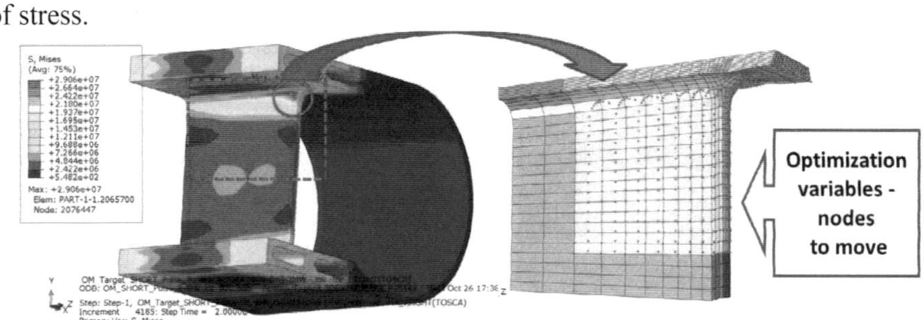

Figure 4. Local shape optimization at inner baffle area.

Problem was resolved by artificially pulling the mercury surface to the corresponding surface of steel after each consecutive optimization iteration as it is illustrated in figure 5. The stresses at the best, 2nd optimization iteration, are presented in figure 6.

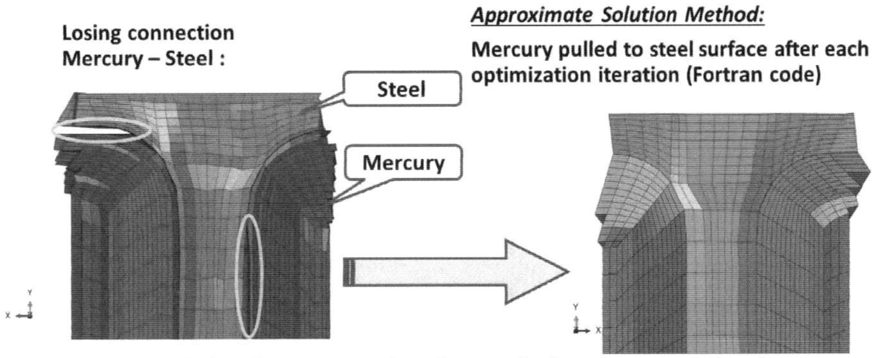

Figure 5. Steel – mercury detachment during shape optimization.

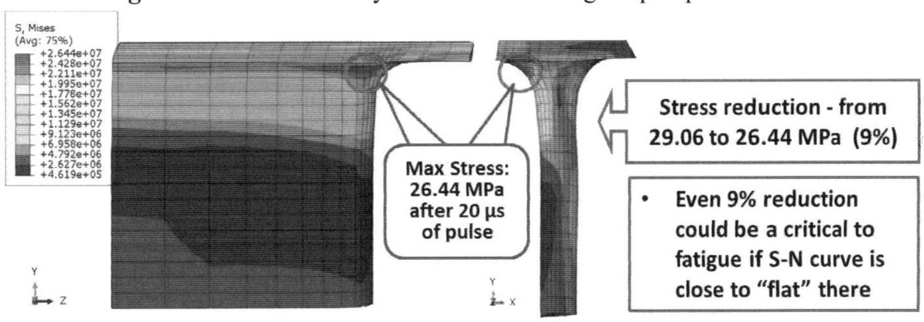

Figure 6. Stress contours after shape optimization.

It has to be noticed the complicacy of the problem. On one hand, the proton beam pulse wave passing through the target is changing the levels and locations of maximum stresses at steel. On the other hand, intervention into optimization process between iterations disorients the optimizer in its searches for direction to the optimal solution. It creates the jumps in objective function history. The best iteration with minimal stress may not be the last one in such case.

5. Conclusions

- Harsh operating conditions of SNS mercury target and increasing levels of power require to achieve the high robustness and reliability of the mercury steel vessel.
- Traditional "Design-Analysis-Evaluation" path is not sufficient to achieve the required design characteristics.
- Design Exploration and Optimization technics are necessary and capable to provide the best possible design and address the increasing demands.

References

[1] Kaminskas S 2016 SNS mercury target fatigue life prediction with Fe-Safe/Verity *Proc. Int. Conf. on High Power Targetry Workshop 2016* Oxford UK
[2] Naoe T et al 2016 Gigacycle fatigue behaviour of austenitic stainless steels used for mercury target vessels *Journal Of Nuclear Materials* **vol 468** pp 331-338
[3] Riemer B et al 2012 Correlation between simulations and observed cavitation damage erosion in operated Spallation Neutron Source target modules *Journal Of Nuclear Materials Proc. IVSMT-11* Ghent Belgium

IOP Conf. Series: Journal of Physics: Conf. Series **1021** (2018) 012053 doi:10.1088/1742-6596/1021/1/012053

ISIS TS1 Project Summary

Stephen Gallimore & Matt Fletcher

STFC - ISIS, Rutherford Appleton Laboratory, Harwell Campus, Didcot, OX11 0QX, UK
E-mail: Stephen.gallimore@stfc.ac.uk

Abstract. Following a feasibility phase which began in 2012, the TS1 upgrade project entered its implementation phase in April 2016. The project is, in essence, a circa £16.25M sustainability project for target station 1 (TS1) of the ISIS facility, Rutherford Appleton Laboratory (RAL), which has been in operation since 1984. Its aims are: to secure the future of TS1 and enable it to operate for many more years, to improve flexibility and future upgradability, to make operation of the target station easier, a neutron performance increase of up to a factor of 2 (on some instruments), as well as providing a strong technical foundation on which the facility can continue to build and develop in the coming years.

This paper will provide a summary of the project structure and progress to date, as well as covering some of the encountered challenges and successes. It will further detail the motivations behind the project's inception, including the critical success factors.

The implementation phase includes design progression, procurement, verification and validation testing, pre-installation testing, along with a large number of other supporting activities. It is currently envisaged that a long operational shutdown of ISIS will occur in a window of 2019-2021, with the project looking to exploit this opportunity for the strip-out, installation and commissioning work required to bring to fruition the planned changes.

1. Introduction

1.1. Background

Target station one (TS1), first produced neutrons in December 1984. Originally designed with a depleted uranium target and with a proton beam repetition rate of 50Hz. With the construction and operation of Target station 2 (TS2) in 2009, TS1 now takes 4 out of 5 proton beam pulses. The many years of operational experience from both target stations, new design techniques (simulation tools for example) and with TS1 being optimised for a uranium target, it was felt that some significant improvements could be made with only moderate expenditure (relative to an accelerator upgrade needed to achieve the same increases).

The project was initiated in 2012 with the creation of a project brief stating the general aims and providing initial estimates of timescales and costs. A 3.5-year definition phase followed, leading to the presentation of a business case to the ISIS management committee in October 2015, who recommended the project be implemented. The first project board meeting, was held in April 2016, where the project management plan was formally approved and moved the project into the implementation phase.

1.2. Project overview

The TS1 project is part of ISIS' commitment to ensuring the sustainability of the facility, utilising the operational experience from TS1 and TS2 as well as other spallation facilities, coupled with the latest simulation techniques and engineering best practice to secure the future of TS1 and enable it to operate for many more years. The project also aims to; provide improved flexibility for future target or moderator changes, make operation of the target station easier, provide a neutron performance increase (of up to a factor of 2, on some instruments), provide confidence in the ongoing operation of TS1 to enable future instrument upgrades and finally to improve the knowledge and skills of staff for future projects, such 'ISIS 2'.

The project was costed using both 'bottom up' and 'top down' approaches, incorporating actual delivery experience of similar projects; phase II instruments and the construction of TS2. The final value has value added tax (VAT) at 20%, a contingency of 10% (which is expected to be spent as part of the project) and no overhead on labour. For the cost of the project to be easily compared to a 'do nothing' option, cost estimates accounted for the disposal of active components deemed to be outside of normal operations (for instance the reflector assembly). This led to an overall total cost of £16,254k at 2015 (Pre-Brexit) prices.

The project includes a complete refurbishment of the internals of the target station, incorporating: the target, reflector and moderators (TRAM) assembly; the fluid systems (water cooling, cryogens for the moderators and inert purge gases) which sit behind the TRAM, the connected instrumentation and controls systems needed for correct operation of the target station. The scope of this work is visually demonstrated by Figure 1, below. The project does not include any significant changes to the TS1 neutron or muon instrument suite, although the development of instruments will carry on in parallel. Several other supporting activities are also required to bring the vision of the project to fruition.

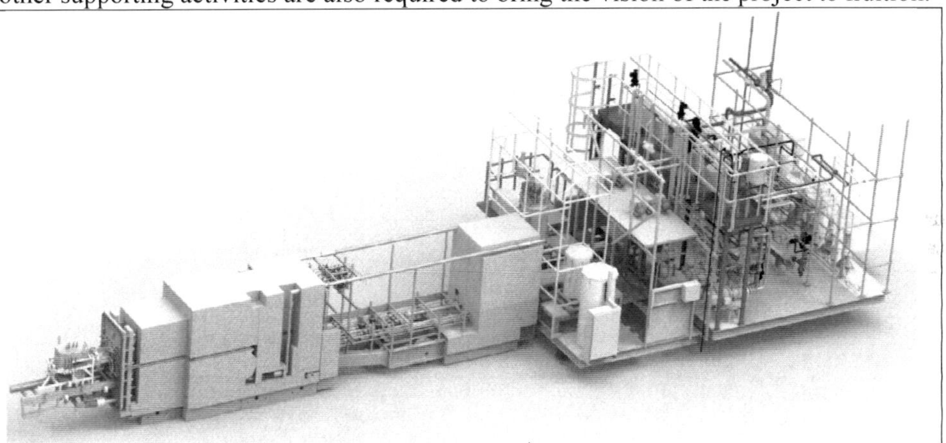

Figure 1 - A 3D CAD model showing the TRAM assembly on the complete target trolley, with the surrounding shielding and infrastructure removed for clarity. This is a representation of the current ISIS TS1 arrangement

The implementation phase is scheduled to be completed in the latter part of 2021, with the hand over to operations. This can be crudely segregated into several sub-phases; design (2016-2017), procurement and delivery (2018-2019), on-site testing and pre-installation checks (2019-2020), strip-out of existing equipment (2020-2021), installation (2020-2021), pre-beam commissioning (2020-2021), with-beam commissioning (2021) and close-out (2021). The long operational shutdown is constrained to be no more than 12 months in duration, during this time there is the potential to have some period of beam operations to TS2 only.

1.3. Project constraints

During the definition phase the high-level constraints were identified and these are as follows; optimise the design (neutronic and engineering) for a beam power 200μA on TS1 target, maintain normal operating parameters currently adhered to for beam position or profile , maximum allowed operational shutdown duration is 12 months, low-risk solution (no increased technical risk during operation), no instrument should be worse off in terms of performance after the project when compare to the present and the existing infrastructure (current beam line inserts face where they do, no significant changes to the biological shielding, accelerator height is unchangeable and the irradiation tube which was installed but never used and now back filled with steel shot as shielding, is to remain untouched. With these constraints in place, it applied bounds for the subsequent neutronic and engineering design work. Although some of these constraints proved challenging to navigate, ultimately it was fundamentally important to have them clearly defined, in order to focus the limited design efforts in the most effective manner.

2. How the project is structured

2.1. Work breakdown structure

The project has been broken down into 20 work packages (WPs) which are managed by 13 work packages managers (WPMs). The often highly-interlinked WPs were selected in such a way that all aspects of the project are covered, from the design of key components to the safety activities and on to the project hand-over and transition to operations.

The project manager, Stephen Gallimore is supported by a technical lead engineer, David Jenkins, and a lead scientist, Rob Bewley. The main bodies providing oversight, challenge and support are; a project board, a technical advisory board and a safety working group. The composition of these bodies was chosen as to provide a broad range of experience and knowledge in scientific, technical and administrative fields from within ISIS, greater STFC and the external community the facility serves.

Over 50 ISIS and other STFC staff have been directly involved with the project so far, with this number likely to grow further as it progresses. This resource has been and will continue to be supplemented with contractors in several areas, such as design, planning and waste management. The project has a dedicated radiological protection advisor (RPA) embedded in the project team.

3. Key areas of work and progress to date

The TRAM assembly, shown in Figure 2, represents the key components of the project in terms of scientific performance. The technical details of these components are covered in more detail by [1],

Figure 2 - A schematic 3D cut-away model showing the target at the centre, surrounded by the different moderators, which are in turn surrounded by the reflector

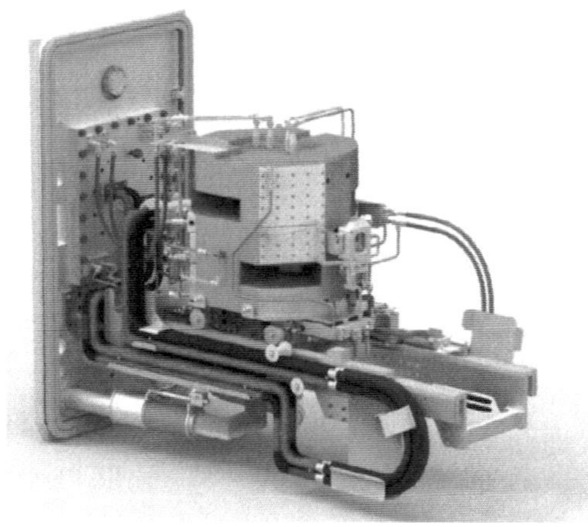

Figure 3 - A CAD model of the proposed TRAM assembly, support frame and void vessel back door for the TS1 Project. The reflector assembly is in its closed position, so the internal components are not visible.

[2], [3] and [4]. The TRAM assembly is supported by a frame which is cantilevered from the 'back door', as shown in Figure 3. This support frame is the only item that operates inside the void vessel (pressure vessel surrounding the TRAM assembly that contains the helium atmosphere during operation) that has not been replaced or modified in some way since its initial installation. The analysis work carried out as part of the design phase is covered by [5].

As previously mentioned, the TS1 project is far more than just the re-optimisation and redesign of the TRAM assembly it also includes a significant amount of work in many other supporting areas, such as controls and instrumentation, as well as large scale replacement and redesign of the services (inert gas, vacuum, water and cryogenic fluids) [6] that feed the TRAM assembly and allow it to operate under its desired parameters. All of this is supported by neutronic and scientific evaluation; better understanding the current situation [7][8][9], often including empirical measurements and assessing the impact of the proposed changes using suite of simulation tools. In addition, there are a number of other activities, for instance, overseeing the construction and utilisation of the planned full-scale mock-up of the TRAM inside the remote handling cell (RHC) [10], communications (both internal and external to ISIS), safety related work (for example production of prior risk assessments and overseeing HAZOPs) and planning and preparation for the dealing with and disposal of the active waste that will be generated.

4. The long shutdown

The long operational shutdown will be when the vast major of strip-out and installation work is carried out. The plan is currently to have approximately 9 months with the beam off and then 3 months of with beam commissioning. During the long shutdown, other projects across ISIS will be carried out, so good communication and robust planning will be essential to minimise conflict in terms of resource utilisation and requirements for running. Systems are being designed in a modular fashion (where possible) and allow for as much testing and commissioning prior to installation as possible.

Another critical aspect during the long shutdown will be the management of radioactive waste. The TS1 project will generate a significant volume of material that will need to be size-reduced, categorised and containerised prior to either storage and/or disposal.

5. Challenges and benefits

5.1. Challenges

As with any project of this technical complexity, duration and budget, there are numerous challenges faced through-out its life; this project is expected to be no different. Below are some of the challenges already experienced or expected:

- Staff resources – e.g. availability of operational staff, constraints on staff numbers
- Procurement & Production – e.g. production delays and contract placement issues
- Physical space (for testing and the management waste)
- Waste management (how to deal with the various waste streams and volumes of waste)
- Scheduling of the long shutdown
- 30+ year old surprises (with the age of the facility these are almost inevitable)

5.2. Benefits already realised

Many of the benefits from this project will only be realised after the successful completion of the long shutdown and some will require years of post-project operation to truly assess. However, the project has already delivered some benefits:

- Better understanding of what we have than probably at any time since TS1 was built
- Have already up-skilled staff – knowledge sharing as well as technical exposure
- Have been able to diagnose and correct existing issues – an example of this is detailed in [11]
- Better records of decisions and the reasons behind them

The project will hopefully continue to maintain these benefits and deliver more throughout its lifetime.

6. Summary

The TS1 project is a major project being undertaken by ISIS to secure the future of the facility.
By 2021 it aims to have delivered on time and on budget:

- Extended target station lifetime
- Improved operations
- Increased performance for some beam lines
- Increased the knowledge, skill and experience level of staff
- Provide confidence for the future

The project is currently on track to complete on schedule and on budget. The project is a major activity for the ISIS design division with significant support from a wide variety of staff across ISIS and STFC.

Acknowledgements

As this paper and its associated presentation is a project management summary, I would like to thank the large number of people involved in the project so far and especially to those who have provided information and direct input to this paper.

References

[1] Jones L 2017 ISIS TS1 project target – design for manufacture *ICANS XXII proceedings*
[2] Coates D 2017 ISIS Target station one project – an overview of the development work being undertaken to improve the target, moderator and reflector TRAM support systems *ICANS XXII proceedings*
[3] Wilcox D 2017 Design and optimisation of the ISIS TS1 project target *ICANS XXII proceedings*
[4] Capstick M 2017 Design and analysis of the high pressure cryogenic moderator vessels for the ISIS target station one *ICANS XXII proceedings*
[5] Williams O 2017 Challenges, analysis and design solutions for a bolted target support frame within a remote handling area *ICANS XXII proceedings*
[6] Dean R, Jenkins D, Probert M & Burridge R 2017 Process filtration of liquid methane radiation products using centrifugal separation *ICANS XXII proceedings*
[7] Probert M, Skoro G, Rudic S, Romanelli G, King S, Bewley R & Webster J 2017 Spin isomers in the ISIS TS1 cryogenic hydrogen moderator *ICANS XXII proceedings*
[8] Bewley R, Skoro G, Lilley S, Smith R, Demmel F, Imberti S & Dalgliesh R 2017 The neutronic characterisation of ISIS TS1 *ICANS XXII proceedings*
[9] Romanelli G, Rudic S, Krzystyniak M, Fernandez-Alonso F, Formalski D, Kibble M, Goodway C, Bones J, Probert M and Skoro G 2017 Direct measurement of proportion of H2 para/ortho isomers in liquid hydrogen moderators at ISIS *ICANS XXII proceedings*
[10] Thomas S, Jenkins D, Haynes D, Russell C, Coates D, Jones L & Gallimore S 2017 ISIS TS1 project mock ups and testing *ICANS XXII proceedings*
[11] Skoro G, Bewley B, Lilley S, Ewings R, Romanelli G, Gutmann M, Smith R, Rudic S & Ansell S 2017 A tale of two foils - upgrade of ISIS TS-1 water moderators *ICANS XXII proceedings*

IOP Conf. Series: Journal of Physics: Conf. Series **1021** (2018) 012054 doi:10.1088/1742-6596/1021/1/012054

The neutron irradiation module at the European Spallation Source ESS

R Senesi[1,2], F Masi[3] , G Gorini[3] , G Scionti[4], C Vasi[2], Y Bessler[5], M Kickulies[6], Y Lee[6], R Linander[6], D Lyngh[6] , V Santoro[6] and L Zanini[6]

[1] Università degli Studi di Roma "Tor Vergata", Dipartimento di Fisica, Centro NAST, Roma, Italy
[2] CNR- IPCF, Sezione di Messina, Messina, Italy
[3] Università degli Studi di Milano-Bicocca, Milano, Italy
[4] Università della Calabria, Dipartimento di Fisica, Rende, Italy
[5] Forschungszentrum Jülich GmbH, Jülich, Germany
[6] European Spallation Source ERIC, Lund, Sweden

Abstract. The neutron Irradiation Module at the European Spallation Source will make use of the high intensity fast neutron spectrum to study the behaviour of the materials used in the facility, set within the ESS research and development program for the target station . By studying how these materials are affected by radiation, estimates of the material degradation in irradiated bespoke samples will allow to optimise of the design and lifetime of regularly replaced target components. The general design and a set of results from the neutronics calculations, aimed at estimating proton and neutron flux distributions, displacement damage, heat deposition and activation for the radiological hazard analysis of the Irradiation Module, set within the Italian contributions to the ESS construction phase, are reported.

1. Introduction

In order to obtain experimental information available on the behaviour and lifetime of spallation target components, ESS has set a test and development programme for structural materials used in radiation environments of spallation facilities [1]. In particular, knowledge of the changes of mechanical properties and microstructure induced by the high neutron flux from the 5 MW ESS spallation source, will allow to define improvements in lifetime and reliability of structural components of the ESS target systems. The outcome of such programme is expected to contribute to the relevant databases for the research and development on high–power spallation target assemblies [2]. The design and construction of the neutron Irradiation Module (IM) at the European Spallation Source is set within the Target In-Kind TIK 4.3 contribution agreement between ESS and CNR. Its primary function is for the study of the materials used in the facility. By using the fast neutron spectra, samples and components contained in the IM can be irradiated and used to support ESS own research and development program for the target station. Samples to be irradiated shall be located as close as possible to the target material, in a position offering both a representative radiation spectrum and an acceptable disturbance of moderator performance. Based on criteria and boundary conditions set by the spallation source construction phase, the IM will be located in the ESS pre-moderator and will be passively cooled by the thermal moderator

IOP Conf. Series: Journal of Physics: Conf. Series **1021** (2018) 012054 doi:10.1088/1742-6596/1021/1/012054

water flow. This set-up is intended to provide a proof-of-concept module, paving the way towards next generation neutron irradiation modules at ESS with active temperature control and large irradiation volume. The positioning of the IM inside the moderator reflector plug is reported schematically in Figure 1 below. With reference to the ESS Target Coordinate system (defining the centre of the Target monolith) the module is located in (x=-16; y=90÷150; z=+96) mm.

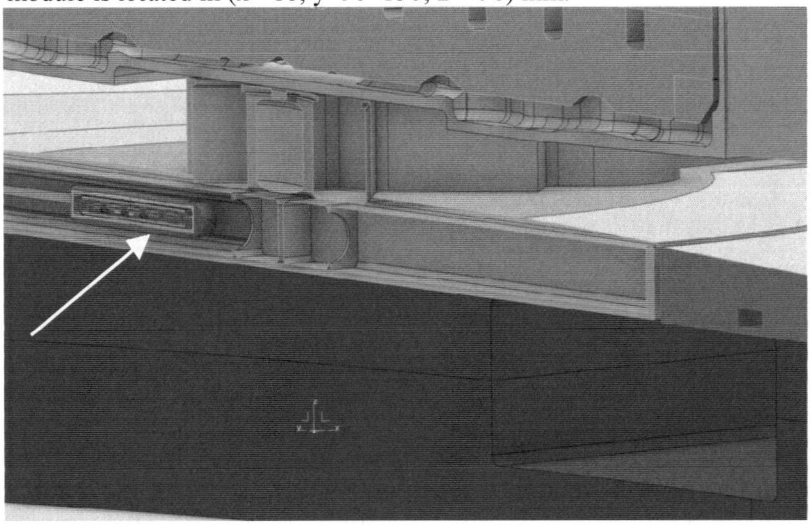

Figure 1. Positioning of the Irradiation Module (indicated by the arrow) into the ESS water pre-moderator

The module consists of a concentric cylinder-shaped container (vessel) enclosing ferrous and non-ferrous metal samples packed and assembled under helium atmosphere. Figures 2 and 3 report the sketch of the vessel and sample assembly, composed of 168 mini, mm sized samples. The external length of the module is 12.5 cm.

The post irradiation examination on the irradiated samples will make use of reference comparison with an identical set of samples that will not be irradiated. The characteristics of the ESS neutron spectrum, with energies up to 2 GeV and power on target up to 5 MW, from the initial to the design operation stages, respectively, will represent a unique opportunity to obtain data from a spallation neutron environment, that is beyond what is available from thermal or fast reactors. In particular, the flux of fast neutrons (integrated between 0. 1 MeV and 2 GeV) on the module will be in the order of 10^{16} neutrons cm^{-2} s^{-1}[3].

Figure 2. Schematics of the outer vessel containing the samples.

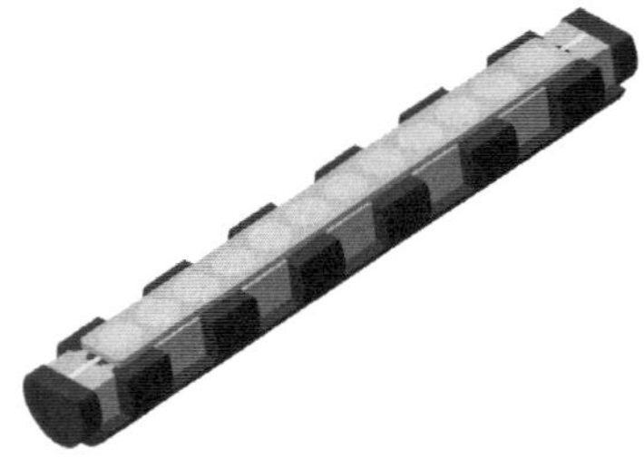

Figure 3. Schematics of the samples assembly

This manuscript reports in section 2, as an example of the expected damage on representative samples, the results of the neutronics calculations for the estimated displacement damage for Al alloy tensile samples. A complete report of the neutronics calculations in terms of neutron and proton fluxes, source brightness losses, gas production, heat deposition, and samples activation will be presented in a dedicated publication [3]. In particular, the design is tailored to guarantee that: 1) the ferrous metal samples will get the same damage dose that is equivalent to the displacement per atom value of approximately 6 dpa in the stainless steel SS316 L, that is of the target vessel's steel -lifetime reference values; 2) the gas production will be in the range of approximately 15 He-appm/dpa, similarly to the damage occurring in fusion related materials [2]; 3) the neutron flux at the irradiation module position is at least 10^3 times more intense than the proton flux.

2. Neutronics calculations

The development of the neutronics model and the IM design was based on inspection and optimization, using neutronics calculations, of the following main parameters:

- Neutron flux in the range of interest
- Impact on moderator's performance
- Radiation damage (dpa)
- Gas (H, He) production
- Heat load
- Activation

The calculations were carried out by making use of the MCNPX code [4]. We report here only an example of the estimated radiation damage on a set of Al alloy samples which will be used for tensile tests. Figure 4 reports the simulation geometry with respect to the primary proton beam.

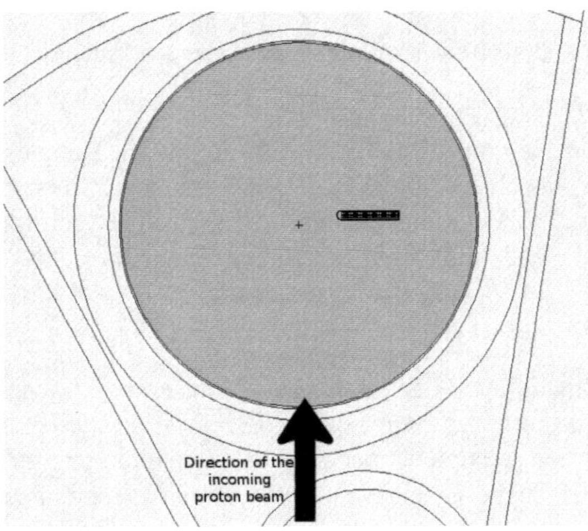

Figure 4. Input geometry for the MCNPX calculations. The Irradiation module is placed downstream with respect to the incoming proton beam. It is about 1.5 cm apart from the centre of the top moderator disk along the beam direction.

Figure 5 reports, as an example, the calculated radiation damage for Aluminium 6061-T6 alloy tensile samples, as a function of the distance from the moderator centre. These calculations shown for proton energies of 570 MeV (30.0 mA average current) and 2 GeV (46.0 mA average current), the former representing the parameters for the initial ESS operation phase, and the latter representing the parameters for the design operation. Both calculations were carried out per "effective year", corresponding to 5400 hours of operation per year. Therefore , the total dpa achievable in a typical - two years irradiation time, will be slightly less than the corresponding dpa from similar irradiation times at PSI [2], but with a much lower gas production per dpa, since the present module will receive an almost pure neutron irradiation [2,3].

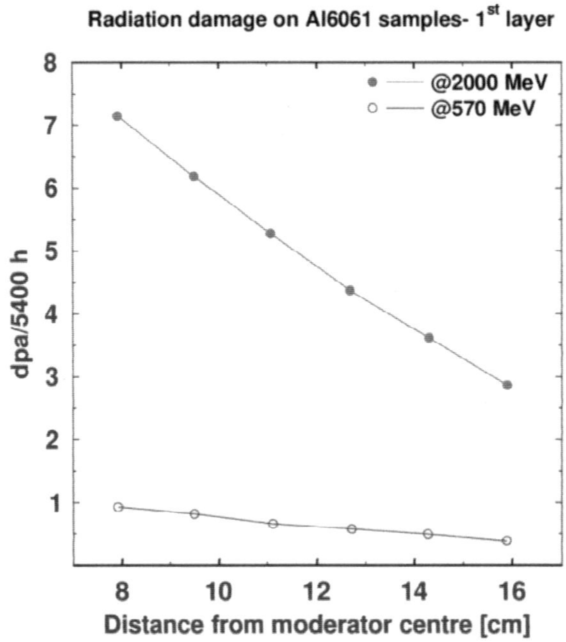

Figure 5. Radiation damage rates for the 6061-T6 alloy tensile samples in the first layer of the module, which is the closest to the target, as a function of their distance from the centre of the moderator. Calculations are reported for primary proton energy of 570 MeV (blue open circles), and 2000 MeV (red full circles), respectively.

3. Conclusions and outlook

The fast neutron irradiation module at ESS will provide relevant knowledge of target materials degradation from the intense neutron radiation field at ESS. The manufacturing of samples, holder, vessel and their assembly has been completed and the module will be integrated into the ESS thermal moderator at the beginning of 2018. Extensive details of the design, neutronic calculations and tomographic analysis of the assembled module will be presented in a forthcoming publication [3].

4. References

[1] Filges D and Goldenbaum F 2009 *Handbook of Spallation Research* (Weinheim: Wiley– VCH Verlag GmbH & Co.)

[2] Dai Y, Brun R, Gao W, Geissmann K, Hahl S, Hou H, Huang Y, Linder H P, Long B, Pahr A, Vontobel P, Wagner W, Wang H L, Zanini L, 2012 *J. Nucl Mater.* 431 2

[3] Scionti G, *et al.*, in preparation

[4] Gorley T, *et al.*, 2012 *Nuclear Technology* 180 298

Acknowledgments
This work is supported within the Target In-Kind TIK 4.3 contribution agreement between ESS and CNR.

IOP Conf. Series: Journal of Physics: Conf. Series **1021** (2018) 012055 doi:10.1088/1742-6596/1021/1/012055

Robust measurement of para-ortho H_2 ratios to characterise the ISIS hydrogen moderators

Giovanni Romanelli[a,b,*], Svemir Rudić[a], Maciej Krzystyniak[a,c], Felix Fernandez-Alonso[a,d], Damian Fornalski[a], Mark Kibble[a], Chris Goodway[a], Jon Bones[a], Molly Probert[a] and Goran Škoro[a]

[a] ISIS Facility, Rutherford Appleton Laboratory, Didcot OX11 OQX, UK
[b] Centro NAST, Università degli Studi di Roma Tor Vergata, Via della Ricerca Scientifica 1, 00133 Roma, Italy
[c] School of Science and Technology, Nottingham Trent University, Clifton Campus, Nottingham, NG11 8NS, UK
[d] Department of Physics and Astronomy, University College London, Gower Street, London, WC1E 6BT, UK

E-mail: giovanni.romanelli@stfc.ac.uk

Abstract. We discuss an experimental approach to determine the concentration of para-H_2 in the hydrogen moderators serving the Target Stations 1 and 2 at ISIS. The outcome of the measurements based on this approach is crucial for the Target Station 1 project, where a detailed characterisation of the present performance is needed to test neutronics simulations. The approach described here is based on the measurement of neutron transmission spectra over a wide energy range of samples extracted from the moderators. Pilot experiments were performed on the VESUVIO spectrometer at ISIS and were combined with measurements of thermal conductivity of the same samples using the ISIS *para*-H_2 rig.

1. Introduction

During the preparation for the ISIS Target Station 1 (TS-1) project, special attention has to be paid to the optimisation of the hydrogen (H_2) moderator, where the suitability of any potential new design strongly depends on the assumed concentrations of *para*-H_2 (pH_2) and *ortho*-H_2 (oH_2) nuclear-spin isomers in the moderator. The very same is true for a potential upgrade of the Target Station 2 (TS-2) coupled moderator. Attempts to characterise the concentrations of pH_2 in TS-1 H_2 moderator include the analysis of the data from the LOQ [1, 2] incident beam monitor over the last 10 years; the diffraction set-up added to CRISP [3, 4] to measure pulse widths over a cycle of experiments; and a comparison between experimental time-of-flight data from OSIRIS [5, 6] with corresponding simulations. On the basis of these results, it is currently assumed that the pH_2 concentration, p_α, should be around 80% in the current TS-1 H_2 moderator and that the concentration in TS-2 should be qualitatively higher.

2. Experimental set-up

To determine p_α experimentally, Neutron Transmission (NT) measurements were performed on the VESUVIO spectrometer at ISIS [7, 8, 9] using the pH_2 Rig (pHR) [10]. The experimental set-up is sketched in Figure 1. Time-of-flight spectra were measured by ^{6}Li monitors placed between the neutron source and the sample (a) and after the sample (b). Normal H_2 (nH_2) was commercially available; a high-purity pH_2 sample was prepared within the pHR; and the mixtures from TS-1 and TS-2 moderators were collected at the end of the 2016/03 ISIS user-round cycle. The experiment was performed ca. 2 weeks after the collection of the samples and these were stored in PTFE-coated bottles. Gas mixtures were loaded in a flat Al container with sample thickness $d = 0.5$ mm. The temperature of the sample during the experiment was set to 15 K in the liquid phase. The gas-loading procedure was controlled by using 316-type stainless-steel buffers attached to the pHR, and by monitoring the stability of transmission and forward-scattering spectra.

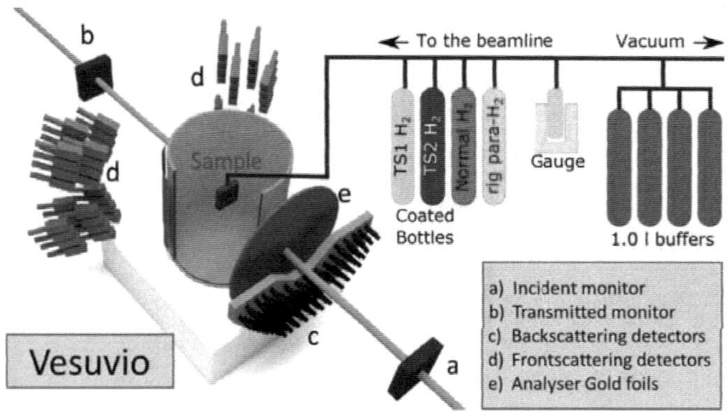

Figure 1. Schematic design of the VESUVIO instrument at ISIS; the additional set-up for sample loading using 1.0 l buffers; and the gauge for resistance measurements using the pHR.

3. Measurements of pH_2 concentrations

Neutron cross sections (σ_α) [11] and heat capacities (C_α) [12, 13] largely differ for pH_2 and oH_2, as shown in Figure 2. At equilibrium conditions and room temperature, nH_2 has $p_\alpha = 25\%$, as a consequence of the 3:1 ratio of nuclear-spin states of oH_2 (triplet) to pH_2 (singlet).

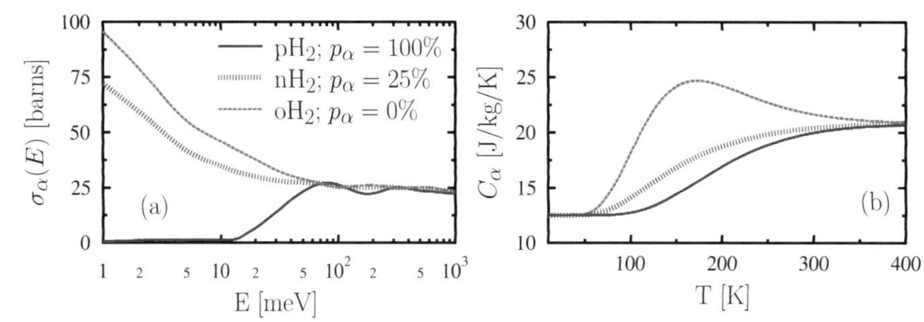

Figure 2. Neutron cross sections [11] as functions of the incident-neutron energy E (a) and isochoric heat capacities as function of temperature T [12, 13] (b) of H_2 mixtures.

NT spectra $T(E)$, as functions of the incident-neutron energy E, are related to the neutron cross section $\sigma_\alpha(E)$ of the sample according to the following expression

$$T_\alpha(E) = \frac{S_\alpha(E) - B(E)}{C(E) - B(E)} = \exp\left(-n\sigma_\alpha(E)d\right), \tag{1}$$

where $\sigma(E)$ is affected by the nuclear dynamics and Pauli exclusion principle [14, 15], $S_\alpha(E)$ is the spectrum from the sample α in the container, $C(E)$ is the empty-container spectrum and

$B(E)$ is the empty-instrument background. Spectra were further normalised to incident-monitor counts. One can express the cross section of the sample as the following combination of cross sections of the two nuclear-spin isomers, $\sigma_\alpha(E) = p_\alpha \sigma_p(E) + (1 - p_\alpha)\sigma_o(E)$. It is worth noticing that pH_2 and oH_2 neutron cross sections, $\sigma_p(E)$ and $\sigma_o(E)$, present the largest differences for incident-neutron energies below 50 meV [11].

Similarly, the calibration gauge in the pHR was used to obtain a relative value of p_α based on the differences in the thermal conductivity of the mixture α with respect to nH_2. A similar approach had already been attempted in the past, as for example in Refs. [16, 17] The gas thermal conductivity is inferred from a measurement of the electrical resistance R of a Pt wire at constant current and voltage. When a non-negligible current flows within the wire, its temperature T and its resistance R increase because of the electrical energy released. The external wall of the gauge is maintained at the temperature of liquid N_2 and the temperature and resistance of the wire are higher for lower thermal conductivity of the gas mixture. Moreover, for an ideal gas, the thermal conductivity can be considered proportional to the isobaric heat capacity. For the steady system, the heat release in the wire per unit of time is equal to the heat absorbed by the N_2 bath. Assuming that the heat capacity can be expressed as $C_\alpha(T) = p_\alpha C_p(T) + (1 - p_\alpha)C_o(T)$, one can determine the value of p_α from the measured resistance R_α of the wire. Alternatively, one can interpolate the measured resistances between two values related to known mixtures, say nH_2 and a pure pH_2 samples.

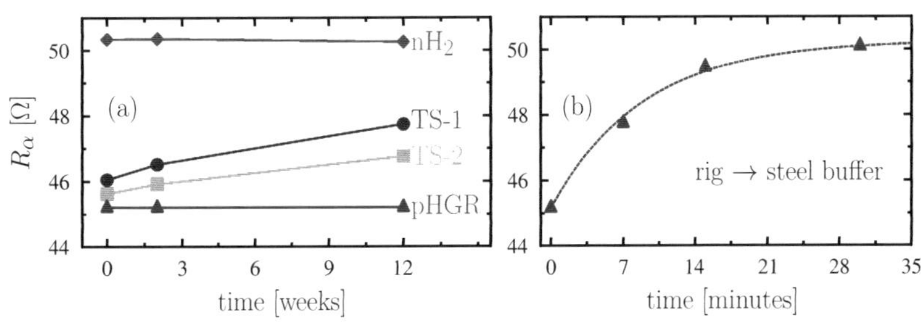

Figure 3. Conversion of pH_2 in the PTFE-coated bottles (a) and in the rig buffers (b) compared to the rate $f(t)$.

4. Results and Discussion

Samples from the H_2 moderators can be collected at the end of an ISIS cycle only. Therefore, a time of the order of 2 weeks is expected before NT measurements can be performed. To take into consideration this delay, the stability of p_α in the samples over a period of time of 12 weeks has been characterised. Figure 3a shows the change of the resistance measured by the gauge in the pHR as a function of the time the samples spent in PTFE-coated steel bottles. A slow decrease of p_α of the order of 2%/per week was observed, and the conversion process was approximately linear over the 12 weeks of monitoring. The values R_n for nH_2 were found approximately constant over the same period. Moreover, the values R_p for pH_2 produced with the pHR several times during the 12 weeks were found to be consistent, thus confirming that our pH_2 generation process is reproducible. On the other hand, a quick decrease of p_α was measured for pH_2 that stayed in the steel buffer for about 30 minutes. The conversion rate is of the form $f(t) = R_p + (R_n - R_p)(1 - \exp(-t/\tau))$ with $\tau \sim 9$ minutes (see Figure 3b).

NT measurements were performed on the nH_2 sample and on the samples from TS-1 and TS-2 moderators partially converted in the buffers. Figure 4 shows the comparison of experimental transmission of the nH_2 sample (8 hours of measurement) to MCNP-x simulations [18] of 0.5 mm thick H_2 mixtures with p_α = 25%, 70%, 75% and 80%. A good agreement is found between experiment and simulation in the known case of the nH_2 sample (green error bars). The case of the partially-converted sample from TS-1 is also shown (black error bars, 2 hours of measurement) as an example of statistical uncertainties for a shorter measurement. By comparison to the simulated predictions with p_α = 70%, 75% and 80%, one can assess the

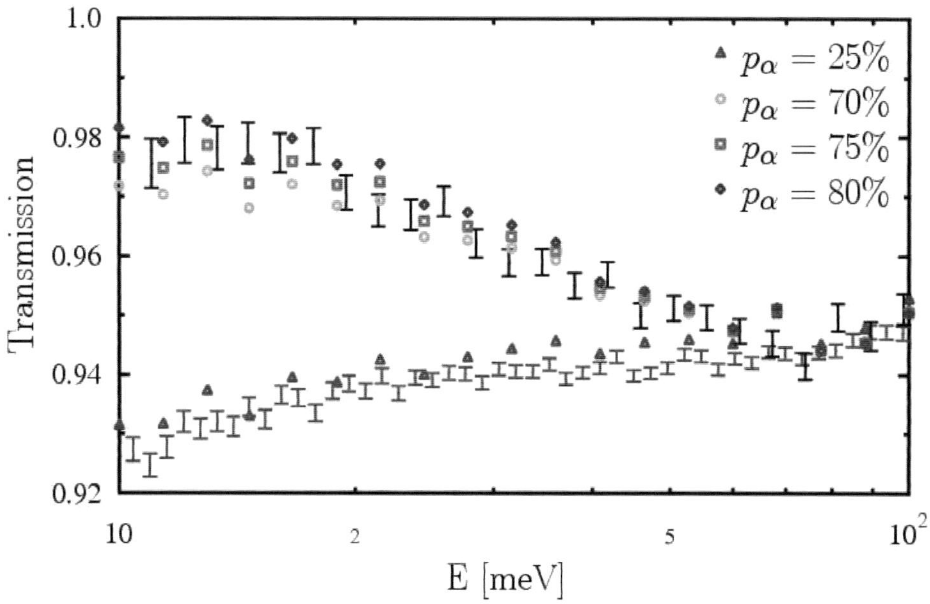

Figure 4. Simulated transmission spectra (markers) for a sample thickness of 0.5 mm. Experimental error bars correspond to nH_2 (green) and partially-convertedTS-1 (black) samples.

experimental uncertainty on the pH_2 percentage within the sample after 2 hours of data acquisition below 10%, that would improve to about 5% after 8 hours of acquisition time.

5. Outlook

In this preliminary study, we have successfully tested an experimental set-up to carry out neutron transmission measurements on VESUVIO to determine the concentration of pH_2 in the ISIS H_2 moderators. We have characterised the conversion of pH_2 in the samples stored in the PTFE-coated bottles and steel buffers. This work paves the way for future transmission experiments.

References

[1] http://www.isis.stfc.ac.uk/instruments/LOQ
[2] G. Skoro *et al*, *Talk at Advanced Moderators for Intense Cold Neutron Beams in Materials Research* 2016.
[3] http://www.isis.stfc.ac.uk/instruments/CRISP
[4] R. Bewley *et al*, *Talk at 3rd ISIS TRAMSNEG Meeting* 2016.
[5] http://www.isis.stfc.ac.uk/instruments/OSIRIS
[6] R. Bewley *et al*, *Talk at 2nd ISIS TRAMSNEG Meeting* 2016.
[7] http://www.isis.stfc.ac.uk/instruments/VESUVIO
[8] Seel A G, Krzystyniak M and Fernandez-Alonso F 2014 *Journal of Physics: Conference Series* **571** 012006
[9] G. Romanelli *et al*, 2017 *Measurement Science and Technology* **28** 095501.
[10] Evans E, Bones J and Goodway C *Para-hydrogen gauge*, internal technical report
[11] Grammer K B *et al.*, 2015 *Phys. Rev. B* **91**(18) 180301
[12] Jensen J E *et al.*, https://www.bnl.gov/magnets/staff/gupta/cryogenic-data-handbook/coverpage.pdf
[13] Leachman J W *et al.*, 2009 *Journal of Physical and Chemical Reference Data* **38** 721
[14] Fernandez-Alonso F *et al.*, 2012 *Phys. Rev. B* **86** 144524
[15] Granada J R *et al.*, 2012 *Physics Procedia* **26** 108
[16] A. T. Stewart and G. L. Squires, 1955 *Journal of Scientific Instruments* **32** 26
[17] T. W. Bradshaw and J. O. W. Morris, 1987 *Review of Scientific Instruments* **58** 83
[18] M. B. Chadwick *et al*, 2011 *Nuclear Data Sheets* **112** 2887 – 2996.

IOP Conf. Series: Journal of Physics: Conf. Series **1021** (2018) 012056 doi:10.1088/1742-6596/1021/1/012056

ISIS TS1 Project target – design for manufacture

L G Jones and D Wilcox

Science and Technology Facilities Council, Rutherford Appleton Laboratory, Harwell Campus, Didcot, OX11 0QX, UK

leslie.jones@stfc.ac.uk

Abstract. The planned upgrade for the ISIS Spallation Neutron Source First Target Station requires a complete redesign of the Target, Reflector and Moderator (TRaM) system whilst working within the constraints set by the existing Void Vessel and target station infrastructure. This talk focusses on converting the optimised simulated design for the TS1 project target into a physical reality that will be robust and give a reliable service life equal to or greater than the existing TS1 Target (>5 yrs). ISIS has the advantage of having its own 'Target Manufacturing Facility' (TMF). The experience it has gained from manufacturing ISIS current TS1 & TS2 targets has been invaluable in identifying potential manufacturing challenges that must be overcome at the design stage. The Target Design Group (TDG) has worked very closely with the TMF to test design ideas and develop techniques. This has culminated in a target design suitable for manufacture. The target design presented here is a work in progress that has now reached the prototype stage and closely resembles the final product. It is the result of a close collaboration between the Neutronics, Engineering Simulation, Engineering Design and Target Manufacturing Groups. The aim is to deliver a target that can be readily manufactured and is reliable in operation without compromising neutronic efficiency.

1. Background

The ISIS spallation neutron source was officially opened in 1985, operating with one target station (TS1). Whilst the target material has changed from the original depleted uranium to tantalum, in the early 1990's, and to tungsten clad with tantalum from 2001 to present the overall design of the target has remained unchanged. The addition of a second target station, which opened in 2008, was an opportunity to design a target that was optimised for tungsten clad in tantalum and therefore neutronically more efficient. Other features were incorporated in the Target, Reflector and Moderator (TRaM) design to reduce time needed for regular maintenance tasks and improve remote handling operations.

With ISIS TS1 having been in service for 32 years and TS2 for almost 9 years, the time is right to develop a new Target, Reflector and Moderator for TS1 using modern software and analysis techniques to increase useful neutron output without the need to increase beam power. The proton beam will remain at 200μA, 800MeV energy.

ISIS has a well-established user community and continues to operate reliably so it is critical that any changes to the TRaM don't compromise the high level of reliability that it currently achieves.

IOP Conf. Series: Journal of Physics: Conf. Series **1021** (2018) 012056 doi:10.1088/1742-6596/1021/1/012056

To arrive at a design that is optimised for neutronic performance, whilst being practical to manufacture and will be reliable in operation, has called for a careful balance between the needs of the various stakeholders. As the TS1 Project is restricted to changing the TRaM components only, it is imperative that the new TRaM assembly integrates well with the existing accelerator system and instruments. ISIS Neutronic Group and Engineering Simulation have co-operated to define the physical attributes for the target, such as size and number of target and position within the reflector and moderator. On the practical side the Target Design Group and Target Manufacturing Facility have worked closely to design the components and develop the special techniques and procedures required for manufacture of the target assembly.

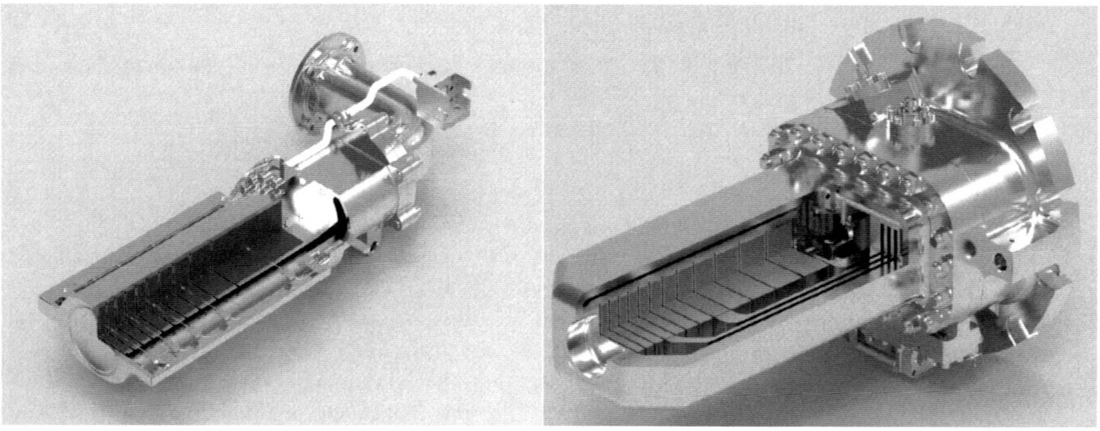

Figure 1. TS1 Project Target compared to TS1 current design (not to scale)

Table 1. Material reduction in TS1 project target.

Mass of material (kg)	Current TS1 target	TS1 Project target	Reduction
Stainless Steel	73.8	8.1	89.0%
Tantalum	32.7	6.9	79.0%
Tungsten	47.3	46.0	2.8%
Total	153.8	61.0	60.3%

2. Overview of TS1 Project target

The current target designs in use on TS1 and TS2 are very different from each other. The older TS1 target was originally designed and optimised for depleted uranium. It included a secondary cooling circuit and thick walled stainless steel vessel to satisfy stringent safety requirements. In the 32 years of service history the only major changes to the design have been to replace the target material for tantalum and later to tungsten clad in tantalum. The newer TS2 target design used only tungsten and tantalum which produced a very efficient neutron source. In the case of the TS2 target this was the obvious choice as there is, effectively, only one target plate. As the TS1 Project target will have a stack of 10 target plates things are not so straight forward, so two options were put forward for consideration. Option 1 was for a TS2 style target with only neutron producing materials used in the construction of the target assembly. Water cooling manifolds were tantalum and EB welded to the target stack. Option 2 proposed that the stack of 10 target plates would be contained within a thin walled stainless steel vessel incorporating the cooling water manifolds in the design.

When weighing up the two options (see Figure 2 below) it was concluded that Option 2 was the most viable as it would reduce the time to manufacture whilst eliminating the risks posed from EB welding the tantalum manifolds onto the finished stack of target plates. EB welded joints have proved troublesome in the construction of TS2 targets on a couple of occasions, and although these problems

IOP Conf. Series: Journal of Physics: Conf. Series **1021** (2018) 012056 doi:10.1088/1742-6596/1021/1/012056

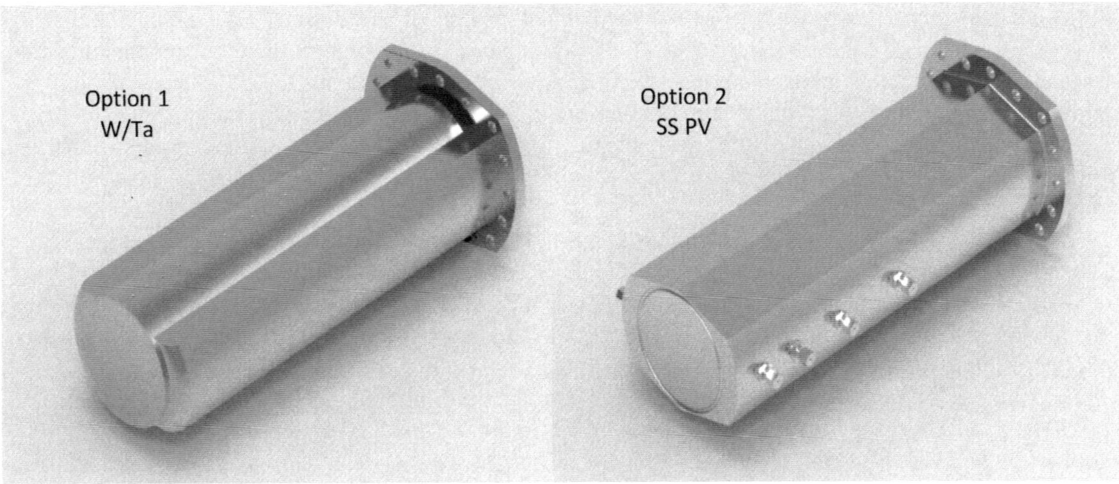

Option 1 – Tungsten/Tantalum (only)	Option 2 – Stainless Steel Outer Vessel
Maximum neutron output	<5% fewer neutrons
Higher material costs	Lower material costs (£50K)
EB welds in tantalum - challenging	Standard material for EB welding
Risk of water leaks from external EB welds	Less external weld length – less risk
Risk to target stack when EB welding manifolds	Target Stack protected in vessel
No stainless steel	8 Kg stainless steel only (89% < TS1)
Longer to manufacture	Stain.St. Vessel mfr in parallel with target plates

Figure 2. Target option merit chart

3. Design for Manufacture

3.1. Target Core

The Target Core comprises a stack of 10 tungsten plates of varying thickness from 11mm to 143mm, each encased in a protective layer of tantalum cladding. The cladding thickness is 1.5mm. This is 0.5mm thinner than the present TS1 target and 0.5mm thicker than TS2. Decay heat calculations and decay heat temperature measurements from our TS1 target suggest this is an acceptable amount of tantalum and will give a good level of protection to the tungsten core whilst providing a practical thickness for EB welding.

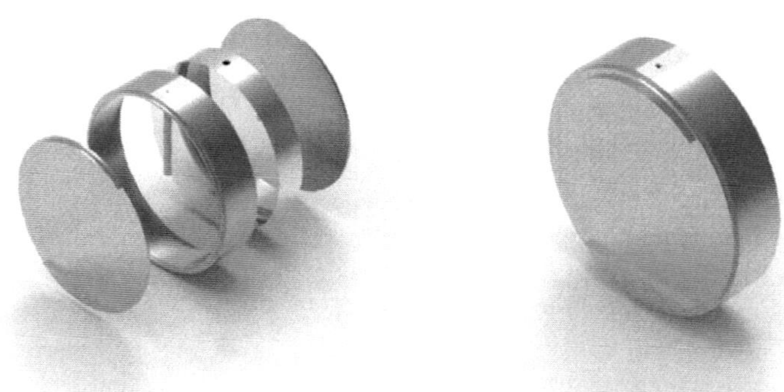

Figure 3. Target plate construction

IOP Conf. Series: Journal of Physics: Conf. Series **1021** (2018) 012056 doi:10.1088/1742-6596/1021/1/012056

Moving to a circular profile, from the rectangular profile of the present TS1 target plates has reduced the number of EB welds from 13 to 3. This is seen as a positive step, as it reduces the number of set-ups, but in practice it has been observed that a lot more heat is deposited into the plates during welding due to the length of weld when completing a 360º + overlap weld @ 3mm/sec. This can lead to distortion of the plates making it more difficult to align for the next welding stage. To help overcome this problem the material thickness at the weld joint is reduced to 1.2mm ensuring full penetration with a less intense beam.

After HIP and NDT each target plate has location features machined to ensure correct orientation of the plates and positioning of the thermocouple holes. The 10 plates are EB welded together to form a connected stack. The reason for this is to ensure that the plates don't vibrate against each other in the fast moving cooling water allowing small particles of tantalum to be rubbed off and travel around the cooling system leading to radioactive contamination in the circuit.

The number 10 plate is located and bolted to the 'Target Spacer Plate' so that the stainless steel vessel does not have to support the weight of the Target Core. This has the added benefit of simplifying the assembly process, because the vessel remains independent, and will make it possible to remove at end of life for PIE or separation of materials for waste disposal. A thermocouple will be inserted into each target plate to monitor the core temperatures. This is similar to the arrangement for TS1 target, except the thermocouples exit through the vessel. Compression fittings will seal onto the thermocouples, which are a swaged design allowing 3mm diameter to seal onto and reducing to 2mm for insertion into the plates. A similar arrangement has been in operation on TS2 manifold since start-up in 2008 with no leakages.

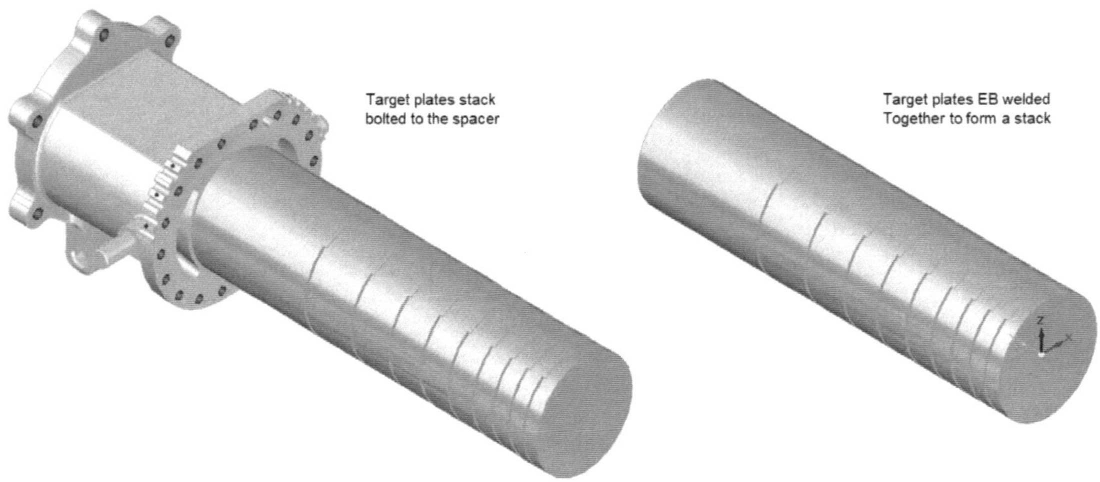

Figure 4. Target plates EB welded together and bolted to Target Space

3.2. Pressure Vessel

Material for the Pressure Vessel will be 316L due to the many years of experience with this grade of stainless steel at ISIS and other synchrotrons plus the abundance of experimental data confirming suitability for use in similar radioactive environments. To reduce the number of external welds it is proposed that the Flange will be machined as an integral part of the vessel with only the Window and thermocouples compression fittings being welded on. The Pressure Vessel cross section is an oval shape which does not lend itself readily to a single pass radial EB weld because of the changing focal point. However, successful development trials at The Welding Institute (TWI) have shown that this is possible and will be our preferred method of construction. The rest of the vessel will be machined using a combination of conventional milling (external profile), Wire EDM (integral water manifolds) & Die sinking (internal water cooling slots).

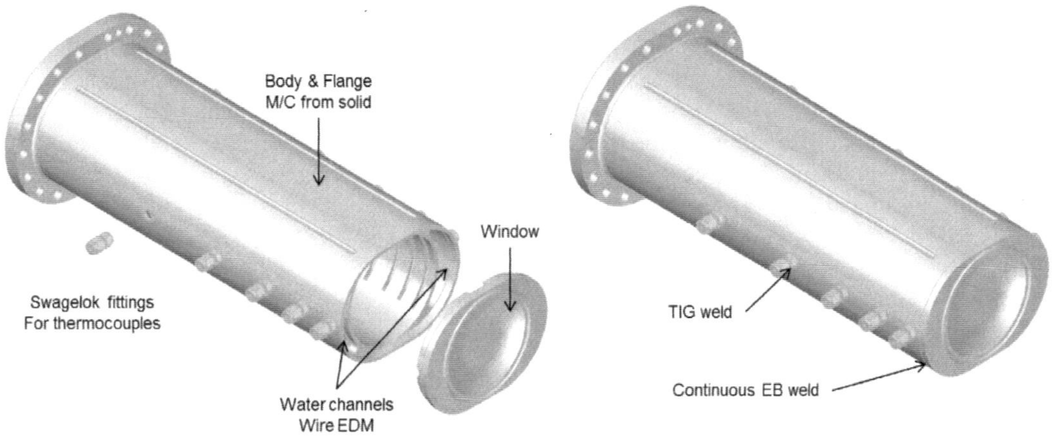

Figure 5. Pressure Vessel construction

3.3. Thermocouple Protection

As the thermocouples exit through the Pressure Vessel before being routed along to the Flange, where they are fixed, they are exposed and liable to damage when the Reflector is open and the Moderators are being worked on with the remote manipulators. A simple guarding system has been designed to cover them whilst allowing a partial view of the compression fittings in case a water leak develops. They are also designed to be easily removed, with the target still in position, should any access to the thermocouples be required.

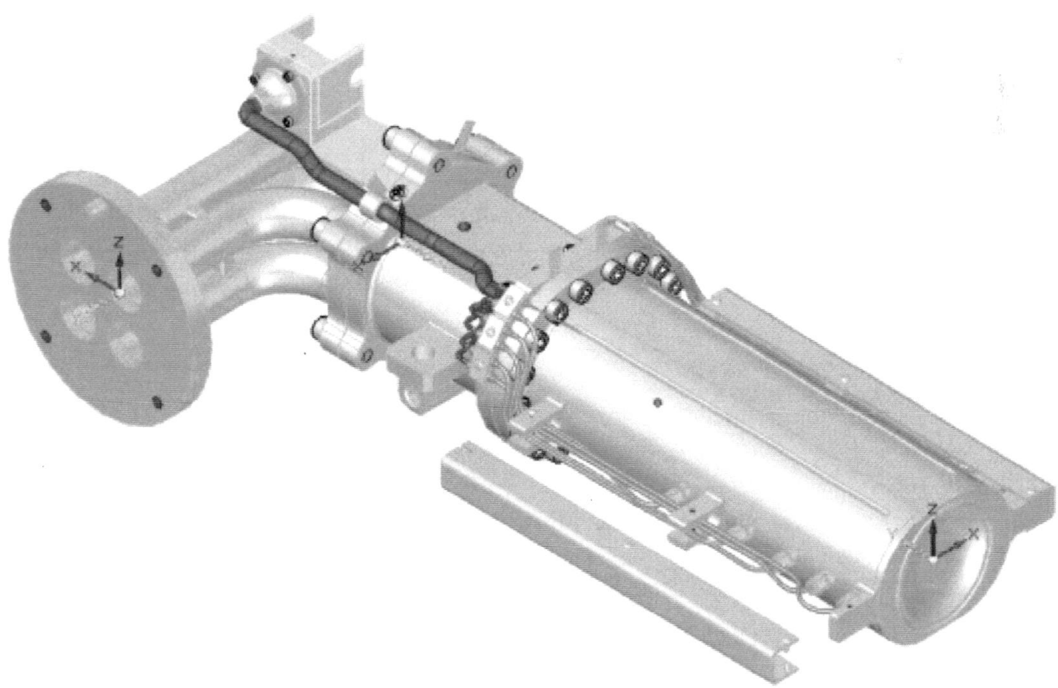

Figure 6. Thermocouple protection

IOP Conf. Series: Journal of Physics: Conf. Series **1021** (2018) 012056 doi:10.1088/1742-6596/1021/1/012056

4. Conclusions

The TS1 Project Target has many similarities to the current TS1 Target. It is consists of multiple tungsten plates clad in tantalum and housed in a stainless steel vessel. However, all the redundant features that were designed for the original uranium target and not required for a W/Ta target have been removed. This has resulted in a compact target that will integrate into the new TRaM, maximising the amount of Beryllium in the reflector and minimising the amount of stainless steel.

Progress with prototyping the target plates and vessel have shown this to be a practical design that can be manufactured within the project timescale.

References

[1] ISIS TS1 Project – Design Philosophy ver 3.
[2] S.Lilley, G.Skoro - MCNPX model 7B
[3] D.Wilcox – Engineering Simulation

IOP Conf. Series: Journal of Physics: Conf. Series **1021** (2018) 012057 doi:10.1088/1742-6596/1021/1/012057

Spin isomers in the ISIS TS1 cryogenic hydrogen moderator

Molly Probert, Goran Škoro, Svemir Rudić, Giovanni Romanelli, Robert Bewley, Stephen King, David Haynes, John Webster, Felix Fernandez-Alonso and Maciej Krzystyniak

ISIS Neutron and Muon Source, Science and Technology Facilities Council, Rutherford Appleton Laboratory, Didcot OX11 0QX, United Kingdom

E-mail: molly.probert@stfc.ac.uk

Abstract. As part of the ISIS TS1 project, work has been carried out on the cryogenic hydrogen moderator system, the main focus being understanding and improving the ratio of spin isomers in the liquid hydrogen moderator.

1. Introduction

Hydrogen comes in two spin states: ortho (o-H_2) and para (p-H_2). Of the two spin states the para state is preferable for the neutronics performance of the moderator [1, 2, 3]. At ambient temperature (so-called normal, n-H_2) hydrogen the ortho/para ratio is 3 : 1, which should convert to an equilibrium state of nearly 99.8% para at the 24K temperature of the moderator. However, the conversion process is very slow without a catalyst and the effects of radiation on the back conversion are relatively unknown (see [4] and the references therein). Catalysis of the relaxation process is necessary as the process is slow with rate of approximately 2% per hour [5] taking weeks to happen naturally with no external encouragement. Catalysis can take the form of magnetic centres on the containment material such as stainless steel, or other materials that decouple the hydrogen molecule [6], as it will naturally tend to recombine in the lowest energy state for the temperature of the environment.

2. Current ISIS TS1 hydrogen cryogenic system

The hydrogen cryogenic system at ISIS Target Station 1 (TS1) was designed and built in the 1980's. The moderator itself started out as a supercritical liquid hydrogen moderator with a cadmium decoupler but the moderator was never run super critical and the decoupler was removed in 2006. Some of the key parameters of the system as it operates currently are displayed in Table 1. There are five instruments that view the hydrogen moderator: CRISP, IRIS, LOQ, OSIRIS and SURF [7]. The hydrogen system is purged and filled with low pressure helium after each user cycle, and purged and refilled with a new batch of n-H_2 at the end of the shutdown. This generally happens about a week before the user run starts as the moderators need to be cold during beam physics. This allows a certain amount of time for the hydrogen to convert naturally with some help from the stainless steel pipework it flows through (the cool down and liquefaction takes between 6 and 8 hours). The original design of the hydrogen cryogenic system

IOP Conf. Series: Journal of Physics: Conf. Series **1021** (2018) 012057 doi:10.1088/1742-6596/1021/1/012057

Table 1. Current Hydrogen system parameters

Parameter	Value	Units
LH_2 Volume in Moderator	1	Litre
Flow Rate	0.5	L/s
Mass Flow Rate	36.5	g/s
Beam Heat	200	W
Delivery Temperature	23	K
Delivery Pressure	8	bar
Catalyst Volume	1.2	cm^3

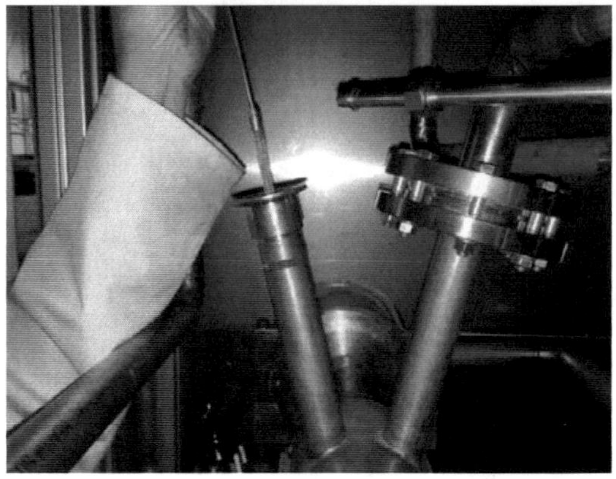

Figure 1. Ferric Oxide catalyst mesh holder inspected in December 2015.

had a very small amount of Ferric Oxide (Fe_2O_3) catalyst that was not of sufficient size to convert the volume of hydrogen in the system at any noticeable rate (see Figure 1). There has also been no reactivation or increase in volume of the catalyst during the lifetime of the cryogenic cold box. Now we are at the stage of designing a replacement system to incorporate a properly sized catalyst and hopefully an in-situ measurement device that should fit in with the existing transfer lines and cold box footprint.

3. Spin isomers ratio determination

The ortho/para ratio affects each of the instruments in a different way due to their operational differences. They all also have their own method of normalising out any effects a change in p-H_2 concentration would have on any of their experiments. This has led to the ortho/para ratio being a known problem but never a high priority to control or investigate the nature of how it changes. Due to the ongoing ISIS TS1 Project the need for better understanding of the ratio of spin isomers in the liquid hydrogen moderator has become greater. A detailed study has been performed in an attempt to characterize the ortho/para H_2 fraction using data from the LOQ incident beam monitor collected over the last 10 years [8, 9]. The main idea was that the variations of count rates (and wavelength dependence of these variations) in a single cycle, if combined with Monte Carlo simulation results, could give an indication about the ortho/para conversion. The simulations are important for an interpretation because neutron intensity fluctuations as a function of wavelength depend strongly on the starting/final values of the para-hydrogen fraction (especially at long wavelengths). So, the cycles where the variations are relatively large or where something happened (for example, the hydrogen system was purged and restarted during a cycle) were of special interest (see Figure 2). The analysis of those special events and comparison with Monte Carlo simulations show that the hydrogen behaviour is compatible with the so-called natural conversion scenario and that the maximal value of the H2 para-fraction during the last 10 years was around 80%. This value is in nice agreement with the value obtained from comparison of the shape of the experimental time-of-flight data from the OSIRIS instrument with corresponding simulation results [10]. In addition to these, the diffraction setup added to the CRISP instrument has been used to measure pulse widths as a function of wavelength over a cycle. The full width at half maximum (FWHM) values of the time pulse lie between the simulated values from 80% and 100% para hydrogen content (Figure 3), suggesting a maximal para-fraction value of 90% [11].

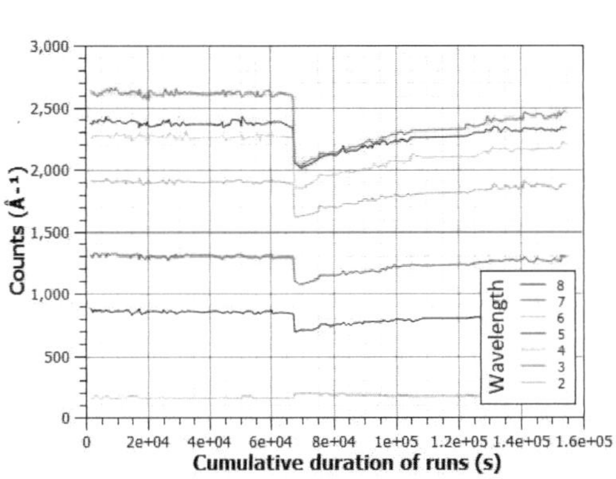

Figure 2. Count rates from LOQ instrument incident beam monitor as a function of wavelength (given in Angstroms in the box at the bottom of the plot) during cycle 2013/05 where the hydrogen cryogenic system was purged and refilled mid-cycle.

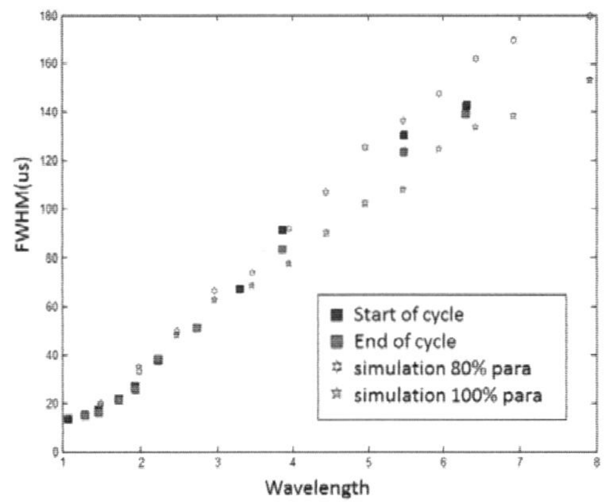

Figure 3. Comparison between measured and simulated time resolution of CRISP instrument as a function of wavelength (in Angstroms) over a cycle (see text for more details).

3.1. Grab sample measurements

A system was devised for taking ambient temperature grab samples from the TS1 and TS2 H_2 systems after a cycle. To alleviate the possibility of back conversion due to contact with stainless steel, cylinders with internal PTFE coating and brass fittings were used as they have been proved to be effective elsewhere [12]. Various methods, including Cavity Ringdown Spectroscopy [13], thermal conductivity measurements and neutron transmission measurements have been attempted to obtain the para hydrogen fraction from these samples. A method that is currently being implemented is based on the direct measurement of the neutron transmission through thin hydrogen samples (with different para-hydrogen fractions) using the VESUVIO instrument [14] at ISIS, combined with measurements of thermal conductivity of the same hydrogen samples using a para-hydrogen gauge (and a para-hydrogen generating cell) [15] originally built for use by the TOSCA instrument [16]. The samples from the hydrogen moderators at TS1 and TS2 were taken at the end of user cycle just before shutdown and measured with the thermal conductivity rig, then when the next user cycle started a couple of weeks later the measurements were repeated and compared to a neutron transmission measurement on the VESUVIO instrument. The initial results of these tests have provided a sound proof of concept and further tests are planned for the next few cycles [17].

4. Upgraded hydrogen system

The TS1 Project aims to increase flux, where possible, but not decrease the performance of the instruments viewing the LH_2 moderator. The expected gain in neutronics performance [18] is due to the presence of water premoderator, increased thickness of the moderating material (from current value of 8 cm to 10 cm) and increased para-hydrogen fraction (from current value of about 80 % to 99 %). As can be seen, the functionality of the associated cryogenic system will be broadly similar. The moderator itself will increase in volume hence the flow speed might change if the heat loads are significantly altered but the needs of the system to have an abundance of p-H_2 is still the same. A catalyst (Ionex, Fe_2O_3) of around 5 litres will be added to the system, which is enough to convert the total LH_2 inventory within an hour. Whilst it is simple to make

IOP Conf. Series: Journal of Physics: Conf. Series **1021** (2018) 012057 doi:10.1088/1742-6596/1021/1/012057

this larger to convert the whole inventory in a single lap, in practice this is never necessary, as the cool down of a new charge of hydrogen takes upwards of 6 hours, mostly due to the heat load of the stainless steel transfer lines. The back conversion from the radiation in the moderator head is minimal compared to the initial conversion from n-H_2. Installation of an in-situ measurement device to determine the ortho/para ratio during a cycle is also in the process of being designed. From several options, the easiest to implement is a relative measurement of the temperature change across the catalyst (a robust calibration will be needed) to monitor its "activity". Access to the catalyst will also be designed to allow for reactivation/replenishment (if necessary), as well as, a facility for hydrogen sample extraction so future results can be compared to the current experiments.

5. Conclusions

The work to characterise the ortho/para ratio in the LH_2 moderator on the ISIS TS1 is ongoing. We have worked from various angles to determine the ortho/para ratio of the hydrogen, from numerical simulations via flux measurements to taking hydrogen samples from the moderator and using different techniques to perform measurements of ortho/para hydrogen fractions. As the system currently stands the catalyst is severely undersized for the volume of liquid hydrogen. Therefore, introducing a suitable catalyst to the system as part of the upgrade and techniques for in-situ measurements during future running of the system are proposed.

Acknowledgments

The work presented in this paper is the combination of a great deal of time and effort by a large number of people. We would like to acknowledge the main contributors to the project here (listed in alphabetical order): John Chapman, Rob Dean, Beth Evans, Stephen Harrison, David Jenkins, Mark Kibble, Andy Robinson, Chris Russell from ISIS and Michele Gianella and Grant Ritchie from the Theoretical Physical Chemistry Laboratory at Oxford University.

References

[1] Harada M et al 2005 *Nucl. Instr. and Meth. in Phys. Res.* A **539** 345362.
[2] Mezei F et al 2013 https://arxiv.org/pdf/1311.2474.
[3] Grammer K B et al. 2015 *Phys. Rev.* B **91** 180301.
[4] Iverson E B and Carpenter J M 2003 *Proceedings of ICANS-XVI, Dusseldorf, Germany* p. 707-717.
[5] Enss C and Hunklinger S 2005 *Low-Temperature Physics* (Berlin: Springer).
[6] Matthews M, Petitpas G and Aceves S 2011 *App. Phys. Lett.* **99** 081906
[7] http://www.isis.stfc.ac.uk/
[8] Škoro G and Probert M 2015 TS1 Hydrogen moderator: ortho/para conversion puzzle *ISIS Internal Report.*
[9] Škoro G, Probert M et al 2016 Cold moderators at ISIS *Talk at 2nd RCM IAEA Coordinated Research Project on Advanced Moderators for Intense Cold Neutron Beams in Materials Research, Vienna, Austria.*
[10] Bewley R 2016 TS-1 baseline: experiment vs. simulatons *Talk at 2nd ISIS TRAMSNEG Meeting.*
[11] Bewley R 2016 TS-1 baseline: experiment vs. simulatons *Talk at 3rd ISIS TRAMSNEG Meeting.*
[12] Tom B et al 2009 *Rev. Sci. Inst.* 80, 016108.
[13] Gianella M 2016 Hydrogen ortho/para ratio *The Ritchie group report.*
[14] http://www.isis.stfc.ac.uk/instruments/vesuvio/
[15] Evans B, Bones J and Goodway C 2010 Para-Hydrogen Gauge Technical Report *ISIS Internal Report.*
[16] http://www.isis.stfc.ac.uk/instruments/tosca/
[17] Romanelli G et al. 2017 Direct measurement of proportion of H2 para/ortho isomers in liquid hydrogen moderators at ISIS *This Proceedings.*
[18] Škoro G, Lilley S and Bewley R 2017 Neutronics analysis of target, moderators and reflector design for the ISIS TS-1 project *Physica B* (to be published).

IOP Conf. Series: Journal of Physics: Conf. Series **1021** (2018) 012058 doi:10.1088/1742-6596/1021/1/012058

Design and optimisation of the ISIS TS1 Project target

D Wilcox and L G Jones

Science and Technology Facilities Council, Rutherford Appleton Laboratory, Harwell Campus, Didcot, OX11 0QX, UK

dan.wilcox@stfc.ac.uk

Abstract. The ISIS spallation neutron source is developing a new target, reflector and moderator assembly for its first target station (TS1). The TS1 Project will increase useful neutron output without increasing the beam power, by using modern software and analysis techniques to redesign these components for improved neutronic efficiency. The design and development of the TS1 Project spallation target is presented here. The final design represents a trade-off between the conflicting demands of neutronic efficiency, engineering reliability and manufacturability. It is essential that the TS1 project does not compromise the reliability of the facility; detailed finite element analysis has been carried out to demonstrate that the TS1 Project target will operate within specified design limits. The new design bears many similarities to the current TS1 target; it is still a plate target, made from tungsten clad in tantalum, cooled by heavy water and contained in a stainless steel vessel. However, the total mass is 60% lower than the current TS1 target. This will improve neutron output, while reducing material cost and the volume of radioactive waste produced.

1. Background

The ISIS spallation neutron source operates two target stations, TS1 and TS2. TS1 has been in operation since 1984, originally with a depleted uranium target. The target material was changed to tantalum in the early 1990's, then to tantalum-clad tungsten from 2001 to present. TS2 was opened in 2008, again using a tantalum-clad tungsten target. The TS1 target consists of 12 plates and operates at 160kW beam power, whereas the TS2 target is a solid rod target operating at 32kW beam power. The TS1 Project refers to ongoing efforts to develop a new target, reflector and moderator (TRAM) assembly for TS1. The project will increase useful neutron output without increasing the beam power, by improving neutronic efficiency using modern software and analysis techniques, along with lessons learned from the development of TS2. This will include the development of a new spallation target, referred to here as the 'TS1 Project target'. Figure 1 shows a comparison of the targets.

The scope of the project is limited to changing the TRAM components only; the new TRAM assembly must be compatible with the existing accelerator, beam transfer lines, neutron beamlines and instruments. The proton beam will remain at 800MeV energy, 200μA, with a circular Gaussian beam profile with 17.9mm sigma. The possibility of changing to an elliptical beam profile was investigated but found to be incompatible with the existing beam transfer infrastructure. The new target must be compatible with the existing target infrastructure including diagnostics and coolant pumps. It is critical that the TS1 Project does not compromise the high level of operational reliability which TS1 currently achieves. The new target must also be manufacturable at a reasonable cost. The ISIS neutronics and manufacturing groups were consulted throughout the design process to ensure their requirements were met. In many cases a trade-off was required between the conflicting demands of neutronic efficiency, engineering reliability and manufacturability.

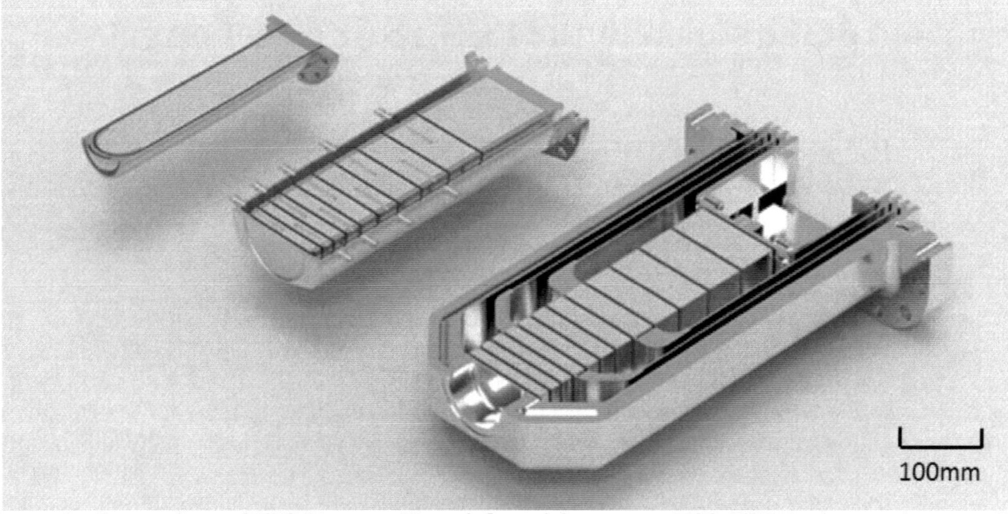

Figure 1. Comparison of ISIS Targets. From left to right: current TS2, TS1 Project, and current TS1.

2. Overview of TS1 Project target

A detailed view of the proposed TS1 project target is shown in Figure 2. The primary neutron producing material is tungsten, clad in tantalum to improve corrosion resistance. Both materials have a high neuron yield plus good mechanical properties at elevated temperatures, and have been used extensively at ISIS and at other spallation neutron sources around the world [1] [2]. The target core is split into plates to improve cooling and reduce thermal stress. The coolant is heavy water, which is contained in a stainless steel pressure vessel and flows between the target plates in parallel. The materials and basic geometry bear many similarities to the current TS1 target, as the initial design study concluded that these were still the best options based on the ISIS proton beam parameters.

Compared to the current TS1 target, the TS1 Project target is much smaller and lighter. The total mass was reduced by 60%, as shown in Table 1. Neutron production only occurs in the core of the target, which interacts with the proton beam, plus a narrow region around this where secondary interactions take place. Any material away from the core such as the pressure vessel and cooling manifolds will absorb more neutrons than it produces, and should be minimised. Therefore the TS1 Project target cross section is circular, to match the circular cross section of the proton beam. The circular cross section is also a standard shape which is easy to manufacture. Reducing the volume of target material also reduces material cost and the volume of radioactive waste produced at the end of the target lifetime.

Table 1. Material reduction in TS1 project target.

Mass of material (kg)	Current TS1 target	TS1 Project target	Reduction
Stainless Steel	73.8	8.1	89.0%
Tantalum	32.7	6.9	79.0%
Tungsten	47.3	46.0	2.8%
Total	153.8	61.0	60.3%

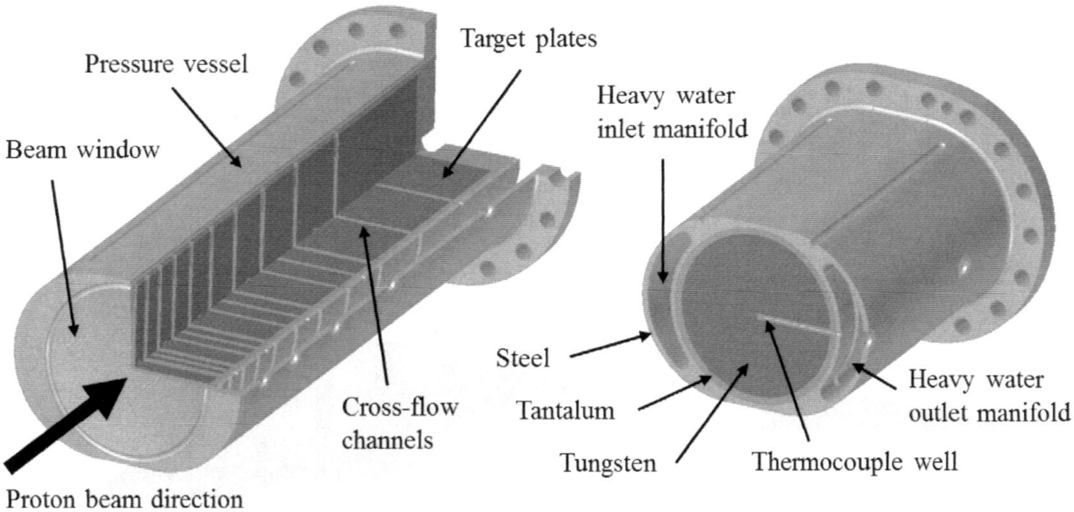

Figure 2. Sectioned views of the proposed TS1 Project target.

3. Design and optimisation

3.1. Target core

The target core must be split into plates to manage temperature and thermal stress. However each additional plate increases the volume of tantalum and heavy water, which is bad for neutron production, and also increases the number of manufacturing operations. Therefore the number of plates should be minimised. ANSYS Workbench finite element analysis (FEA) software [3] was used to optimise the design of the target core. A parametric model of the target core was built, which allowed the simulated plate thicknesses to be easily varied, and output the peak core temperature, surface temperature, and the peak stress in tantalum and tungsten for each plate. The current ISIS targets were simulated using a similar process to give a basis for comparison. Table 2 shows the results for each target, and the design limits which were set. The core temperature is unlimited; it is only a problem due to the thermal stresses it creates. The surface temperature is limited to 100°C to avoid damaging adjacent components such as the moderators. Tungsten stress is limited to half the tensile strength, which is estimated to be 550MPa. Tantalum stress is limited to the simulated stress level in the current TS1 target, as we know this target is reliable. Core temperature, heat flux and tungsten stress for the TS1 Project target are slightly higher than for current TS1, but lower than current TS2 and well within design limits. The optimisation goal was to use the smallest possible number of target plates without exceeding any of the limits. The analysis concluded that 10 plates were required, compared to 12 plates in the current target. The limiting factor was tantalum stress for the front 4 plates and surface temperature for the remaining 6 plates.

Table 2. Summary of finite element analysis results for ISIS targets.

Parameter	Current TS1	Current TS2	TS1 Project	Design Limit
Peak Core Temperature (°C)	184	249	196	N/A
Peak Surface Temperature (°C)	35	35	88	100
Peak Heat Flux (MW/m^2)	2.0	2.4	2.1	3.0
Peak Tungsten Stress (MPa)	89	157	96	275
Peak Tantalum Stress (MPa)	116	90	112	116

3.2. Fluid flow
The target coolant is heavy water, as it absorbs fewer neutrons than regular water. Nonetheless, heavy water anywhere in the target has an adverse effect on neutronic performance, so the volume should be minimised. Heavy water between the target core and the moderators (or pre-moderators) has a particularly large effect and should be avoided entirely if possible. The TS1 project target design achieves this by supplying the heavy water from manifolds either side of the target, with no direct cooling on the moderator-adjacent faces. CFX software within ANSYS Workbench was used to simulate fluid flow through the target. Various manifold geometries were tried before the current version was adopted. It is designed to distribute flow as evenly as possible between all plates, while minimising the volume of coolant and complying with the pressure drop and flowrate limits imposed by the existing pumps and water circuit.

3.3. Pressure vessel
The pressure vessel should safely contain the pressurised coolant, but in order to improve neutron output the volume of steel should be minimised. Stainless steel 316L was chosen as the pressure vessel material, as it has good corrosion resistance, weldability and structural properties. The current TS1 target pressure vessel was designed to cope with the original uranium target plates overheating, so is very over-engineered for tungsten plates and would not be a suitable comparison when designing the new vessel. To ensure integrity without making comparison to the previous target, the new vessel was designed to meet the requirements of British pressure vessel design code PD5500 [4].

3.4. Beam window
The beam window, located at the front of the target, is an integrated part of the pressure vessel and made from the same stainless steel 316L. It must withstand both heat deposition by the proton beam and pressure from the heavy water. The window is curved so as to be thin in the centre where the beam passes through, but thicker at the edges to resist the fluid pressure. A very thick window would absorb a lot of energy from the proton beam and experience high thermal stress, whereas a very thin window would not be able to withstand the fluid pressure. A design study was carried out to determine the optimum window dimensions, using a parametric model coupled to a genetic algorithm in ANSYS DesignXplorer. The current TS1 target has a similar thickness beam window, so unlike the rest of the vessel, the beam window stress can be compared to the current TS1 target, and is found to be lower in the TS1 Project target. The window also meets the requirements of PD5500.

4. Conclusions
The new design is fundamentally quite similar to the current TS1 target; it is still a plate target, made from tungsten clad in tantalum, cooled by heavy water and contained in a stainless steel vessel. The target core contains a similar volume of tungsten to the current TS1 target. However, the volume of additional material (e.g. pressure vessel and cooling manifolds) has been substantially reduced. This will reduce neutron re-absorption in the target, material cost, and the volume of radioactive waste produced. Detailed finite element analysis has been carried out to demonstrate that despite this material reduction the target will operate within all required design limits, ensuring reliability.

References
[1] Kawai M, Kikuchi K, Kurishita H, Li J and Furusaka M 2001 *J. Nuc. Mat.* **296** 312-20
[2] Nelson A T, O'Toole J A, Valicenti R A and Maloy S A 2012 *J. Nuc. Mat.* **431** 172-84
[3] ANSYS® Mechanical Enterprise, Release 17.1, ANSYS, Inc.
[4] British Standards Institution 2011 *PD5500:2009+A3:2011*

IOP Conf. Series: Journal of Physics: Conf. Series **1021** (2018) 012059 doi:10.1088/1742-6596/1021/1/012059

ISIS Target Station One Upgrade Project – An overview of the development work being undertaken to improve the Target, Reflector and Moderator (TRaM) support systems

D Coates

Mechanical Design Engineer

Science and Technology Facilities Council, Rutherford Appleton laboratory, Harwell Campus, Didcot, OX11 0QX, UK

Daniel.coates@stfc.ac.uk

Acknowledgement:- Mathew Capstick, Owain Williams, Andrew Atherton, Daniel Blanco Lopez, Goran Skoro, Steven Lilley.

Abstract. ISIS Spallation source is currently undergoing a new project to develop the Target, Reflector and Moderators (TRAM) of its existing Target Station One (TS1) facility. This project is being undertaken with a view to improve the neutronic performance combined with a gain in output and at the same time, engineered to increase flexibility for upgrades and easier remote operations. The second ISIS target Station (TS2) was completed in 2008. Its design enables the remote handling operator to access the methane moderator in minutes, without having to handle the target with the requirement to drain and split water circuits. With the opportunity to replace much of the structure within TS1, some of the design features of TS2 could be incorporated in the development of the TS1 TRAM.

1. Background

TS1 has been in operation from 1984 and has undergone multiple target changes, the reflector had been replaced in 2004, a replacement hydrogen moderator in 2007 numerous ambient moderator changes and 47 methane moderators. The TRAM is situated at the front of a long trolley system that consists of shielding plugs and services that provide cryogens and cooling for the moderators, reflector and target as shown in *Figure 1.1.*

Figure 1.1, ISIS TS1 Target trolley, The services are located on the yellow trolley to the left. The TRAM is located to the right end of the trolley in front of the forward shielding plug.

IOP Conf. Series: Journal of Physics: Conf. Series **1021** (2018) 012059 doi:10.1088/1742-6596/1021/1/012059

The trolley travels to and from the centre of the target station to the remote handing hot cell for maintenance. The TRAM is too active for human intervention and therefore maintenance operations must be carried out remotely using manipulators while using a zinc bromide window and CCTV cameras as visual aids. The methane moderator is regularly changed twice a year due to clogging. The procedure to change this moderator involves draining down the entire cooling system into holding tanks and drying out the system. The reflector and target are disconnected and handled to their storage locations to enable access to the moderator. The moderator connections are located out of sight below the support beam; this location can make remote handling operations tricky. The entire procedure can take two weeks to change the moderator with risks of damage to the reflector, target, leakages and contamination. The aim is to try to minimise the handling procedures and to reduce the risks.

2. The Neutronic Spec

The new TRAM specification was developed by the ISIS neutronics group and was supplied as a MCNPX file converted to import into Computer Aided Design software (CAD). The latest design being developed for manufacture is the Mark 7. The internal geometry is restricted by the requirements and the positions of the existing beamlines. As shown in *Figure 2.1*, the sizes and positions of the ambient water moderators are unchanged from the current TRAM. The target has been developed to reduce the stainless inventory. The hydrogen moderator head and moderator position has been developed for 99% para operation. The methane moderator has gone from a single poison foil to a dual foil, but is located in the same position. Both the cryo-moderators have the addition of the pre-moderator with a view to extend the moderator life time. Construction of the reflector has progressed from a stainless box filled beryllium structure to a solid beryllium unit.

- *Dual water moderators.*
- *Stainless cased. Tantalum cladded tungsten plate target.*
- *Water pre-moderator for hydrogen moderator.*
- *Water pre-moderator for methane moderator.*
- *Hydrogen moderator.*
- *Dual poison methane moderator.*
- *Surrounded by a solid Beryllium reflector*

Figure 2.1, ISIS TS1 Project TRAM MCNPX Model.

3. Engineered for Remote Handling

The current TS2 TRAM has a clever support frame system that divides the reflector and positions it away from the target and moderators without spliting the water circuits. An attempt to develop the same system for the TS1 project wasn't achievable as it was difficult to overcome the problem of the target being fixed to the interface door blocking moderator access. Therefore, a solution to this problem was to remove the current target support from the door and mount the target directly within the reflector. This then offers enough space to divide the reflector to reveal the cryo-moderators. See *Figure 3.1* for details.

Figure 3.1, The TRAM concept to illustrate a method of splitting the reflector to access the cryo-moderators.

However, this generated a new problem of providing flexible services adequate to supply sufficient cooling for the target. *Figure 3.2* illustrates the complexity of delivering services to the target, reflector and moderators. This concept also highlighted the limititations of the current cantilevered support structure, a new support structure is required.

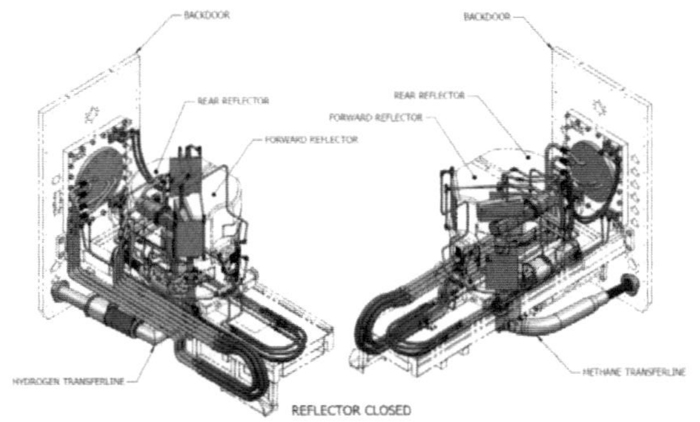

Figure 3.2, Diagram to illustrate the various cooling circuits thoughout the TRAM.

4. Hydrogen Moderator Development

The hydrogen moderator design utilises the current coupling and stem design, but with an optimised moderator head designed for the required 99% para hydrogen, as shown in *Figure 4.1*.

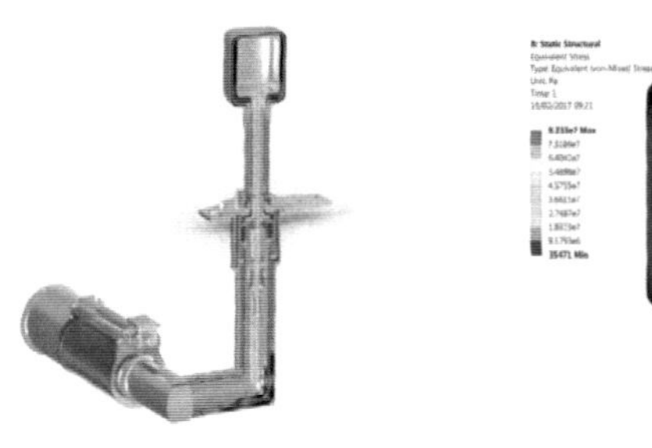

Figure 4.1, Hydrogen moderator design

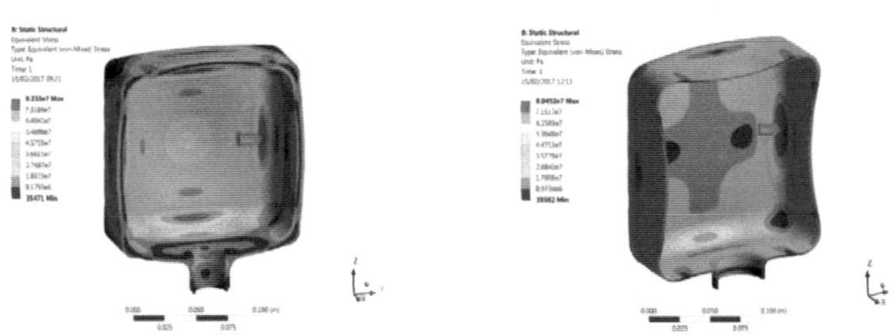

Figure 4.2, FEA plots of the hydrogen moderator head optimisation

The moderator head is constructed from Aluminium 5083 alloy and has been optimised to withstand the operating pressure of 8 Bara and a temperature of 20K. The surrounding vacuum and tertiary vessels have been developed for both pressure and vacuum situations as shown in *Figure 4.2*.

5. Methane Moderator

The proposed methane moderator as shown in *Figure 5.1* has been developed with a dual gadolinium poison layer. The pressure vessel to be manufactured from Aluminium 5083 alloy, has been designed to contain liquid methane at 3 Bara and 110k. The pressure vessel is surrounded by a vacuum vessel to provide thermal insulation. The vessel have been accessed by using FEA modelling as shown in *Figure 5.2*, to produce a robust design without compromising the thickness of the can that can reduce neutronic performance.

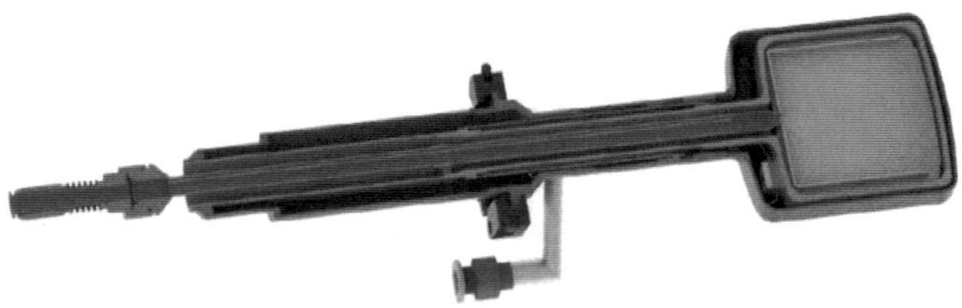

Figure 5.1 Dual Foil Methane moderator development, The cryo- coupling is shown to the left of the image, the moderator pressure and vacuum vessels are shown to the right. The pressure clamp is shown around the moderator in the middle of the image.

The cryo-coupling for this moderator has been designed to position the pressure clamp from beneath, to above the cantilever support frame to provide better access and visibility to aid remote handling during moderator change.

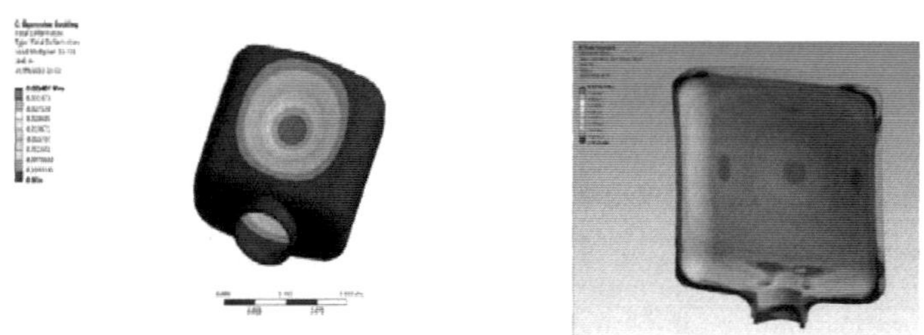

Figure 5.2, illustrates the effect of internal pressure on the methane vessel and external pressure acting on the vacuum vessel.

6. Reflector Development

The technical advancement of a solid beryllium reflector as fitted to the ISIS second target station, is a feature that would be beneficial on the TS1 project to replace the current stainless box filled with beryllium rod design. However, this does have its challenges as TS2 reflector is subjected to 4kw of thermal load whereas TS1 is around 11kw. The reflector is divided into multiple components to provide flexibility to accommodate future upgrades rather than a complete reflector replacement to reduce waste, costs and remote handling operations. The reflector is also designed to support the target, ambient water moderators and pre-moderators. Access to these components is also provided. The design requirement was to maintain the reflector core temperature to below 100°C to protect the aluminium moderators and prevent the deterioration of the neutronic performance. Cooling to the

IOP Conf. Series: Journal of Physics: Conf. Series **1021** (2018) 012059 doi:10.1088/1742-6596/1021/1/012059

reflector is provided by conduction to the outer edges using water flowing through cooling pads. As shown in *Figure 6.1*, FEA analysis was carried out to optimise the structure and by addition of more cooling pads a core temperature of less than 100°C was achieved. As shown in *Figure 6.2* the reflector is broken down into 13 component parts. The size of the entire reflector is 800 X 700 x 700mm and will be manufactured from Beryllium S200FH.

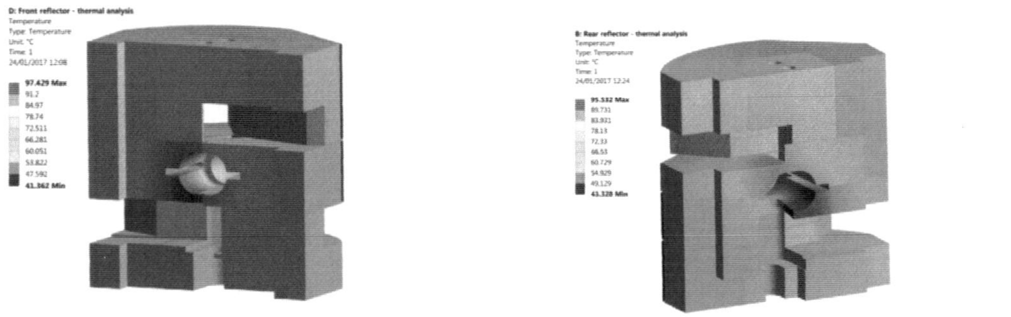

Figure 6.1, FEA plots of the thermal analysis of the reflector halves. The maximum expected temperature measured is 97.5°C.

Figure 6.2, Illustration of the reflector breakdown, The complete reflector assembly as shown on the right.

7. Conclusion

Although the neutronic design is not too different to the current TRAM. A lot of development has already been undertaken to produce a design that will improve remote handling during moderator changes with the advantages of reducing down time along the risks of contamination and damage. The design has taken more focus towards the eventual decommissioning utilising structures that can be broken down to fit into standard flasks to eliminate cutting radioactive material. The solid beryllium reflector has been successfully optimised with a sectional construction for upgrade flexibility while maintaining a critical internal temperature of less than 100°C. Water pre-moderators have been successfully incorporated into the reflector to reduce the damaging effect of the high energy neutrons acting on the cryo-moderators which should improve moderator life.

References

[1] G. Skoro, S. Lilley and R. Bewley, 'Neutronics analysis of target, moderators and reflector design for the ISIS TS-1 project, Physica B, (submitted)

IOP Conf. Series: Journal of Physics: Conf. Series **1021** (2018) 012060 doi:10.1088/1742-6596/1021/1/012060

Implementation of a small-angle scattering model in MCNPX for very cold neutron reflector studies

Kyle B Grammer and Franz X Gallmeier
Oak Ridge National Laboratory, Oak Ridge, TN 37831, USA

Abstract. Current neutron moderator media do not sufficiently moderate neutrons below the cold neutron regime into the very cold neutron (VCN) regime that is desirable for some physics applications. Nesvizhevsky et al [1] have demonstrated that nanodiamond powder efficiently reflect VCN via small angle scattering. He suggests that these effects could be exploited to boost the neutron output of a VCN moderator. Simulation studies of nanoparticle reflectors are being investigated as part of the development of a VCN source option for the SNS second target station. We are pursuing an expansion of the MCNPX code by implementation of an analytical small-angle scattering function [2], which is adaptable by scattering particle sizes, distributions, and packing fractions in order to supplement currently existing scattering kernels. The analytical model and preliminary studies using MCNPX will be discussed.

1. Introduction

Cold neutron (CN) beams with energies of $10^{-4} - 10^{-2}$ eV ($2.86 - 0.286$ nm), for which there are broad scientific applications in fundamental neutron physics and for studying materials, are typically produced through scattering from a cryogenic hydrogenous moderator typically operated at temperatures in the 10-20 K range and with effective cold neutron thermal distributions somewhat warmer. Technologies also exist for producing ultracold neutrons (UCN) with energies below 10^{-7} eV. At present, very cold neutrons (VCN) in the intermediate energy range between UCN and CN is desirable for some physics applications and efficient VCN moderators have not yet been developed.

An intriguing potential material for VCN production taking advantage of small angle neutron scattering from a low absorption, high optical potential material has been proposed [1]. The characteristic diameter of nanoscale materials is on the order of 5.0 nm and is similar in magnitude to VCN wavelengths. A material with a low absorption cross section allows the material to behave as a reflector via multiple small angle scattering events analogous to beryllium and graphite reflectors used in nuclear reactors. In particular, carbon has a small absorption cross section of 0.0035 barns and detonation nanodiamonds can be produced in large quantities

This manuscript has been authored by UT-Battelle, LLC under Contract No. DE-AC05-00OR22725 with the U.S. Department of Energy. The United States Government retains and the publisher, by accepting the article for publication, acknowledges that the United States Government retains a non-exclusive, paid-up, irrevocable, worldwide license to publish or reproduce the published form of this manuscript, or allow others to do so, for United States Government purposes. The Department of Energy will provide public access to these results of federally sponsored research in accordance with the DOE Public Access Plan (http://energy.gov/downloads/doe-public-access-plan).

IOP Conf. Series: Journal of Physics: Conf. Series **1021** (2018) 012060 doi:10.1088/1742-6596/1021/1/012060

through an explosive process [3]. Detonation nanodiamonds form large agglutinates with sizes on the order of microns, and can be purified through sonication followed a milling procedure to isolate primary particles 4.5 nm ± 0.5 in size [4], producing nanoparticles composed of a diamond-like core and a graphite-like shell [5]. Quasi-specular reflection of cold neutrons from diamond nanoparticles has been observed experimentally [6].

Current scattering kernels do not include small angle neutron scattering, which is required to perform transport calculations of powder filter materials. As part of the design effort for the Spallation Neutron Source second target station, an analytical small angle neutron scattering model has been incorporated into MCNPX in order to perform transport calculations and to investigate potential VCN reflector materials. The analytical model describes a Percus-Yevick hard-sphere fluid [2] with the ability to vary the polydispersity, packing fraction, and contrast.

2. Model

The scattering intensity for a medium containing a distribution of particles with diameters σ_i is given by

$$I(k) = \rho \int_0^\infty P_i^2(k) f(\sigma_i) d\sigma_i + \rho \int_0^\infty \int_0^\infty P_i(k) P_j(k) H_{ij}(k) f(\sigma_i) f(\sigma_j) d\sigma_i d\sigma_j, \quad (1)$$

where ρ is the particle number density, k is the momentum transfer, $H_{ij}(k)$ is the pair structure function as derived by Blum and Stell [7]. The particle size distribution function, $f(\sigma)$, is given by the Schulz Γ function,

$$f(\sigma) = \frac{\sigma^z e^{-\frac{\sigma}{b}}}{b^{z+1}\Gamma(z+1)} d\sigma, \quad (2)$$

where z is the Schulz width factor representing the polydispersity and $b = \sigma_\mu/(z+1)$. This distribution has mean particle diameter σ_μ and variance of $\sigma_\mu^2/(1+z)$. The normalized kth moment of the Schulz distribution is

$$\langle \sigma^k \rangle = \int_0^\infty \sigma^k f(\sigma) = \frac{(z+k)!}{z!(z+1)^n} \sigma_\mu^k, \quad (3)$$

which gives an expression for the particle number density, ρ,

$$\rho = \frac{6\eta(1+z)^2}{\pi\sigma_\mu^3(2+z)(3+z)}, \quad (4)$$

where η is the volume packing fraction. The scattering amplitude for hard spheres with uniform scattering length density is given by

$$P_i(k) = 4\pi p k^{-3} [\sin(k\sigma_i/2) - \frac{1}{2} k\sigma_i \cos(k\sigma_i/2)], \quad (5)$$

where $p^2 = (p_0 - p_{\text{med}})^2$ is the contrast between the medium and the particles. Note that two errors in the equations of the original publication [2] were found while verifying the results before implementing the analytical model in MCNPX. There is a factor of k^{-3} that is missing from equation 3 and equation 20 is missing an overall multiplicative factor of 9 and also contains an extra factor of ρ.

The analytical model is invoked via a material modification card, `ms`, on which the model parameters are specified. The macroscopic scattering cross section for the small angle neutron scattering mode is then calculated as a function of $k\sigma_\mu$ using the analytical function in logarithmic bins over 10 decades with 1000 bins per decade in order to provide sufficient

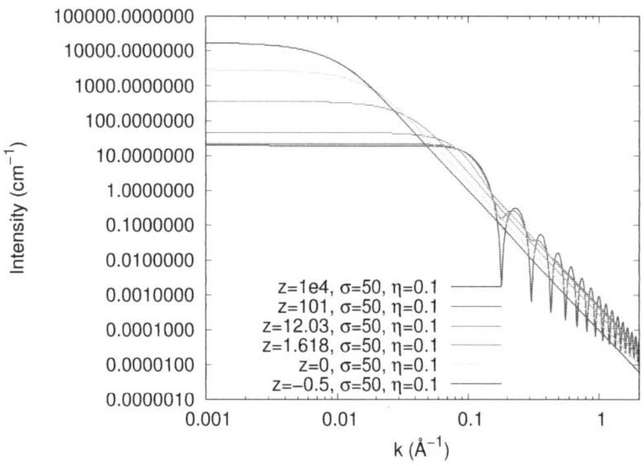

Figure 1. Scattering intensity for $p = 8.09422 \times 10^{-6}$Å^2, $\eta = 0.1$.

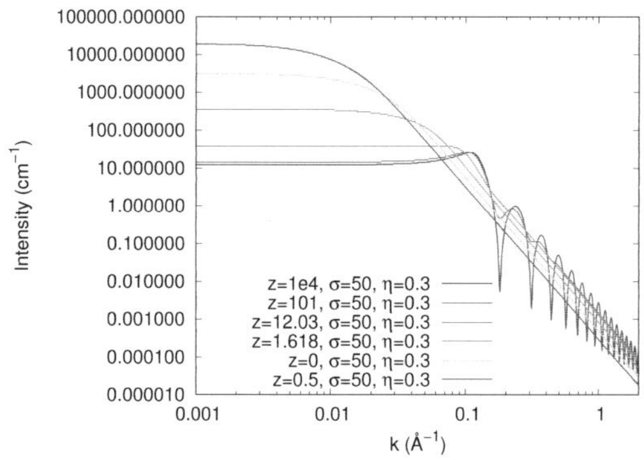

Figure 2. Scattering intensity for $p = 8.09422 \times 10^{-6}$Å^2, $\eta = 0.3$.

resolution of features in the cross section. Numerical precision becomes an issue for $k\sigma_\mu < 0.1$ and the analytical function and a series expansion in k is used below this region. The calculation transitions from using the analytical function to the series expansion in a region where each agrees at the 10^{-6} level. Figures 1 and 2 depict the calculated scattering cross section as a function of momentum transfer for 5.0 nm detonation nanodiamonds in 100% water at two different packing fractions and varying degrees of polydispersity with $z = 10^4$ corresponding to an essentially monodisperse sample. The calculated cross section is integrated as a function of $k\sigma_\mu$ in order to produce a total macroscopic cross section which is then converted to a microscopic cross section in units of barns through the particle number density, ρ.

3. Results

Simulations of simple representations of a SANS instrument and a reflectometer instrument were done in order to demonstrate the behavior of the SANS mode in MCNPX.

The SANS configuration consists of a white beam of cold neutrons collimated to a radius of 0.5 mm incident on a 1 mm thick volume containing a mixture of water and carbon representing nanodiamonds with diameter of 50 nm. The neutron flux is tallied 50 cm away from the scattering volume using a segmented surface tally with 1600 cylindrical segments with radii varying in increments of 0.625 mm to a maximum radius of 100 cm. The neutron flux in each annulus is binned by neutron energy. Figure 3 shows the scattered profile for neutrons of two different wavelengths and two different packing fractions.

The reflectometer configuration includes 0.5 mm beam with divergence of 0.4 mrad incident on the 5 cm thick sample volume composed of pure nanodiamonds with a density of 0.4 gcm^{-3}, meant to simulate a configuration similar to that of Cubitt [6]. The angle of incidence, θ, is varied from 1 to 10 degrees with the detector at an angle of $2\theta + 10$ degrees. The neutron spectrum is tallied at the detector position in order to determine the reflectivity of the sample. The sample is composed of monodisperse nanodiamonds with a diameter of 50 nm. The calculated reflectivity is shown in figure 4.

4. Discussion

First simulation results agree qualitatively with expectations. Experimental verification of the model in order to provide a quantitative benchmark is in the early planning stages.

The reflectivity simulations suggest that a thick nanodiamond powder could serve as a reflector of CN, with a reflection probability approaching 50% for grazing incidence. A nanodiamond reflector could increase CN fluxes provided that it could withstand the high

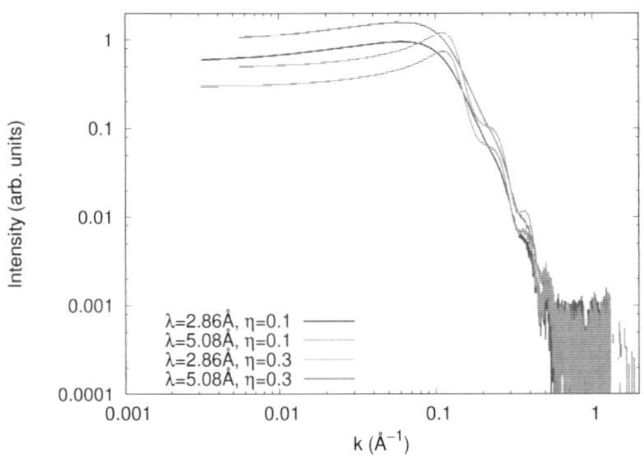

Figure 3. Calculated scattering profile for $p = 8.09422 \times 10^{-6}\text{Å}^2$, σ_μ=50 nm, and z=10^4.

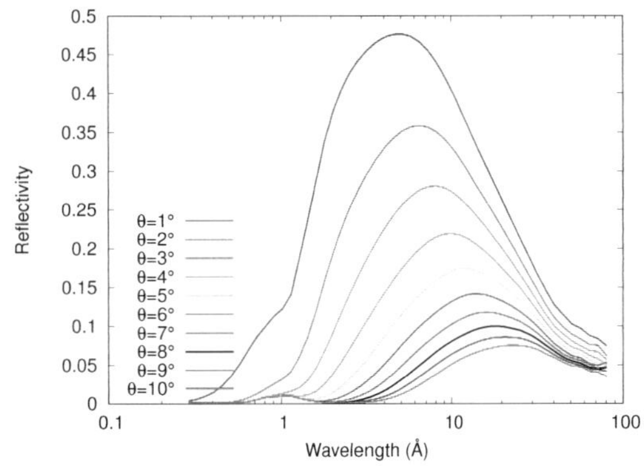

Figure 4. Calculated reflectivty for varying angle of incidence.

radiation environment close to a reactor core where traditional supermirror neutron guides cannot reach.

The seams between individual neutron guide elements give rise to artifacts in the beam profile and beam divergence at the exit of the neutron guide and nanoscale powders are often used to smear these artifacts. The implementation of SANS represents a powerful tool for conducting transport simulations involving nanoscale powders within MCNPX, for which conventional scattering kernels do not presently exist.

This model will be used in a design and optimization campaign for a VCN source, in which the nanoscale material acts as a VCN reflector. The neutronics performance of a moderator with a VCN reflector will be investigated as a function of polydispersity, packing fraction, and particle size.

Acknowledgments

This material is based upon work supported by the U.S. Department of Energy, Office of Science, Office of Basic Energy Sciences, under contract number DE-AC05-00OR22725.

References

[1] Nesvizhevsky V, Lychagin E, Muzychka A, Strelkov A, Pignol G and Protasov K 2008 *Nuclear Instruments and Methods in Physics Research Section A: Accelerators, Spectrometers, Detectors and Associated Equipment* **595** 631–636 ISSN 01689002 (*Preprint* 0805.2634) URL http://linkinghub.elsevier.com/retrieve/pii/S0168900208011261

[2] Griffith W, Triolo R and Compere A 1987 *Physical Review A* **35** 2200–2206 ISSN 10502947

[3] Danilenko V V 2004 *Physics of the Solid State* **46** 595–599 ISSN 1063-7834 URL http://link.springer.com/10.1134/1.1711431

[4] Osawa E 2008 *Pure and Applied Chemistry* **80** 1365–1379 ISSN 0033-4545

[5] Avdeev M V, Aksenov V L, Tomchuk O V, Bulavin L A, Garamus V M and Osawa E 2013 *Journal of Physics: Condensed Matter* **25** 445001 ISSN 1361-648X URL http://www.ncbi.nlm.nih.gov/pubmed/24055978

[6] Cubitt R, Lychagin E V, Muzychka A Y, Nekhaev G V, Nesvizhevsky V V, Pignol G, Protasov K V and Strelkov A V 2010 *Nuclear Instruments and Methods in Physics Research, Section A: Accelerators, Spectrometers, Detectors and Associated Equipment* **622** 182–185 ISSN 01689002 URL http://dx.doi.org/10.1016/j.nima.2010.07.049

[7] Blum L and Stell G 1979 *The Journal of Chemical Physics* **71** 42 ISSN 00219606 URL http://scitation.aip.org/content/aip/journal/jcp/71/1/10.1063/1.438088

IOP Conf. Series: Journal of Physics: Conf. Series **1021** (2018) 012061 doi:10.1088/1742-6596/1021/1/012061

Present fabrication status of spare moderators and reflector in J-PARC spallation neutron source

M Teshigawara, M Harada, M Ooi and H Takada.
J-PARC Center, Japan Atomic Energy Agency, Japan
E-mail: teshigawara.makoto@jaea.go.jp

Abstract. Invar joint and Au-In-Cd decoupler were implemented to prepare spare moderators and reflector. Invar joint was developed to improve manufacturing procedure of the hydrogen transfer line. A low activation Au-In-Cd alloy also could be implemented in the neutron extraction beam hole of reflector as a substitute of Ag-In-Cd decoupler by taking into account of enlarged heat capacity according to actual size in the HIPing process.

1. Introduction

The structural material of moderators and reflector, such as an aluminum alloy, is going to reach to the design value (20 DPA) around 2020 by an accumulation of irradiation-damage at the spallation neutron source of J-PARC. Fabrication of the spare moderators and reflector was started in 2013 for the replacement. The fabrication of coupled moderator, reflector and reflector plug were almost completed.

In the original moderator design, a material combination of aluminum alloy (A6061-T6) and stainless steel (SS316L) were adopted[1, 2] to make the moderator vessel and the hydrogen transfer line with the cryogenic multi-layered pipe of 5th annular geometry. However, some troubles such as welding leak, deformation and contact with other pipes, etc, were occurred in the actual fabrication process[3] due to the thermal shrinkage measure in the hydrogen transfer line.

An idea was proposed to use an invar alloy into the hydrogen transfer line, because of very low thermal expansion, which is one order of magnitude lower than that of aluminum alloy or stainless steel. This would improve the fabrication procedure much simple to eliminate complicated asymmetrical preset considering a certain amount of dimensional change at a cryogenic temperature at an elbow-shaped bend. It was also utilized to make the H_2 transfer line in SNS ORNL[4].

Another new idea is that Gold(Au) was replaced to Silver of the Silver – Indium - Cadmium (Ag-In-Cd) composition as the decoupler material to reduce the residual radioactivity of the used components significantly without sacrificing neutronic performance[5]. The bonding between Au-In-Cd[6] and aluminum alloy (A5083), which is a part of structural material of reflector, was developed by adopting a Hot Isostatic Pressing (HIP) for small-sized-test-pieces (c.a. 25mm in diam.)[7]. However, there were critical engineering issues to make large sized bonding between Au-In-Cd (AuIC) and A5083 alloys required for the actual reflector. We have investigated the HIPing process in terms of the surface condition, size and heat capacity of large-sized AuIC. Finally, we successfully implemented AuIC decoupler to the actual reflector assembly.

In this report, we report these results.

2. Spare coupled moderator fabrication

2.1 Coupled moderator fabrication

In order to utilize the invar alloy in the hydrogen transfer line, we developed the invar-joints for the conversion of invar alloy to aluminum alloy and stainless steel and evaluated the mechanical strength of invar-joints[5], which was also requirement for the Japanese High Pressure Gas Safety Law. These results were reflected the design of the hydrogen transfer line[8]. The invar joints are installed at parallel lines after converted from multiple pipes of hydrogen transfer line. The external shape is as same as original design except for including invar material. The length of multiple pipes from moderator head to the conversion position is c.a. three meters shorter as compared with original design. The thermal shrinkage could be also reduced from 20 mm to 2 mm. These might provide the benefit of eliminating the difficulty in the multiple pipe fabrication with asymmetrical setting.

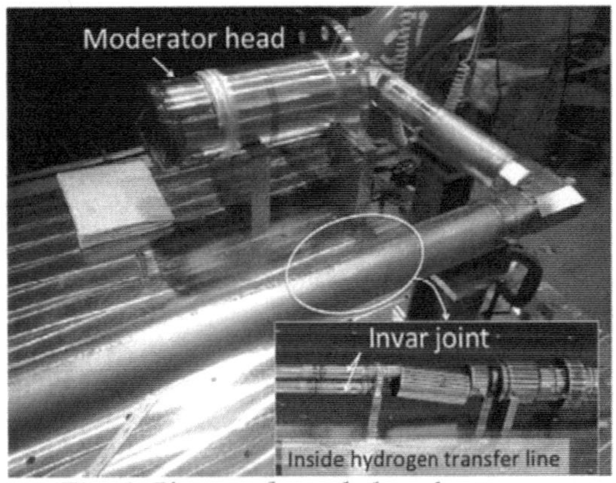

Fig. 1 Picture of coupled moderator manufacturing process

The coupled moderator was fabricated based on the idea. The invar-SS316L joint was installed at parallel lines after converted from multiple pipes as shown in Fig. 1. The results of R&D[5] was also reflected the fabrication process of invar joint.

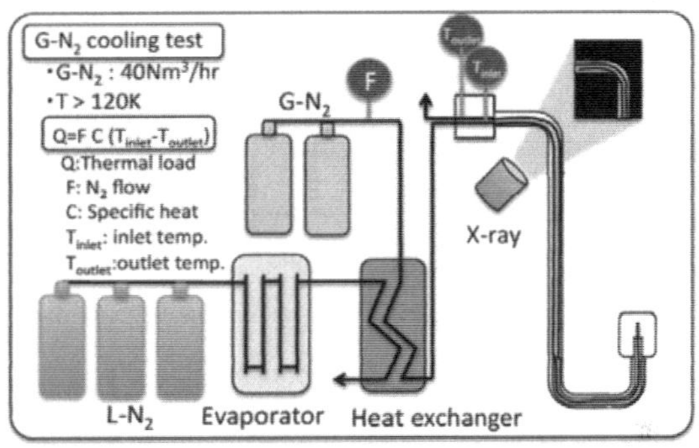

Fig. 2 Layout of cooling test of moderator

2.2 Cooling test of coupled moderator

After finished coupled moderator fabrication, the cooling test was conducted to confirm the design.

We measured the thermal load into the hydrogen transfer line and X ray transmission image to confirm inside hydrogen transfer line, such as the thermal shrinkage and welding defect. The layout of the cooling test is shown in Fig. 2. The gas-nitrogen (G-N_2) was applied to cool down hydrogen transfer line. Figure 3 shows the trends of the temperature and G-N_2 flow rate for case of 150K of inlet temperature. The inlet and outlet temperature was saturated c.a. 150 and 157K, respectively, with G-N_2 flow rate of 36Nm^3/hr in c.a.

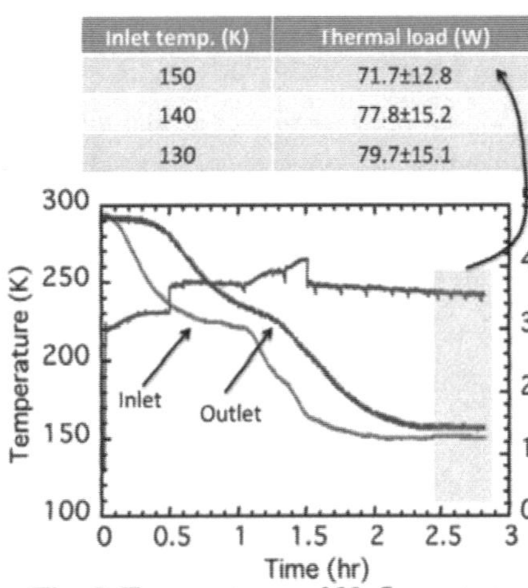

Inlet temp. (K)	Thermal load (W)
150	71.7±12.8
140	77.8±15.2
130	79.7±15.1

Fig. 3 Temperature and N_2 flow rate trends for measurement of thermal load

two hours from beginning of $G\text{-}N_2$ flow, giving c.a. 71.7 W of thermal load. The measured thermal load for other temperature cases are also shown in Fig. 3. These measure thermal load met with the design value [8].

Thermal shrinkage was measured at the elbow pipe mostly away (c.a. 4 m distance) from the fixed position of room and cold temperature, which is installed at inside coupled moderator head as shown in Fig. 1. Figure 4 shows X ray transmission image of hydrogen pipes for case of 140K as compared with room temperature case.

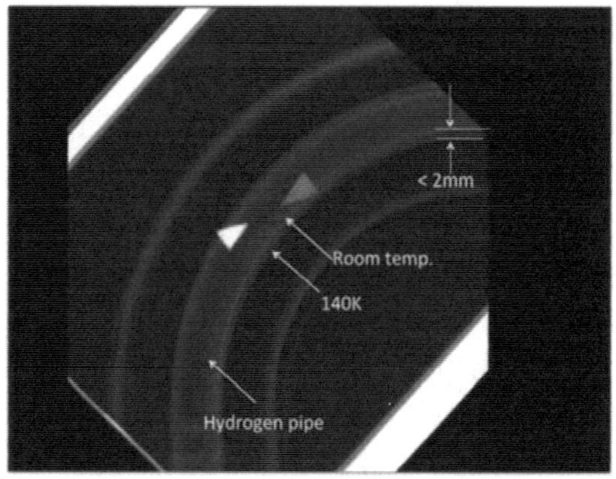

Fig. 4 X ray transmission image of hydrogen pipes for case of 140K as compared with room. temperature case

The measured thermal shrinkage was c.a. 2 mm, which was approximately as same as estimated one in the design calculation. It was also fond that the no welding defect was detected in X ray transmission, which was taken along the hydrogen transfer line.

3. Spare reflector fabrication

As described in Sec. 1, previous study[7] gave an optimal HIPing condition for small sample (c.a 100g in total). The decoupler is installed along the inner surface of beam extraction holes in a reflector assembly as shown in Fig. 5. HIPing method was applied to make the neutron extraction beam holes with decoupler. Before HIPing, the AuIC decoupler material is contained in canning container together with A5083 and SS304 blocks, resulting in 224 + 307 kg in total. This enlarged heat capacity due to actual size might affect on inside temperature profile during HIPing process, especially for bonding strength between AuIC and A5083 alloys. We considered that it takes longer time to reach the required HIPing temperature due to the enlarged heat capacity of the actual HIPing volume. We found the suitable HIPing condition (Pressure: 100MPa, Temperature: 535℃ and Holding time: 2 hours) for actual sized HIPing by numerical simulation and experiment. This detailed result will be discussed somewhere.

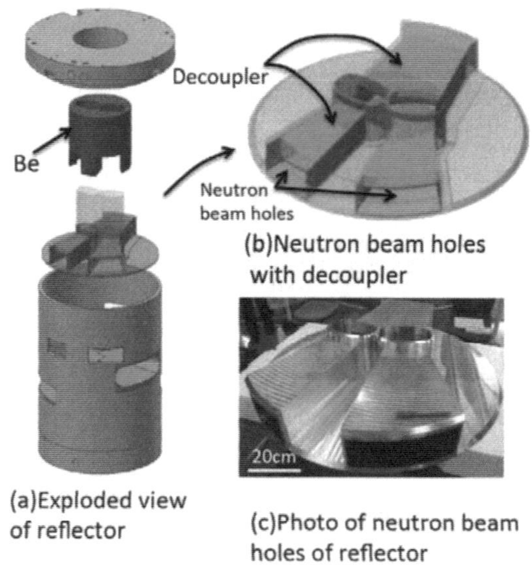

Fig. 5 Exploded view of reflector(a), Neutron beam holeswith decoupler(b) and photo of neutron beam holes of reflector(c).

HIPing was conducted under this condition. HIPed one was machined to parts of reflector with beam holes as shown in Fig. 5 (c). Ultrasonic inspection was conducted to detect the separation of bonding area between AuIC and A5083. No separation was detected at whole area including AuIC. Typical ultrasonic inspection data is shown in Fig. 6. The reflections were detected for each boundary, such as A5083, AuIC and A5083. The thickness for each boundary was 15.5, 2.7 and 1.3 mm, respectively, which was satisfied the design condition. The thickness (2.7mm) of AuIC was converted using sound velocity of AuIC. Finally, the machined parts were integrated into a middle part of the reflector by the welding.

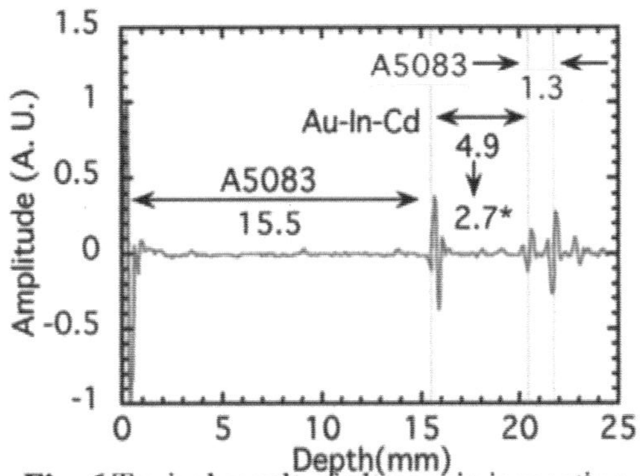

Fig. 6 Typical result of ultrasonic inspection between AuIC and A5083.

4. Assembling test

All moderators, reflector and reflector plugs are connecting at locating pins using remote bolts by remote control machine in hot cell for the installation in target station[9]. The installation error is less than 1.5 mm for each component in our design to satisfy the required performance. We performed the assembling test of coupled moderator and reflector by using the crane and hands on as shown in Fig. 7. We confirmed that each component could combine at locating pins without any rattling and stress and we also confirmed the installation error satisfied with required design.

Fig. 7 Insertion of coupled moderator into reflector in assembling test

5. Summary

Invar joint and Au-In-Cd decoupler were implemented to prepare spare moderators and reflector for the replacement around 2020. Invar joint was developed to improve manufacturing procedure of the hydrogen transfer line. A low activation Au-In-Cd alloy also could be implemented in the neutron extraction beam hole of reflector as a substitute of Ag-In-Cd decoupler by taking into account of enlarged heat capacity according to actual size in the HIPing process.

References

[1] Teshigawara M et al 2003 Proc. ICANS-XVI, May 12-15, Dusseldorf-Neuss, Germany 601-611
[2] Teshigawara M et al 2005 Proc. ICANS-XVII, April 25-29, Santa Fe, NewMexico USA 365-373
[3] Teshigawara M et al 2007 Proc. ICANS-XVIII, April 25-29, Guangdong, China 469-475
[4] McManamy T et al 2008 JNM 377 1-11
[5] Harada M et al 2010 Journal of Nucl. Mat. Vol. 398 pp. 93-99
[6] Ooi M et al 2012 Journal of Nucl. Mat. Vol. 431 issue 1-3 pp. 218-223
[7] Ooi M et al 2014 Journal of Nucl. Mat. Vol. 450 issue 1-3 pp. 117-122
[8]Teshigawara M et al 2015 Proc. ICANS-XVII, Sep. 29-Oct. 3 Mito, Japan 111-115
[9] JAEA-Technology 2011-035

IOP Conf. Series: Journal of Physics: Conf. Series **1021** (2018) 012062 doi:10.1088/1742-6596/1021/1/012062

Raman Spectroscopy as an ortho-para diagnostic of liquid hydrogen moderators

R. C. Gillis[1,†], T. Bailey[2,3], F. X. Gallmeier[1], M. A. Hartl[2], E. B. Iverson[1]

[1] Oak Ridge National Laboratory, Oak Ridge, Tennessee, 37831, USA
[2] European Spallation Source ERIC, Box 176, S-221 00 Lund, Sweden
[3] North Carolina State University, Raleigh, North Carolina, USA
[†] Email: gillisrc@ornl.gov

Abstract.
The intense radiation environment of a neutron moderator provides a mechanism for significant up-conversion of parahydrogen to orthohydrogen inside the moderator, as well as intrinsically catalyzing relaxation of orthohydrogen to parahydrogen. It is plausible that the steady-state orthohydrogen fraction of a moderator in a radiation environment such as at the Spallation Neutron Source (SNS) or the European Spallation Source (ESS) is as high as 30 % without supplemental catalysis. Direct measurement of the orthohydrogen fraction in the liquid hydrogen flow itself is essential to predict and monitor moderator performance, especially for thick or flat moderator concepts such as the ones that have been proposed for the ESS and for upgrades to the SNS. Raman spectroscopy provides a well-known method for directly measuring the hydrogen make-up in an unambiguous way. We describe our tests of Raman spectroscopy for application to the measurement of the orthohydrogen fraction of the hydrogen moderators at SNS and at the ESS. As part of this work, we have additionally developed a sample holder that has been used to perform simultaneous Raman and neutron vibrational spectroscopy on the VISION spectrometer at SNS. We discuss our plans to incorporate such a system as a diagnostic for liquid hydrogen moderators at SNS and at the ESS.

1. Relevance of the orthohydrogen fraction to hydrogen moderators

The neutron scattering cross section of orthohydrogen is much greater than that of parahydrogen at neutron energies less than 15 meV (by about a factor of about 50 at 10 meV). [1] As a result of this fact, it is likely that the optimization of essentially any hydrogen moderator design will depend on an optimal orthohydrogen fraction. In the case of the flat or thick moderators that are being proposed for the ESS and for upgrades to the SNS, it is expected that the difference between < 1 % and 30 % orthohydrogen content would be a factor of two or three

This manuscript has been authored by UT-Battelle, LLC under Contract No. DE-AC05-00OR22725 with the U.S. Department of Energy. The United States Government retains and the publisher, by accepting the article for publication, acknowledges that the United States Government retains a non-exclusive, paid-up, irrevocable, worldwide license to publish or reproduce the published form of this manuscript, or allow others to do so, for United States Government purposes. The Department of Energy will provide public access to these results of federally sponsored research in accordance with the DOE Public Access Plan (http://energy.gov/downloads/doe-public-access-plan).

IOP Conf. Series: Journal of Physics: Conf. Series **1021** (2018) 012062 doi:10.1088/1742-6596/1021/1/012062

in the moderator brightness (prediction relating to flat moderator brightness courtesy of ESS Neutronics). If the freedom does not exist at the design stage to specify an orthohydrogen fraction determined by the figures of merit of instruments served by the moderator, then compromises are made in the design that give up significant performance gains. A system for controlling and monitoring the moderator orthohydrogen content should pay for itself many times over in terms of enhanced facility performance.

The rate of self-conversion [2] between ortho- and parahydrogen is slow, both on the time scale at which hydrogen flows through a moderator loop and on the time scale over which catalyzed conversion [3] can be made to occur. In the absence of other conversion mechanisms, it is therefore possible to control the orthohydrogen fraction by continuously flowing it through a chamber filled with ortho-para conversion catalyst. In this scenario, the entire volume of hydrogen assumes an asymptotic orthohydrogen fraction that corresponds to thermal equilibrium with the catalyst. However, the environment of a moderator introduces uncertainties associated with the fact that the intense radiation field is expected to give rise to radiolysis of the hydrogen. [4] The energy released during recombination of the radiolysis products is such that recombination occurs in a 3:1 orthohydrogen:parahydrogen ratio, and it has been estimated that the orthohydrogen fraction of a liquid hydrogen moderator at an advanced spallation source may in fact be as high as 30 % in the absence of a catalytic chamber.

2. A system for monitoring the orthohydrogen fraction in the moderator loop

Given the uncertainties associated with the orthohydrogen fraction in the moderator loop of an advanced spallation source, it is important to put in place a system that directly monitors the orthohydrogen fraction. Sampling the hydrogen to transport it to a lab for analysis is not the most favorable option as extreme care is required to account for all sources of catalysis when doing so. Furthermore, a real-time monitoring system would allow us to watch the moderator performance evolve as a function of the orthohydrogen fraction, and therefore to develop insight into the physics of moderators and of radiolysis and catalysis of the hydrogen. Raman spectroscopy is interesting as it allows for the signature of molecular species to be identified, and would therefore allow us to see whether the trihydrogen ion (a hypothesized radiolysis product) is present to a significant extent.

With the objective in mind of building a Raman system to monitor the orthohydrogen fraction of the hydrogen in the moderator loop, we have built a sample holder for performing simultaneous Raman and neutron vibrational spectroscopy on the VISION spectrometer at SNS. This equipment, which was designed and built as a collaborative effort between the ESS and the SNS, holds material samples in the neutron beam within the cryogenic environment of the VISION spectrometer, allowing for samples to be studied at temperatures as low as 5 K. An assembly of optical components for directing and filtering light (shown in Fig. 1) is located inside a vacuum chamber within the cryogenic volume at VISION, in physical proximity to the sample. For the measurements discussed in this paper, the sample container was coupled optically to the optical assembly in the vacuum chamber by a 13 mm diameter fused quartz rod. In the case of our studies with hydrogen, the background from the quartz rod was separable from the signal; as such it allowed us to normalize the hydrogen Raman signal to the transmitted laser power.

Raman data of a sample of liquid hydrogen at 17 K taken on VISION using this sample holder are shown in Fig. 2. The ratio of areas of the two peaks can serve as a measurement of the orthohydrogen fraction. With our setup we were able to measure this ratio to within about 3 % uncertainty. This uncertainty was limited by nonstatistical electronic noise which we expect will be substantially improved when we refurbish the optical spectrometer. We used an approximate method of preparing orthohydrogen mixtures by combining hydrogen from a room-temperature cylinder with hydrogen that had been sitting in contact with Oxisorb ortho-para conversion catalyst at 17 K for an extended period of time. We saw 15 % agreement with the expected

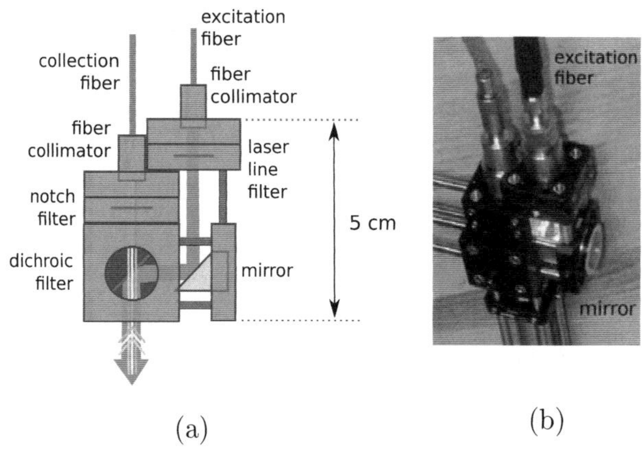

Figure 1: (a) A schematic of the optical assembly, showing the path of the excitation and collected light. (b) A photograph of the optical assembly before it was incorporated into the sample stick. The steel rods emerging at bottom and left were used during alignment and were removed before incorporating the assembly into the sample stick.

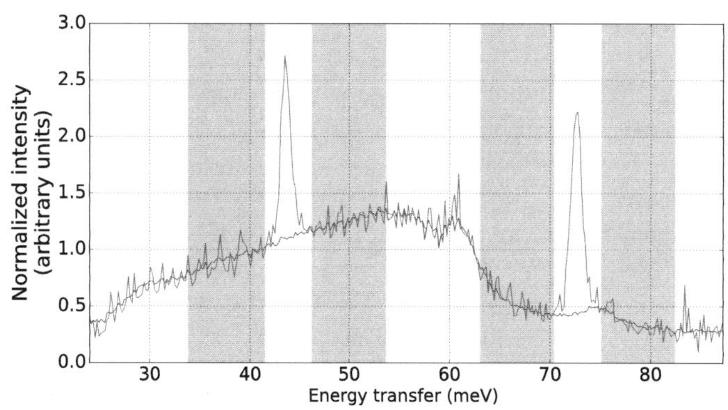

Figure 2: A raman spectrum of hydrogen including the background due to scattering in the quartz rod. A relative normalization has been done of the data to the background within the shaded region. The nominal orthohydrogen fraction of this sample as determined by the sample preparation procedure was 72 %.

peak intensity ratio (based on equation 5.5 of [5]). This difference could be explained by the approximate method of preparing the sample mixtures or by a wavelength-dependent feature of the optics that we have yet to study.

A technical challenge associated with the sample holder on VISION has been the transmission of an optical signal to and from the sample in a cryogenic environment across vacuum-tight boundaries. We do not expect the technical difficulties associated with monitoring the moderator loop to be greater than the ones associated with building the sample holder at VISION. We were able to observe the Raman signal of Fig. 2 with the optics held at 50 K while viewing a liquid sample about 2.5 cm deep, without employing any schemes to amplify the signal. Plans for the hydrogen line involve an installation of optical viewports to allow the Raman probe to view the hydrogen at a stage of the hydrogen loop outside of the moderator.

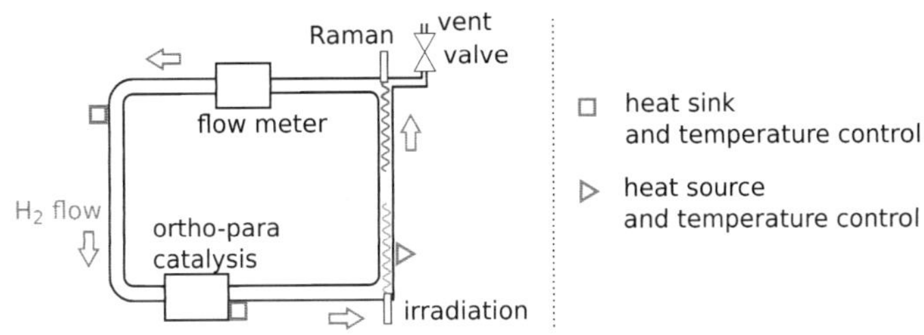

Figure 3: Concept of a benchtop setup for studying the effect of radiolysis and catalysis on the orthohydrogen fraction of a volume of circulating liquid hydrogen.

3. An apparatus for studying radiolysis and catalysis of flowing liquid hydrogen

In addition to monitoring the hydrogen at a stage of the moderator loop that is outside of the moderator, we would like to directly view an irradiated volume of hydrogen to get a better understanding of the nature, relative populations, and lifetimes of the hypothesized radiolysis products. Fig. 3 shows a concept of a benchtop setup that simulates important aspects of the hydrogen loop. It includes a catalysis section, a radiolysis section, and flow of hydrogen through a loop that connects the two sections. This setup bears some similarities to the NPDGamma liquid hydrogen target that is explained in Ref. [6]. Given the low viscosity of liquid hydrogen, [7] and the temperature dependence of the mass density of liquid hydrogen (1.2 % density change per K at 17 K [8]), the hydrogen flow would be generated or assisted by a temperature gradient across the loop. The catalysis section could consist of a chamber or portion of tubing filled with ortho-parahydrogen conversion catalyst, and the radiolysis section could consist of an irradiated chamber or portion of tubing. UV light could serve as a plausible radiation source, and a Raman measurement would allow us to directly view the irradiated hydrogen.

Acknowledgements

This work was sponsored by the Laboratory Directed Research and Development Program of Oak Ridge National Laboratory. It used resources at the Spallation Neutron Source, a US Department of Energy Office of Science User Facility that is operated by Oak Ridge National Laboratory. Oak Ridge National Laboratory is managed by UT-Battelle, LLC, for the U S Department of Energy.

The authors would like to thank Luke Daemen of VISION for his expertise and assistance in developing the Raman sample stick. The authors would also like to thank Neelam Pradhan, June Safieh and John Wenzel of the SNS Sample Environment Research and Development team for producing design drawings of the vacuum chamber and sample container used on the Raman sample stick.

References

[1] K. B. Grammer *et al.*, *Phys. Rev. B*, **91**, 180301(R) (2015)
[2] Yu. Ya. Milenko, R. M. Sibileva and M. A. Strzhemechny, *J. Low Temp. Phys.*, **107** 77–92 (1997)
[3] D. H. Weitzel, W. V. Loebenstein, J. W. Draper and O. E. Park, *J. Res. Natl. Bur. Stand.*, **60**, 221–227 (1958)
[4] E. B. Iverson and J. M. Carpenter, *Proceedings of the ICANS-XVI*, Volume II, pp. 707–717, 2003
[5] I. F. Silvera, *Rev. Mod. Phys.*, **52**, 393–452 (1980)
[6] S. Santra *et al.*, *Nucl. Instr. Meth. Phys. Res. A*, **620**, 421–436 (2010)
[7] B. B. Eselson, Yu. P. Blagoi, V. N. Grigorev, V. G. Manzhelli, S. A. Mikhailenko, N. P. Neklyudov, *Properties of Liquid and Solid Hydrogen*, Israel Program for Scientific Translations, Jerusalem, 1971
[8] J. W. Leachman, R. T. Jacobsen, S. G. Penoncello and E. W. Lemmon, *J. Phys. Chem. Ref. Data*, **38**, 721–748 (2009)

IOP Conf. Series: Journal of Physics: Conf. Series **1021** (2018) 012063 doi:10.1088/1742-6596/1021/1/012063

Numerical flow simulation of the neutron source SINQ of PSI

Sven Jollet, Raffaello Sobbia, Michael Wohlmuther

PSI – Paul Scherrer Institute, 5232-Villigen PSI, Switzerland

sven.jollet@psi.ch, michael.wohlmuther@psi.ch

Abstract. The Swiss spallation source, SINQ, is in operation since 1996. Since its start-up a constant target development program has been pursued, which culminated in the operation of the liquid metal target MEGAPIE. However, the work-horse target consists of a bundle of Zircaloy-II rods filled with Lead, the so-called Cannelloni target. As the high intensity proton accelerator, HIPA, at the PSI constantly delivers higher beam currents to SINQ it is important to perform numerical simulations to understand the response of the target to the power deposition and to ensure the proper D_2O cooling of the Cannelloni rod bundle. Steady-state 3D fluid simulations of the heavy water coolant flow are coupled with the heat source terms in the Cannellonis. The source terms are described as UDFs (User Defined Function) for all Cannellonis with its Zircaloy-II tube and Lead filling. The energy deposition functions for the heat source terms are calculated using MPCNX 2.7.0. Major results from the CFD simulations are the flow structure with the resulting pressure field and the temperature distribution in the target. The results show good agreement with respect to the measured values during operation of SINQ. Additionally the local convective heat transfer coefficients (HTC) are calculated.

1. Introduction

The simulations presented here investigate the D_2O cooling of the Cannelloni target of the Swiss spallation neutron source SINQ. The target consists of a bundle of Zircaloy-II tubes filled with the Lead, the so-called Cannelloni target. The volumetric fraction of the Lead inside the tubes is 90% to allow for the thermal expansion of the Lead during heating up and melting. It is of paramount importance to correctly characterize the flow behaviour in order to predict the heat dissipation from the Cannelloni to the coolant fluid, D_2O. In the present study, the turbulent flow is approximated using the RANS (Reynold-Averaged-Navier-Stokes) formulation with the Shear Stress Transport (SST) turbulence modelling, together with temperature dependent material property values. In the simulations temperature dependent material properties for Zircaloy-II [1][2], Lead [1][3][4][5][6] and heavy water [7] were used. A mass flow rate of 10 kg/s (full target) at 40°C is assumed as starting condition.

2. Basics

A common method for the calculation of turbulent flows is the Reynolds-Averaged-Navier-Stokes (RANS) modelling. In this method, the Navier-Stokes equations are solved for time-averaged flow behaviour and the magnitude of turbulent fluctuations. A hexagonal mesh was set up using ANSYS ICEM for the fluid and the solid part of the simulation. Great care was taken to resolve the near wall region, i.e. the fluid region close to the solid of the Zircaloy-II tube. For the simulation a Coupled-Solver-Algorithm is used with Second Order Formulation. This pressure-based coupled algorithm obtains a robust and efficient solution for single phase steady-state flows [8].

2.1 Turbulence Model

The Shear Stress Transport (SST) Model is used in this study [9]. It is a robust and accurate formulation of the k-ω-Model in the near-wall region with the freestream independence of the k-ε-Model in the far field.

2.2 Heat Transfer Equations

In the simulations the heat flux is computed from two sides. One description used the heat flux to a wall from a fluid cell.

$$q = h_f(T_W - T_f) + q_{rad}$$

Here q is the local heat flux, q_{rad} is the local radiative heat flux and h_f is the fluid-side local convective heat transfer coefficient. h_f is calculated using the local flow-field conditions (e.g. turbulence, temperature, velocity, etc.). The temperature difference is calculated using the wall surface temperature T_W and the local fluid temperature T_f [8].

From the solid cell the heat flux is calculated using the thermal conductivity of the solid k_S, the local solid temperature T_S and the distance between wall surface and the solid cell center Δn [8].

$$q = \frac{k_S}{\Delta n}(T_W - T_S) + q_{rad}$$

With these two equations the boundary conditions at the wall can be calculated. In the simulations for the SINQ target the radiative heat flux q_{rad} was assumed to be zero.

3. Geometry

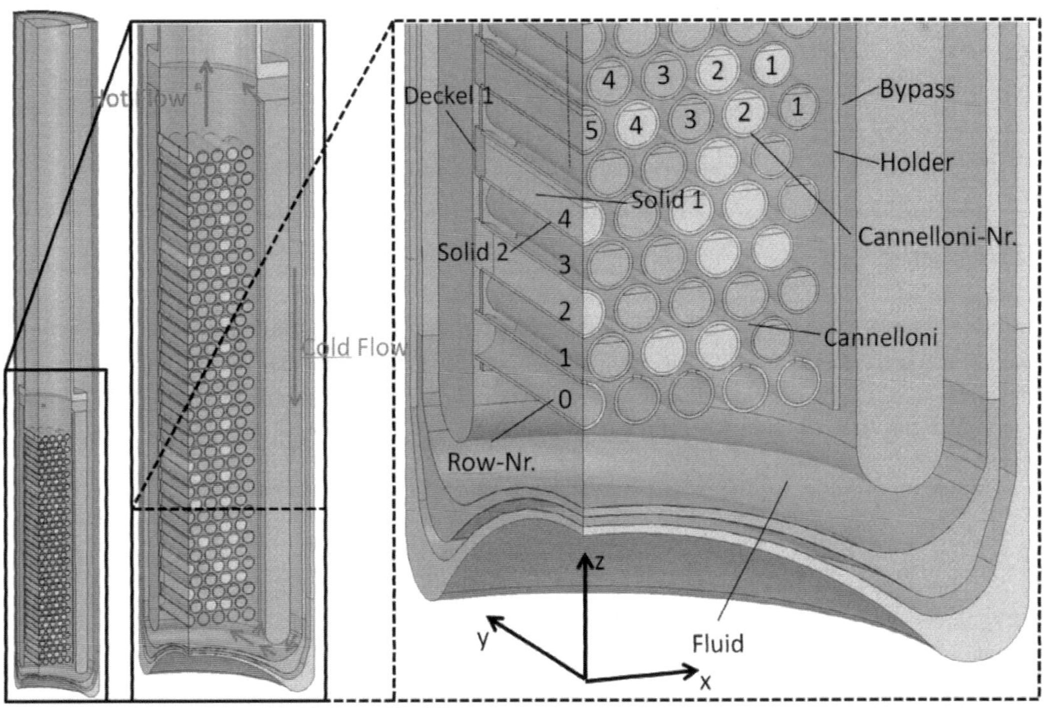

Figure 1 3D geometry of the SINQ Cannelloni Target.

In order to reduce the computational time and due to the inherent symmetry of the SINQ target structure, only a quarter of the 3D geometry is used (Figure 1). In order to prevent numerical reverse flow at the outlet the length of the upper part of the geometry has been extended to a length, L, of about 10 times the hydraulic diameter, D: $L_{t,turbulent} \approx 10 \times D$ [10].

The Cannellonis outer diameter is 10.75 mm, their inner diameter 9.25 mm. The Target holder hydraulic diameter is D = 122 mm. The incoming cold flow (40°C) is split into two parts: the by-pass and the main flow around the Cannellonis. In the model the Cannellonis are built by three parts: Solid 1, Solid 2 and "Deckel" (lid) 1). Solid 1 represents the 90 % Lead filling, the Solid 2 the Zircaloy-II tube and the "Deckel" 1 the end cap of the Cannelloni made from Zircaloy-II (see Figure 1). At the bottom of the target (the target is mounted vertically in SINQ) the incoming flow is split in two parts – the main flow, cooling the rod bundle, and the by-pass cooling the blanket structure of the target.

4. Heat Source (UDF – User Defined Function)

The incident proton beam and secondary particles produced deposit energy in all parts of the target, the Cannellonis, the blanket and of course also the coolant –D_2O. For the moment the heat deposition in the coolant is discarded as it is low compared to the energy deposited in the solid parts of the target. The incoming proton beam is approximated by a double Gaussian distribution. As a consequence also the heat source can be defined by a double Gaussian distribution, described by $f_1(y)$ and $f_2(x)$ in y- and x-directions.
Units are rescaled from MCNPX ($kW/cm^3/mA$) to Fluent (W/m^3) and a proton current of 1.5 mA is used. f_1 and f_2 read:

$$f_1(y) = \frac{A_{1,i}}{s_{1,i} \cdot \sqrt{\frac{\pi}{2}}} \cdot e^{-\frac{2y^2}{s_{1,i}^2}} \qquad \text{and} \qquad f_2(x) = \frac{A_{2,i}}{s_{2,i} \cdot \sqrt{\frac{\pi}{2}}} \cdot e^{-\frac{2x^2}{s_{2,i}^2}}$$

where $A_{1,i}$, $A_{2,i}$, $s_{1,i}$ and $s_{2,i}$ are values from the MCNPX calculations for Lead and Zircaloy-II [11]. To take into account the variation of the power deposition along z-direction, energy is calculated for each Cannelloni separately. The energy deposition of Zircaloy-II and Lead is described by the functions $f_1(y)$ and $f_2(x)$ and a scaling factor c for convert the units to Fluent.

$$f(y,x) = c \cdot f_1(y) \cdot f_2(x)$$

5. Results

The simulation (Figure 2) shows the z-component (the main component of the velocity field), where z-direction points upwards. For the following discussion only the magnitude of z-component will be considered. The positive and negative sign give the direction due to the position of the reference system. The velocity increases at the top of the main inlet (a), where the channel was narrowed from the top very wide section to around 5 m/s. The z-velocity component has a maximum value of about 15 m/s in the inlet region of the bypass (b). The simulation shows also a large mass flow stream along the area between the outer Cannelloni and the holder wall (c). This causes an asymmetric flow in the target with high fluid velocities in the rim areas of the target. The inlet channel widens at the bottom end (d) inducing a decrease of the velocity. Due to the large expansion in point (e) the velocity reaches a relative minimum value.

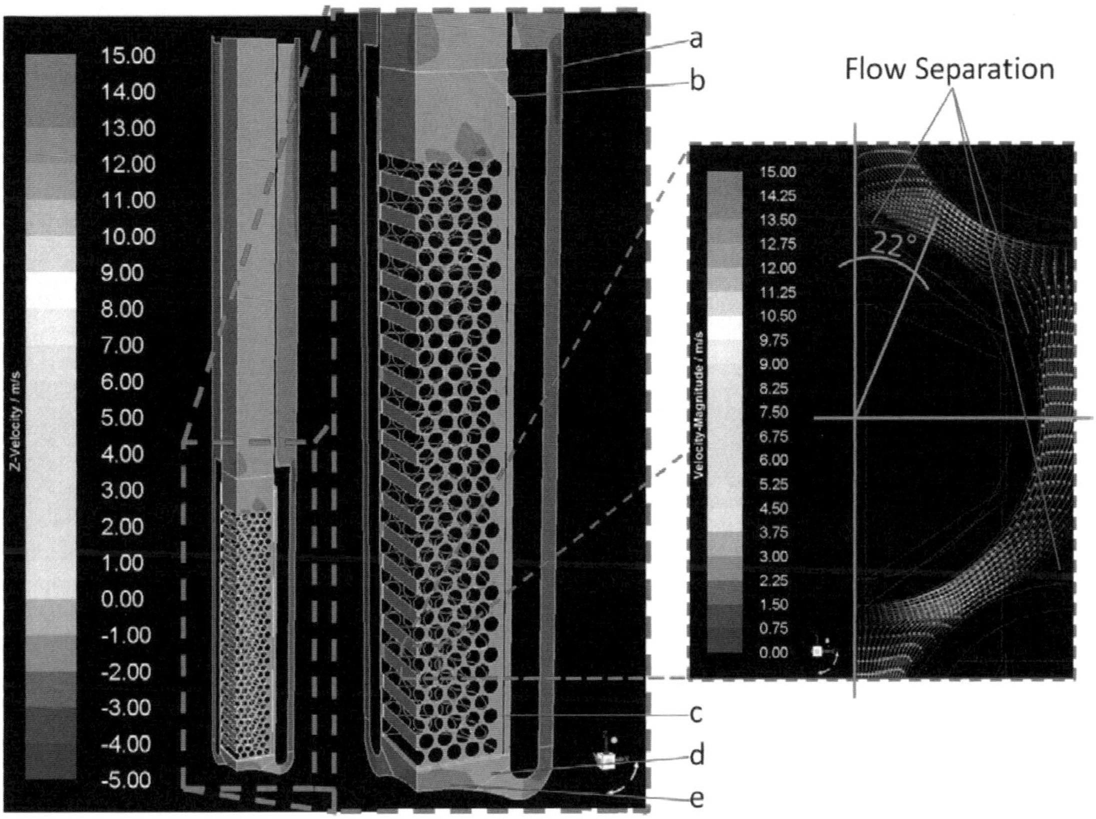

Figure 2 z-Velocity maps obtained in the flow simulation.

A detailed view with velocity magnitude vectors (Figure 2, right) shows the flow behaviour in an elementary cell composed of 3 Cannelloni located at the center of row 6. A flow separation zone in the upper region of the Zircaloy-II tube with respect to the flow direction is visible. The flow separation angle is found to be 22°, see Figure 2, everywhere in the target. This dead water volume results in a bad cooling of the Zircaloy-II rod. In addition to the flow separation at 22° flow fluctuations at larger angles are found. However, these are not present everywhere and are believed to appear/disappear in time.

Figure 3 (detailed view) shows the temperature distributions in the Cannellonis and the heavy-water, as well as the static pressure in the system. The simulation uses the temperature-dependent heavy water properties from [11]. The temperature distribution (Figure 3, left) reach a maximum temperature of 536°C in the central Cannelloni of row 2 (Figure 1, Row 2 Cannelloni 5). The simulation (Figure 3, middle) predicts a heavy water temperature increase from 40°C (inlet conditions) to around 60°C in the outlet region. During operation D_2O temperatures of 70°C are measured in this region. The calculated temperature distribution indicates that from row 14 upwards the Lead is below its melting point of 327°C.

Figure 3, right shows the distribution of the static pressure. The simulation with the inlet-profile and the temperature-dependent heavy water properties give a pressure drop very similar to the measured one of around 1.25 *bar*.

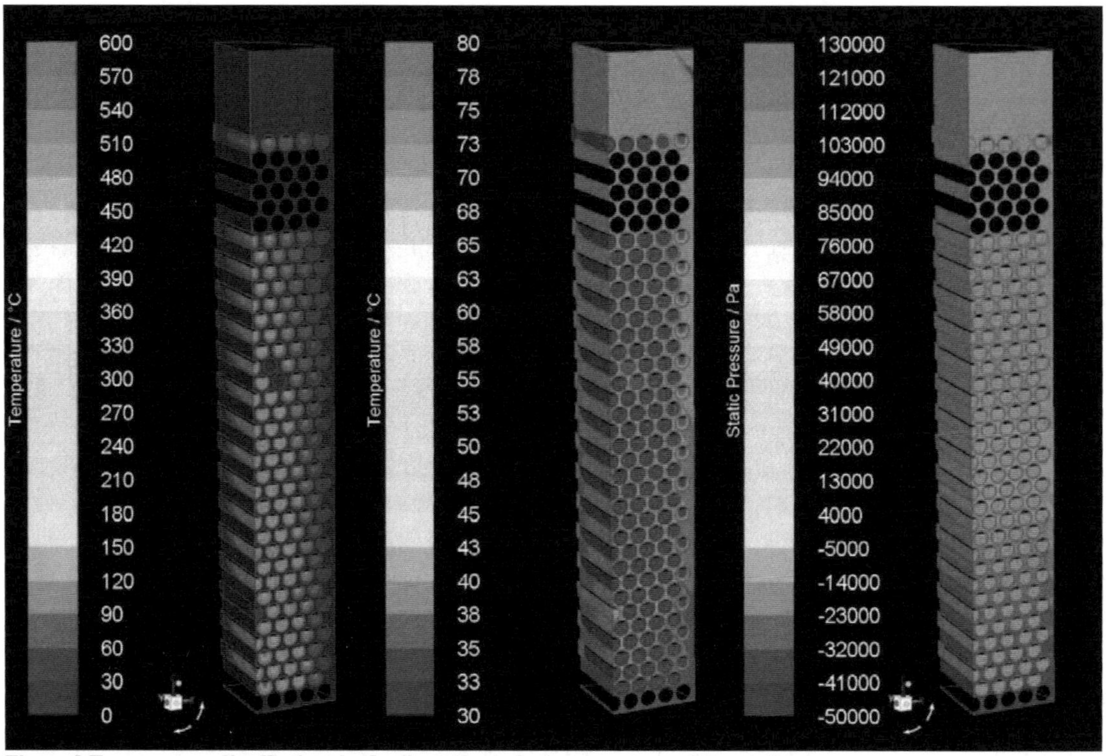

Figure 3 Temperature distributions in the Cannelloni, the heavy water and the static pressure

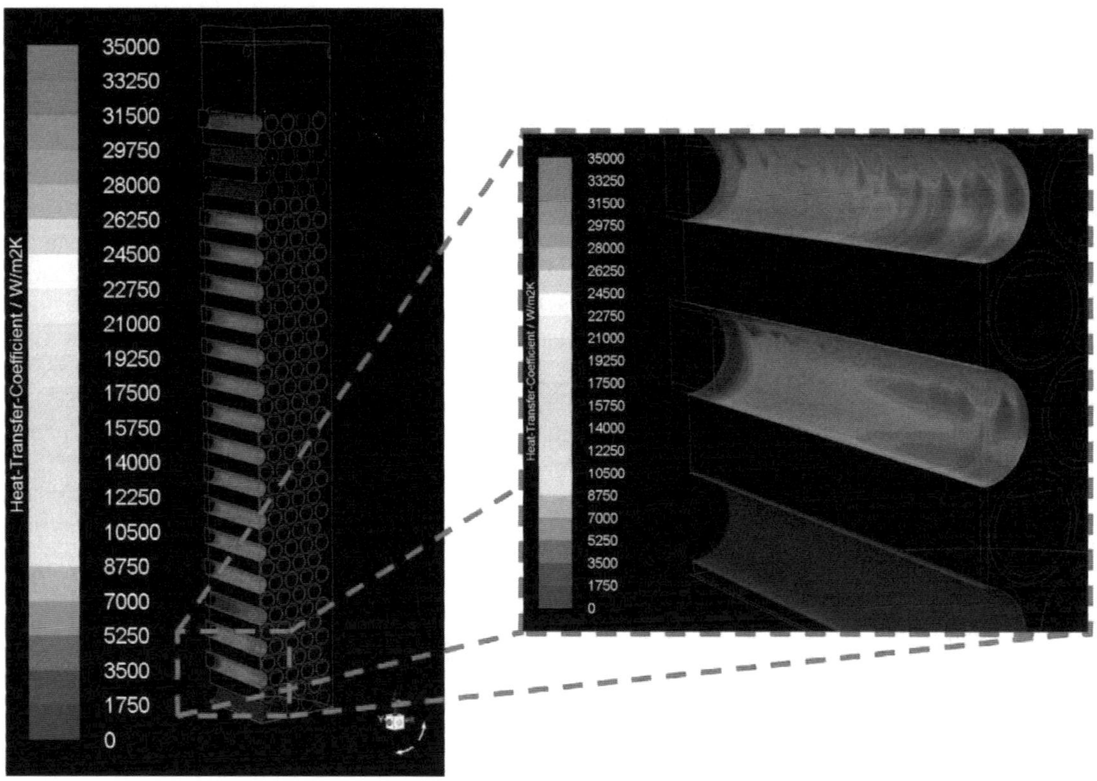

Figure 4 Evaluated Heat Transfer-Coefficients of the SINQ Target

IOP Conf. Series: Journal of Physics: Conf. Series **1021** (2018) 012063 doi:10.1088/1742-6596/1021/1/012063

The fluid-side local convective heat transfer coefficient (HTC) is computed based on the local flow-field conditions, which vary from point-to-point. Thus there is not a constant heat transfer coefficient. The right picture in Figure 4 shows the distribution of the local convective heat transfer coefficient at the Cannelloni surfaces for the central Cannellonis of row 2 and 4. A decrease of the local heat transfer coefficient along the flow direction from 35000 $W/m^2\,K$ (central Cannelloni in row 2), to 25000 $W/m^2\,K$ (central Cannelloni in row 14) and 18000 $W/m^2\,K$ (ccentral Cannelloni in row 24) is observed. The highest local heat transfer coefficients are always found at the bottom side of the Cannelloni with respect to the flow (z-)direction. This behavior is due to the dead water zone and void region above the Lead (90% filling) in each Cannelloni.
The evaluated local heat transfer coefficients serve as an input to FEM-simulations to evaluate the Cannellonis response to the thermal stresses induced by the proton beam.

6. Conclusion

Simulation have performed to characterise the flow and cooling behaviour of the SINQ target. A hexagonal ICEM mesh has been used together with the SST model in Fluent. Predicted D_2O temperatures and the pressure drop agree quite well with measured values during operation. The evaluated local heat transfer coefficients serve as an input to FEM simulations for stress analyses. Besides the ICEM mesh other, semi-automatically produced meshes have been used in the simulations, but are not presented in this paper. These mesh study showed that fine enough, semi-automatically produced tetrahedral meshes are as well suitable to solve the current model giving similar results.

References

[1] R. OTHMAN, Steady State and Transient Analysis of Heat Conduction in Nuclear Fuel Elements, Stockholm: Department of Numerical Analysis and Computer Science, Royal Institute of Technology, 2004.
[2] Nuclear Power Technology Development Section, International Atomic Energy Agency, Thermophysical properties database of materials for light water reactors and heavy water reactors, Wien: International Atomic Energy Agency, 2006.
[3] R. Milenkovic, „Thermal-hydraulics and coupled Thermal-structural analysis of the SINQ water cooled solid target," PSI - Paul Scherrer Institut, Villigen, PSI, TM-85-09-15.
[4] MatWeb, LLC, „MatWeb Material Property Data," MatWeb, LLC, [Online]. Available: http://www.matweb.com/search/DataSheet.aspx?MatGUID=ebd6d2cdfdca4fc285885cc4749c36b1&ckck=1. [Zugriff am 30 07 2012].
[5] S. Guruswamy, Engineering Properties and Applications of Lead Alloys, New York, Basel: Marcel Dekker, Inc., 1999.
[6] OECD/NEA Nuclear Science Committee, Working Party on Scientific Issues of the Fuel Cycle, Working Group on Lead-bismuth Eutectic , Handbook on Lead-bismuth Eutectic Alloy and Lead Properties, Materials Compatibility, Thermal-hydraulics and Technologies, NUCLEAR ENERGY AGENCY ORGANISATION FOR ECONOMIC CO-OPERATION AND DEVELOPMENT, 2007.
[7] Oak Ridge National Laboratory, Thermophysical Properties of Satured Light and Heavy Water for Advanced Neutron Source Applications, Martin Marietta Energy Systems, Inc. 1993.
[8] ANSYS, Inc. ANSYS FLUENT User Guide, Release 13.0 – 2010.
[9] F. R. Menter, „Two-Equation Eddy-viscosity Turbulence Models for Engineering Applications," AIAA Journal, pp. 32(8):1598-1605, August 1994.
[10] Y. A. Cengel, Heat Transer - A Practival Approach, New York: McGraw-Hill, 2002.
[11] PSI – Paul Scherrer Institut, T.Reiss, M.Wohlmuther, S.Dementjevs, R.Sobbia, "Energy deposition calculation with MCNPX of the SINQ target", TM-85-13-13_rev.1, Villigen, Juni 2015.

IOP Conf. Series: Journal of Physics: Conf. Series **1021** (2018) 012064 doi:10.1088/1742-6596/1021/1/012064

ESS Target Water Cooling, Purification and Radiolysis Gas Handling

Håkan Carlsson, Allan Lundgren, Leif Emås, Per Nilsson, Björn Eriksson

P.O Box 176, SE-221 00 Lund, Sweden

hakan.carlsson@esss.se

Abstract. The goal for the ESS facility is to construct an accelerator with the ability to deliver a 5 MW proton beam on the tungsten target. Most of the power is deposited as heat in the fluids and structures in and around the target. To be able to run the facility the nuclear heating from the spallation process need to be cooled in a safe and reliable way. The tungsten target in itself is helium cooled, but this system is in turn water cooled and there are several other water cooling systems, e.g. for moderators, reflectors and structural components. In this paper the water cooling systems are described and what challenges and requirements the systems needs to deal with.

1. Introduction

Unlike fission, spallation is an endothermic nuclear reaction that converts kinetic energy into mass. Of the 5 MW of beam power incident on target, about 89% gets deposited as sensible heat that must be removed, 9% is converted to mass in the liberation of neutrons (56 neutrons per proton incident on target), and about 2% of this power is carried off by neutrinos.

From the above calculations, the cooling needs are roughly 4450 kW where 3000 kW will be deposit in the tungsten blocks in the target wheel and cooled by the helium cooling system. The remaining heat will be deposited in the components within the monolith structure (i.e. moderators, reflectors, plugs) and in the shielding blocks surrounding the target wheel.

The water used to cool the components and the shielding blocks will be highly activated due to the spallation process in the water and the intense neutron radiation. Due to the neutron/proton radiation there will also be a decomposition of the water molecules and eventually there will be hydrogen gas in the system. If this is not handled within the systems it could lead to an uncontrolled hydrogen combustion or detonation.

2. Primary Water Cooling

The primary water cooling systems are divided into three main loops (Thermal Moderator, Reflector and Shielding & Plugs) according to table 1. For the moderator and reflector loops the maximum pressure in the components are set to 5 bar. For the Shielding & Plugs no such requirement exists but the pump will be able give a pressure of around 7 bar(g).

System name	Cooled component	Cooling demands, (kW)	Flow (kg/s)	Inlet temperature (°C)	Outlet temperature (°C)
Thermal Water	Thermal water moderators	172	2.4	20	37.1

Moderators	Proton Beam Window & Plug	6	0.4	20	23.6
Reflectors Cooling	Reflectors	210	1.77	20	48
Shielding & Plugs	Steel shielding	436.5	0.1-1.5	35	35-60
	Moderator/Reflector Frame	489	3.2	35	70
	Target Monitoring Plug	0.6	0.015	35	45
	Proton Beam Instrumentation	5	0.12	35	45
	Neutron Beam Inserts	12.6	4.0	35	<38

Table 1 Primary water cooling (best estimate)

All primary loops are designed in a similar way, fig 1, where the main components are the pump, control valves, delay tank, Gas Liquid Separation (GLS) tank and a heat exchanger. The delay tank is designed for a 90s delay mainly to let the short-lived isotope ^{16}N decay, $t_{1/2}$=7.1s. The GLS tank is designed for a retention time of 300s to allow for de-gassing of the generated gases formed in the spallation process.

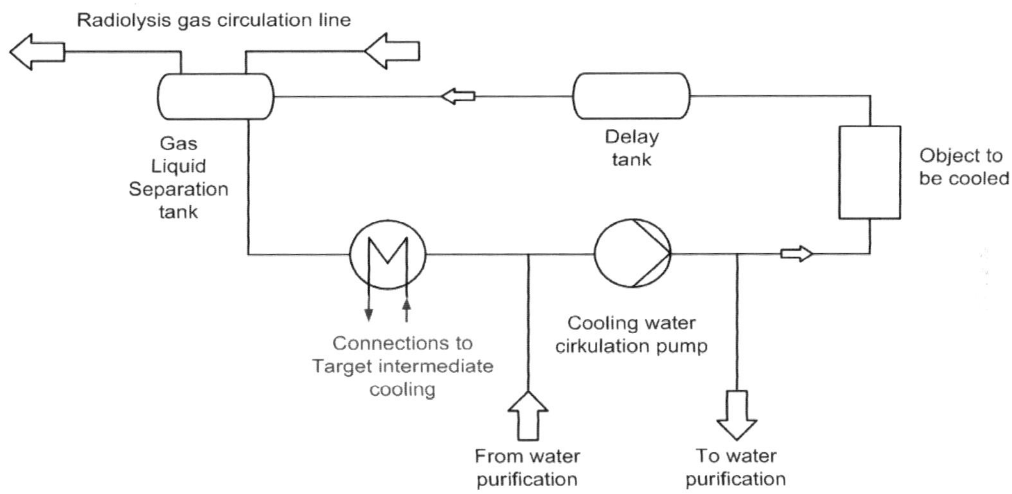

Fig 1 Schematic view of primary water cooling loop

A partial flow from each cooling loop will also be directed through the water purification system consisting of ion-exchange columns.

Within the monolith, the circulating water will be activated due the intense proton and neutron radiation. The main radioactive isotopes in the water systems are the short-lived isotope ^{16}N, ^{3}H and ^{7}Be. All primary water cooling loops will be placed in rooms non-accessible during operation and for approximately 4 h after beam shut-down. The very high energy gammas from ^{16}N decay (6.1 MeV and 7.1 MeV) also requires extra shielding around the outgoing pipes and the delay tanks (5-15 cm Pb). The shielding is needed to protect electronics close to the pipes as well as to be to keep low dose level in adjacent rooms.

IOP Conf. Series: Journal of Physics: Conf. Series **1021** (2018) 012064 doi:10.1088/1742-6596/1021/1/012064

3. Intermediate Water Cooling

The intermediate water cooling systems cools the primary cooling systems and delivers heated water to the Central Utility Building (CUB) where the heat is re-used for district heating or other applications.

There are three intermediate systems cooling different primary systems. One of these, the Target Intermediate Cooling system, then consists of three independent loops. Two of the loops cool the helium in the target primary cooling loop from about 230°C down to maximum 60°C which is a limit for the helium circulator. The third loop will cool the helium after the circulator to a temperature of maximum 40°C at full power before it enters the target wheel.

The primary requirement is to make the systems stable and robust and the second requirement is sustainability meaning returning the water at as high temperature as possible to the CUB.

System name	**Cooled system**	**Cooling demands, (kW)**	**Flow (kg/s)**	**Inlet temperature (°C)**	**Outlet temperature (°C)**
Target Intermediate Cooling	Target Helium Cooling	1500	11.2	50	85
		1600	7.0	25	85
		400	3.8	25	55
15 °C intermediate Cooling	Thermal & pre-moderators + Reflectors	388	3.2	8	40
30 °C Intermediate Cooling	Shielding & Plugs	944	12.4	25	45

Table 2 Intermediate water cooling (best estimate)

4. Water Purification

Ion exchange systems are used for the water purification of the primary cooling systems since those systems are affected by radiolysis process, which generates aggressive compounds that can cause corrosion and that lead to the transport of activated products through the system. Corrosion products have to be removed in a controlled and continuous way in order to keep their concentration as low as practically possible. High concentration of these products will cause a higher activation and radiation levels in the system, and an increased risk for hot spots, due to precipitation of corrosion impurities in valves, pumps and piping systems. This is an obstacle for efficient and safe maintenance of the systems and will lead to increased access times and more complicated maintenance of equipment.

Based on these measurements performed at LANSCE [1], [2], a conservative value of 1 μm/y corrosion rate in the water-cooled austenitic steel at the ESS target station is chosen for the design basis. For the other material present in the systems an extensive literature study and comparison with other facilities lead to the following figures:

- Steel, 316 L 1 μm/y
- Aluminium 10 μm/y
- Beryllium 25 μm/y

Each primary water cooling loop will have two ion-exchange columns, but only one in operation. When resin in one of the column is full the other one will be put in operation. This means it will be in operation for one year and then stored for another year before replacing the ion exchange resin in the column. It is not planned to be operated for longer times, as there might be problems with emptying the column if the operation and storage time are increased to several years

5. Radiolysis Gas Handling

While the water in all three primary water cooling systems is exposed to radiation during operation, radiolysis and spallation gases are formed in the circulating water systems and must be taken care of. The most dominant reaction is the decomposition of water (radiolysis), although formation of other gases does exist (spallation of water and corrosion products). However, water does not decompose into hydrogen and oxygen directly, but rather into hydrogen, hydrogen peroxide, hydrogen radicals and assorted oxygen compounds. The consequence of this is that the recombination of the formed radiolysis products to water is not a simple, stoichiometric problem.
Due to this there will be a build-up of radiolysis and spallation gases in the circulating gas phase, where hydrogen and oxygen, besides the carrier gas, will be the dominating gases. If not treated or vented, the radiolysis and spallation gases will eventually increase the pressure in the circulating gas loop with unpredictable consequences. Also, the concentrations of hydrogen and oxygen may move into the flammability region, which may lead to uncontrolled hydrogen combustion or detonation. A Radiolysis Gas Treatment System is therefore implemented. Such a system should be designed to:

- recombine the hydrogen formed
- keep the concentration of hydrogen below the lower flammability level
- keep the non recombinable gases contained until decay allows release to the atmosphere

With the above requirements, a Radiolysis Gas Treatment System has been designed utilizing a commercial recombiner catalyst for recombining hydrogen and oxygen to water and at the same time ensure a residence time allowing sufficient decay of active non-recombinable spallation gases before release to the atmosphere. Figure 2 below shows a flow diagram indicating the basic design. The system consists mainly of a fan circulating a carrier gas through the Moderator, Reflector and Shielding & Plugs GLS tanks and then through the recombiner catalyst. The condensed water will be returned to the primary cooling loops making a closed loop of the systems.

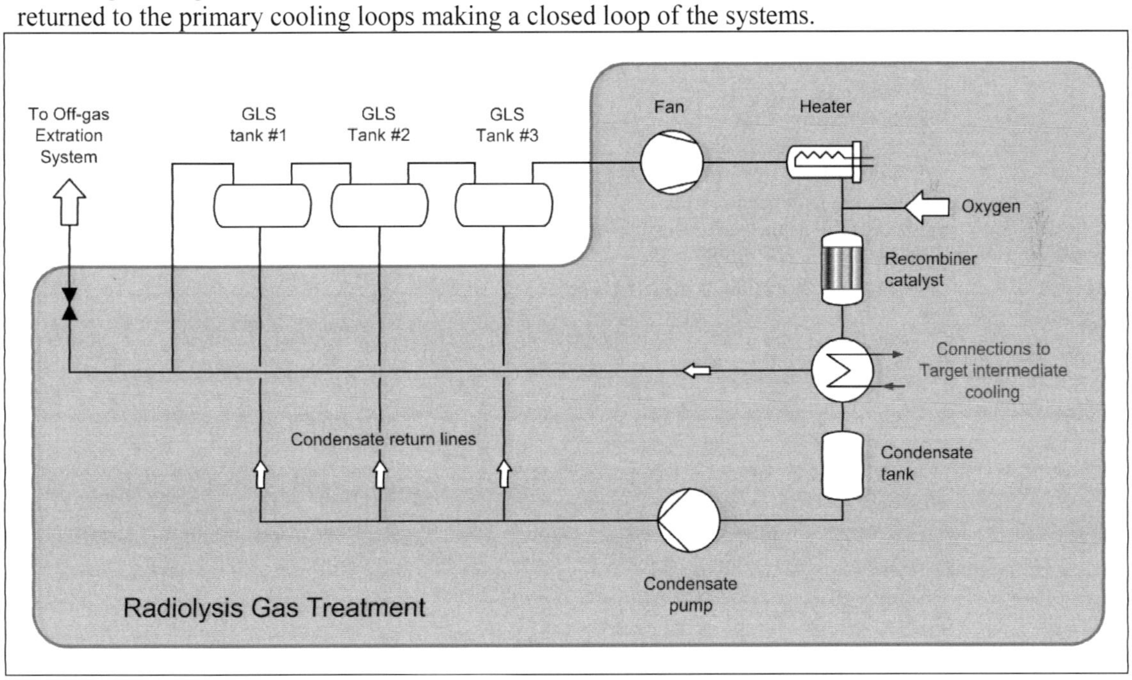

Figure 2 Radiolysis Gas Treatment System.

References

[1] R. S. Lillard et al, Corrosion Measurements on APT Prototypic Materials in the LANSCE High-Power Proton Beam and Applicability to Other Systems, LA-UR-02-2853, May 2012.
[2] R.S. Lillard et al, Journal of Nuclear Materials react-text: 55 278(s 2–3):277–289 April 2000

IOP Conf. Series: Journal of Physics: Conf. Series **1021** (2018) 012065 doi:10.1088/1742-6596/1021/1/012065

ESS Proton Beam Window Design Update

R. Vivanco, T. Mora, J. Aguilar, M. Magán, I. Herranz, A. Aguilar, L. Mena, M. Mancisidor, G. Bakedano, P. Luna, F. Jiménez-Villacorta , M. Wilborgsson (ESS), F. Sordo, J.M. Perlado, J.L. Martínez

Abstract.

ESS target station is devised inside a large vessel that serves as confinement for the atmosphere surrounding main components of the neutron source. Such an environment can not be in contact with the accelerator one, where ultra high vacuum is needed to guarantee a good transporting media for the 2.0 GeV protons.
The Proton Beam Window (hereon PBW) of the European Spallation source is an intricate component that behaves as a physical separator between the accelerator atmosphere and the target one; permitting to work with both environments and thus selecting the optimal vacuum level at each side.
As the proton beam has to cross the window, scattering interactions will take place inside, releasing a considerable amount of energy of around 6 kW, and distorting the protons' path. A thin and robust model has been designed while having to accomplish the crucial requirement founded on the desire to intrude the beam in the lowest way possible; besides a cooling system is conceived to evacuate all the power deposited.

Recent discoveries on materials analyzed in other facilities have motivated a design change from the former beam wind model, together with the aim of improving the design and enhance the cooling capabilities to handle the power deposited. Therefore, the former helium-based cooling system has been replaced by light water, providing larger cooling rates and reducing the high design pressure in the old model. The new water-cooled PBW has a thinner cooling channel, as water can lead to undesired beam distortion, what results in a double cylindrical plate shape of 1 mm each, that leaves another 2 mm cooling channel inside. Solution presented includes the thermo-fluid and mechanical analysis of the PBW that led to the above design.

Prototype manufacturing has started, and progress is presented.

Together, the sealing system that will be mounted on to guarantee the requested leak rates has been design ad-hoc for this component, with a robust pneumatic system to actuate remotely and inflatable belows to bring the desired vacuum levels at both sides. Sealing system has also allowed for beam measuring instrumentation to control the beam parameters, and the mounting device and its cabling have been introduced in a modular way to facilitate the operation inside such aforementioned vacuum requirements.
Moreover, a shielding analysis has been also performed to arrive to the final model, which includes neutron streaming evaluation through the cooling pipes, and thermal and mechanical analyses to assess whether an active or passive cooling should be implemented.

IOP Conf. Series: Journal of Physics: Conf. Series **1021** (2018) 012065 doi:10.1088/1742-6596/1021/1/012065

1. General description

More than 500 m long of accelerator tube will lead 2 GeV protons to a tungsten target, depositing up to 5 MW of energy in a rotating helium cooled wheel.

UHV is needed inside the accelerator tube to avoid any interaction between protons and gases. The target wheel and surrounding components will be confined into a big vessel, and under-pressurized to avoid any undesired leakages.

To separate both atmospheres, a double plate PBW design, inside cooled by water, is presented.

Protons will enter this atmosphere through the PBW, which aims to preserve beam shape and features by reducing scattering. Such a purpose results in two plates of 1.0 mm thickness in aluminum.

As a result of a 5 MW proton beam crossing the material, around 0.1% of its power is deposited there. In such small piece of aluminum, the challenge is to evacuate all that power, while ensuring a resistant structure to the many loads present in the system.

Several systems will be arranged together not only to guarantee proper cooling, but also to ensure that high level sealing, extremely precise alignment, and complete shielding to avoid neutron streaming, are perfectly gathered in a robust and compact plug that follows RCC-MRx nuclear code regulation rules. A thorough description of this system is presented within this document.

1.1. The PBW system

A vessel is designed to wrap the shielding and pipes, and makes connection with a port block place below. Rough vacuum is made inside the vessel to reduce leakages at the seal.

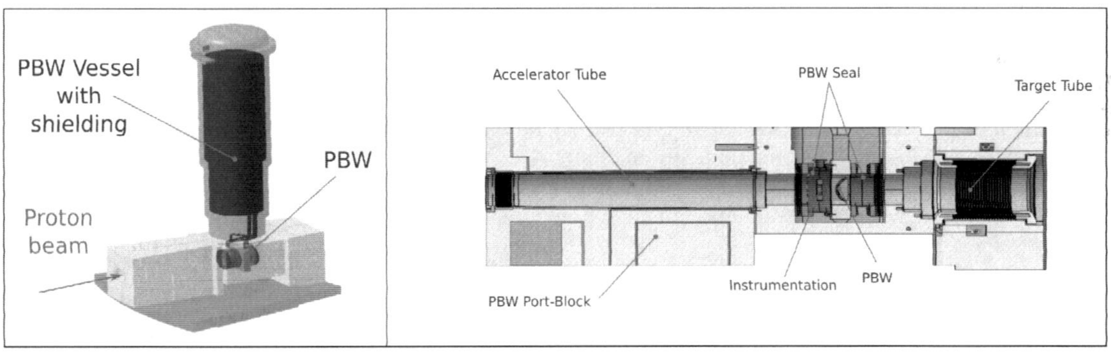

Figure 1. General overview of the PBW assembly inside the port block and vessel.

1.2. The PBW and its cooling structural frame

The PBW consists in 2 thin plates separated by a 2mm cooling channel. This plates are 1.25mm and 1.00mm each. It is located at 3.750 m from the target. Cooling channels are machined in a structural Al6061-T6 block, hereon the "frame", and the PBW is welded to this frame. This system is split into this two components to ease manufacturing procedures.

A 0.3 kg/s water flow rate at 20° is circulated from the pipes above inside the cooling channel, and guided to the PBW plates.

Side supports are made in 1.00 mm thickness to permit deformations and allocate stresses. The outer part of the PBW will be welded to the frame.

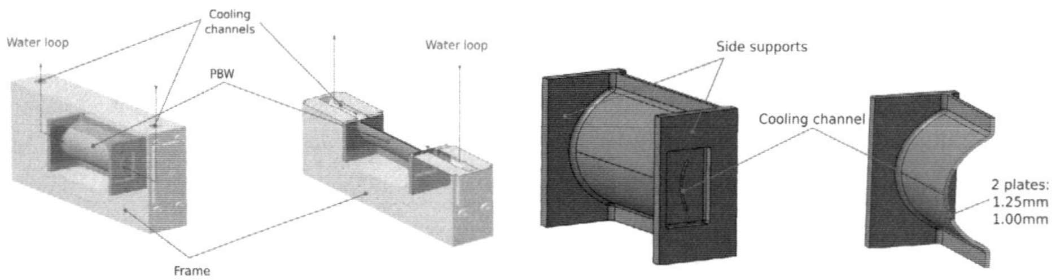

(a) PBW (b) Structural frame

Figure 2. PBW detailed

2. CFD and FEM analyses

MCNPX© nuclear code has been used determine the nuclear deposition in materials, and that power source has been implemented in Computer Fluid Dynamics (CFD) programs. CFD analyses have been performed in water flow, to guarantee that a small 2 mm cooling volume can remove all the heat deposited. Ansys Fluent© with $\kappa - \omega$ turbulence models have been selected to evaluate the fluid behavior.

Once obtained thermal conditions, and ensured the cooling capability, FEM analyses have guaranteed that mechanical stresses are under material limits, the structure won't suffer from buckling or fatigue, and deformations are acceptable.

2.1. *Thermal analyses*

Operating conditions of the beam are quite demanding: 14 Hz, 2.857 ms, 67.5 mA peak eventuating in 5 MW power released. Al6061-T6 will be irradiated, unable to accommodate more than 2100 appm of He due to nuclear gas generation; therefore, no more than 60°C are allowed, to minimize diffusion and prevent from swelling. Every six months of full power operation, the PBW will be removed and completely changed. Coolant pressure needs to be raised up to 5 bar.

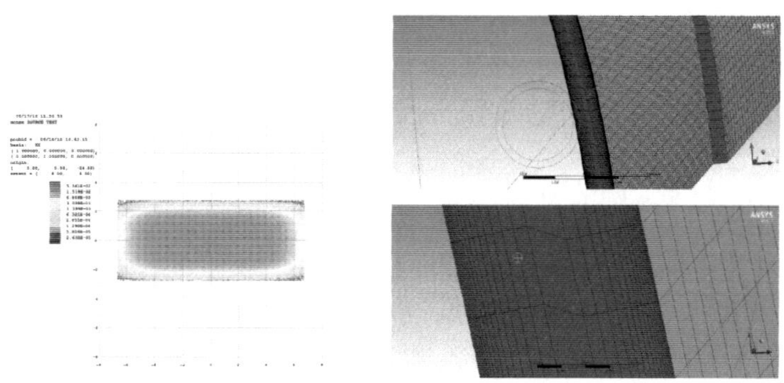

(a) MCNPX©power deposition. (b) Fluid mesh

Figure 3. Integrated thermal analyses performed.

IOP Conf. Series: Journal of Physics: Conf. Series **1021** (2018) 012065 doi:10.1088/1742-6596/1021/1/012065

2.2. Mechanical analyses

Maximum stresses and deformations are calculated by coupling CFD results to mechanical FEM analysis. RCC-MRx methodology for Class 3 (N3Rx) is applied to the analysis. Model is developed to be resistant to type P, type S and buckling damages.

Mechanical and thermal buckling are studied, and modal analysis are also carried out. Results prove that with such geometry and dimensions, safety factor of more than 3 is achieved in the system, what makes it possible to pass all nuclear regulations while pushing the design to its physical limits.

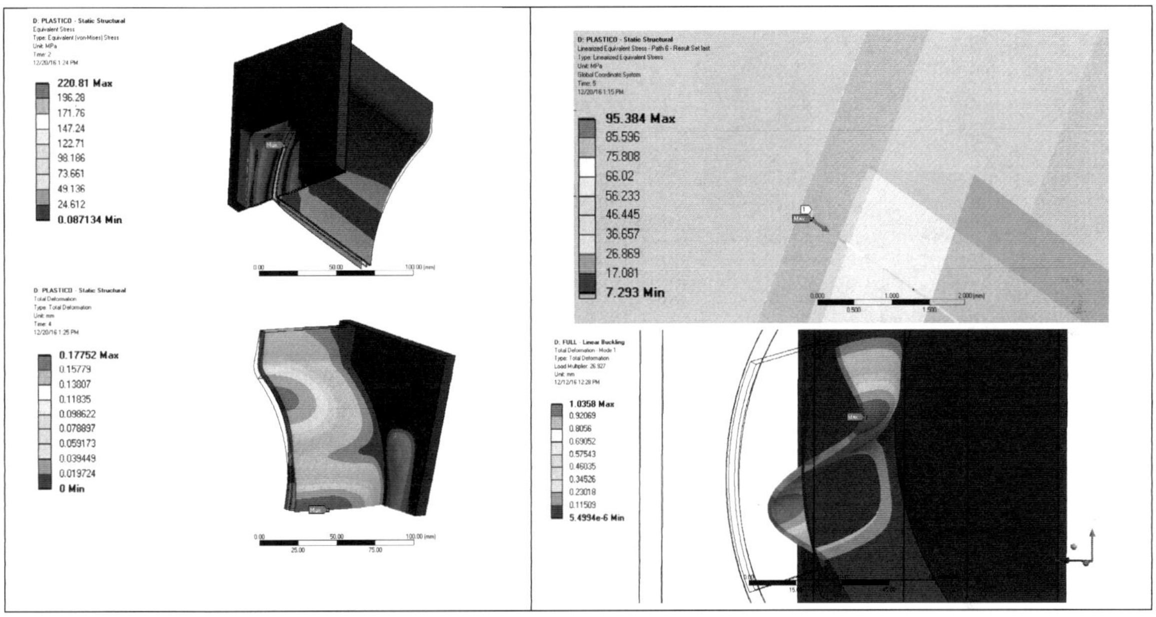

Figure 4. Complete mechanical analyses.

3. Seal design

Different facilities as J-PARC [1], KEK [2] and PSI have been studied as reference for the sealing design. They make use of a pillow seal to guarantee proper contact between surfaces.

An adapted design for ESS has been developed together with spanish company AWGE© [3].

The seal system consists of a grooved cylindrical surface (Figure 5(a)), with membranes between the grooves. Two inner and outer belows (Figures 5(b) and 5(c)) are inflated to expand the system and approach the sealing faces to their counterparts. Besides, different levels of vacuum are applied to the grooves to generate a vacuum resistance for the flow to pass through. Membranes are inflated with same belows' pressure, enhancing the contact between surfaces.

Leak rates are expected to be less than $2 \cdot 10^{-5} \frac{mbar-l}{s}$, considering a UHV atmosphere at the accelerator side, and rough vacuum at the target one.

The full system is mounted as shown in Figure 5(d). Two large flanges will cover the distance between the PBW and the seal, while allocating instrumentation for the proton beam. An aperture monitor will be fed by a vacuum feedthrough from above, cabling the signal up through the 4 m height vessel.

Gaseous nitrogen or helium will be used to inflate the belows, with capability to move the sealing surface ± 3 mm in beam direction, and hence permitting to seal by totally remote operations. To remove it, some pressure can be applied to the grooves, as long as a slight vacuum is applied to the belows, retracting them.

The alignment of the system will relay in several guides placed along the PBW frame.

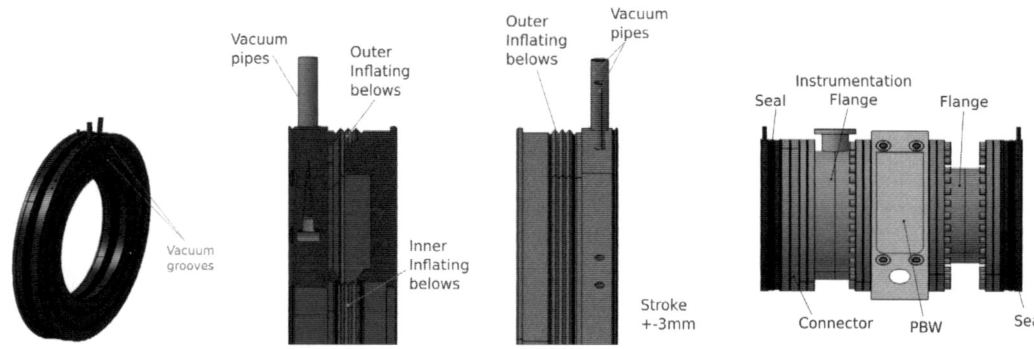

(a) Vacuum grooves (b) Belows - in (c) Belows - out (d) PBW & Seal assembly.

Figure 5. Design of the seal for the PBW.

4. Nuclear shielding, activation and decay heat analyses

Protons are partially scattered in the PBW, releasing around 0.1% of the beam energy. These interactions produce secondary particles, including neutrons and photons, that escalate dose rates above the system, and have to be properly shielded. Planned removal and replacing operations every six months for the PBW force to unplug and plug the cooling media and auxiliary systems placed above the vessel, where hands-on maintenance is envisaged for such operations. Decay dose rates need to be assessed to guarantee integral safety for workers.

Together, heat deposition analyses are performed to check the needs of active cooling in the seal system.

Isotopic composition is of vital importance for the hot cell storage and segregation operations. Activation has been calculated by ACAB© code, and time dependent isotopic contribution to activity is evaluated. Volume independent values permit direct evaluation of future changes in component shapes and geometry.

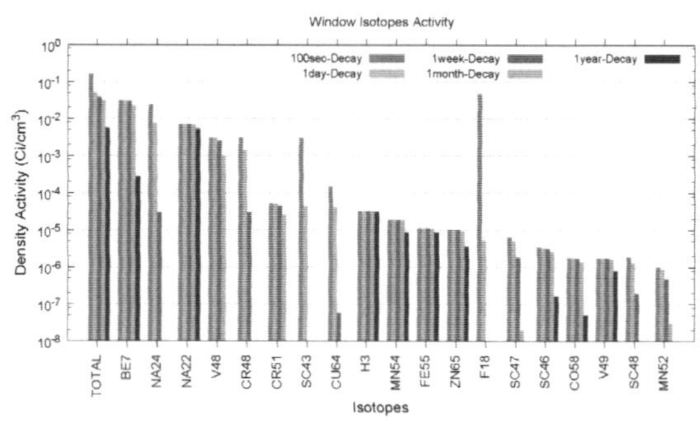

Figure 6. Time dependent isotopic activity in one of the seals.

5. Prototyping

Manufacturing of two thin plates for PBW design is a challenge that entails precise operations. High requirements imposed by ESS force to find a high quality manufacturing process which

IOP Conf. Series: Journal of Physics: Conf. Series **1021** (2018) 012065 doi:10.1088/1742-6596/1021/1/012065

guarantees every thickness described in the design drawings.

Alongside, a prototype for the seal is envisaged in order to learn from experimental values the most suitable vacuum levels and pressures that will have to be applied to make the leak rates dwindle to lower values possible.

5.1. PBW prototype

A bulk of Al6061-T651 is used for the starting operation. Electrical Discharge Machining (EDM) is used to perform a thin and 200 mm length hole in the place where cooling channel will be machined. Afterwards, Wire EDM (WEDM) is sequentially employed to achieve the final 2 mm thickness and channel shape.

External milling is therefore applied in next step, giving the rounded surface needed without the use of any welds.

Measurement techniques while machining, as 3D probes and ultrasonic measurements, are utilized to control thickness down to 10 microns [4].

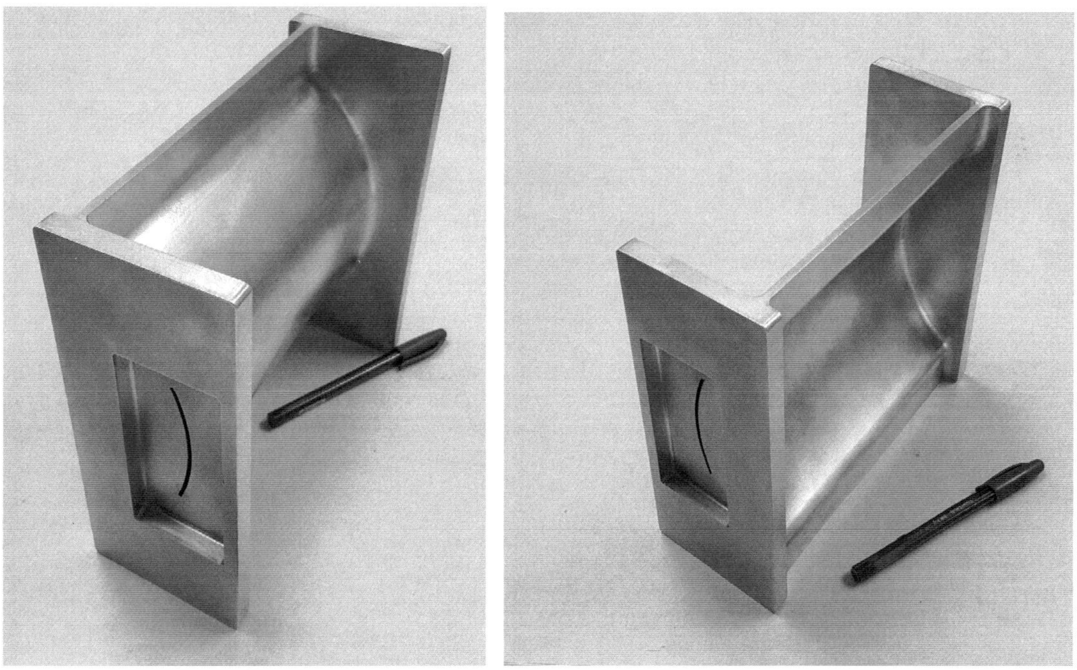

(a) Final prototype 1 (b) Final prototype 2

Figure 7. WEDM operations and final manufactured prototype.

WEDM operations and manufactured prototype are presented in Figure 7. Final geometry has been achieved avoiding any welds, and removing any weak point in the structure, as it eases RCC-MRx regulation.

6. References

[1] Meigo S and Teraoku T 2015 Proton Beam Window Assembly Tech. rep. J-PARC
[2] 2015 New milestone at MUSE Tech. rep. KEK,MUON Science Lab
[3] Mirapeix F, Guerra D, Pérez N *et al.* 2017 Vacuum Seal design for the PBW Tech. rep. AWGE
[4] Rey A 2017 PBW Manufacturing preliminary proposal Tech. rep. Leading

IOP Conf. Series: Journal of Physics: Conf. Series **1021** (2018) 012066 doi:10.1088/1742-6596/1021/1/012066

The neutron moderators for the European Spallation Source

L. Zanini[1], K. Batkov[1], E. Klinkby[1,2], F. Mezei[1], T. Schönfeldt[1,2] and A. Takibayev[1]

[1]European Spallation Source ESS ERIC, Box 176, S-221 00 Lund, Sweden
[2]Technical University of Denmark, DTU Nutech, Risø Campus, Frederiksborgvej 399, Bldg. 201, 4000 Roskilde, DK

E-mail: `luca.zanini@esss.se`

Abstract.
The European Spallation Source will have 42 beam ports as a grid available for a variety of instruments, mostly neutron scattering experiments. Bi-spectral extraction for thermal and cold neutrons must be available to all the beam ports. The moderator design to deliver such neutron beams was driven by the low-dimensional moderator concept. The adopted design, consisting of one *flat* (3 cm high) moderator placed above the spallation target was considered valid for the initial instruments suite. ESS will however have a beam port system designed such that it will be possible to extract neutrons from moderators above and below the target. With all initial instruments pointing to the top moderator, this opens the possibility to have different types of moderators at the bottom, so that other neutron beams of different intensity, or spectral shape, with respect to the ones delivered by the top moderator, could be envisaged, adding additional scientific opportunities to the facility.

1. Introduction

The ESS beam extraction layout contains 42 beam ports, arranged in two sectors covering a total angle of $2\times120°$, pointing to the center of the monolith where the thermal and cold moderators are placed. The points of beam port origin at the moderators are called *focal points.*

This grid is available for placing of instruments, adding flexibility over the life time of the facility for placing of instruments. The layout for the fifteen instruments funded by the construction budget has been fixed. Additional instruments are planned to be installed in the operation phase of ESS.

The moderators to provide thermal and cold beam to the instruments were designed with the requirement to provide a bi-spectral neutron beam to all the beam ports. The thermal and cold moderators are distinct units containing different materials (light water and 20 K pure parahydrogen, respectively).

In order to provide a high thermal and cold flux, moderators should be placed above the *hot spot* of neutron production, which is an area of approximately 15×20 cm^2 where most of the evaporation neutrons are emitted from the target in the spallation process.

2. Low-dimensional butterfly moderators

2.1. Design selection

Different designs were considered for the ESS moderators since design started in 2012.

- The design presented in the Technical Design Report [1] consisted of two volume moderators, with two openings of 60 degrees each.
- The major design change was the decision to follow the low-dimensional moderator concepts [2, 3] The TDR design was replaced in 2014 by the pancake moderator [4] which consisted of a disk-shaped (3 cm height, 20 cm radius) cold moderator. The thermal neutrons were extracted from the sides. This concept was eventually abandoned in favour of the more compact butterfly moderators, with higher brightness (especially in the thermal spectrum) and easier bi-spectral extraction. Different options were considered for the bottom moderator [4], but no decision was taken at the time.
- The butterfly *BF2* design in 2015 [5] consisted of two individual hydrogen vessels, separated by a water cross. Two identical moderator sets were foreseen above and below the target, with different height: 3 cm at the top, 6 cm at the bottom. This choice was officially adopted as baseline for ESS in 2015.
- Finally the design of the top moderator was modified to the *BF1* type, presented here.

Butterfly moderators[5, 6] are the design solution to provide bright bi-spectral moderators for the whole instrument suite. Such moderators (Figs. 1,2) have the following advantages:

- Both thermal and cold moderator have a height of 3-cm for increased brightness, exploiting the advantage of flat moderators [2]. To some extent, also the concept of tube moderators [3] is exploited, thanks to the geometry of the cold moderator.
- both cold and thermal moderators are placed on the hot spot, providing high thermal and cold brightness for a required extraction area of at least 3 (height)$\times$6 (width) cm^2 for both thermal and cold moderators.
- Such moderators are ideally fit for beam extraction in the two 120° sectors; the brightness variation across the sectors is within 15%.
- Their relatively compact shape is an advantage for beam extraction: for all the 42 beam ports, the thermal and cold extraction surfaces lie next to each other, being placed on the two sides of the focal points, allowing instruments to see the brightest part of thermal and cold moderators.

3. Moderator performance

Fig. 3 shows the integrated time-averaged brightness distribution along the 42 ESS beam ports. Results are given for 3-cm and 6-cm viewed horizontal widths, as well as for the (beamline dependent) maximum horizontal widths (the vertical width is 3 cm in all cases). For the 3-cm horizontal widths, by looking at the brightest parts of the moderator extraction surfaces, it is possible to maximize thermal, and especially cold, brigthness, especially for those angles where it is possible to exploit the full depth of the cold moderator in the direction of beam extraction. Selection of the extraction surfaces should be carefully done by means of McStas simulations [7].

Fig. 4 shows the peak wavelength spectra, averaged over the 42 beam ports.

A detailed study of systematic uncertainties in the MCNPX calculations was presented in [1]. On the basis of those results, and considering the more mature level of the present design, we can expect a relative uncertainty of 15% on the present brightness values.

The performance of the ESS source is usually compared with the official ILL brightness values from the *yellow book* [8]. The original design goal of ESS was to achieve a cold peak brightness

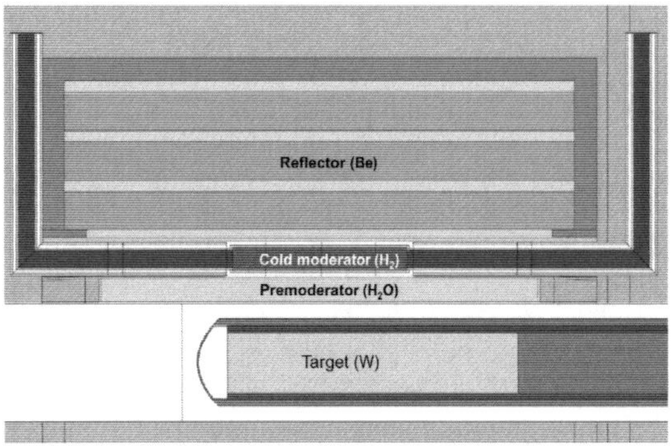

Figure 1. MCNPX [9] geometry, side view. Proton beam comes from the left impinging on the tungsten target. Tungsten (green) has a density of 15.1 g/cm^3 instead of the nominal density of 19.3 g/cm^3 to account for the fraction of helium coolant in the target according to which the filling factor of tungsten is 78 vol%. The beryllium reflector (light blue) includes water channels (green) according to engineering drawings. The reflector is contained in stainless steel (red). The outer reflector (orange) is made of stainless steel, with 10% volume fraction of water, for cooling

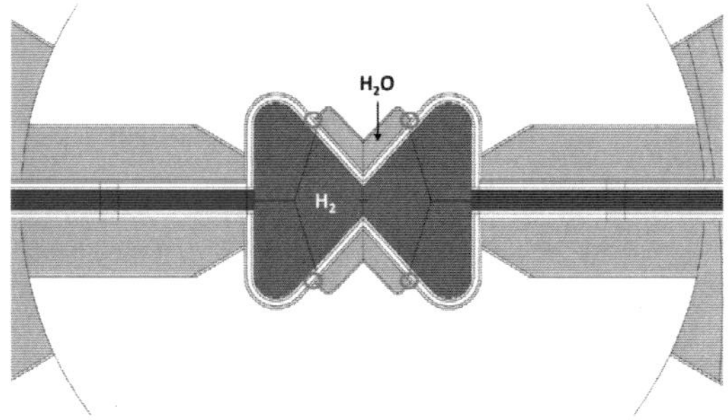

Figure 2. MCNPX geometry of the moderator, top view. The parahydrogen (blue) contains 5 vol% of Al in the model, to account for the presence of Al flow channels, not yet designed. On the sides of the cold moderator, inlet and outlet hydrogen pipes, including vacuum gaps, are modelled. Water (light blue) is placed around the pipes to serve as premoderator and increase the brightness of the cold moderator. Externally, part of the outer reflector (orange) is shown. The four focal points (origin of the beam ports) are circled in red.

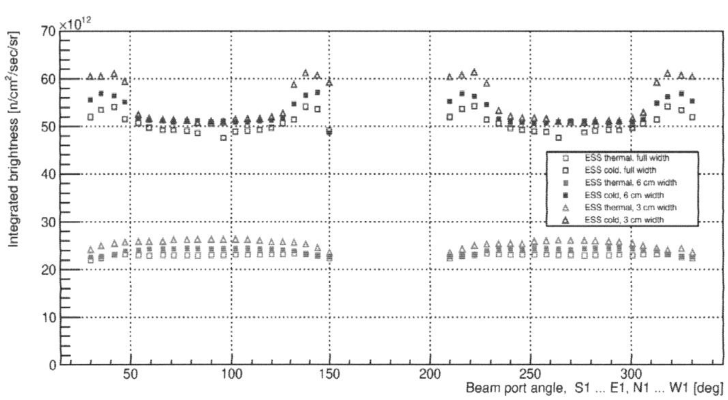

Figure 3. Time-average integrated thermal ($20 < E < 100$ meV) and cold ($E < 20$ meV) brightness for the 42 beam ports, for different horizontal projected widths: full, 6 cm and 3 cm.

30 times the average ILL brigthness [1]. With the use of low-dimensional moderators, we are far above this goal. The cold brightness shown in the present design is (at 4 Å) nearly 140 times higher than the yellow book value. The thermal brightness at 1.5 Å is about 10 times higher than ILL. Considering integral values, the integrated peak cold brightness above 4 Å for the butterfly is of 4.2 $\times 10^{14}$ n/cm^2/s/sr, which is 125 times the ILL average integrated brightness (3.3 $\times 10^{12}$ n/cm^2/s/sr). For the cold neutrons between 2 Å and 4 Å, the ESS brightness is of 9.1$\times 10^{14}$ n/cm^2/s/sr, which is 200 times the ILL brightness (4.5 $\times 10^{12}$ n/cm^2/s/sr) in the same range. For thermal neutrons, from 0.9 Å to 2 Å, the ESS peak thermal brightness is of 6.0 $\times 10^{14}$ n/cm^2/s/sr which is about 10 times higher than ILL (6.2 $\times 10^{13}$ n/cm^2/s/sr).

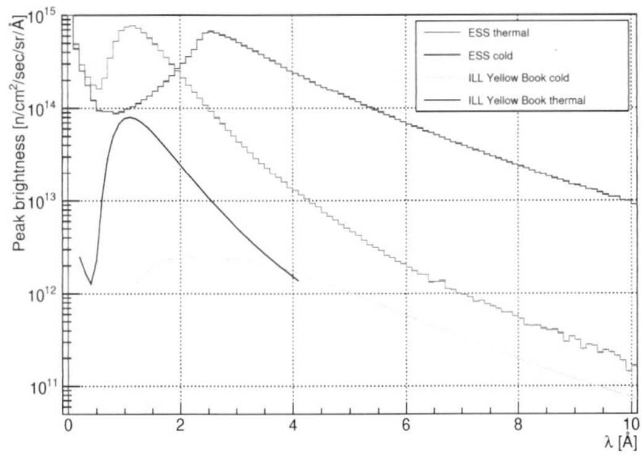

Figure 4. Brightness spectra averaged over 42 beam ports for 3 cm high moderator, compared with ILL official curves [8].

4. Future possibilities

The chosen configuration, where all initial instruments foreseen for ESS point to the top moderator, leaves open several attractive possibilities for future use at ESS. With a high-brightness moderator at the top, the bottom area could be reserved in future for a source of neutrons which should be complementary to the top one. Several possibilities could be explored, such as:

- Ultra Cold or Very Cold neutron source.
- High-intensity moderator, such as a large D_2 moderator, for specific experiments like the nnbar [10].
- A high-brightness moderator, even surpassing the brightness of the top moderator, which could be achieved either by using tube moderators (good for a few beam lines only, due to their strong directionality), or by further reducing the height with respect to the top moderator [3, 11].

References

[1] ESS Technical Design Report, S. Peggs editor, ISBN 978-91-980173-2-8, 2013, http://europeanspallationsource.se/scientific-technical-documentation

[2] K. Batkov, A. Takibayev, L. Zanini and F. Mezei, *Unperturbed moderator brightness in pulsed neutron sources*, Nuclear Instruments and Methods in Physics Research Section A: Accelerators, Spectrometers, Detectors and Associated Equipment 729 (2013) 500 - 505. http://dx.doi.org/10.1016/j.nima.2013.07.031.

[3] F. Mezei, L. Zanini, A. Takibayev, K. Batkov, E. Klinkby, E. Pitcher, and T. Schönfeldt, *Low dimensional neutron moderators for enhanced source brightness*, Journal of Neutron Research 17 (2014) 101 - 105.

[4] L. Zanini, K. Batkov, E. Klinkby, F. Mezei, E. Pitcher, T. Schönfeldt, A. Takibayev *Moderator Configuration Options for ESS*, proceedings of the ICANS-XXI conference, JAEA-Conf 20-15-002, 2015.

[5] L. Zanini, K. Batkov, F. Mezei, A. Takibayev, E. Klinkby, T. Schönfeldt, *The neutron moderators at the European Spallation Source*, proceedings of the AccApp'15 conference, http://accapp15.org/wp-content/data/index.html.

[6] T. Schönfeldt, *Advanced Neutron Moderators*, PhD thesis, Technical University of Denmark, 2017.

[7] P. Willendrup, E. Farhi and K. Lefmann, *McStas 1.7 a new version of the flexible Monte Carlo neutron scattering package*, Physica B, 350 (2004) E735.

[8] Institut Laue-Langevin. 'ILL Yellow Book 2008.' http://www.ill.eu/?id=1379 , 2008.

[9] D. Pelowitz, editor, MCNPX Users Manual, Version 2.7.0, Number LA-CP-11-0438, 2011.

[10] E. B. Klinkby, K. Batkov, F. Mezei, T. Schönfeldt, A. Takibayev, L. Zanini, *Voluminous D2 source for intense cold neutron beam production at the ESS*, in: arXiv physics eprints, No. arXiv:1401.6003, 2014.

[11] L. Zanini, F. Mezei, K. Batkov, E. Klinkby and A. Takibayev, *General use of low-dimensional moderators in neutron sources*, these proceedings.

IOP Conf. Series: Journal of Physics: Conf. Series **1021** (2018) 012067 doi:10.1088/1742-6596/1021/1/012067

Characterization of a Liquid Ammonia Moderator

E B Iverson[1], D V Baxter[2], F X Gallmeier[1], R C Gillis[1], T Hügle[1], W Lu[1], T C McClanahan[3], I Remec[1] and T C Rinckel[2]

[1]Spallation Neutron Source, Oak Ridge National Laboratory, Oak Ridge, TN USA
[2]Center for the Exploration of Energy and Matter, Indiana University, Bloomington, IN USA
[3]Department of Nuclear Engineering, University of Tennessee-Knoxville, Knoxville, TN USA

E-mail: iversoneb@ornl.gov

Abstract.
Liquid ammonia is an attractive intermediate temperature moderating material, offering a high hydrogen density, low tendency toward radiation-initiated polymerization, and broad accessible temperature range. The LENS collaboration has characterized decoupled poisoned liquid ammonia moderators in an attempt to validate scattering kernels we are generating. Additionally, we have measured a borated water moderator in order to test a model for predicting detailed moderator performance at one level of intrinsic homogeneous neutron capture based on measurements at other levels of such capture. We describe both sets of measurements, as well as use them to illustrate some of the reproducibility challenges associated with moderator characterization measurements on compact neutron sources.

1. Introduction

Liquid ammonia (NH_3) is an attractive intermediate temperature moderating material, offering a high hydrogen density, low tendency toward radiation-initiated polymerization, and broad accessible temperature range relative to other intermediate temperature moderators. [1, 2, 3] Solid ammonia offers the same advantages at a broad range of cryogenic temperatures. Both forms offer many low-energy rotational modes providing effective neutron energy loss. A complication for the use of ammonia as either an intermediate temperature moderator or as a cold moderator is the significant neutron capture cross section for natural nitrogen; more than five times that of hydrogen on a per atom basis, nearly tripling the macroscopic capture cross section for the molecule. Ammonia enriched to high levels of ^{15}N would avoid this additional capture, while retaining similar neutron scattering characteristics. Isotopically enriched ammonia would be practical for use at a production facility if merited, although it would likely require a recovery and re-use system.

This manuscript has been authored by UT-Battelle, LLC under Contract No. DE-AC05-00OR22725 with the U.S. Department of Energy. The United States Government retains and the publisher, by accepting the article for publication, acknowledges that the United States Government retains a non-exclusive, paid-up, irrevocable, worldwide license to publish or reproduce the published form of this manuscript, or allow others to do so, for United States Government purposes. The Department of Energy will provide public access to these results of federally sponsored research in accordance with the DOE Public Access Plan (http://energy.gov/downloads/doe-public-access-plan).

At the Spallation Neutron Source facility at Oak Ridge National Laboratory (SNS), we are considering the replacement of the ambient water moderator with one composed of liquid ammonia. Various previous studies concerning ammonia moderators have concerned solid ammonia at relatively low temperature. [1, 2] This is not our interest here: the ambient water moderator currently in use at SNS [4, 5] serves instruments that would benefit from an intermediate temperature spectrum, but not necessarily from a low temperature spectrum, and the power densities deposited in the moderator systems are not easily handled with a solid moderator material. In order to predict the practical performance of a liquid ammonia moderator, and thus evaluate its suitability for our use, we require two things: a validated scattering kernel with which to predict neutron thermalization, and a method for predicting the performance of a moderator of ^{15}N-ammonia based on characterization and validation measurements performed with ^{14}N-ammonia. Our group is in the process of measuring the inelastic and total neutron scattering from ammonia at various temperatures in order to generate scattering kernels. These efforts will be reported elsewhere. Here we describe integral measurements of prototypical liquid ammonia moderator performance intended to validate those scattering kernels. We have characterized ammonia moderators at temperatures of 220 K (liquid) and 185 K (just below the freezing point); while the solid ammonia measurements are not directly relevant for our desired application, they provide additional points of comparison for measurements reported in the literature (which were performed at 77 K for differently sized moderators), [1, 2] and were practical with relatively little additional effort. On a separate occasion, we performed comparable measurements of room temperature water and various concentrations of boric acid, with which we hope to test a model for extrapolating from measurements of natural ammonia to performance of ^{15}N-ammonia with its lower capture rate. The details of those measurements, and the testing of that extrapolation model, will be reported elsewhere—however, the nature of the two experimental campaigns provided an opportunity to perform measurements on what should nominally have been the same moderator (distilled water in identical geometries). The comparison of these measurements provides insight into challenges for consistency and reproducibility in moderator characterization.

2. Measurements Performed

We have measured neutron wavelength spectra and wavelength-dependent emission time distributions (pulse shapes) from moderators composed of natural ammonia at 220 K and 185 K (liquid and solid, respectively). Measurements at these temperatures were repeated for moderators of size 100 mm by 100 mm by 25 mm, and size 100 mm by 100 mm by 15 mm. For each moderator vessel, the measurements were repeated with room temperature distilled water to provide a relative control. These control experiments were accomplished by removing the ammonia from the moderator vessel and replacing it with water, without removing or disturbing any physical components. Details of the LENS moderator test facility, our moderator performance characterization equipment, and our data analysis methods can be found elsewhere, [6, 7, 8, 9] and will not be repeated here. All measurements described were performed with cadmium decoupled moderators, using accelerator pulse widths of 13.5 μs.

The detailed measurements we have performed provide integral quantities we will use as a first level of comparison to calculations based on scattering kernels currently under development. Ammonia is a thermalizing moderator; robust and convenient integral metrics include the spectral temperature of the emitted neutron beams, the fundamental decay constant of the long-wavelength emission time distributions, and (somewhat less integral in nature) the wavelength-dependent metrics for emission time distribution width; mean emission time, full-width half-maximum (FWHM), and root-mean-square (RMS) emission time.

The spectral temperature T_n of the neutron beam is estimated by fitting a linearized efficiency-weighted Maxwell-Boltzmann distribution in the range of 1–3 Å (beyond 3 Å the spectra depart

IOP Conf. Series: Journal of Physics: Conf. Series **1021** (2018) 012067 doi:10.1088/1742-6596/1021/1/012067

from the Maxwell-Boltzmann distribution). The fundamental decay time is estimated as the average of the decay times obtained by fitting the appropriate portion of each long-wavelength emission time distribution to an exponential decay. Mean emission times, RMS emission times, and FWHM were determined after subtraction of a linear background, and the RMS emission times σ_t and FWHM δ_t corrected for finite rectangular source pulse width $\Delta t_s \approx 13.5\mu s$ and data acquisition channel width $\Delta t_c = 2\mu s$ as

$$\sigma_t^2 = \sigma_m^2 - \frac{1}{12}\left((\Delta t_s)^2 + (\Delta t_c)^2\right) \tag{1}$$

and

$$\delta_t^2 = \delta_m^2 - \frac{(1.96)^2}{12}\left((\Delta t_s)^2 + (\Delta t_c)^2\right). \tag{2}$$

While these corrections are approximate, and rely on a square primary source pulse and a particular form for the emission time distribution, they are reasonable—perhaps more importantly, they are the same corrections used in earlier studies [2] and are therefore justified absent significantly better alternatives.

3. Results

Wavelength spectra measured for the 25 mm moderator are shown in Figure 1, and for the 15 mm moderator in Figure 2.

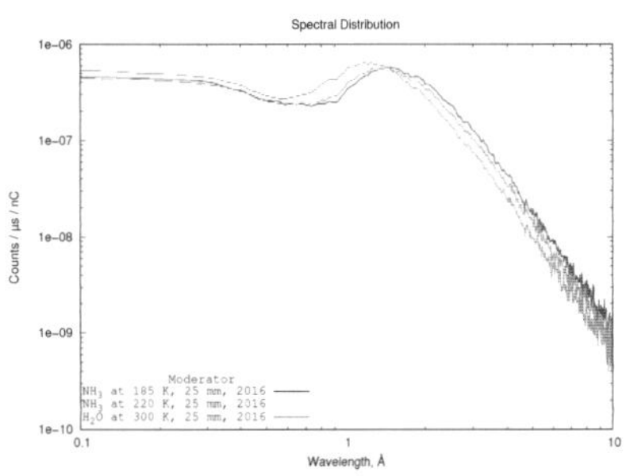

Figure 1. Neutron wavelength spectra measured for 25 mm thick moderators of water and ammonia.

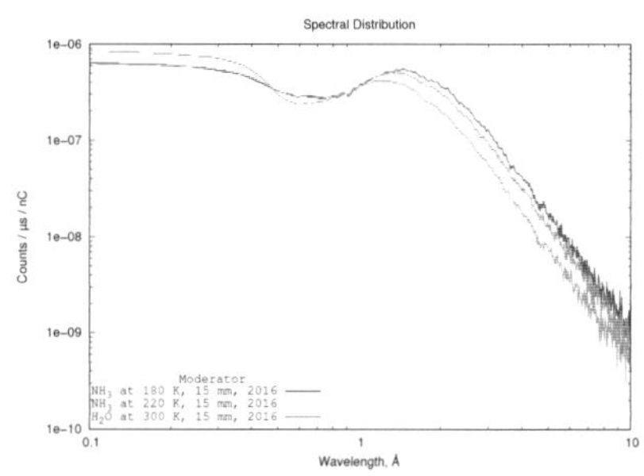

Figure 2. Neutron wavelength spectra measured for 15 mm thick moderators of water and ammonia.

Emission time distributions at a single wavelength for a single moderator geometry are shown in Figure 3. Similar distributions were measured at other wavelengths, and for both moderator geometries, but are not shown here. The apparent fundamental decay time τ_1 can be seen in Figure 3 between 50–100 μs. A second, longer-lived decay time τ_2 can be seen beyond 150 μs, and is assumed to represent leakage from the reflector through an imperfect cadmium decoupler.

Spectra taken from the distilled water control moderators in both experimental campaigns (using the same moderator vessels, but with significant reassembly of the moderator-reflector system) are shown in Figures 4 and 5.

Numerical values for various metrics are shown in Table 1. The apparent total flight path length (to the emission time analyzer detector) $\tilde{L}_T$ is determined by fitting the arrival time of the analyzed pulses. Since the analyzing crystal and the detector do not move between measurements, the variation in that fitted length can be assumed to reflect changes in the moderator position along the beamline upon cooling and reassembly. We make the following observations:

IOP Conf. Series: Journal of Physics: Conf. Series **1021** (2018) 012067 doi:10.1088/1742-6596/1021/1/012067

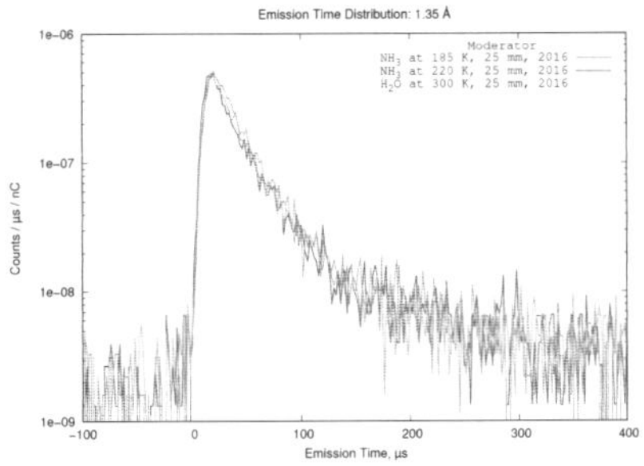

Figure 3. Emission time distributions measured at 1.35 Å for 25 mm thick moderators of water and ammonia.

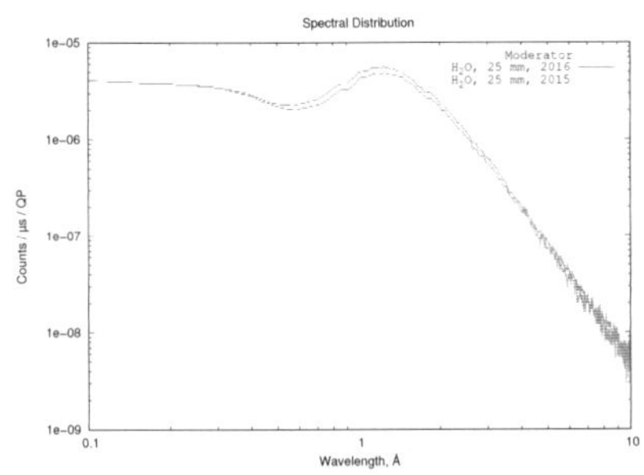

Figure 4. Neutron wavelength spectra measured for a 25 mm thick water moderator on two separate occasions.

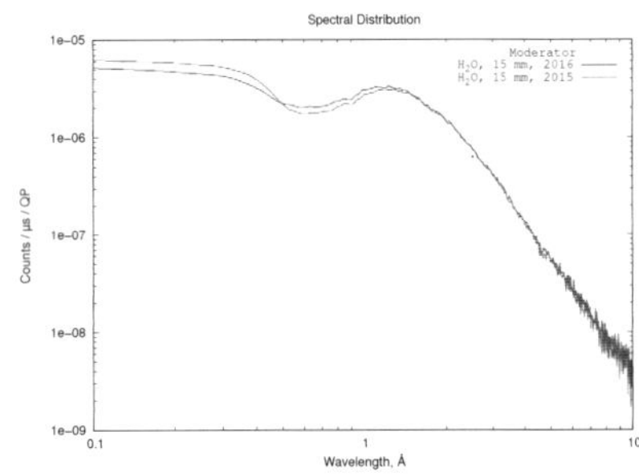

Figure 5. Neutron wavelength spectra measured for a 15 mm thick water moderator on two separate occasions.

Table 1. Measured parameters for ammonia and water moderators. All moderators were 100 mm by 100 mm transverse to the viewing direction.

Material	Temperature (K)	Depth (mm)	T_n (K)	τ_1 μs	τ_2 μs	FWHM (1.08 Å) μs	$\tilde{L}_T$ (m)
NH_3	220	25	280	18	68	14	9.2625
NH_3	185	25	253	20	78	12	9.2625
H_2O	302	25	357	21	108	23	9.2661
NH_3	220	15	294	13	141	14	9.2755
NH_3	185	15	270	14	139	9	9.2755
H_2O	302	15	346	18	165	14	9.2786

- The intensity at the shortest wavelengths is slightly smaller for an ammonia moderator than for a water moderator of the same dimensions. This is as expected for the slab moderator configuration at LENS; the higher hydrogen density means that the viewed moderator surface is neutronically farther from the primary neutron source.
- The spectral temperature for the ammonia moderators is significantly higher than the physical temperature (20–50%), as might be expected for small moderators (also the case for water) and high capture cross section (specific to ammonia).

- The ratio between the spectral temperature and the physical temperature is larger for the solid ammonia than for the liquid ammonia, consistent with the loss of thermalizing efficiency associated with recoil.
- The moderator vessel appears to shift approximately 3–4 mm upon cooling from 300 K to 200 K. This is not unreasonable. The viewed face of the moderator vessel appears to have moved some 12.4 mm upon removal and reinsertion. While we would expect a movement of 5 mm (due to the different size of the moderator vessel), the additional 7.4 mm motion provides an indication of the reproducibility of the moderator position.
- There is a long-lived tail on the emission time distributions for all moderators tested. It is significant in both length (i.e,. the large decay time) and magnitude. It appears to vary quite a bit, possibly in response to the apparent shift in moderator position. We believe that it is due to in-leakage from the reflector through a geometrically imperfect cadmium decoupler. While its effects can be dealt with in assessing emission time distributions, we are concerned that it may significantly perturb apparent spectral temperatures and thermalization ratios.
- The reproducibility of the absolute intensity of the water moderator spectra from the two different campaigns is not as good as we would like. While the 25 mm data shown in Figure 4 match at short wavelengths as we might expect, the 15 mm data indicate a gain of 15% for the subsequent measurement. Even more troubling than this absolute change is the fact that the thermalization ratio changed for both geometries; a loss of 15% for the 25-mm geometry and 30% for the 15-mm geometry. Finally, the spectral temperature appears to have changed for the 15 mm case, being nearly 10 K lower in the second measurement. These discrepancies will be addressed further below.

Our ammonia moderator measurements, if taken at face value, provide the integral quantities, shown in Table 1, needed to validate scattering kernels used in moderator simulations. However, the consistency check performed on nominally identical water moderators are not satisfactory at this stage of data analysis. We have identified the following hypotheses for further consideration:

- The apparent shift in moderator position upon reassembly could result in a significant change in the target-to-moderator distance. We plan to perform a Monte Carlo sensitivity study. This alone would not explain changes in spectral temperature nor thermalization ratio.
- The apparent shift in moderator position upon reassembly could also result in different geometric imperfections in the cadmium decoupling. We will examine emission time distributions to see if consistent effects are seen there, and assess decoupler gaps in Monte Carlo studies.
- The second round of measurements (ammonia and water) took place a year after the first round (water and boric acid in water). The aluminum moderator vessels could have been contaminated with boric acid, causing unpredictable effects on subsequent moderator measurements. We are considering repeating the ammonia measurements with fresh moderator vessels built to the same specifications.
- The process we used to fill the moderator vessels with water or boric acid solution without removing them from the target-moderator-reflector assembly did not provide positive assurance that the material in question all made it to the desired location. We are considering various options for checking the moderator fill level remotely for future measurements.
- The absolute normalization of the spectra is done to delivered proton charge. Unfortunately, the proton beam diagnostics are not particularly well suited to the short-pulse low power operational mode we use for moderator characterization, and variations in beam position

and energy could result in significant changes in primary neutron production. We have implemented a separate neutron detector system within the LENS reflector assembly in order to track primary neutron production independently from proton beam delivery.

- The absolute normalization of the spectra further depends on details of the data acquisition system, which was changed between these two measurements (the changes should affect only the proton normalization, not the neutron measurements). We are currently characterizing the data produced with both systems in order to assess differences in potential systematic errors between the two campaigns.

4. Summary

We have measured integral performance parameters of small decoupled ammonia moderators pertinent to scattering kernel validation. In the process of performing these measurements, we have seen inconsistent performance of water moderators used as control samples. The inconsistencies raise questions as to the reliability of these measurements, and thus offer insight into the qualities necessary for robust and reliable moderator characterization measurements:

- Consistent moderator positioning within the reflector assembly.
- Consistent and complete decoupling from the reflector assembly.
- Consistent and monitored moderator fill level.
- Consistent and monitored primary neutron source production.
- Robust moderator vessels immune to the effects of tested moderator materials, and/or characterization of the vessels themselves.
- A well-understood and self-consistent data acquisition system.

We expect that we will repeat this series of moderator characterization measurements, both to test our ideas for improving reliability and reproducibility and to verify that the data we have taken for ammonia are suitable for scattering kernel validation.

Acknowledgments

This material is based upon work supported by the U.S. Department of Energy, Office of Science, Office of Basic Energy Sciences, under contract number DE-AC05-00OR22725.

References

[1] Fluharty R G, Simpson F B, Russell G J and Menzel J H 1969 *Nuclear Science and Engineering* **35** 45–69
[2] Graham K F and Carpenter J M 1972 *Nuclear Science and Engineering* **49** 418–437
[3] Carpenter J M, Iverson E B and Littrell K C 2003 *Journal of Neutron Research* **11** 41–49
[4] Iverson E B, Ferguson P D, Gallmeier F X and Popova I I 2003 *Transactions of the American Nuclear Society* **89** 673–674
[5] Lu W, Gallmeier F X, Ferguson P D, Iverson E B and Popova I 2012 *Proceedings of the Tenth International Meeting on Nuclear Applications of Accelerators—AccApp'11* p 203
[6] Iverson E B, Baxter D V, Muhrer G, Ansell S, Dalgliesh R, Gallmeier F X, Kaiser H and Lu W 2014 *Nuclear Instruments and Methods in Physics Research Section A: Accelerators, Spectrometers, Detectors and Associated Equipment* **762** 31–41
[7] Muhrer G, Schönfeldt T, Iverson E B, Mocko M, Baxter D V, Hügle T, Gallmeier F X and Klinkby E B 2016 *Nuclear Instruments and Methods in Physics Research Section A: Accelerators, Spectrometers, Detectors and Associated Equipment* **830** 454–460
[8] Baxter D V, Ansell S, Ferguson P D, Gallmeier F X, Iverson E B, Kaiser H, Lu W, Muhrer G, Rinckel T C and Steinbach T 2012 *Physics Procedia* **26** 153–160
[9] Baxter D V, Leung J, Kaiser H, Ansell S, Muhrer G, Iverson E B and Ferguson P D 2012 *Physics Procedia* **26** 117–123

IOP Conf. Series: Journal of Physics: Conf. Series **1021** (2018) 012068 doi:10.1088/1742-6596/1021/1/012068

Status update on the design and construction of the Active Cells Facility and Remote Handling Systems

Magnus Göhran, Lennart Åström, Paul Erterius, Srdjan Vareskic, Eldin Mukovic

European Spallation Source ERIC; Lund, Sweden

magnus.gohran@esss.se

Abstract. The Remote Handling Systems at ESS are responsible for making sure that the radioactive waste from the target station monolith operation can, in a safe way for operators and public, be lifted and transported from the monolith vessel to the Active Cells Facility, adequately treated for interim storage and eventually shipped off site. Connected to the target station building, a mock-up and test stand area will be available that will be used for operator training as well as for monolith systems cold testing.

This paper will report on the construction progress of the Active Cells Facility starting with first concrete casting quarter one of 2017. The design concepts and chosen technical solutions of the Active Cells Facility have previously been presented at ICANS XXI (The ESS target station hot cell facility and associated logistics).

The internal casks and associated lifting devices as well as the mock-up and test stand facility are still in preliminary design. The casks are used for monolith maintenance and for transport of activated components from the monolith to the Active Cells Facility. The mock-up is a steel framework structure designed to support operators in training and system testing. This paper includes a status update on these systems as well as discusses safety classification on cranes and lifting equipment dealing with the radioactive monolith components.

1. Introduction

The Remote Handling Systems (RHS) is a project under the target division at ESS responsible for delivering facilities, systems, structures and components intended for maintaining the target station monolith systems. The main deliverables are the Active Cells Facility (ACF), the Internal Casks and Handling Devices (Casks) and the Mock-up and test Facility (MUTS). All systems will be physically located in the target building (D02) and are shown in Figure 1.

The systems are currently at different stages of design where the structural parts of the ACF is the part that is the most advanced at the moment. The Casks and MUTS systems have undergone Preliminary Design Reviews (PDR). These systems will be posted as possible for in-kind contribution and the detailed design will thus ideally be performed by in-kind partners. UKAEA/RACE at the Culham Science Centre/UK is the in-kind partner for the design of the systems within the ACF.

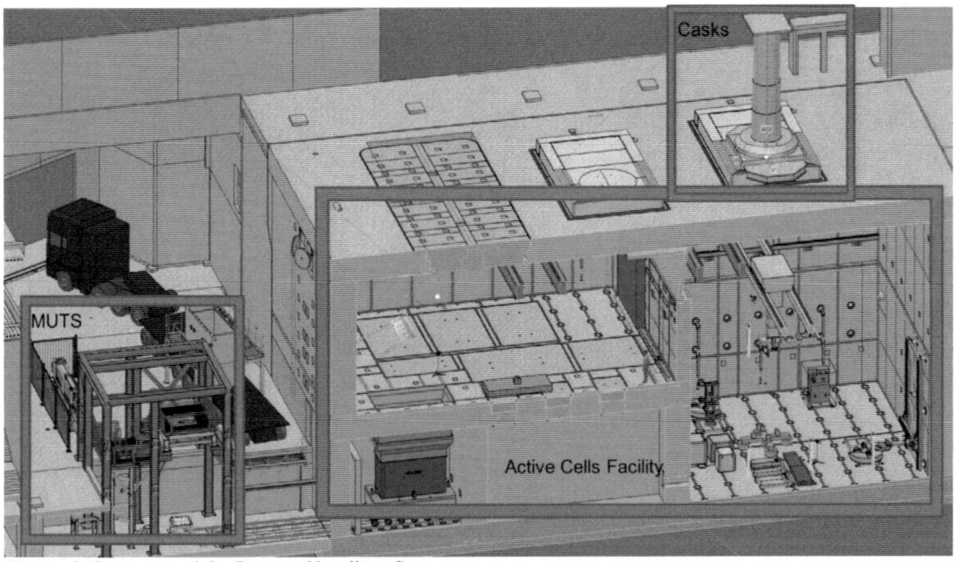

Figure 1. Summary of the Remote Handling Systems.

2. The Active Cells Facility

The construction of the structural parts of the ACF as well as all the cast in items is currently in its final stage. The first 240 m^3 of concrete was poured at the 23rd of March, covering the first 400 mm of the basement slab. Another 1100 mm will be casted in 2 stages where cast in items will be installed at the 400 mm level as well as the 1300 mm level.

2.1. Active Cells Facility – Cast in items

All system specific cast in items (not re-bars) are handled in-house ESS. The main items are liner beams, electrical conduits, 10 inch through wall confinement penetrations as well as anchor plates. The anchors are multipurpose for general use or explicitly designed for a specific task. The design of the cast in items have two major design interfaces, the in-kind partner and the construction company Skanska. The intricate interfaces have to accommodate specific requirements in terms of loads, sizes, tolerances and location in conjunction with the structural design including planning and design of the re-bars and casting sequences. In Figure 2, the installation of the electrical conduits is shown, where the bottom of the Ø114 mm pipe is attached to the top of the first 400 mm casting sequence. Figure 3 is showing a CAD image of the finally installed system where hot- and cold-side junction boxes will be installed and cables will be possible to route between in- and outside of the cells avoiding radiation streaming due to its dogleg through the floor.

In order to facilitate the installation of the 10 inch confinement penetrations as well as the liner beams, Skanska have together with sub-contractors designed a system of in wall steel structures. These structures will be able to support the confinement penetrations as well as other casted in items to fulfil the stringent mechanical tolerances that is required. The structure includes spring loaded rods that will push the liner beams against the formworks during casting in order to get a nice stainless steel surface to weld the liner plates to once the structural design is done. This solution has been iterated and refined through several mock-up tests. The principle is shown in Figure 4 and the final casting mock-up is displayed in Figure 5 where the finished surfaces of the liner beams are visible.

Generally, anchors are placed as generic patterns, both in walls and floors, in order to facilitate a maximum flexibility of in-cell layout and system installation. The ACF will not have any windows and therefore the layout is, from a work perspective, not bound to any specific location in the cells.

Figure 2. Electrical conduits installation with the floor re-bars in place.

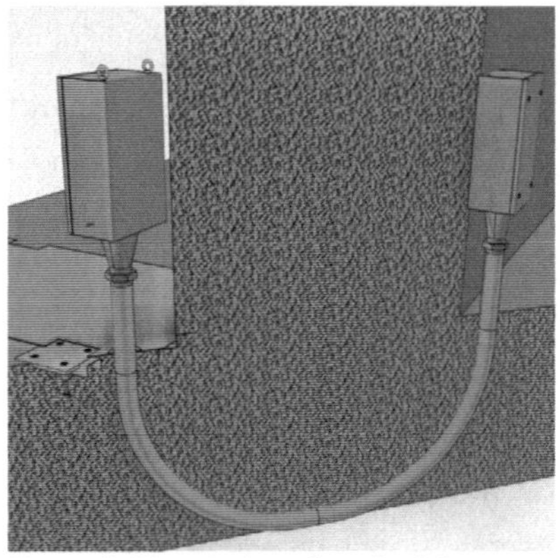

Figure 3. Electrical conduits final design.

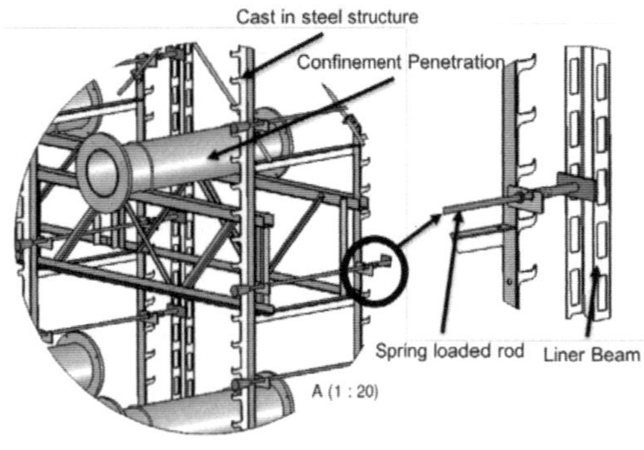

Figure 4. Steel wall structure and systems to accommodate confinement penetration and liner beam cast in items.

Figure 5. Concrete and liner beam mock-up result where the liner beams are visible.

3. Internal Casks

In order to be able to retrieve and transport the radioactive spent components from the monolith operation to the ACF, a set of shielding systems including casks, covers, floor valves and local shielding will form the logistics for these tasks. The main focuses for the design are worker safety and a technically viable solution. The design of the system is currently in a preliminary state where the main focus has been concentrated on getting all requirements defined. A lot of the requirements are derived from the hazard analysis where lifts of the radioactive components has been one of the main drivers for classification of cranes as one example, where the 100 tons high bay crane is rated as a nuclear crane in accordance with Swedish nuclear regulation in order to avoid hazardous risks of drop loads in and around the high bay area. The cask for the target wheel assembly is shown in Figure 6 as one example of how the casks will be designed. The cask for the moderator and reflector assembly will be designed as a single purpose cask as well, then there will be a multipurpose cask for instrumentation plugs and the proton beam window.

The design of the cask system also has specific challenges with respect to shielding. The main casks are designed with their own internal lifting devices and during the lifts, the position of the radiation hazard is changing in relation to where workers are located. Therefore, all angles have to be investigated for sufficient shielding as well as various shine paths along the lifting path. This is true, both for the handling around the monolith as well as when the casks are docked to the ACF and the components are lowered into the cells.

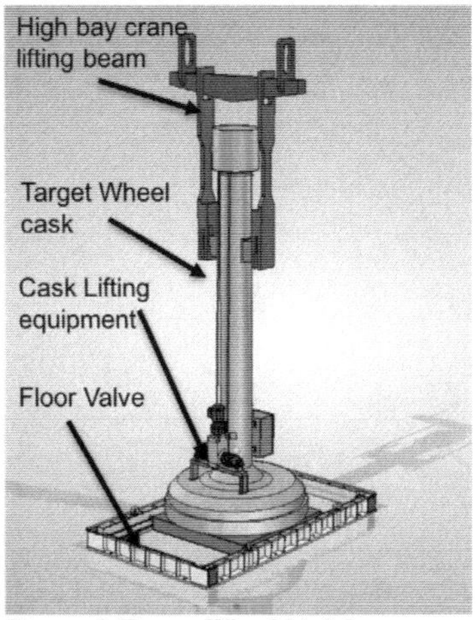

Figure 6. Target Wheel high bay transport cask.

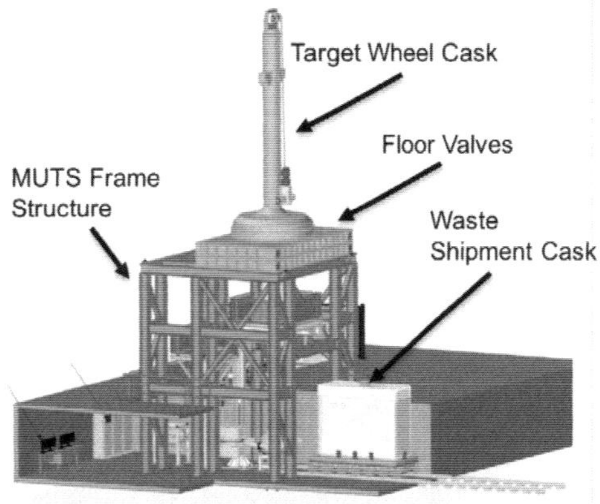

Figure 7. Mock-up and test stand facility within the D02 building.

4. Mock-up and test stands

The MUTS facility will be installed in the back of the transport hall within the target station building. The main functions driving the requirements are focused on the possibility to cold test monolith handling procedures. This include both vertical and horizontal cask operations but it also provides the possibility to cold test components in terms of rotation of the target wheel as one example. As seen in Figure 7, the space allocated is very limited since the waste management have to lift the waste offsite shipment casks from the same area. Also, the high bay crane has a limited coverage towards the short wall in the high bay where the far end of the MUTS only have 50 tons lift capacity.

The primary monolith components facilitated in the stand are the target wheel, the moderator and reflector plug, the proton beam window, the proton beam instrumentation plug and the neutron beam guide inserts. The MUTS facility is currently in a preliminary design state and the aim is that an in-kind partner will be responsible for the detailed design as well as delivery and installation.

Since the construction of the target building is done in conjunction with the component detailed design where the transport hall area of the building is the last part of the facility to be finished, means of installing the MUTS at some other location is currently investigated. This, in order to be able to have the MUTS facility up and running when the monolith components are starting to arrive to ESS. This put other requirements on the construction in terms of being able to disassembly and assembly again in the target building once it is ready and handed over to operations.

5. Conclusion

The various systems within the work package Remote Handling Systems are in various stages of design, stretching from installed at site to preliminary design. Various challenges are addressed spreading from on-site construction interfaces related with installation, to basic design parameters related to requirements of radiation protection around monolith cask operation.

IOP Conf. Series: Journal of Physics: Conf. Series **1021** (2018) 012069 doi:10.1088/1742-6596/1021/1/012069

Lifecycle of the ESS Moderator and Reflector System

M Kickulies[1], Y Beßler[2], Y Lee[1], D Lyngh[1]

[1] European Spallation Source ERIC, P.O. BOX 176, 221 00 Lund, Sweden
[2] Central Institute of Engineering, Electronics and Analytics – Engineering and Technology (ZEA-1) Forschungszentrum Jülich GmbH, Germany

E-mail: marc.kickulies@esss.se

Abstract. The European Spallation Source (ESS) will be a 5 MW class spallation neutron research facility. An important part of the target station is the Moderator and Reflector (MR) System including structure and handling. The primary function of the MR Plugs is to efficiently reflect and moderate fast neutrons from the spallation target to thermal and cold neutrons suitable for the neutron scattering systems. However, the MR System need to fulfil many operational functions as well, which include cooling of radiation heat in liquid and metal bodies, positioning and structural support, capability of handling of active components and confinement, shielding and avoidance of streaming. Depending on the design, the accumulated neutron flux and irradiation induced material degradation mechanism limit the lifetime of the MR Plug. At full beam power of 5 MW, the MR Plugs need to be changed yearly. That imposes high demands on logistics, manufacturing and handling on the system. This paper presents the lifetime criteria of the MR System, in view of radiation induced material degradation. The complete MR System lifecycle is presented, which ranges from purchase of raw materials (aluminium alloy, stainless steel and beryllium), manufacturing, factory acceptance test FAT), transport, delivery control, pre-installation test until a new MR System is ready for installation.

1. Moderator and Reflector (MR) System

The primary function of the Moderator and Reflector System is to reflect and moderate fast neutrons from the spallation target to thermal and cold neutrons suitable for the neutron scattering systems. The MR Plugs are located in close vicinity of the target wheel. One MR Plug is positioned above the target wheel and the other below. The MR System consists of a pre-moderator, a cold moderator, a thermal moderator, a beryllium reflector and the support structure including cooling pipes as shown Figure 1. In order to facilitate replacement of the MR Plugs without interactions with the target wheel, the MR Plugs can be rotated to clear the wheel. They can then be lifted straight up.

2. Motivation

Dependent on the design, the accumulated neutron flux and the different irradiation material effects the lifetime of the MR Plugs is limited. The current ESS design is based on the assumption of exchanging the MR System each year during the summer down time. That puts particular demands on logistics, manufacturing, transport and handling on the system.

IOP Conf. Series: Journal of Physics: Conf. Series **1021** (2018) 012069 doi:10.1088/1742-6596/1021/1/012069

At this early stage the aim is to get an understanding, when it is needed to start final design, purchase and manufacturing of a second MR Systems during initial operation and finally to establish a schedule for sustainable series production of MR Systems during normal ESS operation.

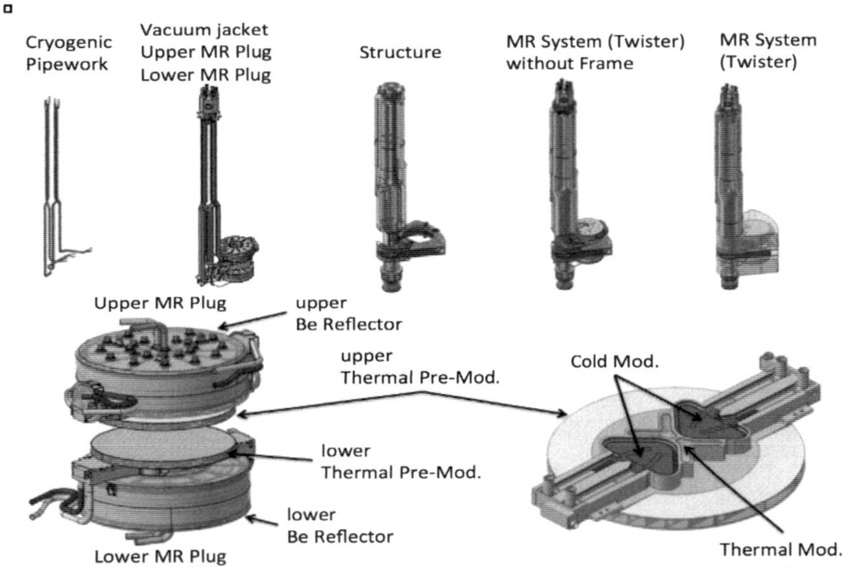

Figure 1. Moderator and Reflector System (Twister) consisting of cold, thermal moderator, beryllium reflector, piping and supporting structure

3. Expected Lifetime of Moderator and Reflector System

The maximum lifetime of the MR System is estimated to be 7000 hours as illustrated in Figure 2. The deterioration leading to the end of lifetime in the MR System is the irradiation damage of the aluminium alloy Al 6061 T6 located close to the target centre. The irradiation damage mechanisms relevant for the aluminium parts of the MR System are thermal neutron induced silicon transmutation caused embrittlement and fast neutron caused radiation damage.

The maximum local thermal neutron flux in the MR Plugs is about $6.8 \cdot 10^4$ $n_{th}/cm^2 \cdot s$, which amounts to the total annual fluence of $1.36 \cdot 10^{22}$ n_{th}/cm^2 for 5400 hours of 5 MW operation at ESS. The green line in Figure 2 is showing the thermal neutron fluence versus time. The maximum value given in the RCC MRx code [1] (A3.2A.32) for negligible irradiation is $1.8 \cdot 10^{22}$ n_{th}/cm^2 @ 20°C. Beyond $1.8 \cdot 10^{22}$ n_{th}/cm^2 rules for irradiation damage must be applied. According to RCC MRx, a maximum fluence of $6.7 \cdot 10^{23}$ n_{th}/cm^2 is allowed. However, the design code focuses on the material degradation in view of thermal neutron induced silicon transmutation caused embrittlement only. Considering that the ESS moderators are exposed to a high level of fast neutron flux, a more stringent lifetime criteria is needed. But, the RCC MRx provides with a robust design guideline, which is complemented with

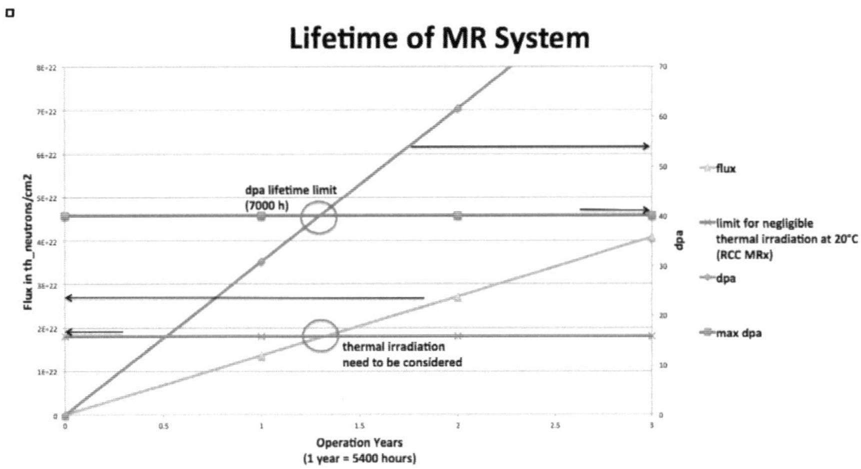

Figure 2. Expected Lifetime of MR System

IOP Conf. Series: Journal of Physics: Conf. Series **1021** (2018) 012069 doi:10.1088/1742-6596/1021/1/012069

additional lifetime criteria based on fast neutron based radiation damage.

As the pre-moderator is located next to the spallation source, the relatively high fast neutron fluences induces a high radiation damage rate of about 28 dpa for 5400 hours of annual operation at 5 MW. For reference, the operating experience at HFIR (High flux isotope reactor), Oak Ridge National Laboratory shows that aluminium alloy Al 6061 T6 is able to withstand up to 40 dpa of irradiation [2]. Taking the nominal lifetime limit of 40 dpa, the ESS water pre-moderator should be replaced every 7000 hours (Figure 2). For the future, as the displacements and the point defects in the aluminium self-anneals at a moderately elevated temperature [3], there is still potential for further relaxation of the dpa-based lifetime limit.

4. Lifecycle Stages of the First Moderator and Reflector System

Until today a long development of the MR System is done. Currently the end of final design and start of manufacturing is reached. The Technical Design Report (TDR) was issued in April 2013 and the MR System ready for installation will be achieved in summer 2018 after 5 years.

The main features of the TDR concept have been: cold and thermal volume moderators and a 25 ton backpack for remote handling of the MR System. After 2 years of extensive iterations the Preliminary Design Review (PDR) was held in June 2015. In order to improve neutronic performance, now the MR System now consists of a butterfly 2 (BF2) flat moderator and a rotating Twister structure for remote handling.

The final design is done by the in-kind partner Forschungszentrum Jülich. With CFX and FEM analysis the outcome of the preliminary design was confirmed and improved. According to manufacturing possibilities and restrictions the design was adapted and detailed. For critical parts or processes mock-ups were built and tests were done. The final outcome of the detailed design ready for manufacturing is shown in Figure 1. The duration of the final design was 16 month for the cold, thermal moderator and reflector. For the Twister the final design lasted 18 + 6 month. 6 month design work was added due to the change to replace the lower MR Plug (still shown in Figure 1) with a water-cooled steel structure.

The make phase consist of material purchase, manufacturing and testing of sub-assemblies and the complete MR System. With the exception of the MR System shaft the manufacturing is done in-house in Forschungszentrum Jülich's own workshops. Because of this, the time to build and test the cold, thermal moderator and reflector can be squeezed to about 6 month and the time to build the complete MR System is about a year. The major procurements for the MR System are the raw materials: aluminium alloy, stainless steel and beryllium. Whereas stainless steel 1.4306 and aluminium Al 6061 T6 are standard materials still with a delivery time of up to 20 weeks, the beryllium with high purity demands is a very special material with limited availability and a time for delivery of about one year.

Testing is done for sub-assemblies and the complete MR System. The final test will be conducted in a factory acceptance test (FAT). Testing comprises: visual inspection, He leak test, dye penetrant test, x-ray examination and pressure test. The cold moderator will even be tested cold with liquid nitrogen, combined with a He leak test before and after the moderator was cryogenic cold. The overall testing is estimated to take about 1 month, assuming that sub-assembly tests have been conducted before.

5. Moderator and Reflector System Replacement

To understand replacement scenarios a very generic sequential schedule for a new built MR System is sketched in Figure 3 based on the experience described above. Without any integration or optimization a new MR System ready for installation could be done: with 6 month final design work, 1 year raw material purchase (due to the long delivery time of beryllium), 1 year of making and testing including FAT and finally 3 month for delivery and pre-installation site testing.

According to the replacement schedule the second MR System is planned for installation in 2024. Already today it is decided to do a major design change. For the second MR Plug the intention is to have a butterfly 1 (BF1) cold moderator design. With the experience from the preliminary and final

IOP Conf. Series: Journal of Physics: Conf. Series **1021** (2018) 012069 doi:10.1088/1742-6596/1021/1/012069

design the change is estimated to take about 6 month of design work before manufacturing can start. The manufacturing time is expected to be the same as for the first MR System, 1 year. Special for the second MR Plug is, that beryllium is available and need not to be purchased. That is due to the late change during construction to remove the lower MR Plug, as the beryllium for the component was already ordered. Under this conditions the second MR System could be built in about 18 to 20 month depending on the purchase and delivery time of aluminium alloy and stainless steel. That would mean the concept to go for final design must be decided latest in December 2022.

The third MR System is planed for summer 2025. That means after one year of operation of the second MR System exposed to irradiation damage corresponding to full 5 MW proton beam. But, even without any design work, the purchase of the beryllium for the third MR System must be started in July 2023, long before the second MR system is installed. About 6 month may be saved if machining would start without having beryllium delivered.

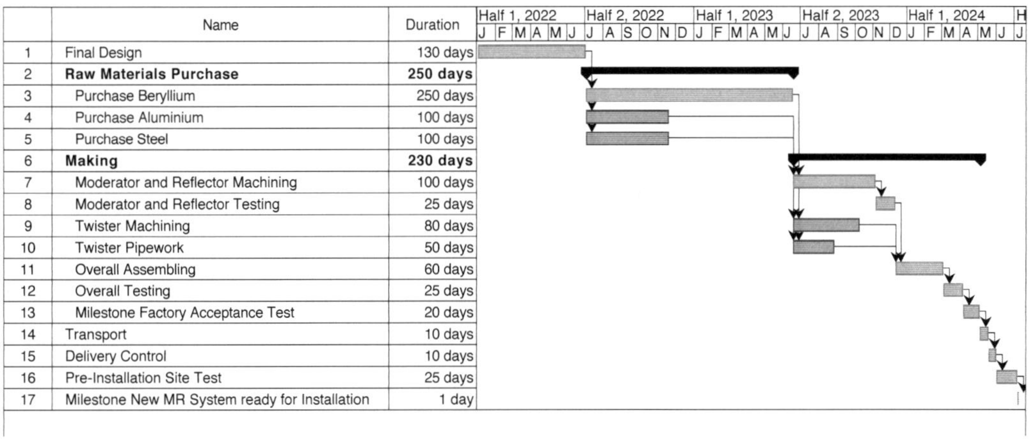

Figure 3. Sequential Schedule for MR System Replacement

6. Conclusions

Still there are possibilities for time optimization for MR System replacement, but it can be concluded that the yearly replacement of MR Systems is very challenging because of beryllium delivery time, machining time and workshop capacities. Furthermore, time for design changes is very limited or need to be done before operation experiences are available.

Regarding the beryllium purchase it seems to be necessary to develop a longer-term strategy to have beryllium available but not restrict flexibility for design changes. The high cost of beryllium needs to be considered as well. The manufacturing workshop needs to establish a concept for series production for MR System manufacturing and testing.

References

[1] Afcen RCC-MRx Code 2012
[2] W. Lu, P. D. Ferguson, E. B. Iverson, F. X. Gallmeier, and I. Popova 2008 Moderator poison design and burn-up calculations at the SNS *Journal of Nuclear Materials, 377:268–274*
[3] K. Farrell 1999 Materials selection for the hfir cold neutron source *Technical Report ORNL/TM-99-208, ORNL*

IOP Conf. Series: Journal of Physics: Conf. Series **1021** (2018) 012070 doi:10.1088/1742-6596/1021/1/012070

Beam power nonlinearity: twice the power, but not twice the neutrons?

T. Hügle, E. B. Iverson, F. X. Gallmeier

Instrument and Source Division, Oak Ridge National Laboratory, Oak Ridge, TN 37831, USA

Email: hueglet@ornl.gov

Abstract. Over the course of SNS operations, the neutronics team has observed indicators of a non-linear relationship between proton beam power and neutron brightness on multiple beamlines. As we prepare to replace the inner reflector plug, we have conducted our largest and most complete tests of this non-linearity in dedicated experiments. We will discuss our findings and subsequent experiments, their impact and our investigation of possible explanations for this behaviour.

1. Introduction

It would be reasonable to assume that increasing the beam power of a spallation neutron source by increasing the proton current by a certain percentage would result in a neutron production increase by the same percentage. It was therefore troubling to find indicators based on several individual measurements that the target/moderator system at SNS exhibits nonlinearity between beam power and resulting neutron yield.[1] Recent dedicated facility wide measurements confirm this for certain moderators.

1.1. Metrics

In order to measure the impact of changes to the moderator system, beam monitor data as well as instrument detectors were used. It is useful to define a metric "neutron yield", which is "detector counts per proton charge delivered to the target". In an ideal case, the neutron yield would be constant, meaning that e.g. doubling the beam power results in twice as many neutrons being delivered to the user instruments.

$$Y_{id} = \frac{instrument\ detector\ counts}{proton\ charge} \sim \frac{count\ rate}{beam\ power}$$

$$Y_{bm} = \frac{beam\ monitor\ counts}{proton\ charge} \sim \frac{count\ rate}{beam\ power}$$

This manuscript has been authored by UT-Battelle, LLC under Contract No. DE-AC05-00OR22725 with the U.S. Department of Energy. The United States Government retains and the publisher, by accepting the article for publication, acknowledges that the United States Government retains a non-exclusive, paid-up, irrevocable, worldwide license to publish or reproduce the published form of this manuscript, or allow others to do so, for United States Government purposes. The Department of Energy will provide public access to these results of federally sponsored research in accordance with the DOE Public Access Plan (http://energy.gov/downloads/doe-public-access-plan).

IOP Conf. Series: Journal of Physics: Conf. Series **1021** (2018) 012070 doi:10.1088/1742-6596/1021/1/012070

The two neutron yields each are sensitive to different parameters and serve different purposes: The instrument detector counts in Y_{id} differ drastically from experiment to experiment and depend on extrinsic effects like the sample in place, the exact sample location and a wide variety of instrument settings such as slit settings and chopper configuration. At the same time, the instrument detectors offer high precision, little intrinsic effects like dead time, and (data being collected in event mode) the ability to reconstruct a finely binned time line.

Beam monitors, on the other hand, are usually built into the flight path far enough upstream to be unaffected by most instrument changes. However, they often suffer from dead time effects, which can introduce a systematic error at increasing beam powers, and usually collect data only as histograms, which means the finest time resolution available is the duration of the experiment.

Therefore, Y_{id} is best used for limited time dedicated experiments where the instrument is left at a certain configuration and the source characteristics (proton beam power, moderator temperature ...) are modified, whereas Y_{bm} is best used to monitor moderator health in the background during ongoing operations on a timescale that stretches months or even years. Attention has to be paid in this case to either comparing only data collected at identical power levels, or to develop a reliable calibration method to adjust for possible dead time issues.

Neither Y_{id} nor Y_{bm} can account for beam line components such as neutron guides, and both are convoluted with a wide range of parameters like detector efficiency, bandwidth selection and instrument background. They can therefore not be expected to deliver absolute neutron production numbers. However, it is reasonable to assume that –within the boundaries just discussed- changes in these metrics reflect changes in the target/moderator performance itself.

2. Experimental setup

In order to measure the impact of proton beam characteristics on neutron yield, we decided to perform a series of dedicated experiments using the neutron scattering instrument suite at SNS as detectors. For that, we chose one or two instruments per moderator surface, and asked the instrument scientists to illuminate the instrument detectors. In most cases, this was achieved by putting a piece of Vanadium into the sample position. Data was then taken continuously in event mode or, if this was not feasible, in five minute increments for the duration of the experiment.

2.1. Experiment 1: beam power dependence

These were tests to investigate any direct beam power dependence on moderator performance. To this aim, the beam power was first ramped up continuously to 450 kW, and following that increased in 30 kW steps (~20 minutes each) up to the final power (1 MW). The accelerator protocol used for this sequence is identical to the one used for foil conditioning routines.

2.1.1. Decoupled poisoned Hydrogen moderator.

The neutron yield of the decoupled poisoned liquid Hydrogen moderator shows strong power dependence. As the beam power was increased, the total number of neutrons detected in the TOPAZ instrument (BL12) decreased (Graph 1a). The same holds true for the opposite side of the moderator (not pictured). The decline seems to proceed in a linear fashion. Upon further analysis, it can be seen that the changes are wavelength dependent (Graph 1b). While the neutron yield in the wavelength range from 1.0 to 1.2 Å declines by approximately 10% when the proton beam power is increased from 100 kW to 1 MW, neutron yields at ~2 Å and ~3 Å fall by closer to 15%. The ramifications of this change in the neutron spectrum are significant: calibration measurements at one beam power are not necessarily valid at other power levels any more.

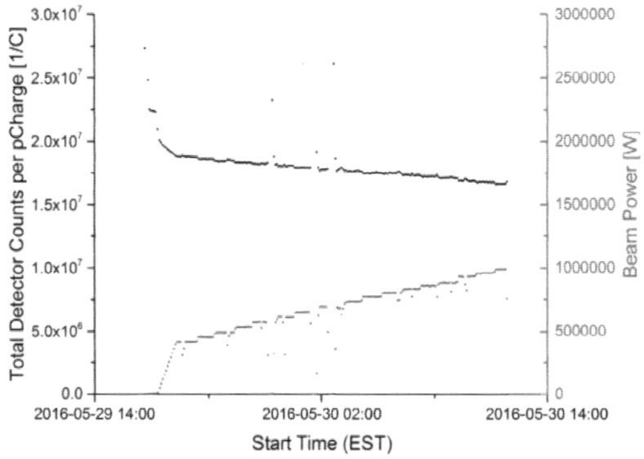

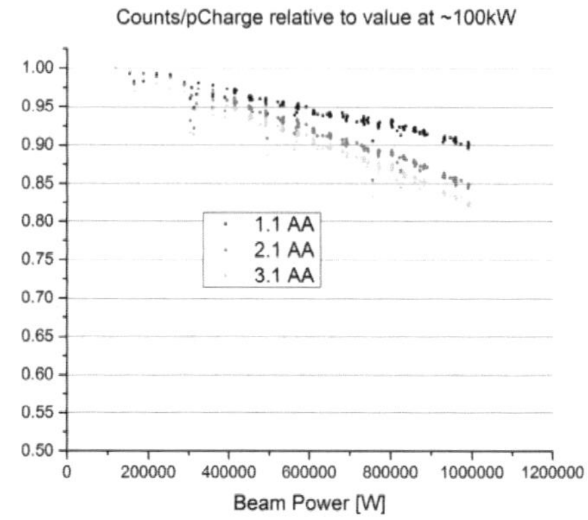

Figure 1. a) (left) Development of neutron yield (red) at TOPAZ during Proton beam power ramp up (black). b) (right) Change of neutron yield at various wavelength bands (100% = neutron yield at 100kW): black: 1.0 Å to 1.3 Å, red: 2.0 to 2.2 Å, green: 3.0 to 3.2 Å.

2.1.2. Ambient water (thick side). An example of a moderator not influenced by proton beam power can be seen in Figure 2. The ARCS instrument on the thick side of the water moderator does not see substantial changes in neutron yield over the course of a 1 MW ramp up. Apparent changes can be seen at low beam powers. Possible explanations for this might be a lack of background subtraction (artificially increasing the detector counts encountered per proton charge), or errors due to the fact that any effective beam power during a rapid ramp up such as here is necessarily averaged.

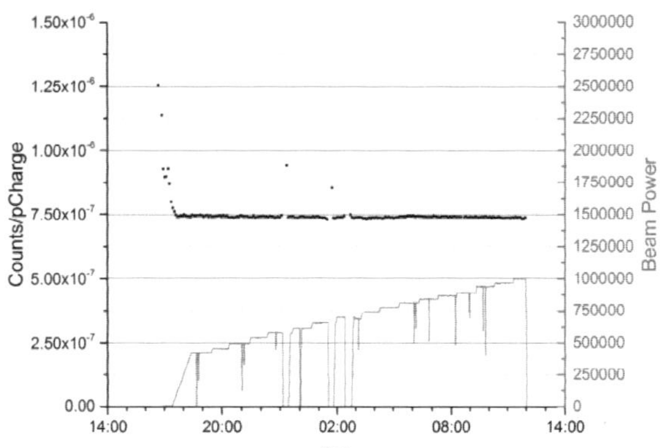

Figure 2. Development of neutron yield at ARCS (black) during power ramp up (red).

2.2. Experiment 2: investigation of repetition rate

To establish whether the moderator changes described above are a result of time averaged power or of the power per individual proton pulse, an experiment changing accelerator repetition rate was conducted. Several beam power levels were reached by adjusting beam current and repetition rate inversely. This led to the accelerator settings of e.g. 500kW@60Hz and 500kW@30Hz, where the latter has proton pulses twice as intense, but only half as often as the former.

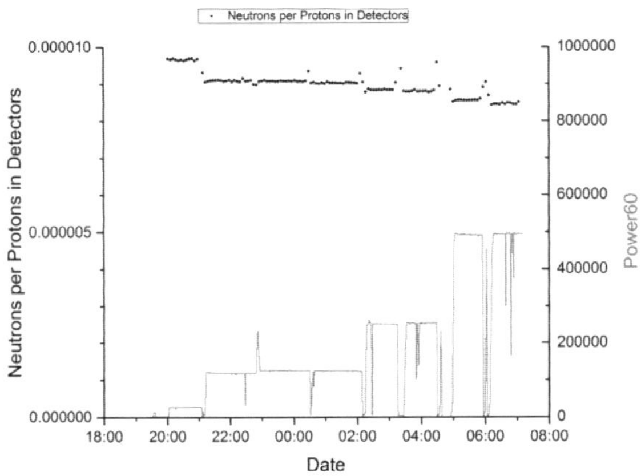

Figure 3. Development of neutron yield at TOPAZ (black) as a function of beam power and repetition rate (red). The individual beam settings (~1h each) were (in order): 125kW@15Hz, 125kW@30Hz, 125kW@60Hz, 250kW@30Hz, 250kW@60Hz, 500kW@30Hz, 500kW@60Hz.

As can be seen in Fig 3, the repetition rate has no influence on neutron yield at low beam powers. At 500kW, a small difference in neutron yields can be seen: 500kW@30Hz produces slightly more neutrons than 500kW@60Hz. This would mean that a more evenly distributed proton beam delivery has a larger detrimental effect than individual, highly intense pulses. However, the substantial differences in proton pulses between these two settings required comparably large changes in proton beam footprint to prevent damage to the target, which might have caused changes in the neutronics characteristics of the overall target/moderator system. More experiments focused on keeping the proton beam footprint consistent will be necessary. However, the extent of the changes suggest that the primary source of the nonlinearity effects is the average bam power, with possibly a secondary effect of pulse intensity.

3. Possible explanations

Several factors might influence moderator performance of a hydrogen system in a radiation field:

- Changes in ortho/para ratio
- Changes to temperature, especially through formation of 'heat pockets' by inconsistent flow patterns
- Changes in density, e.g. by the shockwave of the pulse rippling through the moderator at precisely the time that moderation occurs

Ortho/para ratio and temperature are both mostly dependent on average beam power, whereas density effects would likely be on the time scale of individual pulses. Calculations by Iverson suggest that the effects we see would be consistent with a hydrogen concentration of at least 30% ortho-H_2. Further insight will be gained upon installation of future generations of IRPs, which will offer improved flow patterns in the decoupled/poisoned hydrogen moderator and the incorporation of an ortho/para converter system.

This material is based upon work supported by the U.S. Department of Energy, Office of Science, Office of Basic Energy Sciences under contract number DE-AC05-00OR22725.
This research used resources at the Spallation Neutron Source, a DOE Office of Science User Facility operated by the Oak Ridge National Laboratory.

References

[1] E. B. Iverson, F. X. Gallmeier, W. Lu T. C. McClanahan, M. B. R. Smith, *Measurements of Neutron Beam Performance at the Spallation Neutron Source,* American conference on neutron scattering 2014, Knoxville, USA

IOP Conf. Series: Journal of Physics: Conf. Series **1021** (2018) 012071 doi:10.1088/1742-6596/1021/1/012071

Neutron poison burnout and effects on SNS moderator performance

Franz X. Gallmeier, Wei Lu, Erik B. Iverson

Instrument and Source Division, Oak Ridge National Laboratory, Oak Ridge, TN 37831, USA

Email: gallmeierfz@ornl.gov

Abstract. Detailed neutron absorber burnup calculations were performed for the poison plate and decoupler layers of the decoupled ambient water and liquid hydrogen moderators of SNS. Although local poison burnout was identified for the water moderator at 32 GWh and local decoupler burnout was becomes apparent at the hydrogen moderator at 37 GWh delivered proton beam energy, both moderators see major impacts to the moderator performance by increases of pulse widths and spectral changes only past the stated burnout energies.

1. Introduction

The Spallation Neutron Source (SNS) produces pulsed neutron beams for material research and fundamental physics applications based on a proton beam delivering sub-microsecond pulses into a mercury target at 60 Hz repetition rate and 1.4 MW design beam power. The fast neutrons originating in the spallation reactions in the target are moderated by four ambient water and liquid hydrogen moderators placed in close proximity of the target in the so-called inner reflector plug (IRP). To generate sharp neutron pulses for high resolution measurements, the water moderator and one of the three liquid hydrogen moderators are equipped with a gadolinium poison plate in the center and with a decoupler cadmium layers absorbing thermal neutrons arriving from the reflector.

High-neutron-flux exposure depletes the poison and decoupler materials in those isotopes with high neutron absorption cross section such that the pulse shaping ability is diminished over time. Indeed, the burnout of decoupler and poison is the lifetime limiting effect of the IRP containing the moderators. At the design, the thicknesses of the decoupler and poison layers were chosen to guarantee an IRP life of 30 GWh delivered proton beam on target as estimated by calculations of neutron absorption rates and the criterion that at end-of-life we wished to have a thermal neutron blocking fraction of at least 0.80. Burnout simulations of the poison plates by Lu confirmed the 30 GWh life time[1].

This manuscript has been authored by UT-Battelle, LLC under Contract No. DE-AC05-00OR22725 with the U.S. Department of Energy. The United States Government retains and the publisher, by accepting the article for publication, acknowledges that the United States Government retains a non-exclusive, paid-up, irrevocable, world-wide license to publish or reproduce the published form of this manuscript, or allow others to do so, for United States Government purposes. The Department of Energy will provide public access to these results of federally sponsored research in accordance with the DOE Public Access Plan (http://energy.gov/downloads/doe-public-access-plan).

IOP Conf. Series: Journal of Physics: Conf. Series **1021** (2018) 012071 doi:10.1088/1742-6596/1021/1/012071

Several reasons motivated us to revisit the IRP life time predictions. Lu performed his depletion analyses with the CINDER'90 code [2] fed with neutron fluxes provided by MCNPX [3] calculations for a layered gadolinium plate sliced vertically into 1-2 cm wide stripes.

As local variations of the neutron flux persist across the width of the moderator, neglecting this variation underpredicts the rate of burnup. On the other side, the use of CINDER'90 activation cross section in for the burnup calculation has found to overpredict the neutron absorption in the resonance region because of multi-group representation, and hence overpredict the burnout. Additionally, Lu did not include the burnout of the decoupler material in his analyses. Over the course of burnout, the neutron fluxes in the moderator increase by as much as 20% and hence also speedup the burnup over the IRP life. Also the decoupler may burn out locally before the poison does due to large variations of thermal neutron fluxes in the reflector peaking close to the target in the areas between neighboring moderators. Finally, we want to investigate the evolution of moderator performance, the growth rate of brightness and the changes to the pulse shapes due to poison and decoupler burnup to arrive at a more realistic criterion to establish the end-of-life of the IRP.

As of the time of the ICANS-XXII conference, SNS will approach 35 GWh delivered beam on target and we had plans to start the replacement of the present IRP by an identical replacement component. Because of manufacturing problems with the replacement IRP, the changeout will have to deferred to the end of calendar year 2017 by the time SNS will have accumulated 41 GWh. The accumulated information was essential for making decisions about the feasibility of continuing operations to this time, and about the vulnerability of the various neutron instruments in the SNS science program.

2. Methods of calculation

The neutron absorption property in thermal neutron fields and hence the ability of functioning as neutron poison in and around the moderators is dominated by the isotope ^{113}Cd in the decoupler and by the isotopes ^{155}Gd and ^{157}Gd in the poison plates. All three isotopes are transmuted into stable isotopes. Also, the feed-in into these three isotopes is driven from stable isotopes with lower neutron numbers. For these reasons, involving a transmutation code solving complex systems of decay chains is not required to track the evolution of isotopic compositions of the poison and decoupler zones. The equations governing the nuclide concentrations ${}_Z^AN$ with time t are easily established from isotope production rates ${}_Z^AP$ as

$$ {}_Z^AN(t) = {}_Z^AN(0) + \left({}^{A-1}_{Z}P - {}^{A+1}_{Z}P\right) * t $$

where

$$ {}_Z^AP = {}^{A-1}_{Z}N \int {}^{A-1}_{Z}\sigma(E)\, \phi(E)\, dE $$

with the absorption cross section ${}_Z^A\sigma(E)$ for isotope of mass number A and charge number Z and neutron flux $\phi(E)$ varying with energy E. The isotope production rates are directly calculated by a transport calculation of neutron flux buildup due to the proton beam impact in the SNS target station for a given material composition of poison and decoupler materials.

The time evolution was calculated by a multi time-step Runge-Kutta predictor-corrector method with time steps of 1000-hours length. Numerical problems expressed in large fluctuations in the number densities of the leading isotopes are found when burnout of these isotopes is reached. These are easily dampened by forcing the number density of the leading isotopes to arrive in an equilibrium such the production rate from minor isotopes balances their destruction rate:

$$ \frac{d\,{}_Z^AN}{dt} = -{}^{A+1}_{Z}P + {}_Z^AP = 0 $$

$$ {}_Z^AN = {}^{A-1}_{Z}N \int {}^{A-1}_{Z}\sigma(E)\, \phi(E)\, dE \Big/ \int {}_Z^A\sigma(E)\, \phi(E)\, dE $$

All the analysis was cast into a PERL script, which obtains production rates by MCNPX calculations, performs the transmutation calculations for the time steps, establishes revised material compositions for the next round of calculations of production rates and repeats the loop to a requested value of delivered beam energy on target.
The calculations of isotope production rates were based on a detailed as-built model. All poison and decoupler layers with typical thicknesses of one millimeter were sublayered into 5 zones and split into segments of about 2x2 cm^2 in total 7x8x5=280 pieces. Likewise, the decoupler zones were split into pieces to allow to account for local variations of burnup. The decoupler layer is flame-sprayed on the vacuum vessel and water vessel of the hydrogen and water moderators, respectively, except for the view ports on the front and back faces of the moderators and extended for 50 cm up the supply piping. All decoupler is facing a thin water cooling channel followed by bulk beryllium reflector, except for the hydrogen moderator bottom and water moderator top that are set into a water-cooled aluminum structure towards the target, and a side edge of the hydrogen moderator, which faces a 40-cm thick water column integrated into the neighboring coupled hydrogen moderator. In order to consider uneven decoupler burnup effects of zones facing thicker water zones at the hydrogen moderator, the decoupler was subdivided not only around the perimeter but also along the poison plate plane in BL11-side and BL2-side zones.
At several time steps, detailed moderator performance analyses were conducted for a large number of proton histories to arrive at converged results for moderator brightness spectra, and pulse shape analyses.

3. Results of poison and decoupler burnup calculations

The poison plates and decoupler layers burn out the leading neutron-absorbing isotopes very unevenly as shown in Figures 1 and 2. The hydrogen moderator experiences burnout of the decoupler at the BL11 side at the bottom of the moderator pointing towards the coupled moderator at around 37 GWh delivered proton beam. At his time, the its poison plate has 7% of the neutron poison left in its fastest-burning location and would last for 3 GWh longer.

For the water moderator, the burnout is reached first in the top third of the poison plate at 32 GWh (see Figure 2). Its decoupler would hold up for 11 GWh longer before springing a neutron leak.

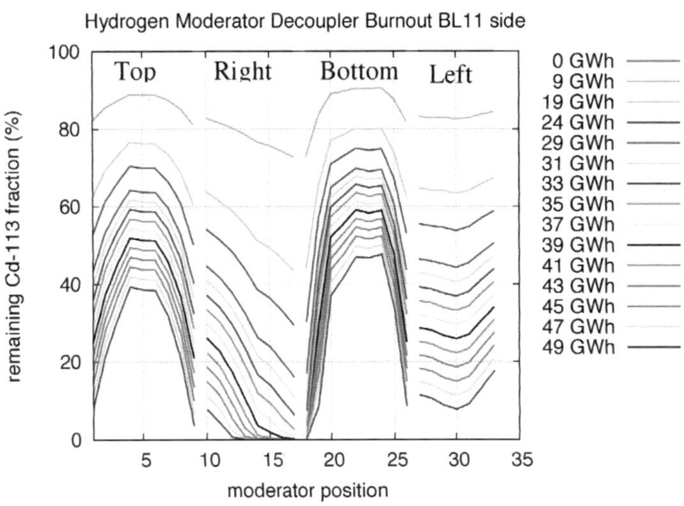

Figure. 1: Evolution of local loading of ^{113}Cd in the decoupler layer in percentage of initial loading of the decoupled poisoned hydrogen moderator at various stages of delivered beam energy on target with the moderator positions running around the circumference of the moderator.

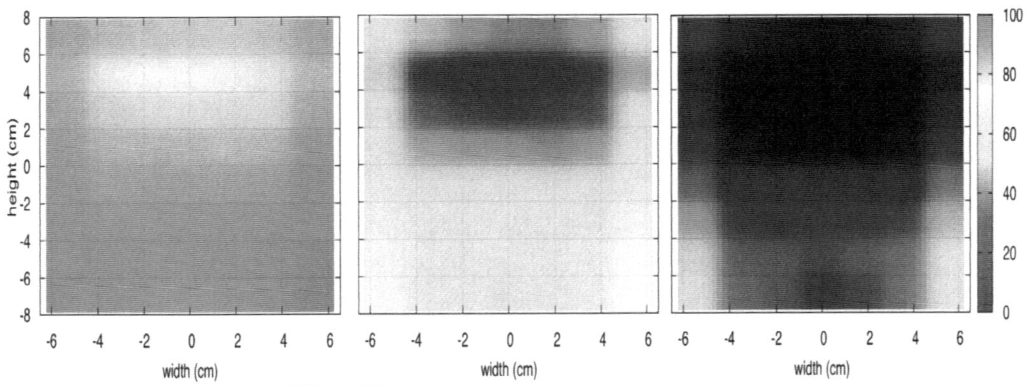

Figure 2: Local loading of $^{155}Gd+^{157}Gd$ in the poison plate of the water moderator in percent of initial loading: 9 GWh-left, 25 GWh-center, 40 GWh-right. The moderator is positioned at the bottom of the target.

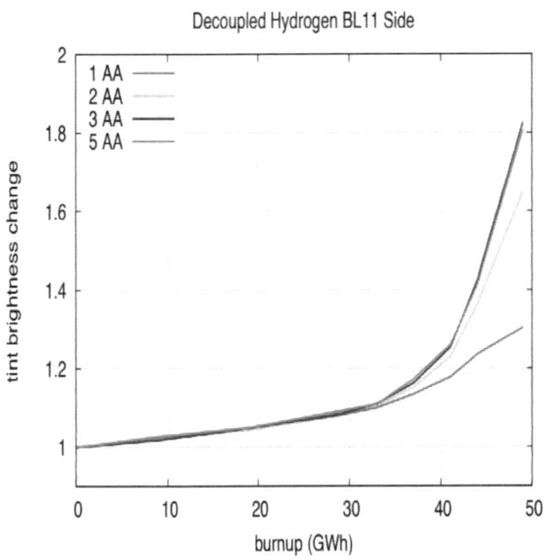

Figure 3: Change in time-integrated pulse brightness of the hydrogen moderator versus delivered proton beam energy as viewed from the BL11 side.

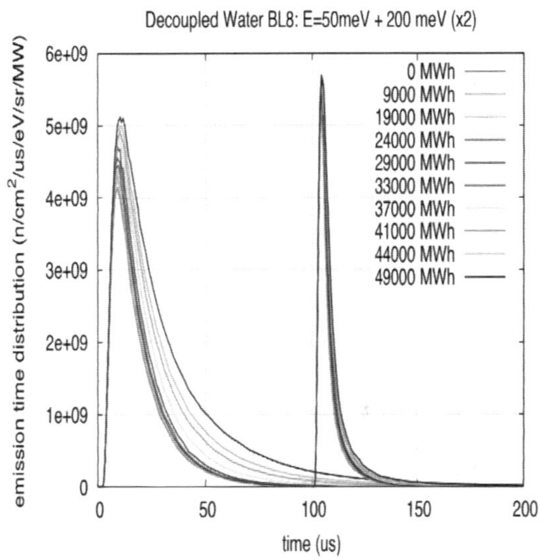

Figure 4: Neutron emission time distribution of water moderator at 50 and 200 meV evolving with delivered proton beam energy as viewed from the BL8 side.

4. Impact of poison and decoupler burnup on moderator performance

With changes of the isotopic composition of the poison and decoupler due to burnup, the moderating processes, pulse-shaping effects, and the moderator performance change with time. The effect on moderator performance was analysed by performing longer transport calculations at various steps of the IRP life with decoupler and poison material compositions resembling the state of burnup of the requested time. A steady increase over integrated delivered beam energy is experienced up to about 15% increase, and much steeper increase thereafter for both the water and hydrogen moderators as shown in Fig. 3. The break-away points signal holes in poison or decoupler. The BL17 side of the water moderator with 25 mm water depth to the poison plate is much less sensitive than the BL8 side with 15 mm depth.

The emission time distributions of the pulses as shown in Figure 4 for the BL8 side of the water moderator for 50 and 200 meV (offset by 100 μs) neutron energies show that the increase in time-integrated brightness after the break-away point pile in not so much in the pulse peak but rather in the tail of the distribution.

As the scheduled IRP replacement in March 2017 (at 37 GWh) has been postponed to beginning 2018 due to late delivery of the replacement IRP, the predicted moderator performance data and confirming performance measurements let us confidently extend the IRP life to 41 GWh.

Acknowledgement
This material is based upon work supported by the U.S. Department of Energy, Office of Science, Office of Basic Energy Sciences, under contract number DE-AC05-00OR22725.

References

[1] Lu, W., Ferguson, P.D., Iverson, E.B., Gallmeier, F.X. and Popova, I., Journal of Nuclear Materials, Vol. 377(1), pp. 268-274 (2008).
[2] W. L. Wilson, T. R. England, K. A. Van Riper, Proceedings of the Fourth Workshop on Simulating Accelerator Radiation Environments (SARE4), Knoxville TN, Sept. 14-16, 1998.
[3] D. B. Pelowitz, MCNPX User's Manual, Version 2.6.0, Report LA-CP-07-1473, Los Alamos National Laboratory, Los Alamos, New Mexico, 2008.

IOP Conf. Series: Journal of Physics: Conf. Series **1021** (2018) 012072 doi:10.1088/1742-6596/1021/1/012072

Target test facility for ADS and cross-section experiment in J-PARC

Shin-ichiro Meigo, Hiroki Iwamoto, Hiroki Matsuda, and Hayanori Takei

Nuclear Transmutation Division Facility and Application Development Section J-PARC/JAEA, Tokai-mura, Naka-gun, Ibaraki, 319-1118

Materials and Life Science Division Neutron Source Section J-PARC/JAEA, Tokai-mura, Naka-gun, Ibaraki, 319-1118, Japan

E-mail: meigo.shinichiro@jaea.go

Abstract. For the development of Accelerator Driven System (ADS), a Transmutation Experimental Facility (TEF) is planned to be built in J-PARC. In this facility, negative hydrogen (H^-) beam with a power of 250 kW will be delivered to a Lead-Bismuth Eutectic (LBE) target, which is one of the powerful candidates target for the ADS, placed at ADS Target Test Facility (TEF-T). On the beam transport to TEF-T, laser charge exchange (LCE) technique will be utilized to Transmutation Physics Experimental Facility (TEF-P) to deliver a very low-intensity beam such as 10 W with high stability from high-intensity H^- beam. To examine this technique, we carried out the experiment at the RFQ test bench. By using the LCE, very stable short pulse beam extraction has been demonstrated for a long duration. We have started the development of beam monitor for the TEF-T by using heavy ion to give high DPA rate. Some experiments for validation of code and nuclear data are planned in J-PARC to obtain nuclear data for the ADS.

1. introduction

For the development of Accelerator Driven System (ADS) to the mitigate hazard of long-life nuclides produced in a nuclear reactor, especially for minor actinide (MA) such as neptunium and americium, a Transmutation Experimental Facility (TEF) shown in Fig. 1 is planned to be built in J-PARC. In this facility, negative hydrogen (H^-) beam with a power of 250 kW and repetition rate of 25 Hz, which is accelerated in a 400-MeV LINAC, will be delivered to a Lead-Bismuth Eutectic (LBE) target. The LBE is one of the candidate targets for the ADS, placed at ADS Target Test (TEF-T). In the TEF-T, multi-purpose use of the facility is planned. To perform experiments such as examination of

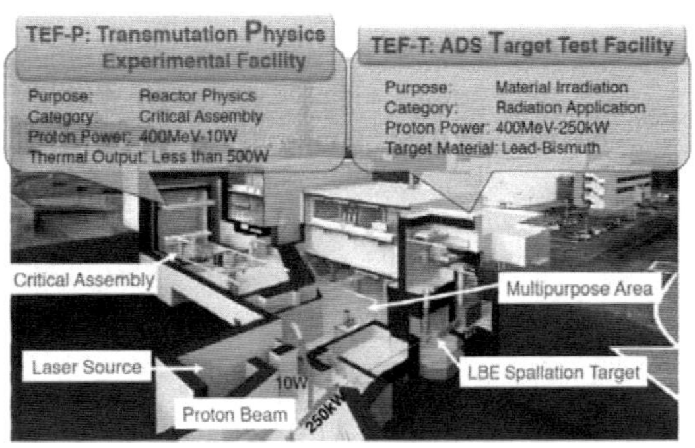

Figure 1. Schematic drawing of Transmutation Experimental Facility (TEF) in J-PARC.

semiconductor devices, spallation neutrons produced at the LBE target will be provided to the experimental neutron hall at TEF-T. At the TEF-T, Isotope Separator On-Line (ISOL) is planned to be built as a candidate for multi-purpose use in the second target station.For a study of reactor physics with spallation neutrons, a Transmutation Physics Experimental Facility (TEF-P) is planned to be built in TEF, which is composed of an MA core with effective neutron multiplication factor (k_{eff}) of 0.95.

To transform significant amounts of actinide nuclides, the accelerator with a beam power of 20 MW as a candidate with extremely high reliability is considered in the design of JAEA. For beam injection to TEF-P with a very low intensity such as 10 W from the high-intensity H^- beam (250 kW), a reliable beam extraction system is required. To realize the ADS, further neutronics studies and reaction cross-sections of reaction are required. In this paper, the present status of the design of the TEF and the research and development for TEF and ADS are described.

2. TEF-T and TEF-P

In the Japan Proton Accelerator Research Complex (J-PARC), a MW-class pulsed neutron source, the Japan Spallation Neutron Source (JSNS), and the Muon Science facility (MUSE) were deployed in the Materials and Life Science Experimental Facility (MLF). In the present beam operation, a LINAC provides 400 MeV H^- beam at 25 Hz to a Rapid Cycling Synchrotron (RCS). After upgrade of the LINAC to deliver beam to TEF, the repetition rate will be increased from 25 Hz to 50 Hz. By adding extraction device into the LINAC, an H^- beam of 25 Hz is derived to TEF-T. Preliminary design of the TEF-T has been completed and is described in a Technical design report[1].

3. Research and development for TEF-T and ADS

For TEF-T and ADS, beam extraction system has been developed. To validate the calculation code and the nuclear data used in the neutronics design for ADS, we have carried experiments in J-PARC. In this section, research and development of a beam extraction system based on laser charge exchange and some experiments to obtain cross-section are described.

3.1. Beam extraction based on laser charge exchange

On the beam transport to the TEF-T, the beam extraction technique involves a very low-intensity beam such as 10 W taken from the high-intensity 250-kW H^- beam[2]. One electron of the H^- ions will be detached by photons of Nd-YAG laser to produce neutral hydrogen beam (H^0). Following a thin foil, the H^0 stripped to H^+ so that eventually protons are introduced into the TEF-T. This technique satisfies the requirement for the safety demand for beam separation. It should be noted that the maximum beam power is physically determined. An alternative extraction technique using a thin wire and magnets could produce an unexpected beam power so that these techniques can not be applied at the TEF. Also, an abnormal vacuum level in the accelerator may cause unintentional beam extraction. Therefore, the interaction point is placed in the middle of bending magnet, which prevents unexpected beam extraction by minimizing the straight length in the interaction section.

To demonstrate this technique, we carried out an experiment by using a 3-MeV RFQ placed on the test bench of the LINAC building by using experimental setup [3] shown in Fig. 2. The interaction point of the laser was put in the center of the dipole magnet, which bends the H^- beam by 23°. Consequently, the charge exchanged beam by the laser is delivered at 11.5°, which is the half angle of bending magnet. At 11.5° direction, we placed a thin foil to strip the remaining electron from the H^0 ions and also deployed a beam position monitor (BPM), a slow current transformer (SCT), and a Faraday cup (FC) to observe the extracted beam position and power. The length between the laser oscillator, which is Nd-YAG with 1.6 J/shot at 25 Hz

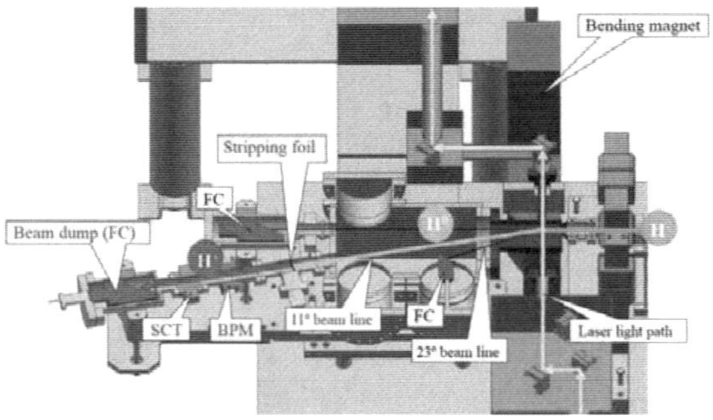

Figure 2. Laser charge exchange experiment using 3-MeV H^- beam at test-stand RFQ.

having a length of 6 ns (σ), and the interaction point is chosen to be similar to the TEF-T condition.

At the extraction channel, an H^+ beam was observed as expected, which showed an efficiency of the charge exchange as 100% during the laser injection. These signals show, the laser light time structure having 6 ns in 1 σ and the signals of Faraday cup, the BPM and the SCT. From the signal of the SC, the extracted beam power was 35 mW at 3MeV. A stable beam extraction with a fluctuation of intensity less than 2% was demonstrated for a duration longer than 7 days. At the TEF-T, 400 MeV will be utilized so that the extracted power was interpreted to be 7.7 W, which satisfied the requirements of the TEF-T. It should be noted that the present successful result means that one of the significant milestones for the TEF construction is achieved.

The current experimental result showed the capability of a small pulse in a macro pulse with the length of 0.5 ms. Using the LCE with a long length laser loop, which makes many interaction events in the macro pulse, many bunches of the beam with short pulse length about several ns with relative high repetition can be created. Those pulses may be used for a spallation neutron source with high repetition rate with short pulse length by introducing an additional target placed in the TEF-T. For the development of the ADS, neutron data around 1 MeV, which has a peak in fission and spallation neutron production, are required. Therefore, the repetition rate of the laser applies to 1 MHz, which gives the length of laser loop 300 m. Consequently, the beam power will be 700 W. Owing to the short flight path of the neutrons to the experimental hall, a relatively high-intensity neutron beam can be obtained. By using a Time-of-flight (TOF) technique, neutron experiments with high energy resolution can be performed to obtain cross section for the ADS.

3.2. Beam profile monitor

During beam operation, the observation of the beam profile on the target is essential for stable beam operation. As a candidate for the beam profile monitor, luminescence monitor is considered. Since the peak current density of the TEF-T is 6 times higher than that of MLF[4], TEF requires a profile monitor that will stand the high-intensity beam. To observe the degradation of the luminescence of chrome doped Al_2O_3, the experiment was performed using an Ar^{+15} (150 MeV) beam at the cyclotron in Takasaki Ion Accelerator for Advanced Radiation Application (TIARA). The result of the degradation light is shown in Fig. 3, which shows 20 % degradation of light yield for 160 h beam operation in the MLF. Similar results were obtained at Spallation Neutron Source (SNS) in Oak Ridge National Laboratory (ORNL)[5]. A new material with less degradation will be developed and confirmed

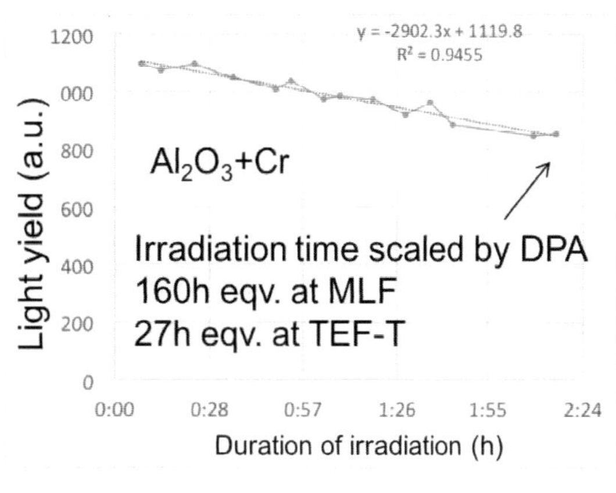

Figure 3. Result of degradation of luminescence during Ar irradiation.

IOP Conf. Series: Journal of Physics: Conf. Series **1021** (2018) 012072 doi:10.1088/1742-6596/1021/1/012072

with the Ar beam irradiation.

3.3. Measurements cross section for TEF-T and ADS

To obtain cross section, we had begun activation cross-section measurements of target and window materials at the beam dump line of the beam dump placed near the extraction channel of the RCS [6]. Since gasses at the LBE loop are continuously purged to prevent erosion on the surface of the target, the gas production cross section is important, especially for rare gas because the rare gas can not be reduced in the off-gas system. Because of its particularly long half lifetime (32 days), an accuracy of estimation of ^{127}Xe is important. In Fig. 4, a cumulative cross section of $Pb(p,x)^{127}Xe$ is shown, which is normalized for the beam power. The yield of spallation neutrons is roughly constant to the beam power irrespective of the proton energy. The cross-section is shown in Fig. 4 and may give the optimum energy for the ADS to reduce the inventory of hazardous nuclides. However, the experimental data is scattered in the energy range between 1 to 2 GeV so that we will carry out further measurements for validation. In Fig. 4, comparison of various calculation using PHITS[7, 8] is also shown, and this drastically underestimates the experiment. In future, modifications will be applied to PHITS.

4. Summary

To evaluate the performance of the LCE, we carried out an experiment by using a 3-MeV RFQ placed on the test bench of the LINAC building. It was found that very stable beam with power of 8 W at 400 MeV can be extracted for a long duration. By using the LCE, a very short pulse length can be created. By using the LCE with long length loop, very short pulse spallation neutron source can be set up at the TEF-T for measurement of cross-sections for the ADS. We had started some experiments for the TEF-T and ADS.

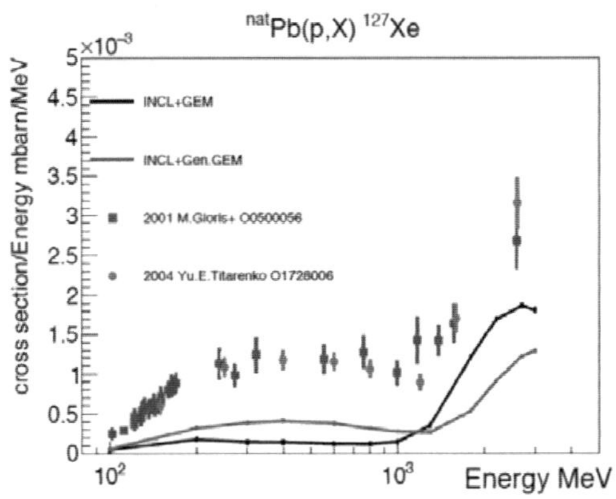

Figure 4. Normalized beam power cumulative cross section of $Pb(p,x)^{127}Xe$.

Acknowledgment

This work is partly supported by the Ministry of Education, Culture, Sports, Science and Technology (MEXT) KAKENHI Grant-in-Aid for Scientific Research (C) Grant no. 26390114. Also, this paper includes results obtained with the Subsidy for Research and Development on Nuclear Transmutation Technology.

References

[1] Nuclear Transmutation Division, "Technical Design Report o J-PARC Transmutation Experimental Facility - ADS Target Test Facility (TEF-T) -", JAEA-Tech. 2017-003 (2017) (in Japanese).
[2] S. Meigo, et al., J. Nucl Mater. **450**, 8 (2014).
[3] H. Takei, et al., Proc. IBIC2016, WEPG45, Barcelona, Spain (2016).
[4] S. Meigo, et al., Nucl. Instrum. Meth. **A600**, 41 (2009).
[5] T. Shea et al., Proc. DIPAC09, MOOC04, Basel, Switzerland (2009).
[6] H. Matsuda, et al., "Measurements of activation cross sections of target and proton beam window materials at J-PARC" Proc. ICANS-XXII held at Oxford Mar 27 - 31 2017 (2017).
[7] K. Niita, et al., JAERI-Data/Code 2001-007 (2001).
[8] A. Boudard et al., Phys. Rev C **87**, 014606 (2013).

IOP Conf. Series: Journal of Physics: Conf. Series **1021** (2018) 012073 doi:10.1088/1742-6596/1021/1/012073

Initial implementation of helium Gas into the SNS mercury target for mitigation of fatigue and cavitation damage

P Rosenblad, M Wendel, L Jacobs, B Riemer, D Winder, C Barbier, M Dayton, G Stephens, R DeCosta, S Parson, and D Freeman

Oak Ridge National Laboratory, USA

Abstract. The short-pulse, 60 Hz beam used at the Spallation Neutron Source (SNS) initiates pressure waves in the mercury target with each pulse that 1) lead to huge numbers of fatigue stress cycles in the target vessel, and 2) pitting / erosion damage to the target vessel caused by mercury cavitation. Both phenomena can lead to target leaks and interruption of neutron production.

The number of beam pulses expected over a successful target life is on the order of 10^9; the number of response stress cycles at a vessel position can exceed 1 per pulse. This number is even beyond the regime typical of high-cycle fatigue. It is important to be aware that contemporary knowledge fatigue damage includes probabilistic behavior for many materials, and that for many materials a true endurance limit does not exist. Obtaining data in the giga-cycle regime is difficult, particularly repeatable data. In this regime, the effects of very small material defects, inclusions and inhomogeneity become relevant as their size approaches that of the material microstructure. Detecting and controlling such defect sizes is problematic in the SNS mercury target module weld assembly.

While considerable efforts are being made to improve the robustness of the SNS target, one thing that can be assuredly stated is that reducing high-cycle fatigue stress magnitude will add margin to its fatigue life. Injecting gas into the target mercury is one way to accomplish this. The degree of pulse stress reduction to be expected from a given gas condition is difficult to predict. There are SNS experimental data and JPARC target gas injection experience to expect stress reduction to ½ ~ ¼ of that without gas throughout much of the target mercury vessel.

As a first step towards reduction of cyclic stresses and mitigation of cavitation damage, SNS will implement gas injection in its mercury target module beginning in the fall of 2018. The project is known as Gas Injection Initial Implementation, or GI3. GI3 will comprise of a once-through system. That is, the injected helium will be processed through the Mercury Off-gas Treatment System (MOTS) rather than recycled for re-injection. GI3 will employ relatively low flow rate of up to 1 slpm of helium gas through a total of sixty orifices, each with a diameter of 8 microns. The upstream supply pressure will be approximately 6.9 atm.

GI3 in envisioned as a first step towards a more complete and complex gas injection system that would include much higher flow rates, gas re-circulation, and the potential combination of small gas bubble and gas wall supplies. Despite this reduced scope, there are significant design,

IOP Conf. Series: Journal of Physics: Conf. Series **1021** (2018) 012073 doi:10.1088/1742-6596/1021/1/012073

installation, operational, and safety related challenges associated with the project. This paper and presentation will describe these challenges and the strategies employed to overcome them.

1. Introduction and Scope

This paper describes the system configuration for the Gas Injection Initial Implementation (GI3) project, which is a hardware and operational modification to the Spallation Neutron Source (SNS) mercury target system.

The short-pulse, 60 Hz beam used at SNS initiates pressure waves in the mercury target with each pulse that (1) lead to huge numbers of fatigue stress cycles in the target vessel, and (2) pitting / erosion damage to the target vessel caused by mercury cavitation. Both phenomena can lead to target leaks and interruption of neutron production. The injection of helium gas within the mercury target could mitigate both phenomena.

The first phase of mercury target gas injection at the SNS is known as GI3. The general parameters and requirements of GI3 are:

- The system shall allow continued neutron production in the event that gas injection is disabled or turned off.
- The system shall prevent backflow of mercury and spallation products out of the service bay via the helium supply line.
- Helium gas shall be injected into the target module at flow rates up to 1 slpm, total, which is an injected mass fraction of 0.1% compared to the mercury flow rate at a pump speed of 350 rpm. The injected volume fraction of helium in the mercury vessel is less than 0.1% because the mercury is under pressure greater than one atmosphere.
- Helium gas bubbles shall be as small as reasonably achievable within the constraints of existing target designs, which is likely to be an average diameter of 150 microns or less.
- The helium gas supply pressure shall be controlled and measured continuously.
- The helium gas flow rate shall be measured continuously.
- The system shall not allow liquid mercury to enter the mercury off-gas treatment system (MOTS) such that it either disables the operation of MOTS or leads to a loss of mercury from the service bay.
- The system shall not negatively impact the ability of the flowing mercury to remove heat from the system.

Following the implementation of GI3, it is anticipated that further upgrades will be made to the gas injection system allowing higher gas flow rates, smaller gas bubbles, and an additional target gas supply dedicated to protective gas wall. The scope of this document is limited to the GI3 phase only.

2. Gas Injection Initial Implementation (GI3) System Overview

The system is shown on the GI3 schematic diagram, which is shown in Figure 1. The schematic diagram shows the three main subsystems that comprise GI3, which are the helium supply system, modifications to the existing mercury loop system, and modifications to the existing mercury off-gas treatment system (MOTS). In addition to equipment necessary for GI3, some equipment will be installed with GI3 that will allow for future upgrades of the gas injection system. Fundamentally, GI3 involves the transport of helium gas through the following chain:

1. A new helium supply system located in the target building high bay, which is fed by the existing building helium system, which is fed by a helium bottle truck located outside the building
2. Through a GI3 gas panel located in the target building high bay
3. Through the a spare penetration of the water system gas liquid separator (GLS) cavity located adjacent to the High Bay
4. Through an existing 2-inch pipe that runs from the GLS cavity to the west end of the target process bay, which houses the mercury process system

5. Through a flexible hose from the 2-inch pipe to the rear (east end) of the target carriage
6. Through an existing gas line running from the rear to the front of the target carriage
7. Into the mercury target where it is injected into the flowing mercury through inlet orifice bubblers (IOB)
8. Through the mercury loop itself until the gas reaches the free surface at the top of the mercury pump sump tank or in the target mercury vent
9. Through MOTS
10. Through the Hot Off-Gas treatment system (HOG)
11. And finally, out of the building stack to the atmosphere

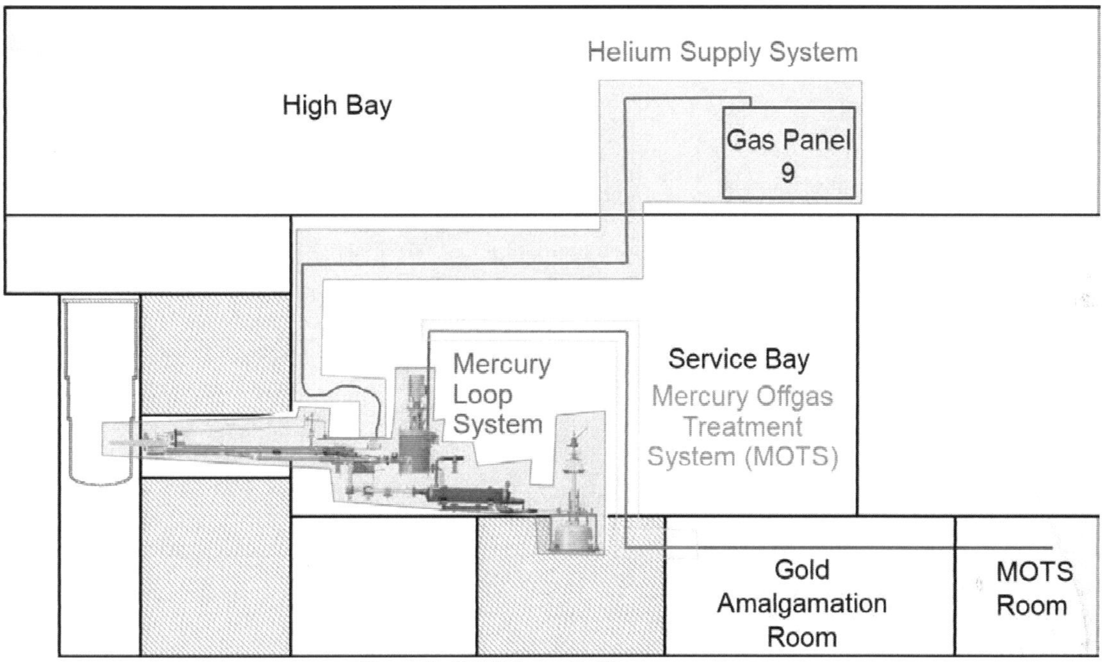

Figure 1 - GI3 Schematic Diagram

3. System Details and Specific Challenges

3.1. Gas Supply System

The most significant feature of the gas supply system is its location. Because the system requires a gas line to enter the high-pressure portion of the mercury loop downstream of the pump discharge, there is a potential safety accident in which mercury and spallation products back flow through the supply line. Because the gas panel is located in the high bay and the supply line enters the service bay from above, there is not enough pressure (even in off-normal cases) to push mercury out of the service bay, which results in an inherently safe design.

The panel for the gas supply system is a relatively standard design. It consists of valves, filters, a flow meter, a pressure regulator a pressure relief valve, and a pressure gauge. The system is designed to maintain a steady helium gas pressure from 0 to 100 psig at all times. From the gas panel, the helium is transported into an existing cavity which contains water system GLS and then into the mercury service bay through an existing 2-inch pipe. Transporting the helium through this pipe, which is about 80 feet long and has five 90 degree bends is particularly challenging. The pipe is used as a conduit through which two flexible hoses will be routed. Only one supply hose is used for GI3, but the second, separate

IOP Conf. Series: Journal of Physics: Conf. Series **1021** (2018) 012073 doi:10.1088/1742-6596/1021/1/012073

hose is required for future upgrades to gas injection that will require an independent gas supply at a different pressure and flow rate.

The 2-inch pipe terminates along the north end of the service bay, which is closest to the monolith. From here, another flexible hose will transport the helium to an existing connection on the back of the target carriage. An existing gas line transports the helium to the target location and a new jumper brings the helium to the target module itself. The jumper is fitted with an additional check valve and filter. The filter prevents contamination of the small injection orifices and prevents backflow of mercury to the supply line.

3.2. Helium Gas Introduction to the Target Module

Helium is introduced to the mercury loop through new hardware added to target modules. The initial target modules have a retrofit design as shown in Figure 2.

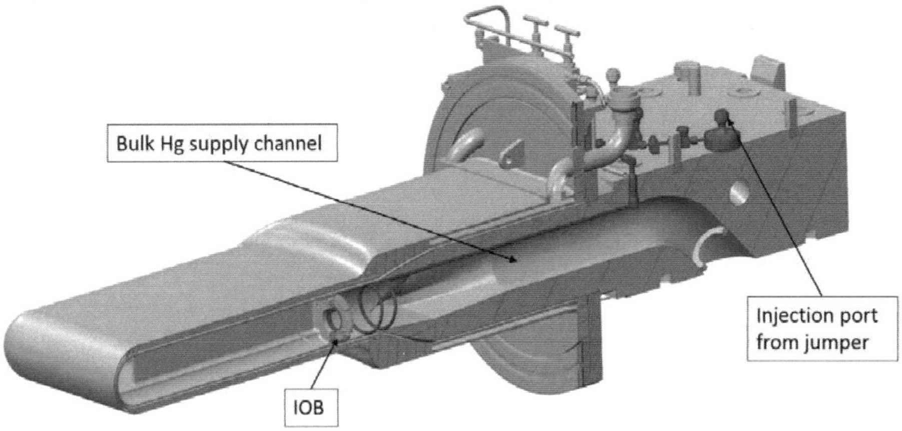

Figure 2 - GI3 Configuration at Mercury Target Module

On the target module, the gas supply is split into two paths feeding an inlet orifice bubbler (IOB) installed in each of the two bulk mercury supply flow paths. Each IOB has a circular array of 30 orifices oriented radially. Each orifice has a diameter of 8 microns (.0003 inches). At 100 psig supply pressure upstream of the IOBs and the mercury pump operating at 350 rpm, the flow is choked and the nominal total flow is 0.7 slpm.

3.3. Helium Gas within the Mercury Loop

Once injected, the helium bubbles interact with the turbulent mercury flow, where they may coalesce and/or be temporarily held up in low-pressure zones within the mercury. Some of the gas exits the target module through the vent line, and some of the gas is swept through the mercury loop, (Figure 3), but eventually, the gas reaches the pump tank and rises to the surface due to buoyancy. From the pump tank, the injected helium joins with the sweeping cover gas flow and is then drawn through the MOTS.

IOP Conf. Series: Journal of Physics: Conf. Series **1021** (2018) 012073 doi:10.1088/1742-6596/1021/1/012073

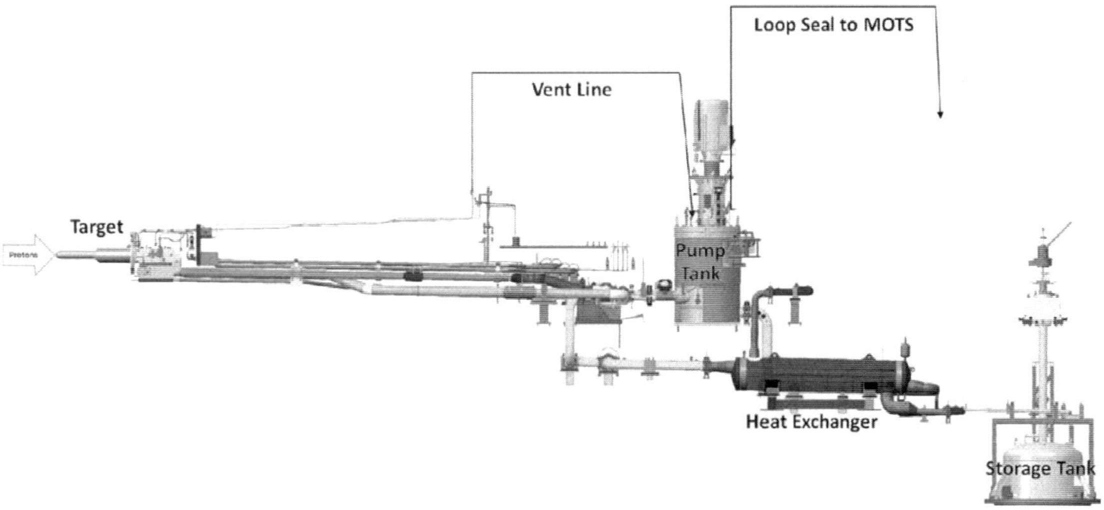

Figure 3 - Mercury Loop with Major Components

Figure 4 highlights potential collection points for gas, which are either local high points or downward turns in the flow path. Gas that is held up in the mercury flow causes the combined fluid volume in the mercury loop to increase, raising the mercury level in the pump tank. The trapped gas is eventually vented when the Hg pump is stopped or when the Hg is drained from the loop.

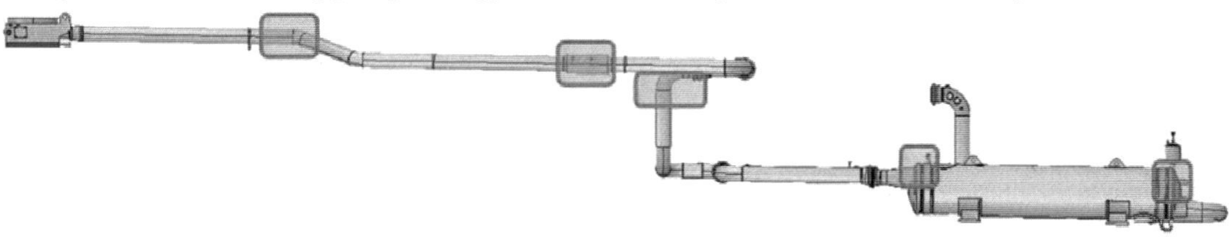

Figure 4 - Likely Collection Points for Injected Helium

3.4. Mercury Off-gas Treatment System (MOTS) modifications

3.4.1. Prevention of liquid Hg flow through MOTS. Figure 3 shows the loop seal, which is the beginning of MOTS. It is paramount to the successful operation of SNS that liquid mercury does not pass through MOTS. A runaway event caused by higher than expected gas injection rate or unexpected transients of capture and release of gas within the loop could conceivably send mercury over the loop seal and towards MOTS. Therefore, GI3 adds two features to protect against these potentials.

First, the set point for the existing burst disk atop the pump sump tank will be lowered so that a static column of mercury will burst the disk before its height could exceed the elevation of the loop seal.

Second, a gas liquid separator (GLS) has been added downstream of the loop seal. The GLS protects against an event in which a pocket of trapped gas is released and moves through the mercury loop towards the pump tank. As the gas rises, it is subject to lower and lower pressure, which allows it to expand. This expansion raises the mercury level in the pump tank, which could occur relatively quickly. If the rising gas pocket pushes the free surface of the mercury against the top of the tank before breaking the surface, then some amount of mercury might be trapped inside the line leading to the loop seal. If the (3/4-inch) line is not large enough to allow the gas to flow past the mercury during this dynamic event, the buoyant gas pocket could push the trapped mercury up the loop seal line and potentially over the loop seal. In effect, the mercury loop would "burp" some volume of mercury. Figure 5 shows a conceptual representation of the GLS. If mercury collects in the GLS, it can be collected and returned to the mercury loop (not shown).

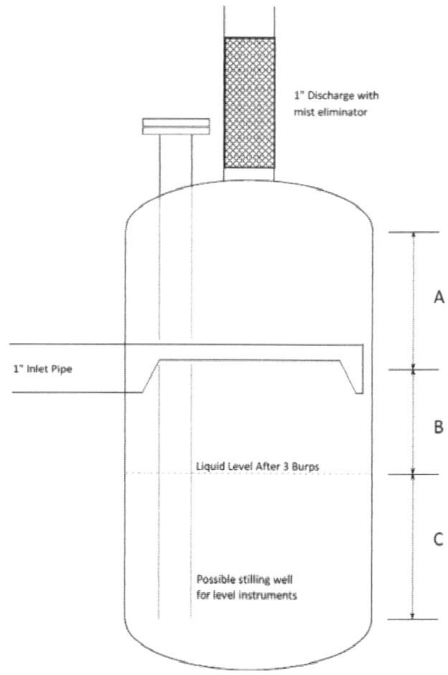

Figure 5 - Conceptual representation of MOTS GLS

3.4.2. Additional Carbon Adsorber

SNS operational experience has demonstrated that gas flow through the mercury loop causes accelerated shedding of radioactive spallation product noble gases from the liquid mercury. This has occurred when the gas leakage rate of the seal between the target module and the target carriage has significantly increased during operation. The fast shedding of noble gases leads to high dose rates adjacent to the MOTS components in the target building basement. As GI3 will be intentionally injecting gas, this phenomenon is expected. Therefore, an additional shielded carbon adsorber is to be installed in MOTS. The purpose of the device is not to add additional absorption capacity. Rather, it provides a simple delay tank so that the radioactive noble gases can decay before reaching downstream MOTS components for which the high does rates are not acceptable due to operational limitations.

4. Schedule

GI3 is planned to be commissioned in October, 2017. However, this completion date will be challenging. In addition to the hardware engineering and installations, extensive controls programing and modifications are required. These systems must be integrated with operations procedures and training. Finally, the system and readiness will undergo significant internal and external review prior to approval.

5. Summary

A comprehensive system for the initial implementation of gas injection into the SNS mercury target has been designed. The primary goal is to increase target module lifetime and reliability by (1) reducing the amplitude of cyclic loading and (2) reducing the damaging effects of cavitation bubbles within the mercury vessel structure. GI3 is expected to provide appreciable improvements and support the near-term goal of reliable 1.4 MW operation. In addition, GI3 provides the first step towards full scale deployment of gas injection.

IOP Conf. Series: Journal of Physics: Conf. Series **1021** (2018) 012074 doi:10.1088/1742-6596/1021/1/012074

Process Filtration of Liquid Methane Radiation Products Using Centrifugal Separation

R Dean, P Harrison, R Burridge, D Jenkins, M Probert

ISIS, RAL, Science & Technology Facilities Council, Harwell, Oxford, OX11 0QX

robert.dean@stfc.ac.uk

Abstract. The radiation chemistry of methane is relatively well known and the creation of hydrogen and long chain hydrocarbons (in the form of oils, waxes and even solid carbon-like deposits) in the presence of a neutron source has presented a challenge to the ISIS liquid methane moderator operation since first start up. The rapid build-up of gaseous hydrogen leads to pressure fluctuation and circulation failure as the pipework becomes gas-locked. Current operational practice sees us replace the methane charge every single day which, while proving effective at removing the built up hydrogen gas, leads to temperature instability and periods where the system is unavailable to the users. Additionally, this approach does very little to curtail hydrocarbon build-up; previous attempts to remove the hydrocarbons using mesh filters have also proved ineffective and instead the system must be regularly flushed with solvent, while the moderator vessel requires complete replacement after only 2 or 3 operational cycles.

We intend to investigate the possibility of using centrifugal separation in order to constantly remove heavy hydrocarbons and hydrogen gas from the process stream of liquid methane. If successful this could realise great benefits in terms of system stability, maintenance costs and beam time available to users. This paper will provide details of the proposed changes to be incorporated as part of the Target Station 1 (TS1) upgrade project.

1. Introduction

The ISIS Target Station 1 uses liquid methane as a moderator material, reducing the energy of fast neutrons created at the target down to thermal neutron levels which are useful to the installed instruments. Within the moderator vessel, the intense neutron radiation causes methane molecules to break apart into atomic hydrogen and free radicals of methane. These then recombine, forming molecular hydrogen (H_2) and a variety of hydrocarbons. As the material is recirculated through the moderator, the hydrocarbon chains progressively increase in length, forming oils and waxes (referred to as 'polymer') and eventually resulting in solid carbon-like deposits. Figure 1 shows a moderator can which has become completely blocked as a result of this hydrocarbon formation.

The molecular hydrogen, which is gaseous at operating conditions, builds up very rapidly within the system. After only a matter of hours of operation, the methane system experiences significant pressure and temperature excursions, resulting in a loss of availability to users. It is believed that this is caused by H_2 concentration reaching a level where the performance of the circulation pump is impaired. No such instability is experienced if the system operates with no beam on the target, and outside of these momentary excursions the control system is comfortably able to maintain steady state conditions with varying beam current.

Fig.1. Moderator blocked with hydrocarbon deposits

2. Previous Work

Although the potential issues with methane radiolysis products were know when the system was first designed, the scale of the problem become apparent only when beam energy was increased after several years' operation [1].

A theoretical analysis [2] was carried out in order to estimate the rate of build-up of the contaminants, and this work was backed up by examination of irradiated samples [3]. The study estimated that, at current beam energies, approximately 2 liters (at operating conditions) of hydrogen and 0.5 grams of polymer are created within the moderator per day.

The system was originally fitted with two 70μm filters, but these were not found to be effective – the filters did not produce any operational benefit but did create a large amount of radioactive waste. When removed, the filters were found to be saturated with oil, but it is suspected that this only occurs as the system warms up and the oils become liquid.

In addition, a small swirl tank was incorporated into the Cryogenerator head with the hope that the resulting region of low methane flow would allow the hydrogen to naturally bubble out of the liquid. The efficacy of this feature is unknown but operational experience suggests it offers no meaningful improvement.

A major modification was undertaken to the coldbox system in order to allow the full methane charge to be replaced at operational temperature. Gaseous methane is liquefied using excess refrigeration from the process stream over the course of a day, and this is then released into the system, pushing the existing charge into a dump tank from where it can be vented in a controlled manner.

In current operations, the methane charge is replaced every single day in order to prevent hydrogen build-up from reaching a level where the system becomes gas-locking. This represents a significant cost to the organization and the charge change itself inevitably creates large temperature fluctuations in the methane, resulting in periods where the system is unavailable to users. Additionally, the methane moderator vessel requires replacement every 2 to 3 cycles, while the transfer lines linking the moderator and supply system are periodically flushed with the solvent MEK to prevent hydrocarbon build up from blocking the system completely.

3. Continuous Filtration

It is therefore hoped that a means can be found to continuously remove contaminants from the fluid as it circulates, as this could allow the methane moderator system to run for a greatly extended period with much improved pressure and temperature stability. It is proposed that, since H_2 will be in a gaseous state and oils are likely to be solid at the operating conditions, centrifugation of the process fluid could utilise the differences in the relative density in order to separate contaminants from the liquid methane. The replacement of the moderator supply system as part of the TS1 Project represents an opportunity to incorporate new equipment, and two options are currently being investigated, a motor-driven centrifugal separator, and a hydrocyclone separator.

IOP Conf. Series: Journal of Physics: Conf. Series **1021** (2018) 012074 doi:10.1088/1742-6596/1021/1/012074

3.1. Centrifugal Separator

A motor-driven centrifuge, in which the methane fluid enters a drum rotating at very high angular velocity, could create large forces, and thus a high degree of separation, in the fluid components. Figure 2 shows how such a unit might be arranged; methane returning from the moderator enters through a central vertical inlet pipe which directs the flow against the wall of the drum. The drum, spinning at very high angular velocity, induces rotation in the fluid and thus subjects it to centrifugal force.

The drum is shaped so that the heavier oils are forced to move outwards and downwards, collecting at the base, while the less dense hydrogen moves towards the top and center. A vent hole allows the gas to collect in a void above the centrifuge, from where it can be periodically released. Clean methane then exits, through outlets at the base, to the Cryogenerator Unit where it is re-cooled to operating temperature.

The design of such a device would share much in common with the pumps currently installed to circulate the methane around the system. Indeed, by angling the outlets of the centrifuge drum, a degree of pumping action could be provided, mitigating the pressure drop created.

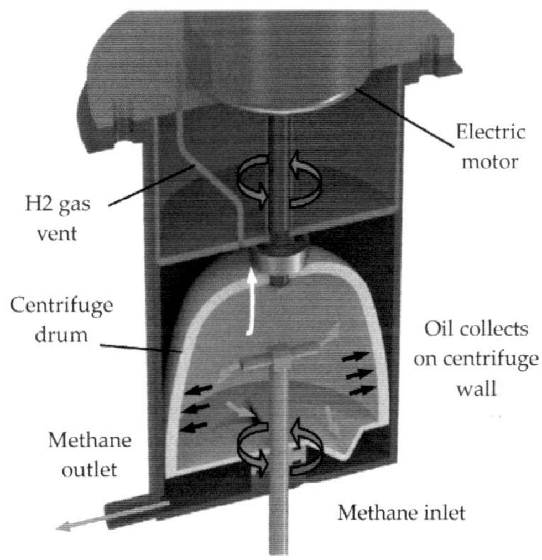

Fig. 2. Motor-driven centrifuge concept

A potential shortcoming for this design would be that the centrifuge drum would need to be regularly removed from the coldbox for cleaning in order to remove accumulated oils, unless a novel solution could be found to clean the drum during operation. Given the radioactive nature of this material, this would represent a hazardous operation.

3.2. Hydrocyclone

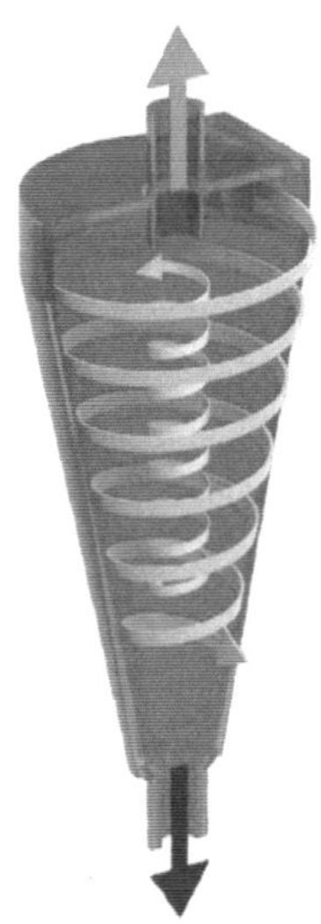
Fig. 3. Flow pattern within the hydrocyclone

A hydrocyclone consists of a conical vessel with outlets at the top and bottom. Fluid enters tangentially into the top of the unit and the geometry induces a vortex, causing heavier components to move towards the outer wall and exit through the apex at the base of the unit. At the same time, the flow pattern sets up a secondary vortex moving vertically upwards through the center of the unit. As lighter components are drawn to the middle, this secondary vortex transport them away, exiting the through the overflow pipe at the top of the separator.

A combination of two hydrocyclone units could be employed in order firstly remove less dense H_2 from the process stream, followed by the higher density long chain hydrocarbons.

This solution has the advantage that it uses no moving parts and so is essentially maintenance free, hydrogen gas can be vented through the normal vent system and accumulated oils can be collected and drained during planned shutdown periods.

Hydrocyclonic separators are well established and widely used in industrial processes. However, it is unclear at present whether the low flowrate in the methane system will be sufficient for separation to occur, or if the particle size would be sufficient to be effectively captured using this type of device.

4. Future Work

A prototype hydrocyclone unit has been produced using 3D printing and a test program, initially using water as the flow medium, will be carried out in order to ascertain suitability of this type of unit at low flow rates and to gauge the likely effect on pressure drop in the system.

Further investigation into material properties of hydrocarbons at cryogenic temperatures is needed. Although hydrocarbon samples are readily available – approximately 100ml of oil is drained from the system after every cycle – the highly radioactive nature of this substance prevents any analysis being carried out. This means that the actual composition of the oil cannot be easily determined and so material properties at cryogenic temperatures are unknown. A representative mixture based on theoretical radiolysis models could provide a starting point to address this requirement.

A motor-driven centrifuge will require significant development work and an industrial or university partner is being sought to collaborate in this work.

5. Conclusion

Cryogenic liquid methane has excellent properties as a neutron moderator, but the operational difficulties associated with its use mean that it is not a favored option for this purpose - ISIS is currently the only spallation source using a liquid methane moderator.

The TS1 Project represents an opportunity to address the long-standing operational challenges inherent in liquid methane and it is hoped that, with some development, a centrifugal separator could provide a successful solution to these problems, providing great benefits in terms of system stability, maintenance costs and beam time available to users.

References

[1] Lucas A T 1988 Radiation damage in liquid methane moderators – *ISIS Document ATL/TN9-88 5/9/88*
[2] Evans D 1995 Irradiation effects in liquid methane used as a neutron moderator *Cryogenics* **35** 763-6
[3] Evans D 2000 Summary of methane moderator runs 10/09/99 - 04/08/00 *ISIS Internal Document*

IOP Conf. Series: Journal of Physics: Conf. Series **1021** (2018) 012075 doi:10.1088/1742-6596/1021/1/012075

ISIS TS1 Project mock-ups and testing

S J Thomas, D Coates, C Russell, D Haynes, L Jones, & S Gallimore

ISIS, Rutherford Appleton Laboratory, Science & Technology Facilities Council, Harwell Oxford, Didcot, Oxfordshire, OX11 0QX

stephanie.thomas@stfc.ac.uk

Abstract. The ISIS TS1 target station has operated with the current tantalum clad tungsten design of target for over a decade with only minor changes to the accompanying moderators and reflector since ISIS's creation in the 1980s. The TS1 upgrade project is currently being undertaken to redesign the target, moderators, and reflector (TRaM), as well as their associated support structure, and also the services located in the target services area (TSA). This paper will detail the work being undertaken to underpin design, verify and validate through an on-site testing programme. As the area that the new equipment will be installed in is already radiologically active, everything must be comprehensively tested for fit and function before installation. Most of the testing will be performed on a full scale representation of the remote handling cell (RHC), complete with the remote handling manipulator arms. There will also be rigs in place for testing the new water pumps and instrumentation prior the installation into the TSA. In addition to these testing mock-ups, there is also work planned to support and verify various aspects of the fluid circuits that are required for operation of the target station. Presently, all ISIS targets are flow-tested on a dedicated rig to ensure that the cooling channels are clear and can achieve the required cooling flow rate. This rig is to be replaced with a dual purpose rig that will also allow cooling testing to be performed while improving the operability of the rig's design. Testing is scheduled to begin in summer 2017 and will continue until installation in 2019/2020.

1. Introduction

As ISIS TS1is now over 30 years old, the TS1 Project is working to improve its operability and efficiency by modifying the target, reflector, and moderators (TRaM), the associated services and support structures. The radiation level in the areas where the new equipment is to be installed restrict or prevent access, and will activate and/or contaminate any equipment that is introduced. To ensure that anything that enters a radiation area is fit for purpose prior to installation, a program of off line testing will be performed, necessitating a number of testing rigs and full scale mock-ups.

2. RHC Mock-Up

The TS1 TRaM is supported within the void vessel on a cantilever frame which attaches to the door of the vessel. This door attaches to the target services trolley, which can be rolled back to move the TRaM components into the remote handling cell (RHC). All operations within the RHC are conducted using four master/slave manipulator arms; two on either side of the cell. As the arms do not have any power amplification, any heavy lifting is performed by the in cell 1tonne crane. The RHC has two operator rooms, one north of the cell and one to the south; each room has a liquid zinc bromide window to view the cell, with a viewing area of 550x550mm^2 [1]. There are also four pan/tilt cameras, one in each corner of the cell, and two flying lead cameras (one on each side of the cell) which can be held in the manipulator arms.

The TS1 project will be replacing all of the TRaM components, as well as the cantilever beam, all of which will have to take place within the RHC. This means that all of the work must be able to be performed with the manipulation equipment already available in the RHC, and with the view restrictions of the operator windows and cameras.

Figure 1 View of RHC from the south manipulator operator room showing 1) void vessel door 2) cantilever beam 3) cryogenic moderators 4) manipulator arms

Prior to installation, the alignment, fit, and operability of the new TRaM components need to be tested. As the design of the cantilever beam is to be altered (see figure 2 & figure 3), the existing full scale mock-up of the cantilever beam and door cannot be used, so a new mock-up is required.

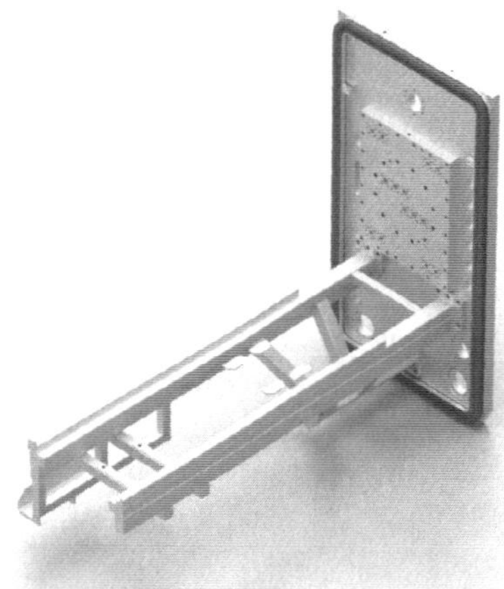

Figure 2 Current TS1 cantilever beam design

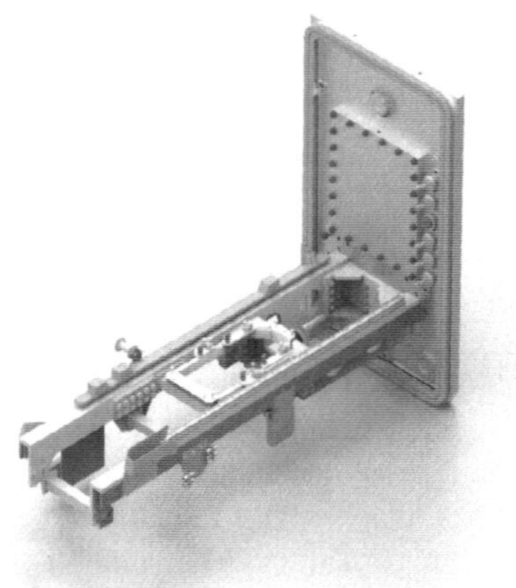

Figure 3 New (Mk 7) TS1 cantilever beam design

The test reflector blocks will be delivered and require testing before the design of the new cantilever beam is finalised, meaning that it would not make financial sense to have a detailed beam manufactured. Due to this, a functional replica will be made, using Bosch Rexroth aluminium profile to form the basic support structure, with machined attachment plates and rails. This will allow the initial testing and alignments to be performed; further testing will be completed once the cantilever design in completed and a true mock-up can be manufactured to replace the 'table' style frame. The mock-up frame will be accompanied manipulator arm frame which will allow manipulator accessibility to be established.

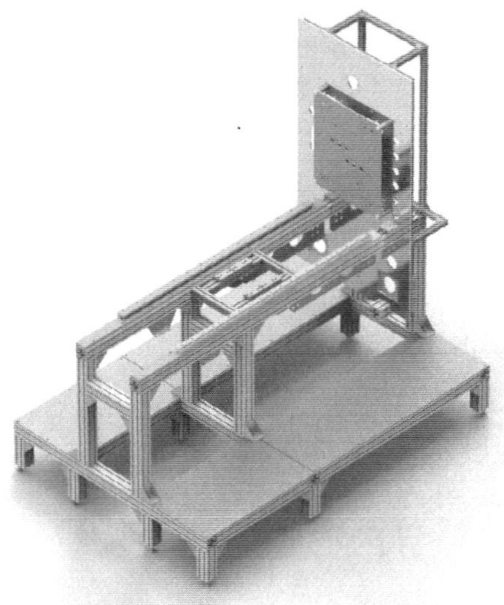

Figure 4 Mock-Up TS1 table frame

3. Target Flow Test Rig

All ISIS targets are flow tested prior to use to ensure that the cooling channels can achieve the required flow rates and that the pressure drop across the target is within acceptable limits. In order to extend the capability of the existing test rig, it will be altered to include heat extraction and improved flow control. The inclusion of a heat exchanger and chiller unit will allow cooling tests to be run to test different target designs and to verify FEA results. The present valve arrangement on the rig does not lend itself to easy or accurate control of the flow; using more appropriate valves in the return branches and at a distance from the flow metres should improve ease of operation.

Figure 5 Fig. 5: Current target cooling water flow test rig

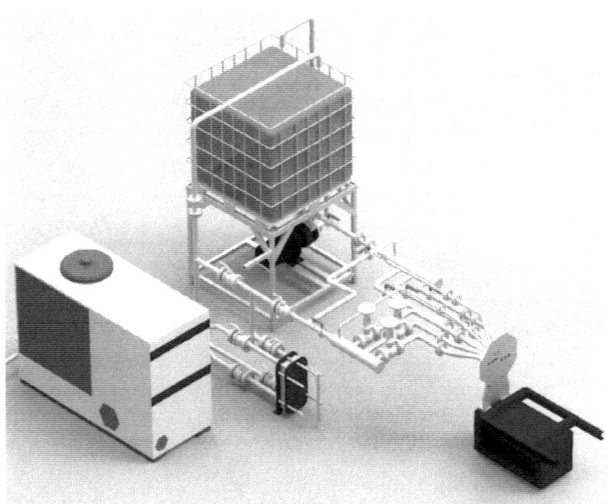

Figure 6 Altered target cooling water flow test rig with heat extraction and improved flow control

IOP Conf. Series: Journal of Physics: Conf. Series **1021** (2018) 012075 doi:10.1088/1742-6596/1021/1/012075

4. Target Services Area

The target services area (TSA) is located behind the RHC and contains the target services trolley which houses all of the services equipment required to operate the TRaM. These services were designed for use with uranium targets [1], some of which has been deemed unnecessary via a HAZOP analysis now that ISIS runs with the more stable tungsten/tantalum targets. Since removing this superfluous equipment will allow easier access for the project installation, and for general maintenance, this task has been deemed important to include as part of the TS1 Project.

4.1. Skids

As a lot of equipment will need to be moved to facilitate this strip out, the opportunity will be taken to rearrange the target services trolley in order to improve future accessibility and to speed up maintenance procedures. The intension is to arrange the trolley such that the equipment associated with each service (water, hydrogen, methane, etc) is allocated its own skid, separate from the others. This should make future works easier as new skids can be assembled and worked on offline, allowing the skid to be introduced to the TSA as a single entity, reducing radiation exposer of workers.

4.2. Pump test area

New pumps are to be installed into the TSA to replace the existing aging pumps. A pump test area will be required, including a representative pipe which will mimic the pressure drop over the pipework connecting the main cooling water circuit pump to the target. This area will allow the new pumps to be run under conditions similar to operating conditions and will allow testing of instrumentation.

4.3. Cryogenic transfer lines

The new cryogenic transfer lines will have to be assembled and inspected prior to installation. A mock-up of the forward shielding will be created to allow the operations personnel to practice feeding the transfer lines through between the TSA and the void vessel door.

5. Conclusion

There is a very real need for the provision of full scale mock-ups for testing to be carried out prior to any installation work commencing. This allows alignment adjustments to be made while access is still available, and means that if any errors are found, they can be rectified before the items have been introduced to a high radiation area. The offline testing also allows the operations personnel to become familiar with the how they will be interacting with the new equipment which will allow for more efficient installation and maintenance work.

References

[1] Carne A and Eaton G H 1983 *SNS target station safety assessment SNSPC/P6/82*

IOP Conf. Series: Journal of Physics: Conf. Series **1021** (2018) 012076 doi:10.1088/1742-6596/1021/1/012076

Investigation of novel moderator geometries at the Necsa accelerator based neutron sources

C B Franklyn

Radiation Science Department, Necsa, PO Box 582, Pretoria, South Africa

cbfranklyn@gmail.com

Abstract. Investigations into novel neutron moderator geometries yielding significant yield enhancements is a challenging exercise when trying to model possible structures, primarily due to a lack in suitable cross-section data, as a function of neutron energy, for various materials at cryogenic temperatures. Current studies at the small accelerator based neutron sources based at Necsa, South Africa, centre around investigations into the effect of unusual geometries for the primary moderator. Due to the inherently low sub-thermal neutron yield of small accelerator based neutron sources, further studies of neutron moderator and reflector configurations that yield a directional bias have been pursued, specifically for a multi-layered polyethylene and silicon wafer concept, maintained at cryogenic temperatures. The status of these investigations, including comparison between experiment and MCNP modelling are presented. Identification of particular deficiencies in neutron cross-section data will also be presented.

1. Introduction

The neutron yield from small accelerators, such as those based at the South African Nuclear Energy Corporation SOC (Necsa), are more than adequate for fast neutron radiography applications, however, the primary neutron flux is considered inadequate, after conventional moderation for thermal neutron based investigation of materials.

With the aim of achieving an enhanced moderated neutron yield with a directional bias, various unconventional moderator geometries have been investigated utilizing the linear accelerators at Necsa to generate primary neutrons in the sub-MeV range. A description of the experimental facility developed and some preliminary results obtained is given. Comparisons with MCNP modelling of the various moderator geometries are also presented.

2. Primary neutron source

By considering the various neutron producing reactions shown in Figure 1, for H^+ or D^+ ion beam energies available from the Necsa 4MV van de Graaff accelerator both Li and Be thick targets are equally suited from a total neutron flux point of view. For the purpose of this research, to generate a directionally biased cold neutron source, it was considered advantageous to select a nuclear reaction with a strong bias to generating as many neutrons as possible of low energy, and even more advantageous if these neutrons are produced with a forward direction bias. This narrowed the

IOP Conf. Series: Journal of Physics: Conf. Series **1021** (2018) 012076 doi:10.1088/1742-6596/1021/1/012076

selection down to the $^{7}Li(p,n)$ reaction, ideally at a beam energy of 2.25 MeV yielding a neutron flux at 0° as shown in Figure 2.

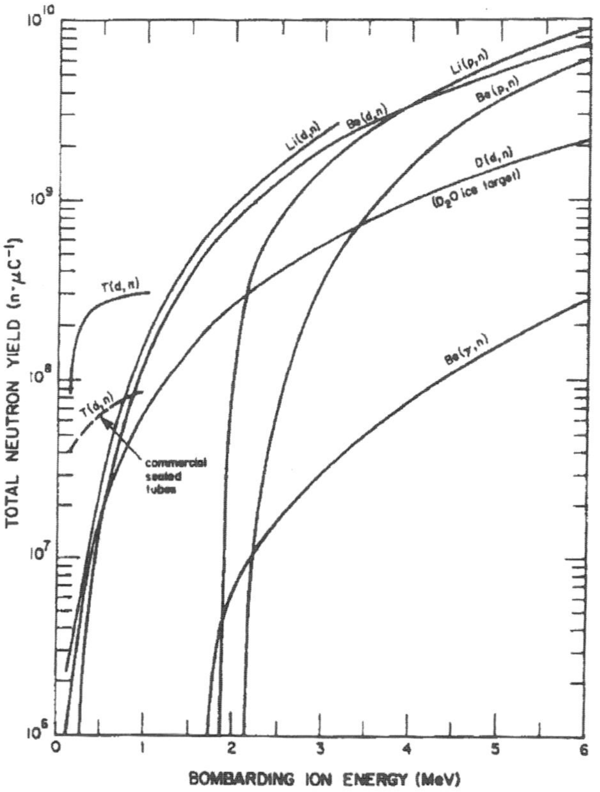

Figure 1. Total neutron flux yields for selected reactions [1].

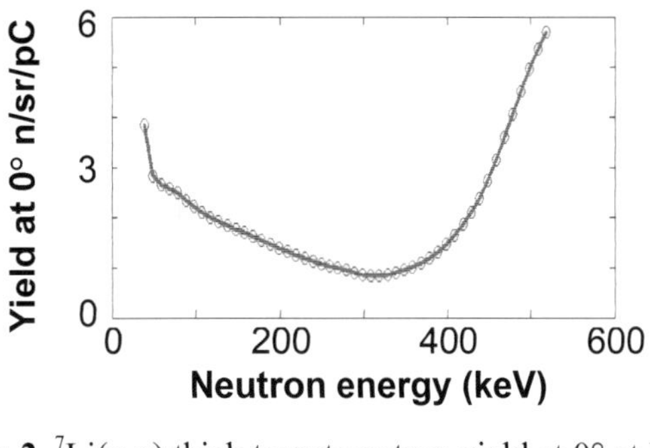

Figure 2. $^{7}Li(p,n)$ thick target neutron yield at 0° at E_p=2.25MeV [2].

3. Neutron moderator

Most large accelerator based neutron sources rely on liquid H_2 or D_2 as the neutron moderator and have proven very successful. Unfortunately the usable cold neutron yield is highly reliant on having a very intense primary neutron source, well outside the scope of the small accelerators at Necsa. Use of alternative moderator material, such as methane, mesitylene, etc have been reported, as well as the use of neutron reflectors to further enhance cold neutron generation.

Recent studies by E.B. Iverson *et al* [3] demonstrated a potential for enhanced cold neutron beam production using multiple layers of moderator and low scatter cross-section material. This concept was adopted in the Necsa studies with the inclusion of a cold finger to cool the moderator material to 70K. Figure 3 is a schematic of the neutron moderator system used at Necsa. The moderator itself consisted of 13 layers of 10 cm diameter polyethylene (2mm thick) and silicon wafers (1 mm thick). The Si wafers (which had been cut normal to the [1 1 1] direction) could be aligned for specific crystal lattice orientations. For this work the alignment was perpendicular to the [2 –2 0] vector. Since the moderator can be cooled to 70K with a cryogenic cold finger, the moderator material was mounted inside a vacuum chamber. Shielding around the vacuum chamber consisted of borated polyethylene with 5 cm diameter openings for neutron beam input and scattered neutron output. Neutrons were detected in the forward direction using a ^{3}He position sensitive detector (PSD).

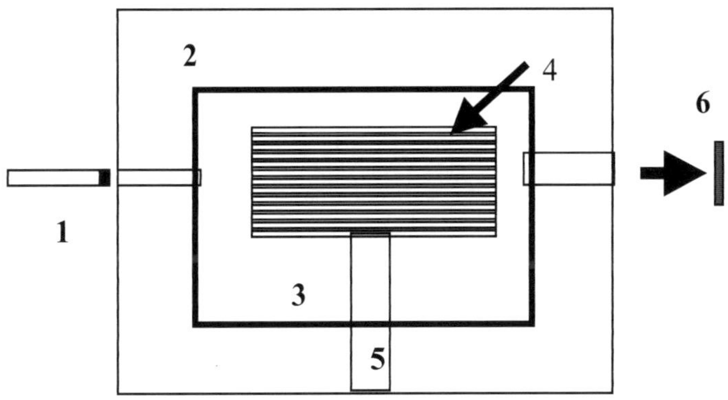

Figure 3. Schematic of the moderator experimental set-up. 1) H^{+} ion beam and thick Li target, 2)shielding, 3) vacuum chamber, 4) Si & polyethylene layers, 5) cold finger, 6) neutron PSD

4. Experiment and results

A 2.25 MeV H^{+} beam from the van de Graaff accelerator was used to conduct the experiments. A 5 mm beam spot was directed onto a 1 mm thick natural Li target deposited on a Ta cup. To minimize beam induced damage to the Li target and maximize beam intensity to typically 2 μA, the Ta cup was cooled with a fine water spray and continuously rotated, off centre, around the beam axis. The PSD, 1 cm diameter and 60 cm long, was placed approximately 50 cm from the moderator at 0° to the incident ion beam. Due to the unavailability of pulsed beam, no meaningful neutron energy information could be gathered at this stage.

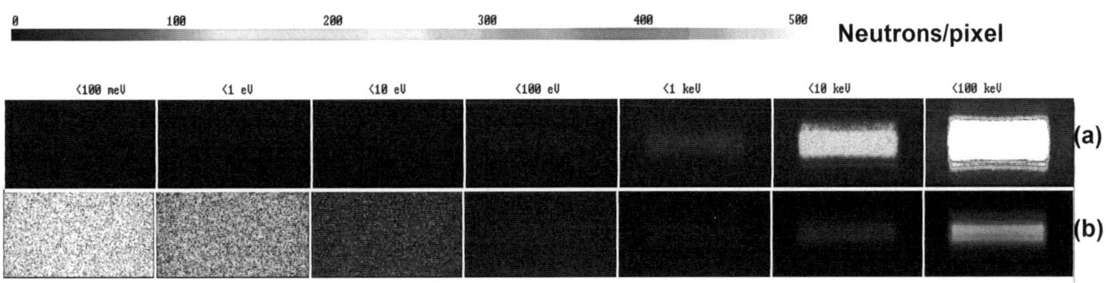

Figure 4. MCNP-X simulation of experiment; a) raw neutron counts and, b) corrected for detection efficiency. Distributions are divided into log scale neutron energy bins from <100 meV to 100 keV. Each distribution covers a 20 cm x 4 cm high area, 50 cm beyond the moderator.

Simulation of the experimental set-up, using MCNP-X, yielded spectra as illustrated in Figure 4a which shows the expected neutron distribution over a 20 cm x 4 cm area in various energy bins from <100 meV to 100 keV. Figure 4b is the same data but corrected for detector efficiency. If one models the experimental set-up with solid Si or polyethylene, their distributions do not compare with the combined Si/polyethylene model, thus demonstrating the enhancement prospects of multi-layered moderator material. Although there is still a significant high-energy component in the distribution it is <100 keV, whereas only 20% of the incident neutron beam is <100 keV.

Figure 5a shows the total neutron counts along 20 cm of the PSD with the target at room temperature (~300 K) and at ~100 K, indicating enhanced transmission at low temperature. Figure 5b is the equivalent MCNP distribution, which could only be calculated for 300 K due to a lack of Si and polyethylene neutron scattering cross-section data at 100 K.

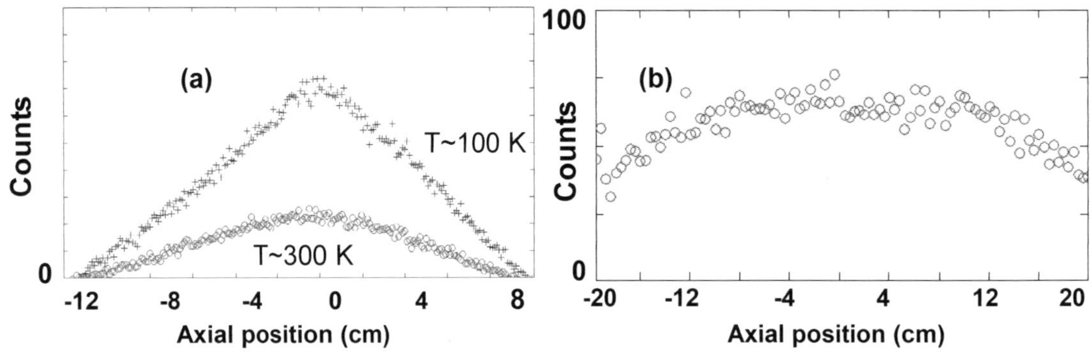

Figure 5. (a) Measured neutron transmission at indicated temperatures, b) MCNP predicted distribution at T=300K.

From Figure 5 it can be noted that the MCNP predicted distribution (detector energy efficiency corrected) implies a greater scattering of neutrons, whereas the experimental data indicates enhanced forward scattering, which is even more evident at 100 K. To facilitate modelling of this experiment it will be necessary to acquire neutron scattering cross-section data at various cryogenic temperatures for polyethylene and for Si, especially for various crystal orientations

5. Discussion

The experiment so far has been proof-of-principle with results correlating with Iverson et al [3]. Further measurements will be conducted wherein the Si crystal alignment will be varied, with respect to the incident neutron beam. Kinematically the H forward scattering dominates from high energy down to ~10 eV, then the scattering predominates perpendicularly. Taking into account that the neutron cross-section for a bound H atom is higher than for a free H, if we consider a biphenol ring ($C_{12}H_{10}O_2$), there are 8 H atoms that move more easily in the plane of the crystal than perpendicularly, thus we will have a medium wherein neutron slowing down is much less perpendicular to the plane compared to within the plane. Similarly if one considers the properties of liquid crystals that display a nematic phase, such as didodecyloxyazoxybenzene ($C_{36}H_{58}N_2O_3$), one can obtain similar preferred alignment of the H and C atoms. Further studies will assess the benefit of the abovementioned crystals to see if even more directionally enhanced neutron production can be achieved.

Acknowledgements

The author wishes to acknowledge funding support by IAEA Research Contract No.18179; access to the Necsa accelerator facilities; and the loan of silicon wafers by Dr G. Török, BNC, Hungary.

References

[1] Hawkesworth M R 1977 *Neutron radiography: Equipment and Methods*. *Atomic Energy Review* 15 169-220.
[2] Drosg M DROSG-2000 *IAEA-NDS-87*.
[3] Iverson E B, Baxter D V, Muhrer G, Ansell S, Dalgliesh R, Gallmeier F X, Kaiser H, Lu W 2014 *Nucl. Instr. Methods* A **762** 31.

IOP Conf. Series: Journal of Physics: Conf. Series **1021** (2018) 012077 doi:10.1088/1742-6596/1021/1/012077

Challenges, Analysis and Design Solutions for a Bolted Target Support Frame within a Remote Handling Area

Owain Williams

Science and Technology Facilities Council, Rutherford Appleton Laboratory, Harwell Campus, Didcot, OX11 0QX, UK

owain.williams@stfc.ac.uk

Abstract. The ISIS Neutron Source TS1 Project will include the replacement of a significant amount of equipment, including the existing support frame for the Target, Reflector and Moderator components. This paper discusses the challenges affecting the design, and a summary of the analyses carried out to verify the safety case. Having established a baseline model, verified using hand calculations, a number of investigations were carried out to inform design decisions. The results were that the new design could support the applied loads for all load cases, with a greater factor of safety than the current installed frame. In addition, it was found that the number of bolts could be significantly reduced, potentially representing a significant time saving for end-of-life disassembly.

1. Background

The ISIS Neutron Source TS1 project (discussed thoroughly in the work of S. Gallimore, D. Coates, D. Wilcox and L. Jones at ICANS XXII) will incorporate a new Target, Reflector and Moderators (TRAM) assembly, along with a full new set of services. All of this will be supported by a new frame, made of bolted stainless steel plates. Due to the high radiation environment around the Neutron Target, the assembly is enclosed in a shielded cell, and only accessible with remote handling equipment. The frame is cantilevered from the door to the remote handling cell to allow for sealing around it during operation. The resulting complex loading situation on the frame and bolt patterns required further analysis to evaluate the safety case, and to inform further design decisions.

1.1. Reasoning for moving to a bolted design

As discussed by D. Coates, one of the significant engineering improvements being implemented in the TS1 project is to move from the current layered reflector (shown in Figure 1) to a design that can be split horizontally, allowing fast access to the moderators (see Figure 2). It was found that this could not be implemented on the existing frame, and the decision was therefore made to replace it with a new design that better suits the new TRAM layout. However, removing the existing welded frame (shown in Figure 3) poses its own problems: the frame is too large to be put in a radiation shielded flask and extracted; instead, it must be cut up into smaller pieces for extraction (tests of this have already been carried out, shown in Figure 5). It has therefore been decided that the new design should be made of bolted sections (see Figure 4), removing this difficulty (and the radioactive material handling considerations that go along with it) for the future end-of-life disposal.

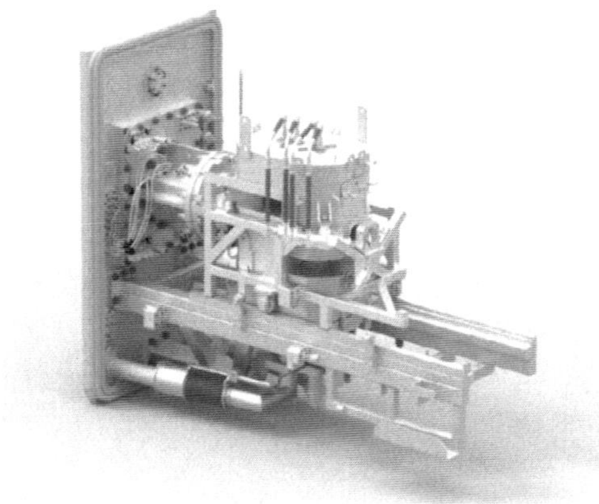

Figure 1: Current installed TS1 TRAM

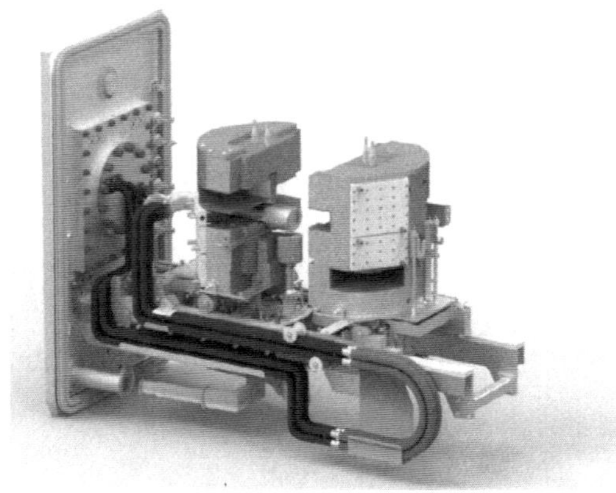

Figure 2: TS1 Project TRAM design, shown in the Open position

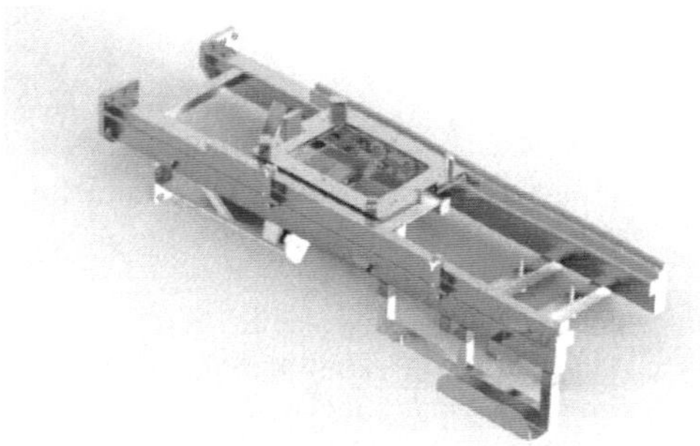

Figure 3: Current installed TS1 Support Frame

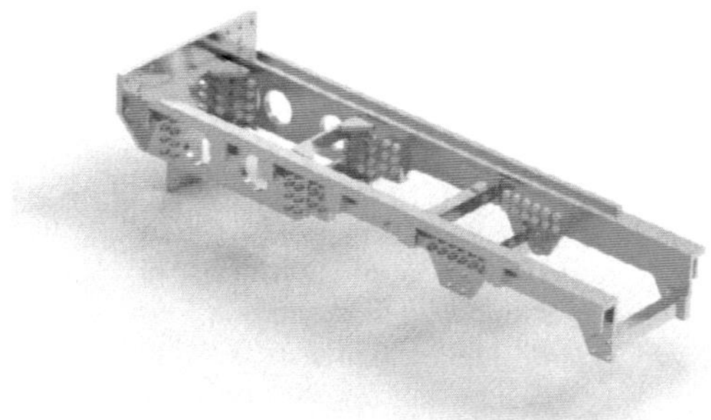

Figure 4: TS1 Project TRAM Support Frame design

2. Design Challenges

Whilst moving to a bolted design facilitates easier end-of-life disposal, the challenging environment and loading state prevent this from being a straightforward transition. To begin with, the combined cantilevered moment and torque on the frame from the TRAM and services loads gives a sufficiently complicated stress state that Finite Element Analysis (FEA) modelling is required in order to accurately assess the safety case for the frame and bolt patterns.

Figure 5: Cutting tests for currently installed welded frame

Figure 6: Current TS1 Remote Handling Area and Manipulator Arms

From a logistical point of view, the extremely high radiation environment around the Neutron Target (> 1 Sievert per hour at a distance of 1m) also represents a significant challenge for the installation, maintenance and end-of-life disassembly and disposal of the frame, and must be considered carefully at the design stage. For example, stainless steel is required for both the bolts, and the plates, requiring specially coated bolts to prevent galling of the screw threads. All interactions with the frame must be done with remote handling "Manipulator Arms" (see Figure 6); these are limited in maneuverability, strength, and, crucially, the torque load that can be applied to bolts.

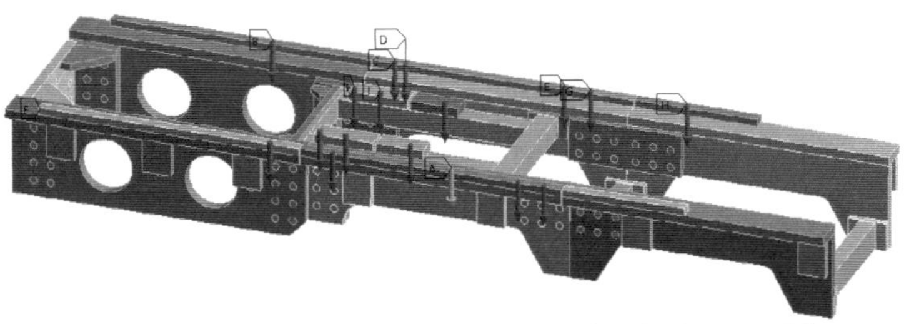

Figure 7: Extracted TRAM and services loads applied to the frame model

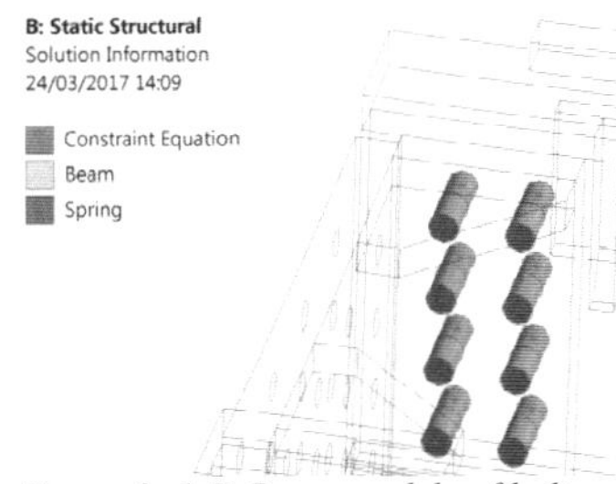

Figure 8: 1-D Beam models of bolts with "Spider" connections to holes

3. Structural Analysis

Ansys Workbench FEA software [1] was used to assess the stress state and maximum deflection in the frame. In order to model the stress state in the frame, the loads for each of the components in the TRAM assembly needed to be extracted and modelled (as shown in Figure 7). A baseline model was then developed, with verification using hand calculations approximating the loading state of the frame. From this, a variety of different load cases were evaluated, and the effect of different design features, or material choices could be evaluated for designers. The deformation and Von-Mises Equivalent Stress results are shown in Figure 9 and Figure 10 respectively.

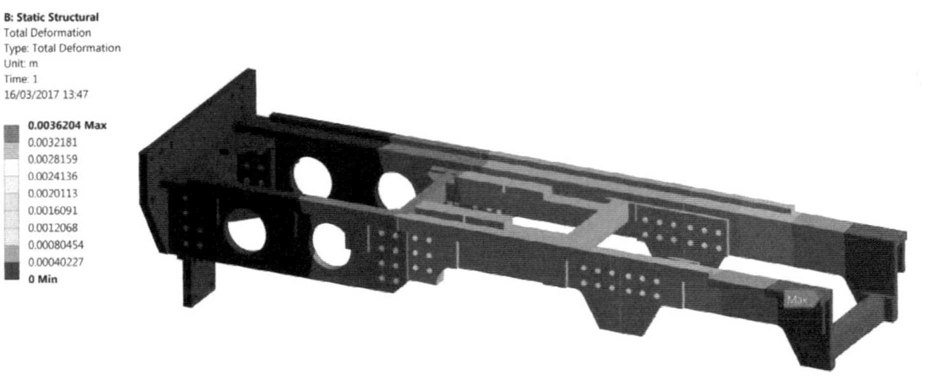

Figure 9: Deformation results, assuming bonded contact

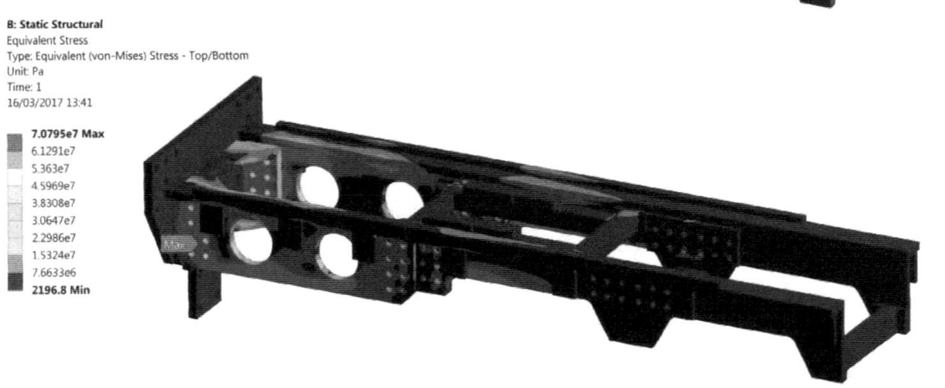

Figure 10: Equivalent Stress results, assuming bonded contact

Further information can be extracted from this model, such as the expected discontinuities in the rails along which the reflector wheels roll at the transitions between the plates (shown in Figure 11).

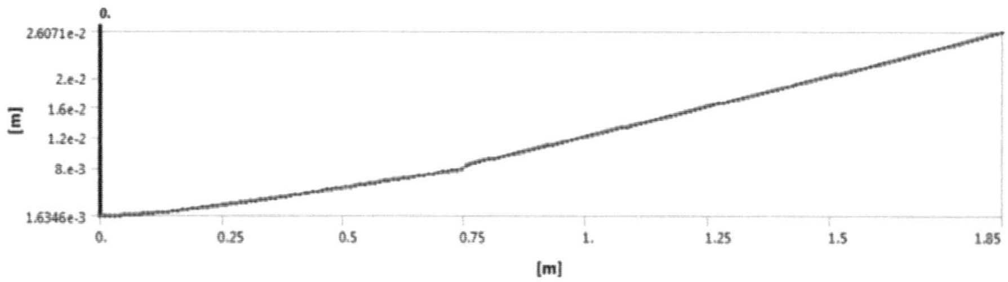

Figure 11: Vertical deflection along the right-hand rail showing discontinuities at the plate transitions

4. Bolt Pattern Analysis

A significant challenge was to quantify the loads and stresses on the bolts and plates due to the combined moment, shear and torsion load on each of the bolt patterns. Hand calculations gave initial estimates for the bolt loads, but only for a simplification of the loading on the frame. Using Finite Element Analysis of the full frame, with realistic load locations, yielded the additional shear and axial bolt-loads associated with torsion due to unequal loads on each side of the frame. This was done using 1-Dimensional Beam Elements, a method of bolt modelling requiring comparatively low computation time; this was crucial to be able to model all 68 bolts on a single workstation. This simplified methodology gives quite poor local stress results due to the unrealistic rigid bonds at the head and the nut of the bolt [2][3], but generates accurate bolt loads; these loads could then be extracted and compared with the recommendations from standards for Stainless Steel bolts and frames [4][5].

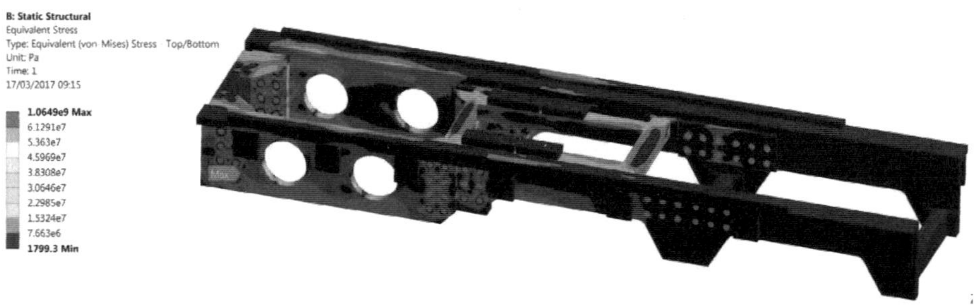

Figure 12: Equivalent Stress results for Bolt Simulation

5. Conclusions and Further Testing

The first (and crucial) requirement of the analysis was to verify that the frame is capable of supporting the applied loads for all load cases; this was shown to be the case, with a significantly higher factor of safety than the current installed frame. Additional investigations were also conducted, exploring design concepts such as alternative materials, different bolt load-transfer methods, and the impact of additional shielding. The results showed that alternative materials (such as Aluminum) could feasibly be used for the frame (although would represent other challenges in manufacture and installation). Another useful conclusion was that the size and number of bolts in each bolt pattern could be reduced, thereby reducing the amount of work required for end-of-life disassembly of the frame.

References

[1] ANSYS® Mechanical Enterprise, Release 17.1, ANSYS, Inc.

[2] Montgomery, J. 2002 *Methods for Modelling Bolts in the Bolted Joint*, Ansys Users Conference.

[3] Abbey, T. 2014 *Joints and Connections in FEA* [Online] http://www.digitaleng.news/de/joints-connections-fea/

[4] Baddoo, N. R. and Burga, B. A. 2012 *Structural Design of Stainless Steel – SCI Publication P291,* The Steel Construction Institute

[5] British Standards Institution *BS EN 1993-1-8:2005 Eurocode 3: Design of Steel Structures*

IOP Conf. Series: Journal of Physics: Conf. Series **1021** (2018) 012078 doi:10.1088/1742-6596/1021/1/012078

Liquid hydrogen for cold neutron production at European Spallation Source ERIC

Jesper Ringnér[1], M. Klaus[2], J.Jurns[1], Y.Beßler[3], D.Lyngh[1]

[1] European Spallation Source ERIC, Sweden
[2] Institute of Power Engineering, Technische Universitaet Dresden, Germany
[3] Forschungzentrum Juelich, Germany

jesper.ringner@esss.se

Abstract. The European Spallation Source (ESS) will be a 5 MW, long-pulsed spallation neutron research facility. One of the key feature is that ESS will use liquid hydrogen as a moderating media for the cold neutrons. The hydrogen operates at cryogenic temperatures at approximately 17K. The challenge for the cryogenic system is to meet the high neutronic heat load and maintain a narrow operational span. To handle the large variations in heat load a special developed pressure control device is designed. To maintain the hydrogen operating temperature, a Helium refrigerator with unique capabilities to control temperature in the hydrogen system is developed. It also has the capacity to react and compensate for a lower heat load than anticipated in the moderator system on short notice. Another key feature is the Ortho-Para catalyst and the in line OP measurement that verifies the high para content at all time, to optimize neutronic performance. This paper describes the process development, planned commissioning and operation of the cryogenic hydrogen and ancillary systems.

1. Development of the Cryogenic Moderator System (CMS)

1.1. The first conceptual design. [1]

The CMS at ESS was more or less a scaled up version of the SNS and J-Parc hydrogen systems [2], [3]. To cope with the larger power of ESS proton beam everything was 5 times bigger than J-Parc (1MW vs 5MW). At this stage none of the components in the system was detailed for the new requirements.

The specification was to keep the liquid para hydrogen at a supercritical pressure (>1.3MPa), and the temperature around 20K for optimal neutronic performance. Design pressure was 20 bara and expected heat load was 20 kW in the volume moderators. Moderator was similar to SNS/J-Parc.

To maintain the pressure in the loop, a pressure control buffer was included, consisting of two parts. The first was an accumulator with steel bellows, back pressured with helium to accommodate volume changes in the system. The second was a heater that compensated for the loss of beam power and maintained the heat load that the Target Moderator Cryoplant (TMCP) would expect. The main idea was to always feed the TMCP with a 20kW heat load regardless of whether or not the beam was on. A simplified schematic of the proposed system is shown in Figure 1.

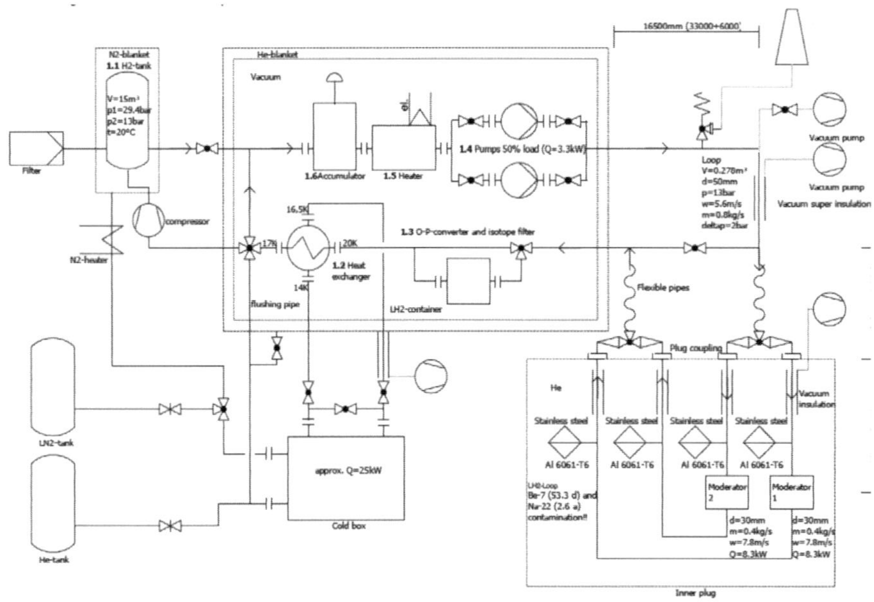

Figure 1, Concept P&ID courtesy of Y. Beßler

The system was set up with two pumps that operated in parallel to ensure availability. Both pumps should provide the full pressure head in the system (roughly 1.2 bar), but only half of the flow (400g/s), each. If one pump failed the other one would speed up to compensate for the flow loss. One drawback with this set up was the complicated design of an impeller to coop with two different flows still meeting a high pressure head.

An Ortho-Para catalyst was foreseen in the system but the measurement method to verify the para content was not specified.

The hydrogen system was intended to be located close to the monolith in the A2T area and the helium cooler located just above inside the high bay as shown in Figure 3.

1.2. Preliminary design to final design

Along the road to a viable preliminary design, some drawbacks appeared while attempting to scale up the design. First of all the intended location of the TMCP was moved to the cryogenic building and co-located with the Accelerator helium cryo plants. This was done to gather all helium cryoplants in the same area. This resulted in a 335m long cryogenic transfer lines to connect to the CMS heat exchanger. The hydrogen room was also moved because of safety regulations. Originally located in the A2T area it was now located on the top floor of the target building as shown in Figure 3.

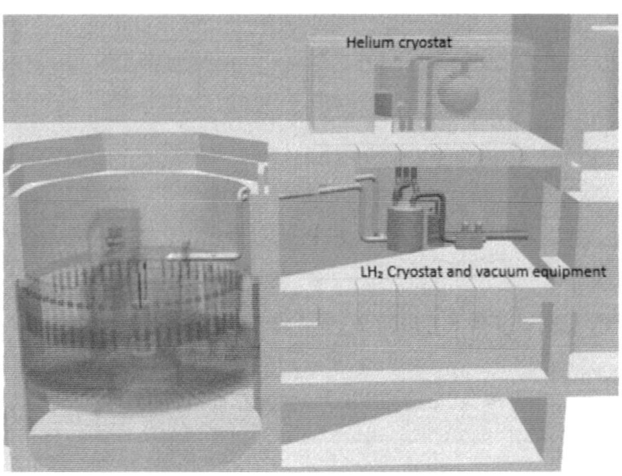

Figure 2, Original layout A2T room with TMCP above.

Figure 3, New routing through the building. CMS connected to the moderator plug. TMCP moved 300m outside.

In the spring of 2014, the neutronic team made some quantum leaps in moderator design and in enhanced brightness. First out was the pancake design which later was replaced with the current butterfly 2 design, see Figure 4.

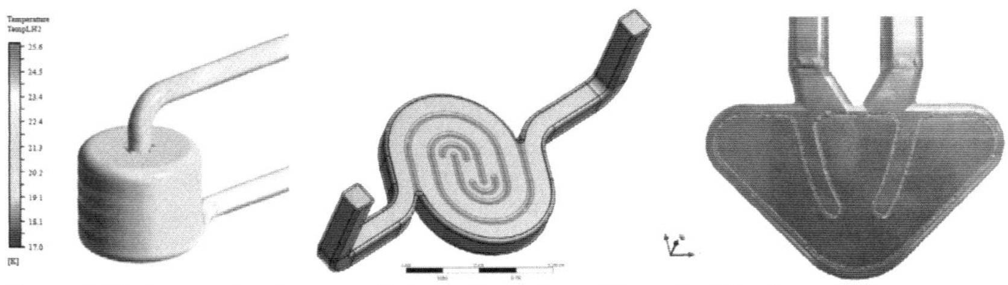

Figure 4, Moderator development: Volume, Pancake and butterfly 2(implemented).

The new refined design resulted in a new neutronic heat load of 19kW (which includes some contingency). For the hydrogen system this meant that when adding static heat loads, pump efficiency etc. the expected heat load/cooling capacity will be increased to roughly 28kW. Calculations showed that the accumulator would be large. The heater that was intended to compensate for beam power, would contain a large amount of liquid hydrogen and have too slow response time. As the overall requirement is to minimize the total H_2 inventory which was increasing due to the move of the equipment, the accumulator/heater was shown to be an unacceptable solution. The possibility to use a buffer vessel was investigated and an actively controlled buffer was implemented as the solution. The heater was removed and the decision to let the TMCP handle the temperature swings in the hydrogen was taken.

Two pumps in parallel was discarded for two pumps in a serial set-up. A benefit is that each pump provides half the pressure head but the full flow. If one pump fails the system has the possibility to speed up the remaining one to maintain the full flow and the full pressure head.

For the Ortho-Para measurement a Raman spectroscopy with an in-line measurement is decided. For the catalyst Oxisorb is the first choice but the design will coop with Ionex as well. Ongoing research on the materials kinetic data is ongoing and a part of ESS scope. Sapphire glass viewports are to be installed in the hydrogen pipeline which enables sensors to look into the hydrogen flow.

High requirements on the moderator vessel in terms of reliability resulted in a change of design pressure to 17bar. We assumed that keeping the system pressure range between 13-17bar at all time was an impossible task. To make sure that the system doesn't blow of the hydrogen unintentional, an investigation started to clarify if it's possible to reduce the system pressure. The biggest risk we foreseen was boiling and two phase flow in the moderator vessel. With an acceptable margin in temperature, the decision was taken to increase operational range to 9-17bar during operation.

1.3. Target Moderator Cryoplant, (TMCP)

The initial design of the TMCP [1] followed the conceptual design of the SNS system, with a fairly simple cryoplant operating at constant heat load of 25 kW at 16 K. Due to the challenges of using a heater in the hydrogen circuit to compensate for beam power, ESS decided to allow the TMCP to perform this function. Additionally, as the required cooling capacity of the TMCP will vary significantly especially during the first years of operation as beam power ramps up, the TMCP design evolved to allow significant turndown in capacity. Compensating for beam power is accomplished by bypassing a small flow from the cold helium supply to the CMS, warming it with an ambient heater, and introducing the warm gas to the helium return line from the CMS. This allows the TMCP to operate at a constant heat load in the case of short-term beam trips. In the case of longer periods with low or no beam power, the cryoplant is designed with two parallel warm compressors and two parallel cold turbines. Turning off one of the compressor/turbines can easily cut the cooling capacity in half. Cooling capacity can be further reduced by discharging a portion of the helium from the cooling loop to warm helium buffer tanks, thereby reducing the system pressure and helium mass, while maintaining a relatively constant pressure ratio between the low and high pressure sides of the refrigerator. The TMCP can operate between 2.3 and 30.3 kW capacity, providing helium at 15 K to the CMS for hydrogen cooling. A simplified schematic of the TMCP and is shown in Figure 5.

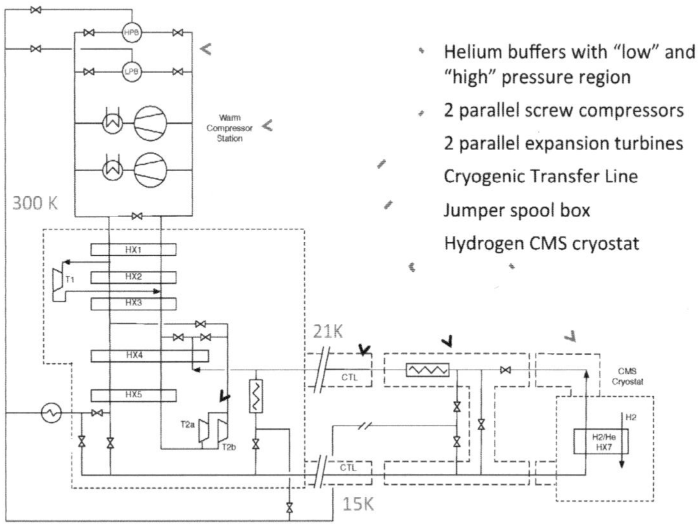

Figure 5. P&ID of the TMCP

2. Manufacturing, assembly and installation

A design review for early procurement was performed in spring 2016 for the pumps and they were purchased early autumn the same year. This was necessary due to the roughly one year long lead time. They are planned for installation into the cryostat August 2017.

The hydrogen transfer lines specification has been finalized and a call for tender was issued spring 2017. Installation and testing are expected early 2019.

A Critical Design Review for the CMS was successfully performed in February 2017 and components and instruments will now be purchased and manufactured.

The cryostat and all designed parts will be produced in-house by our In-kind partner Forschungszentrum Jülich (FZJ). They are also responsible for assembling, testing of components, installation, Factory Acceptance Test and Site Acceptance Test. The process development and cryogenic know-how are developed by Technische Universitaet Dresden (TUD).

Planned access to the Target building/hydrogen room is March 2019 and installation of the cryostat, hydrogen transfer lines and the helium transfer lines will take place at the same time. The TMCP is delivered by Linde and will be ready to perform Site Acceptance Test (SAT), in Q2 2019 and the CMS will be ready for SAT February 2020.

Scheduled beam on target is March 2020.

3. Commissioning and start of operation

Beam on target ends the construction phase and take the project into the commissioning phase. The beam power during commissioning will be low thus also the expected heat load on the hydrogen system. During this time we plan to fine tune the process controls, learn the system behavior and make the system as robust as possible. First users expect us to deliver cold neutrons for the instruments 2023.

References

[1] ESS Technical Design Report ESS-0016915
[2] Technical Design Report of Spallation Neutron Source Facility in J-PARC, 2012 JAEA
[3] Spallation Neutron Source Project Completion Report, SNS 100000000-BL0005-R00

IOP Conf. Series: Journal of Physics: Conf. Series **1021** (2018) 012079 doi:10.1088/1742-6596/1021/1/012079

Cryogenic Moderator Design for the ISIS TS1 Project

M. Capstick[1]

[1.] ISIS Neutron and Muon Source, Rutherford Appleton Laboratory, Harwell Campus, Didcot, OX11 0QX, UK

[1.]matthew.capstick@stfc.ac.uk

Abstract. A project to improve the neutronics and serviceability of the ISIS Target Station One TRAM (Target, Reflector, and Moderators) is being carried out. This includes a redesign of the cryogenic methane and hydrogen moderators to improve functionality with the new target and reflector, whilst accommodating new water pre-moderators. Both moderators consist of a high pressure, low temperature fluid vessel contained within a vacuum vessel. Finite Element Analysis along with Parametric Optimisation has been used to design and optimise all the moderator vessels. Buckling analysis was investigated as a method of verifying the moderator vacuum vessels and used to show that the vacuum vessels are conservatively designed.

1. Introduction

The ISIS Target Station One (TS1) project is currently in the design and implementation phase. The project includes a redesign of the cryogenic methane and hydrogen moderators. This paper will cover the design challenges encountered and the computer modelling used to support the design process. This includes the optimisation carried out on the moderator fluid vessels and an investigation into the potential use of computational buckling analysis.

1.1. TS1 Project Cryogenic Moderators

ISIS TS1 employs two cryogenic hydrogenous moderators; one containing liquid methane, the other containing liquid hydrogen (Figure 1). Both of these moderators consist of a high pressure, low temperature fluid vessel contained within a vacuum vessel. The hydrogen moderator also employs a tertiary containment vessel. As can be seen in Figure 1; both moderators are positioned directly below the tantalum/tungsten target. As part of the project, water pre-moderators are being introduced between the cryogenic moderators and the target. The hydrogen moderator supplies neutrons to instruments on one side of the target station, whereas the methane moderator supplies neutrons to both sides of the target station. To prevent back scattered neutrons passing through the methane moderator, two gadolinium poison foils are placed in the centre of the moderator.

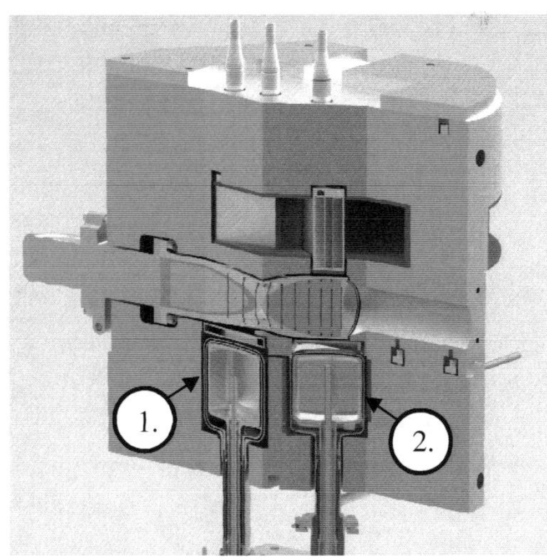

Figure 1: *A section of the ISIS TS1 TRAM showing the hydrogen (1.) and methane (2.) cryogenic moderators.*

1.2. Moderator Design Process

The neutronic design team established ideal moderator dimensions. These dimensions provided a starting point for the engineering design process to work from. The design was adjusted from the ideal moderator dimensions to improve the manufacturability and usability of the moderator vessels before being passed back to the neutronics team. This iterative process can be continued until a final, highly optimised design is produced. As part of the engineering design process, the Finite Element package ANSYS was used to ensure the peak stress and maximum deflection met the specification. Parametric Response Surface Optimisation was also used to further optimise the dimensions of the moderator vessels. Due to the variable wall thickness of the moderator fluid vessels, a hex-dominate volumetric mesh was used for all the analysis of the fluid vessels. The vacuum and tertiary vessels have a near constant wall thickness therefore shell elements were used as a reasonable simplification.

2. Parametric Optimisation

The moderator designs were optimised using Parametric Optimisation with the FEA package ANSYS. This process uses a statistical model between the key geometric parameters (e.g. the wall thickness or internal radius) and the key output parameters; in this case the peak stress and maximum deflection. Figure 2 shows the relative influence each parameter has on the peak stress. As expected, the wall thickness is the most significant input parameter and is inversely related to the peak stress. An allowable wall thickness was agreed with the neutronic design team therefore the internal radii (the second most significant factor shown on Figure 2) became a focus of the optimisation.

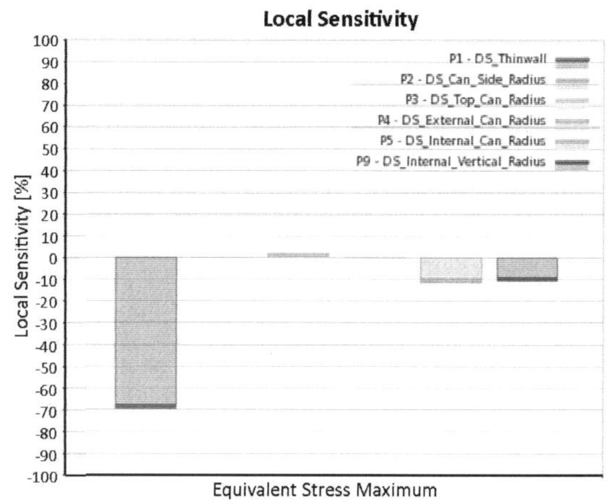

Figure 2: *A bar chart showing the influence each parameter has on the peak stress within the methane moderator fluid vessel.*

During parametric optimisation the statistical model is used to determine the specific set of input parameters which result in the required output, as defined by the specification. Figure 3 shows the theoretical relationship between wall thickness, internal radius, and peak stress according to the statistical model. Figure 4 shows specific design points within this range with the constraints applied; each coloured point represents a feasible set of parameters for the design. By applying the allowable limits to every parameter, a highly optimised design which meets the specification can be selected.

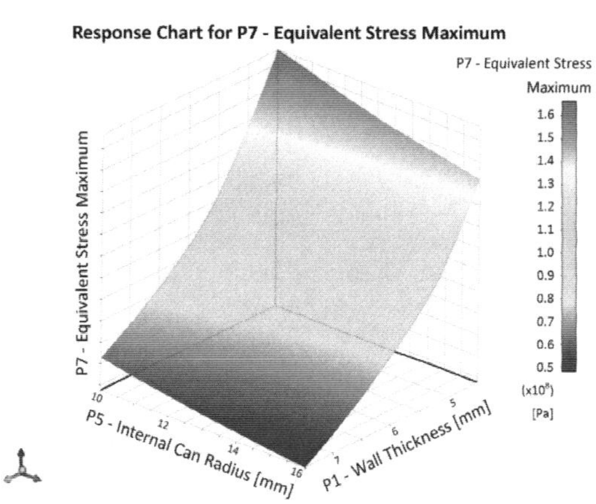

Figure 3: *A surface plot of the relationship between wall thickness, internal radius, and peak stress.*

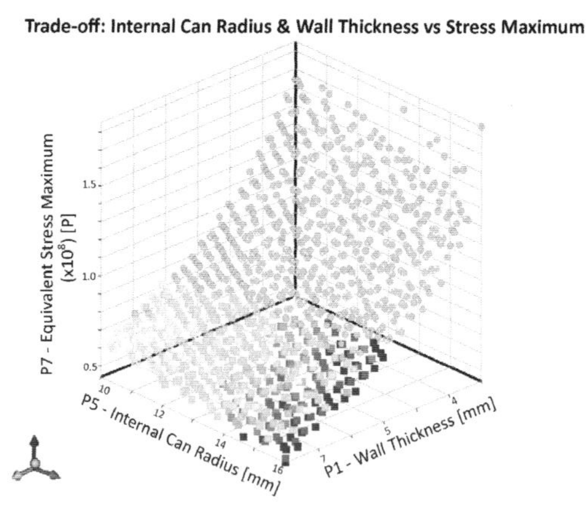

Figure 4: *Specific design points with the wall thickness, internal radius, and peak stress constraints applied; a coloured point represents a feasible design.*

IOP Conf. Series: Journal of Physics: Conf. Series **1021** (2018) 012079 doi:10.1088/1742-6596/1021/1/012079

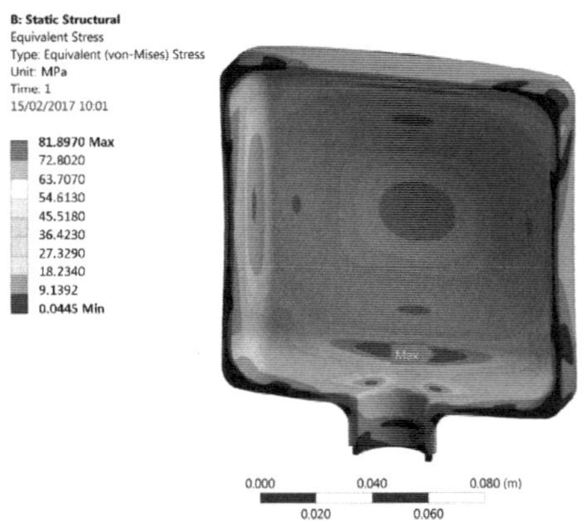

Figure 5: *The stress profile within the methane moderator fluid vessel under peak internal pressure.*

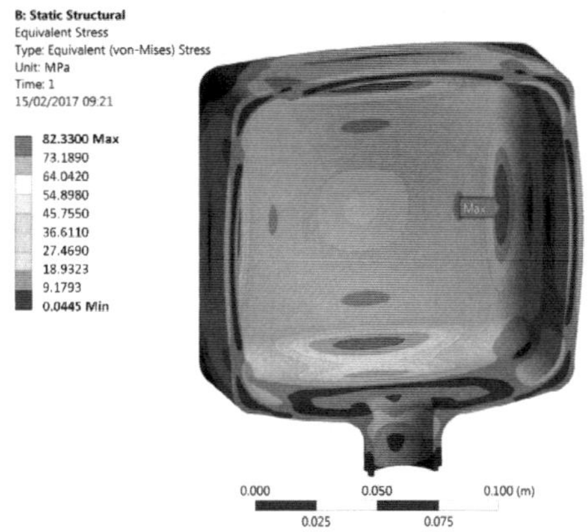

Figure 6: *The stress profile behind the hydrogen moderator fluid vessel under peak internal pressure.*

2.1. Optimised Designs

The process of parametric optimisation (Section 2) allowed the geometry of all the moderator vessels to be optimised while ensuring the design met the specification. The specification stated that the peak stress within all moderator vessels must be lower than 83MPa, in accordance with PD5500 for the Aluminium Alloy 5083-O used for moderator vessels at ISIS [1]. The limit of 83MPa has previously been used for the computational verification of moderator vessel designs at ISIS; and therefore was also deemed suitable for this analysis. As part of the geometry optimisation the internal radii were decreased in order to minimise the apparent wall thickness around the edges of the vessel, and therefore reduce unnecessary beam-loss in these areas. However, as can be seen in Figure 5 and Figure 6, the internal radii of the fluid vessel are also the areas experiencing peak stress. The optimisation process allowed designs to be produced with minimal internal radii without exceeding the maximum allowable stress. This process was repeated for all the internal and external radii, as well as the radius of the top and the bottom of the vessel itself. The curvature of the top of the vessels was also reduced in order to accommodate the water pre-moderators that have been introduced as part of the TS1 project.

The current design provides a nominal 4mm gap between the undeformed cyrogenic vessel and the vacuum containment vessel. The maximum deformation at this point was limited to less than 0.5mm in the centre of the moderator face so as to prevent thermal contact between the internal cryogenic vessel and the surrounding vacuum vessel during peak deformation of both vessels. An example of the predicted deformation of the methane fluid vessel under maximum operating pressure is shown in Figure 7.

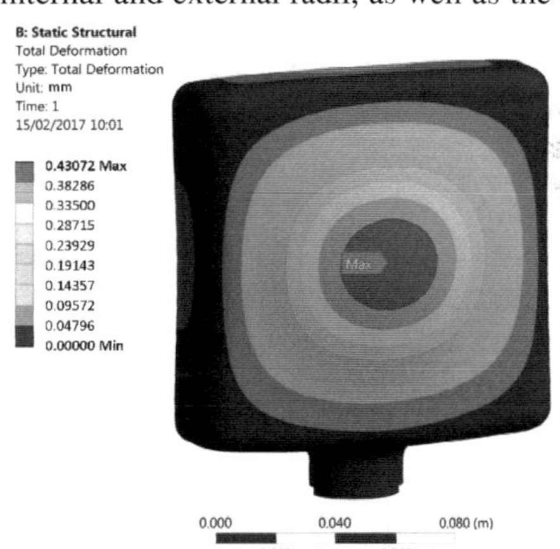

Figure 7: *The deformation of the methane moderator fluid vessel under peak internal pressure.*

3. Buckling Analysis

The vacuum vessels were verified in ANSYS to ensure they would not fail in buckling. Computational analysis has not previously been used at ISIS for buckling analysis of moderator vessels; however it has the potential to allow less conservative designs, therefore has been investigated here.

3.1. Eigenvalue Buckling

The specific buckling analysis used was eigenvalue buckling within ANSYS. This method assumes that the material behaves linearly elastically up until the point of buckling, which generally leads to an overestimating of the load which the moderator can withstand before buckling occurs [2]. Due to the fact that computational analysis has not been used for this purpose at ISIS before, some validation of the method was carried out first.

3.2. Validation

Previous vacuum vessel designs have been verified using PD5500; considering the external surface of the moderator vessel to be a section of a cylinder under external pressure [1]. By ensuring that the critical buckling pressure for this cylinder is greater than the max operating pressure of the vacuum vessel, the vessel has been deemed safe. When this method was replicated by analysing the same cylinders using ANSYS the elastic instability pressure was found to agree with the hand calculations to within 10% for all designs. This was taken as a demonstration that the eigenvalue buckling analysis method can predict the elastic instability pressure consistently with the previously used methods.

3.3. Analysis Results

Figure 8 and Figure 9 show the primary buckling mode for the methane and hydrogen vacuum vessel respectively. In these cases the elastic instability pressure is 7.3Bar and 14.9Bar, which equates to pressures 1.5 times greater than the pressure calculated previously using PD5500. This reassures the previous assumptions that the vacuum vessels are a conservative design, and that less conservative designs could be investigated in the future, however less conservative designs would require further analysis and testing before use due to the overestimation typical of eigenvalue buckling analysis [2].

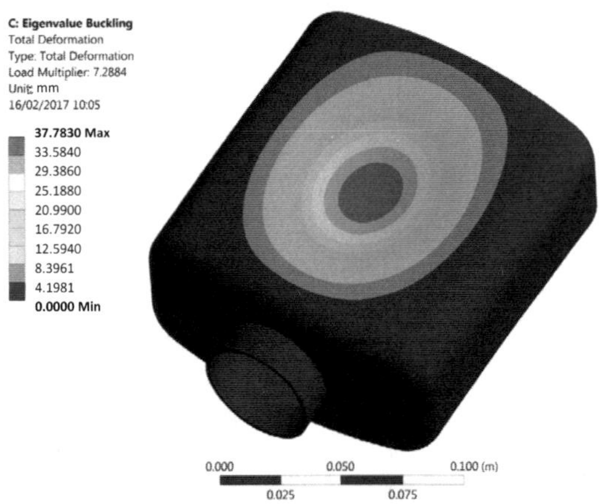

Figure 8: *The primary buckling mode of the methane vacuum vessel.*

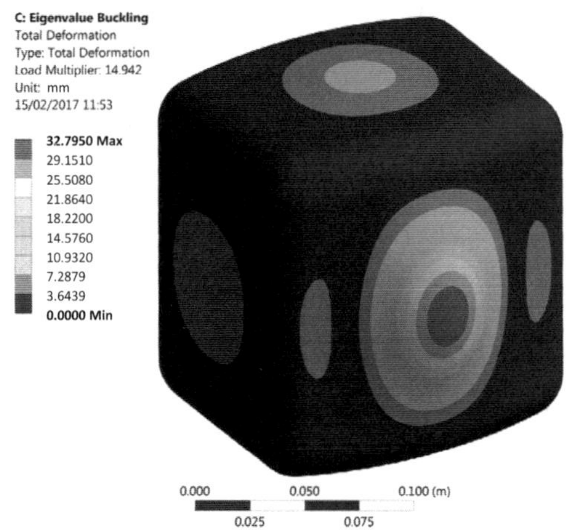

Figure 9: *The primary buckling mode of the hydrogen vacuum vessel.*

4. Conclusions

The cryogenic moderator designs for the TS1 project have been parametrically optimised to closely match the ideal neutronic design while meeting the engineering specification for all the vessels. The peak stress within all the moderator vessels in below 83MPa as required to prevent structural failure. The max deformation of all the vessels will not produce thermal contact between the cryogenic vessel and the surrounding vacuum vessel. Buckling analysis has been investigated and shown to agree with the PD5500 standard when analysing a conservative simplification of the vacuum vessels. Finally, when considering a non-conservative analysis of the moderator vessel geometry, an instability pressure at least 1.5 times greater was predicted. This suggests further optimisation is possible, however it would require physical testing to verify.

References

[1] PD 5500:2015 *Specification for unfired fusion welded pressure vessels* (The British Standards Institution: 2015)

[2] Kurowski, PM. Kurowski, TP. *Buckling Analysis with FEA* (Machine Design, London, Ontario, Canada: 2011)

IOP Conf. Series: Journal of Physics: Conf. Series **1021** (2018) 012080 doi:10.1088/1742-6596/1021/1/012080

The Cryogenic Moderator System Cryostat Design for the European Spallation Source

Y Beßler[1], M Klaus[2], C Henkes[1], G Natour[1] and J Ringnér[3]

[1] Forschungzentrum Juelich GmbH, Juelich, Germany
[2] Institute of Power Engineering, Technische Universitaet Dresden, Germany
[3] European Spallation Source ERIC, Lund, Sweden

E-mail: y.bessler@fz-juelich.de

Abstract. The European Spallation Source in Lund, Sweden, is going to be a neutron scattering research center that aims to provide around 30 times brighter neutron beams than any other existing facility. As one subsystem of the Target Station, the Moderator & Reflector System slows down high energy neutrons from the spallation process. To gain maximum neutron flux intensities along with high system availability for condensed and soft matter research, an optimized liquid hydrogen Cryogenic Moderator System (CMS) has been developed. Hydrogen with a pressure below critical, an inlet temperature of around 17 K, and a parahydrogen fraction of at least 0.995 will be utilized to interact with neutrons in a unique Moderator vessel arrangement. Two turbo pumps are arranged in series and circulate the cryogen. The pressure stabilization is achieved by an active pressure control buffer in a bypass stream between high pressure and low pressure side. Hydrogen conversion from ortho- to parahydrogen will be controlled and monitored using a catalyst bed in a second bypass line between the high and the low pressure branch of the circuit. A helium refrigerator, the Target Moderator Cryoplant (TMCP), continuously recools the hydrogen mass flow.

1. Introduction to the ESS Cryogenic Moderator System

ESS is going to provide a long-pulsed cold and thermal neutron beam with highest brightness for a broad range of experiments. The main task of the Cryogenic Moderator System CMS is to supply the moderating media hydrogen to the cold Moderator vessels (cM) with a circuits average temperature of about 18.5 K (cM inlet controlled to 17.4 K) and an average pressure of 10 bar.abs (min. value at cM outlet). The cM's facilitate neutrons passing through the hydrogen, where they interact, get slowed down and leave the cM at useful energies. Two different moderator types with different geometry (30 and 60 mm LH_2 height, Butterfly 2 design), neutronic features and cooling demands are used at ESS. Subcooled liquid hydrogen, specifically parahydrogen (antiparallel proton spin), needs to be supplied continuously to the four cM's within tight operating boundaries and a high system availability.

The general CMS mission is to reduce the speed of higher energy neutrons set free during the spallation process by supplying a moderation media and producing cold (slow) neutrons, complying with the following key requirements [1, 2]:

- Cryogenic hydrogen at average 18.5 K and elevated pressure of 10 bar.abs (subcritical, subcooled LH_2) with a parahydrogen content of > 99.5 % shall be fed to the cM's,
- The hydrogen shall be kept within operational boundaries at all times to ensure availability,
- The hydrogen shall be provided in a safe and reliable way,

- The average temperature increase over the cM's due to a 5 MW proton beam and spallation neutrons shall be kept ≤3 K.

The overall target cryogenics at ESS [3 to 6] can be separated into the Target Moderator Cryoplant TMCP (helium refrigeration plant with about 30.2 kW cooling power at 15 K) which interfaces to the Cryogenic Moderator System (closed liquid hydrogen loop) via two heat exchangers inside the CMS cryostat. Figure 1 shows the overall system setup of the CMS and its peripherals.

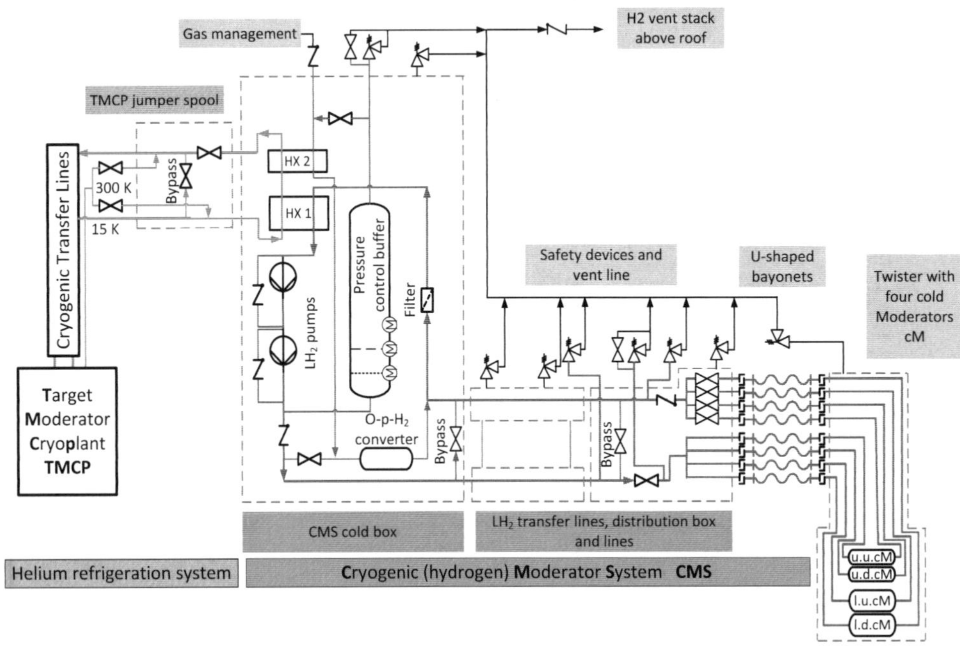

Figure 1: Schematic flow diagram of Cryogenic Moderator System (blue), Target Moderator Cryoplant (green) and vacuum sections (dotted red lines).

The flow diagram shows the two process media: The TMCP and its helium refrigeration cycle in green whereas the closed hydrogen circulation of the CMS is in blue color. The neutron moderation process results in average 17.3 kW as a dynamic heat for a 5 MW proton beam. The absorbed heat caused by neutron interaction with aluminum and hydrogen inside the cM's needs to be extracted by a steady hydrogen circulation. The static heat load as a sum of heat transfer from higher to cryogenic temperatures as well as the power input from the circulation pumps add 5.9 kW to the hydrogen. The total heat load adds up to 23.1 kW.

2. General description of the hydrogen circulation: Cryogenic Moderator System

The task of the CMS is cooling four cold Moderators, take up the neutron heat during the moderation process as well as static heat loads from the CMS loop in a safe and reliable manner. Subcooled liquid hydrogen at average 18.5 K and 10 bar.abs is used as the moderating and cooling media. Thus, the parahydrogen critical pressure of 12.85 bar.abs is not reached and temperatures above the boiling point (ca. 30.2 K at 10 bar.abs) result in vapor formation and phase separation. The design pressure of the system is 17 bar.g constrained by the structural limits of the complicated cM vessel geometry as the CMS most fragile components. The overall heat load gets transferred to the TMCP helium flow in the main H_2-He heat exchanger. Additionally, the system consists of two turbo pumps in series, a pressure control buffer in a bypass and an ortho-parahydrogen converter in a bypass. The CMS cryostat is placed in the upper level of the target building in a separate, ATEX classified room. Transfer lines get routed to the target ground level and close to the Connection Cell. A distribution box, splitting up the single hydrogen supply flow into four streams, sits close the Connection Cell wall. From here four

IOP Conf. Series: Journal of Physics: Conf. Series **1021** (2018) 012080 doi:10.1088/1742-6596/1021/1/012080

supply distribution lines get routed into the Connection Cell and further into the Monolith vessel. Inside the Twister four cM, two of 30 mm LH_2 height above and two of 60 mm LH_2 height below the Target wheel, get cooled by liquid hydrogen and moderate neutrons.

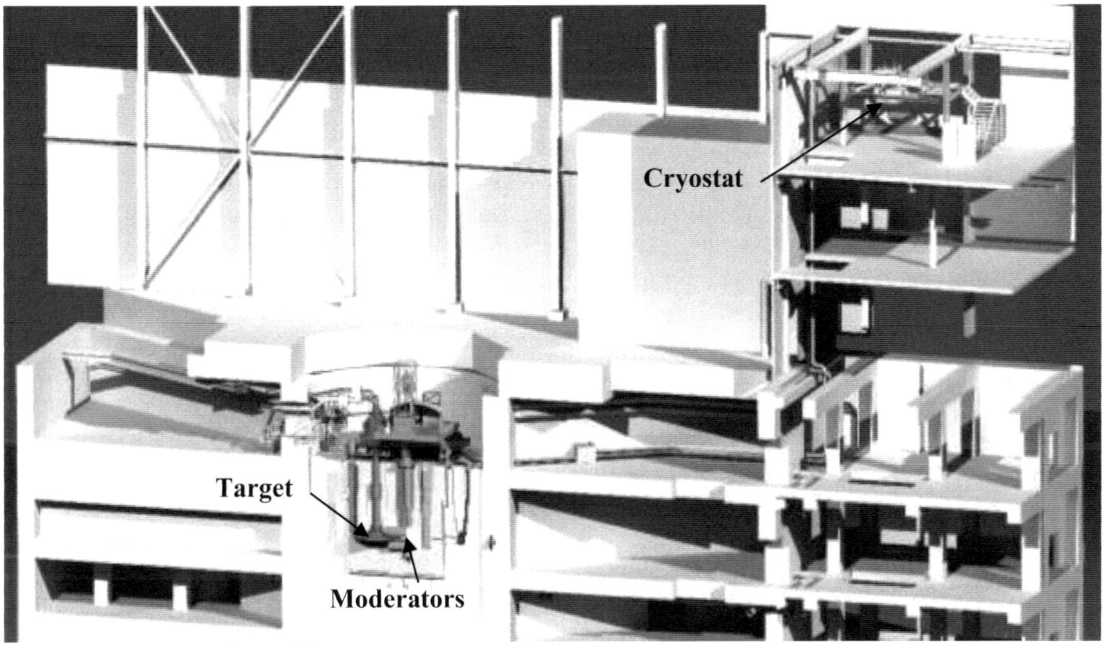

Figure 2: Isometric view of the ESS Target Station

Two identical hydrogen pumps are designed to circulate 1000 g/s hydrogen at 17 K and 9.1 bar.abs inlet conditions to the serial arrangement. A variable frequency drive VFD for each pump allows for speed control. The pump head is 0.94 bar per pump, which adds up to overall 1.87 bar to overcome the circuits pressure drop. Due to required redundancy, the pump design allows to operate the CMS with only one pump by increasing its speed and thereby operating at the same total pump head and mass flow. A bypass with a non-returning check valve over each turbo pump ensures sufficient and directed flow. Continuous circulation is vital for a safe, reliable and stable CMS operation.

The pressure control buffer PCB is a vertical vessel of 45 l volume in a bypass of the high to the low pressure side of the hydrogen circulation. Since the main circuit's subcooled liquid hydrogen behaves like an incompressible fluid, changes in the neutron heat load and fluid temperature would cause severe pressure fluctuations for a closed circuit at constant average density. A pressure control mechanism is required to stabilize the average CMS pressure level. Hence, the PCB is connected to the piping downstream of the second turbo pump and holds a liquid volume in its lower part. Above the liquid level a pressurized vapor fraction acts as gas spring or accumulator for the CMS pressure control system [7]. Liquid hydrogen can be stored or set partially as liquid storage volume to the loop by the lower bottom connecting pipe in case of a pressure increase or decrease because of changing neutron heat loads. A control valve on top of the PCB top enables vapor release to the low pressure side of the CMS. The gas gets liquefied in a coiled fin-tube heat exchanger and passes the ortho-parahydrogen converter before entering the bulk liquid hydrogen flow. Pressure overshoots after neutron heat ramp up can be effectively mitigated by doing so. For the other case of a loss of beam heating, electrical heaters on the PCB at the lower part boil liquid and restore extracted vapor avoiding too low pressures. Thus, the pressure control buffer provides volume compensation to the CMS.

The ortho-parahydrogen converter is a vessel of ca. 35 l volume and filled with a commercial o-p-H_2 catalyst (30 l bed volume), preferably chromium oxide on silica gel or ferrous oxide. It is arranged in a bypass of the high to the low pressure side of the hydrogen circulation and flow control is done by

IOP Conf. Series: Journal of Physics: Conf. Series **1021** (2018) 012080 doi:10.1088/1742-6596/1021/1/012080

a cryogenic control valve. At the moderator inlet, a parahydrogen concentration of $\geq$ 99.5% is required, because the moderator vessel geometry is optimized for max. neutron brightness and pure parahydrogen as moderating media. Orthohydrogen could decrease the moderation performance significantly, absorb neutrons instead of slow them down and let them pass. This reduces the number of useful cold neutrons and increases the dynamic heat load.

Cryogenic transfer lines supply liquid hydrogen from the CMS cold box towards the Twister that holds the four cM. For the first 40 m the bulk flow is carried in a single vacuum insulated pipe and gets split into four distribution lines in a distribution box. From here four lines are routed ca. 10 m inside the Connection Cell and Monolith vessel close to the upper part of the Twister. Removable u-shaped bayonets connect each distribution line with Twister piping to the cM and back. The parallel flow allows operating each moderator with the same temperature and results in less pressure loss. Drawbacks are an increased hydrogen inventory and flow distribution compared to a serial flow through the cM's. Four distribution return lines connect again to the distribution box. Four manual fine control valves adjust the flow so that 240 g/s are being supplied to each of the two upper and 260 g/s to each of the two lower cM's. The four individual flows get united again and take the bulk return line to the CMS cold box. A filter upstream of the plate-fin heat exchanger and turbo pumps prevents any debris or particles to block the heat exchanger, harm the pump impellers or block valves. Passive, spring-loaded safety relief valves protect the CMS process pipes. At least one pressure relief device protects each insulation vacuum space from overpressure. Two actively controlled valves allow H_2 release in case of pressure excursions before the spring-loaded safety relief valve lifts. Every source of hydrogen release is connected to a shared vent line to the outside of the building. The vent line features a small flow inert helium gas with an overpressure because of flap at the end above the building's roof. Explosive mixtures with hydrogen can be excluded at all times.

3. Cryostat design as result of CMS process design

The main components of the cryostat (Figure 3) are the vacuum vessel consisting of three parts (a valve plate, an upper section and a lower part), the pipes for the process gases (helium and hydrogen) and related cryogenic components, i.e. control and check valves, heat exchangers, pumps, a pressure control buffer and an ortho-parahydrogen converter. The weight of the cryostat adds up to 2.9 t with a height of about 3.1 m from the bottom to the upper control valves and its diameter is about 2.5 m.

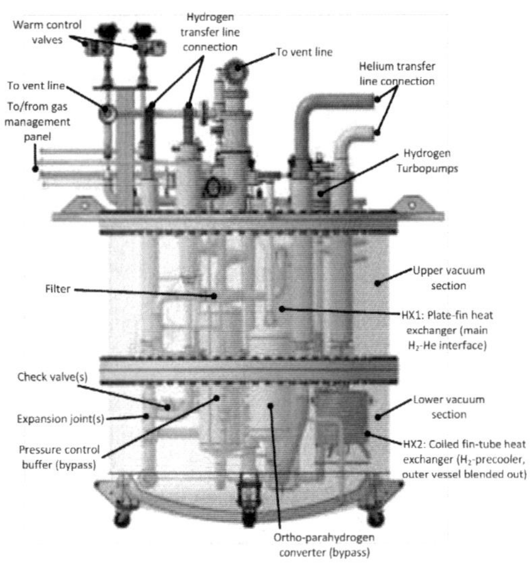

Figure 3: CMS cryostat vessel with most of the hydrogen systems functional components.

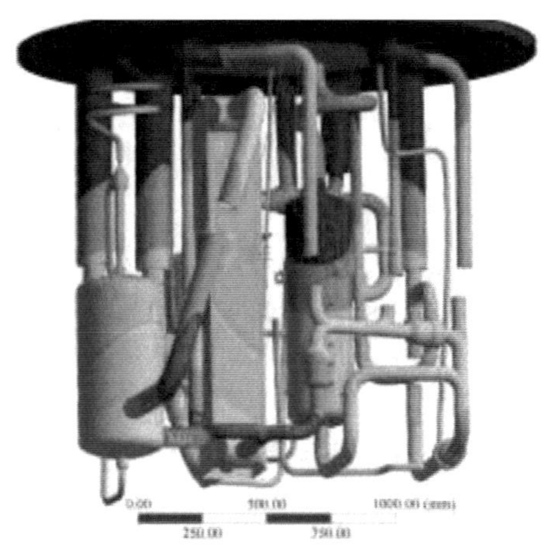

Figure 4: Thermal simulation result of total deformation during cold operation (linear color gradient: blue min. 0 mm to red max. 8.4 mm).

Figure 4 shows an exemplary result of the conducted thermal and mechanical analysis for the verification of the CMS cryostat structural integrity under cold operating conditions.

4. Conclusion and Outlook

Based on the ESS Preliminary Design and the requirements a final Cryogenic Moderator System has been developed and a cryostat containing most of the CMS functional components has been designed.

References

[1] Gallimore S, Nilsson P, Sabbagh P, Takibayev A, Weisend II J G, Beßler Y and Klaus M 2014 *AIP CONF PROC* **1573**, 659 doi: 10.1063/1.4860765

[2] Klaus M, Haberstroh Ch, Quack H, Beßler Y and Butzek M 2015 *MATER SCI ENG* **101** 012050 doi:10.1088/1757-899X/101/1/012050

[3] Arnold P, Hees W, Jurns J, Su T, Wang X L, Weisend II J G 2015 *MATER SCI ENG* **101** 012011 doi:10.1088/1757-899X/101/1/012011

[4] Weisend II J G, Arnold P, Fydrych J, Hees W, Jurns J M and Wang X L 2015 *PHYSCS PROC* **67** 27–34 doi: 10.1016/j.phpro.2015.06.006

[5] Jurns J, Ringnér J, Quack H, Arnold P, Weisend II J G and Lyngh L 2015 *MATER SCI ENG* **101** 012082 doi:10.1088/1757-899X/101/1/012082

[6] Chang H M, Kim S G, Weisend II J G and Quack H 2016 *Proc. Int. Cryocooler Conf.* **19** 463

[7] Klaus M, Eisenhut S, Quack H, Haberstroh Ch and Bessler Y 2015 *MATER SCI ENG* **101** 012050 doi:10.1088/1757-899X/101/1/012050

Acknowledgments

This work was conducted within the In-Kind TIK 3.1/3.2 agreement between ESS and FZ Juelich GmbH (TU Dresden contributing the final CMS process design) supported by the German BMBF.

IOP Conf. Series: Journal of Physics: Conf. Series **1021** (2018) 012081 doi:10.1088/1742-6596/1021/1/012081

Simulation Methods and Results of the SINQ Cold Neutron Source Upgrade Study

R. M. Bergmann, U. Filges, D. Kiselev, C. Klauser, E. Rantsiou, V. Talanov, M. Wohlmuther, M. Yamada
Paul Scherrer Institut, Villigen, Switzerland
E-mail: ryan.bergmann@psi.ch

Abstract.
Changing the configuration of the cold neutron source during a planned extended shutdown of the Swiss Spallation Neutron Source (SINQ) is being considered for improving performance of instruments at SINQ that use cold neutrons. The cold neutron source consists of a 20 L volume of liquid D_2 at approximately 25 K. This study includes making the re-entrant hole external, redesigning the re-entrant hole geometry, adding a Pb-208 reflector, replacing the zirconium safety hulls, adding cryogenic beryllium reflector/filters, and incorporating an active ortho-D_2 conversion loop. The optimization methods and the effects of the best-case modifications at all six neutron guides are presented. The best case design is predicted to yield a maximum gain factor of 1.5 in cold neutron intensity.

1. Introduction

The Swiss Spallation Neutron Source (SINQ) is a spallation neutron source driven by a continuous 590 MeV proton beam at the Paul Scherrer Institut (PSI) in Villigen, Switzerland. The incoming protons impinge on a zircaloy clad lead target cooled with heavy water, producing high energy neutrons. These neutrons are moderated by the tank of D_2O surrounding the target. The cold source (CS) is a 20 liter volume of liquid D_2 at approximately 25 K located within the D_2O moderator tank with its innermost face approximately 35 cm away from the center of the target. It further reduces the neutrons' energies to a colder spectrum more useful to instruments which view it. To enhance the number of emitted cold neutrons, a re-entrant hole (REH) has been attempted in the past. The goal of a REH is to remove scattering and absorbing material between the spot with the maximum cold neutron flux (CS center) and the direction of interest (guide hall). The REH is presently a five-sided box with the open face at the bottom. It sits inside the D_2 volume directly next to the wall facing the guide hall. Voiding this kind of REH relies on radiation-induced heating to boil D_2 which in turn fills the REH cavity with D_2 vapor, pushing the liquid D_2 out. Proposed plans include making the REH external, ensuring that it is not filled with liquid D_2 (or vapor). Calculations done prior to these changes predicted cold neutron fluence gains from 1.2 to 1.9, with gains increasing with wavelength [1]. These increases have not been observed, and it is suspected that the REH is not fully voided [2]. All new designs have an external REH, which will decouple the fill level from radiation-induced heating and ensure that the REH is never filled with liquid D_2.

Neutrons are transported from the CS to the instruments via neutron guides, which are internally coated with neutron supermirrors reflective to neutrons at low energies and low incident angles. In the optimized model used for this study, an approximation was used to model the angle-integrated flux exiting 6 meters of guide, the method for which has been published and presented previously in [3]. The method uses a precomputed energy- and angle-dependent transfer functions (using a patched version of MCNP that can model neutron mirror reflections with the McStas reflectivity model [4]) which correlates the angle and energy of an incoming neutron with one that exits the guide. These transfer functions are multiplied with the angular neutron flux at the entry of the guide and summed over angle to give a good approximation of the exiting neutron flux for rectangular guides.

In the planned upgrade strategies [5], changes will be made to the neutron guides and the CS at SINQ to increase the neutron output to instruments without additional proton current on the spallation target. This paper presents the REH design presented in [3] as well as a new one which has been determined with a genetic optimizer [6]. The effects of adding a Pb-208 reflector, adding cryogenic beryllium windows, using 100% ortho-deuterium, and replacing the zirconium safety hull with aluminum have also been computed. The effects of these changes on the non-REH side of the CS are also shown.

2. Simulation Geometry

Simulations for the optimization study were done using the Monte Carlo N-Particle code (MCNP) developed as Los Alamos National Laboratory (LANL) [11]. Figure 1 shows a zoomed-in view of the important parts of the MCNP model geometry.

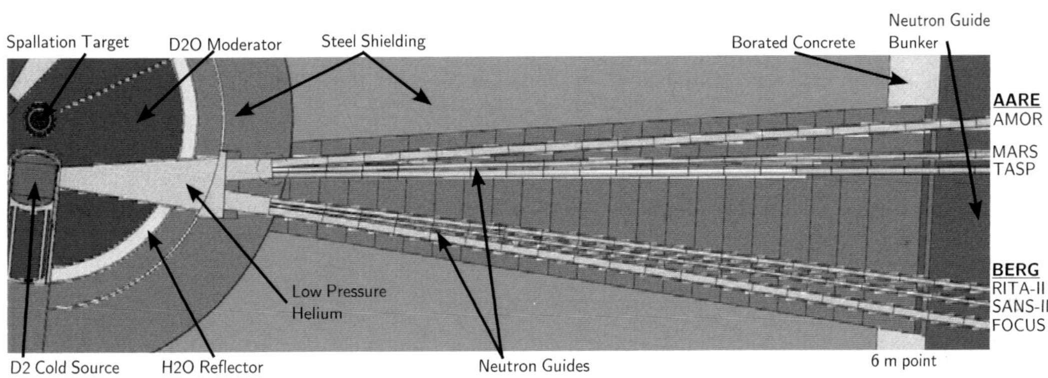

Figure 1. A horizontal slice through the MCNP simulation geometry highlighting the relevant components of the target, moderators, and guides present in the MCNP model.

A CAD model of the neutron guide bundles was converted to MCNP geometry by the Monte Carlo Modeling Interface Program (MCAM) package and integrated into the model [7]. The target is zircaloy clad lead rods cooled by D_2O, the moderator tank is filled with D_2O, the CS is liquid D_2, the nozzles are filled with a low-pressure helium mixture, and the shielding and neutron guide structure is steel. All structural material in the D_2O moderator tank is an aluminum alloy (AlMg-3), which is shown in yellow in Figure 1. The cross sections used in the simulations were from the ENDF/B-VII.1 library. Thermal scattering $S(\alpha,\beta)$ tables (continuous-energy when possible) were used for both moderators and structural materials within the moderator tank [8, 9].

3. Variance Reduction and Acceleration

Three variance reduction or acceleration methods were used to perform the simulations in an amount of time amiable to doing an optimization study. First, mesh-based weight windows were generated based on the response of a point detector located inside the AMOR/SANS-I neutron guide. The generated weight windows serve to increase the number of neutrons in regions that make large contributions to the point detector, which in this case is the CS [10, 11]. A map of the values is shown in Figure 2 (lower values indicate higher sampling).

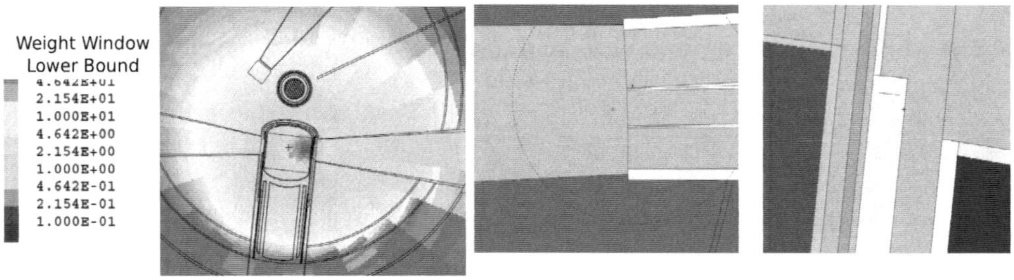

Figure 2. [LEFT] The weight window lower bounds, [CENTER] the Aare-side guide bundle with a DXTRAN sphere, and [RIGHT] the geometry near the corner of the REH.

The second acceleration method used was a DXTRAN sphere. This method serves to preferentially scatter neutrons into the direction of the neutron guide instead of relying on probability to transport them there (uses a weight adjustment to keep the simulation fair). The placement of the DXTRAN sphere around the Aare guide bundle is shown in Figure 2. The last method was using a surface source at the outer surface of the CS insert. In an initial run, a proton source was used on the spallation target and neutron tracks crossing into the CS insert were recorded and saved. In all subsequent runs, this generated surface source was used as the primary particle source instead of protons. With all of the variance reduction and approximations implemented, each case took about 7 minutes to run on a 144-core configuration on a computer cluster at PSI. As a side note, using the surface source alone reduced the runtime by a factor of 10.

4. REH Redesign

If only flat surfaces are considered, the REH geometry can be described by nine (9) variables: the distance of the side walls from the nozzle center (4), the angles the walls make with the centerline (4), and the depth of the back wall (1). These were the parameters varied by the optimization procedures used in the REH redesign study.

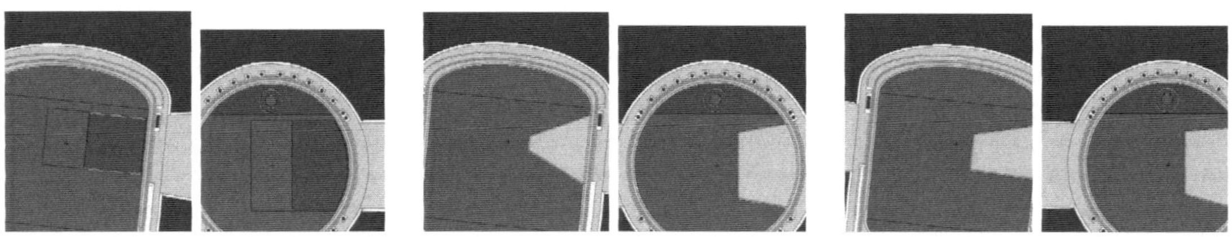

Figure 3. [LEFT] The current REH geometry. [CENTER] The best case REH geometry from the parametric scan. [RIGHT] The best case REH geometry from the genetic optimization. In all cases, the horizontal geometry is shown to the left and the vertical geometry to the right.

The leftmost case in Figure 3 shows the present REH design, and is called the "original" case subsequently. The first new design results, from here on called the "scan REH" case (center case in Figure 3), was determined by taking the best performing combination of a parametric scan of the REH geometry. Two scans were done; one in the horizontal plane and one in the verical plane. In both scans, the geometry was kept symmetric about the centerline. The rightmost case in Figure 3 shows the best case as determined using the DEAP genetic optimization framework [6], which from here on called the "genetic REH" case. The figure of merit (FOM) in all the optimization studies was chosen to be the total current exiting the AMOR/SANS-I guide six meters from the CS surface from 3 to 6 Å – the wavelength band which is useful for most of the instruments in the guide hall [12].

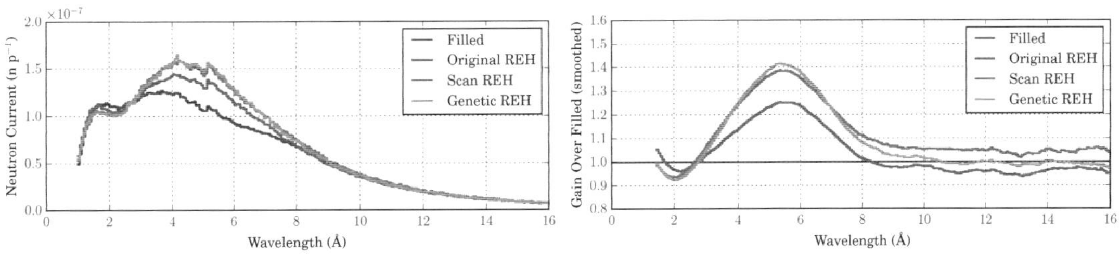

Figure 4. [LEFT] The calculated spectra in the AMOR/SANS-I guide at the 6 m position, and [RIGHT] the gain curves for the original, scan, and genetic REH geometries compared the case where the REH is filled with liquid D_2.

Figure 4 shows the calculated spectra exiting the 6 m length of guide leading towards AMOR/SANS-I and the gains over the "filled" case (which is the original case, but with the REH filled with liquid D_2). It shows that the new designs produce gain curves that are about 60% higher than that of the original design. The gain curve of the genetic case has a slightly higher peak value (1.42 vs. 1.39) and a slightly higher average value from 3-6 Å (1.32 vs. 1.30), but has lower gains for wavelengths longer than 7 Å. The effect that the asymmetric genetic REH has on the other guides is discussed in Section 6.

5. New Materials

Other than changing the geometry of the REH, changing some key materials was also studied. Many options were considered, but many were either unfeasible in the strong radiation field or produced very negative results. In all cases where materials new were studied, the scan REH was included in the model since a similar geometry will almost certainly be present in the upgrade. The short list of cases considered were: 1) Adding cryogenic Be windows at the REH [13] to reflect small wavelength neutrons back into the moderator while allowing long wavelength neutrons to exit freely; 2) Adding a Pb-208 reflector [14], which has very low neutron absorption, to the back side of the CS in order to reduce upscattering on the surrounding warm D_2O; 3) Making the moderator be 100% o-D_2 in order to increase the macroscopic cross section; 4) Replacing the Zr safety hull with AlMg3, which has a lower total macroscopic cross section.

Figure 5 shows that the Pb-208 reflector gives about a 5% gain uniformly, but implementation of such a heavy mass and acquisition of pure material could prove to be difficult. The cold beryllium windows show a slight gain for wavelengths longer than 4 Å, and losses for wavelengths between 2 and 4 Å. Due to the losses in the wavelength range of interest, and the gains for short wavelengths (contribute to background), the beryllium windows are unattractive for the upgrade. Figure 5 also shows that moving to a 100% o-D_2 moderator would give about a 5% decrease for neutrons between 2 and 3.7 Å, a 5% increase for neutrons between 3.7 and 5.5 Å, and a 10% increase for longer wavelengths. Such a moderator could be realized with small recirculation loop in the storage volume constantly flowing the liquid D_2 over a converter material (like iron oxide [15]). Replacing the zirconium hull with the same thickness of AlMg3 gives 7% increase for neutrons less than 5.5 Å. If this is acceptable due to safety implications has yet to be seen.

IOP Conf. Series: Journal of Physics: Conf. Series **1021** (2018) 012081 doi:10.1088/1742-6596/1021/1/012081

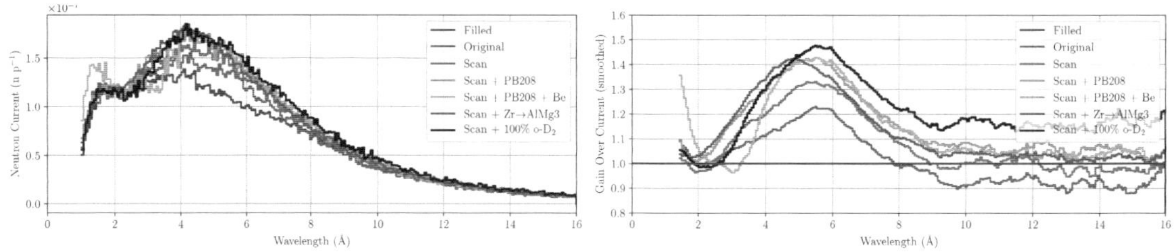

Figure 5. [LEFT] The calculated neutron spectra in the AMOR/SANS-I guide at the 6 m position, and [RIGHT] the gain curves over the filled case from material modifications.

6. Individual Guide Results

The optimization was based on the response of the AMOR/SANS-I guide, but the effect of the modifications were investigated on all guides to make sure the changes don't favor a single guide too heavily. Table 1 shows the peak and 3-6 Å average gains in the 6m neutron current for the proposed modifications for all six guides, and Figure 6 shows the spectra and gain curves for the RITA-II guide.

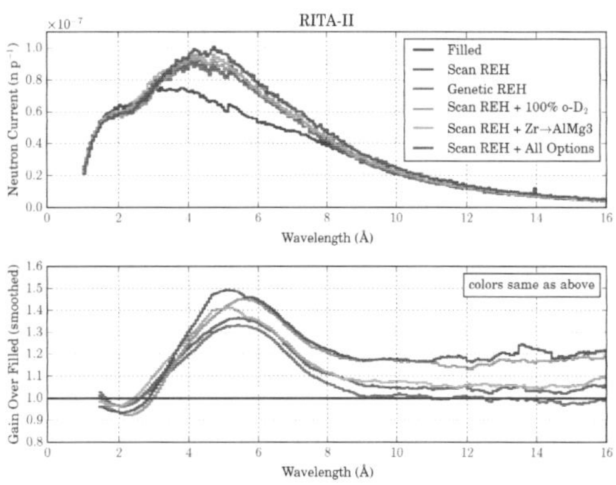

Figure 6. [TOP] The calculated neutron spectra in the RITA-II guide at the 6 m position, and [BOTTOM] the gain curves over the filled case for the original, scan, and genetic REH geometries.

Table 1. Peak gains (bold) and average gains from 3-6 Å due to the proposed CS upgrades at the 6 m position for each of the six guides. The top row are the guides on the Aare side, and the bottom row are the guides on the Berg side.

	AMOR	MARS	TASP
Scan REH	**1.39** 1.30	**1.41** 1.32	**1.40** 1.31
Genetic	**1.42** 1.32	**1.41** 1.31	**1.44** 1.33
100% o-D_2	**1.48** 1.36	**1.49** 1.37	**1.49** 1.37
Zr→AlMg3	**1.46** 1.35	**1.48** 1.36	**1.48** 1.36
All Options	**1.53** 1.41	**1.56** 1.42	**1.56** 1.42
	RITA-II	SANS-II	FOCUS
Scan REH	**1.37** 1.29	**1.40** 1.31	**1.39** 1.31
Genetic	**1.33** 1.26	**1.39** 1.30	**1.40** 1.31
100% o-D_2	**1.46** 1.34	**1.50** 1.38	**1.48** 1.37
Zr→AlMg3	**1.42** 1.33	**1.47** 1.37	**1.46** 1.36
All Options	**1.51** 1.39	**1.56** 1.43	**1.54** 1.41

When compared to the symmetric, scan-optimized REH, it can be seen that the asymmetric, genetically-optimized REH gives about a 2% benefit to guides on the Aare side and either has neutral or a small negative effect on the Berg side. The scan-optimized REH is more uniform across all guides. Figure 6 only shows the RITA-II guide since it has the biggest negative effect from the genetic REH. All the other neutron guide cases have similar shapes and Table 1 gives the magnitude of the differences. Separate simulations were run for each case (DXTRAN centered around the Aare or Berg bundle) in order to keep the DXTRAN sphere small and therefore the underlying tallies efficiently sampled.

7. Conclusions and Future Work

The most important feature to realize in the upcoming SINQ upgrade period is the new REH geometry (scan-optimized is least risky), followed by a 100% o-D_2 CS, and finally replacing the Zr hull with AlMg3 if possible. Incorporating all three attractive upgrade options, neutron intensity is predicted to increase by **1.41** on average (3-6 Å) and **1.54** at peak for all guides. The effect at ICON and BOA will be negative if only the REH is realized, but will be positive if the additional material modifications are realized (these results are not presented here, but important to note). The next step will be doing technical feasibility studies such as thermal analysis of the REH geometry to ensure the integrity of the D_2 vessel is unaffected and the D_2 flow is not impeded by it.

Acknowledgments

This work was supported by Swiss National Science Foundation grant 200021_150048/1. The authors would also like to thank Tibor Reiss for his assistance in creating the MCNP models and advice on performing the simulations.

References

[1] F. Atchison, H. Spitzer, and W. Wagner "New Cold D_2 Moderator Vessel with Re-entrant Beam Window for SINQ", 16th Meeting of the International Collaboration on Advanced Neutron Sources (ICANS-XVI)", 2003, Forschungszentrum Jülich.
[2] K. Thomsen, "Conceptual Proposal for Compound Moderators with Preferential Emission Directions", Physics Procedia 60, pp. 278 - 293, 2014, http://dx.doi.org/10.1016/j.phpro.2014.11.038
[3] R. M. Bergmann, U. Filges, D. Kiselev, T. Reiss, V. Talanov, and M. Wohlmuther, "Upgrades to the SINQ Cold Neutron Source," Journal of Physics: Conference Series, Vol. 746, No. 1, 2016, doi:10.1088/1742-6596/746/1/012035.
[4] F. X. Gallmeier, M. Wohlmuther, U. Filges, D. Kiselev, G. Muhrer, "Implementation of Neutron Mirror Modeling Capability into MCNPX and Its Demonstration in First Applications", Nuclear Technology Vol. 168 No. 3, December 2009, pp 768-772.
[5] Christian Rüegg , "Upgrade plans for SINQ – Moderators, Guides, Instruments, and New Science", 21st Meeting of the International Collaboration on Advanced Neutron Sources (ICANS-XXI), 2014, Mito, Japan.
[6] Félix-Antoine Fortin, François-Michel De Rainville, Marc-André Gardner, Marc Parizeau, Christian Gagné, "DEAP: Evolutionary Algorithms Made Easy", 2171–2175, July 13, 2012, Journal of Machine Learning Research.
[7] Y. Wu, FDS Team, "CAD-based interface programs for fusion neutron transport simulation", Fusion Engineering and Design 84, 2009, 1987-1992.
[8] D. Kent Parsons and Jeremy Lloyd Conlin, "Release of Continuous Representation for $S(\alpha,\beta)$ ACE data", LA-UR-12-00800, Feburary, 2012, Los Alamos National Laboratory, Los Alamos, NM, USA.
[9] Wolfgang Bernnat, Jürgen Keinert, Margarete Mattes, "Scattering Laws and Cross Sections for Moderators and Structure Materials for Calculation of Production and Transport of Cold and Ultracold Neutrons", Journal of Nuclear Science and Technology Vol. 39, sup2,pp 124-127, 2002, http://dx.doi.org/10.1080/00223131.2002.10875056
[10] X-5 Monte Carlo Team, MCNP - A General Monte Carlo N-Particle Transport Code, Version 5 (Revised 2/1/2008), April 24, 2003 Los Alamos National Laboratory, Los Alamos, NM, USA.
[11] Denise B. Pelowitz, John T. Goorley, Michael R. James, et. al., "MCNP6 USER'S MANUAL VERSION 1.0", Rev. 0, May 2013, Los Alamos National Laboratory, NM, USA
[12] T Reiss and U Filges and F Gallmeier and V Talanov and M Wohlmuther, "Bi-spectral moderator for spallation sources optimized for instrument requirements", Nuclear Instruments and Methods in Physics Research Section A, Vol. 728, pp. 117 - 125, 2013, http://dx.doi.org/10.1016/j.nima.2013.06.093
[13] G. Muhrer, E.J. Pitcher, G.J. Russell, "The neutron performance of a pre-moderated beryllium reflector-filter hydrogen moderator system for the Manuel Jr. Lujan Neutron Science Center", Nuclear Instruments and Methods in Physics Research Section A, Volume 536, Issues 1-2, 1 January 2005, Pages 154-164, ISSN 0168-9002, http://dx.doi.org/10.1016/j.nima.2004.07.201.
[14] T. Schonfeldt, Konstantin Batkov, et al., "Broad spectrum moderators and advanced reflector filters using (208)pb", Nuclear Instruments and Methods in Physics Research. Section A, Vol. 769, 2015, http://dx.doi.org/10.1016/j.nima.2014.09.048
[15] M. Hartl, R. C. Gillis, et al., "Hydrogen adsorption on two catalysts for the ortho- to parahydrogen conversion: Cr-doped silica and ferric oxide gel", Phys. Chem. Chem. Phys., Vol. 18, No. 26, pp. 17281-17293, 2016, http://dx.doi.org/10.1039/C6CP01154C

IOP Conf. Series: Journal of Physics: Conf. Series **1021** (2018) 012082 doi:10.1088/1742-6596/1021/1/012082

Implementation of the proton beam instrumentation into the proton beam instrumentation plug

Naja de la Cour[1], Mattias Wilborgsson[1], Thomas Shea[1], Cyrille Thomas[1], Fabien Rey[1], Sara Ghatnekar[1], Rikard Linander[1], Erik Adli[2], Håvard Gjerdsal[2], Martin Jaekel[2], Mark Ibison[3,4] and Raul Vivanco[5]

[1] European Spallation Source ERIC, Lund, Sweden
[2] University of Oslo, Oslo, Norway
[3] University of Liverpool, Liverpool, UK
[4] Cockroft, Daresbury, UK
[5] ESS Bilbao, Madrid, Spain

E-mail: naja.delacour@esss.se

Abstract. The 5 MW proton beam to be delivered to the ESS spallation target must be tuned to its nominal value and controlled at all times. To achieve this, a suite of beam diagnostics has been proposed and is currently been designed. It is located in two plugs, the Proton Beam Window (PBW) plug and the Proton Beam Instrumentation Plug (PBIP). Here we present the design concept for the PBIP imposed by the beam physics requirements as well as by the high power target environment, i.e. material lifetime, remote handling and waste management.

1. Introduction

The ESS accelerator delivers to the target a 125 MW peak power, 2.86 ms pulsed proton beam with a low emittance of 1 mm mrad. Even a fraction of this beam is capable of inducing damage in material it intersects. The chosen beam delivery strategy is to raster the beam in order to reduce the high current density below the permanent damage threshold of the target materials. Simultaneously, the transverse beam extent can be constrained to limit energy deposition outside of the aperture[1, 2]. The required beam control is enabled by a suite of diagnostics, located mainly in the target area[3], and composed of two imaging systems, one multi-wire grid and three aperture monitoring systems. They permit measurement of the beam transverse profile, the beam divergence, the position of the beam with respect to the Target Wheel (TW), and the fraction of the beam outside the nominal aperture that could potentially cause damage, or pollute the neutron flux delivered to the neutron instruments. The two imaging systems are both supported by the PBIP and they provide images of the beam intersecting the PBW at the entrance of the target monolith area and the target wheel. The image analysis provides the location of the beam at the PBW and at the TW in the general ESS coordinate system, and also the current density distribution on both PBW and TW from which a measurement of the raster beam pattern can be compared with the defined nominal distribution of the beam on the TW. The accuracy of the measurement should provide the position of the beam to within ± 1 mm and the flatness of the density to within 20% of the mean peak current density. The optical system design is highly constrained by the required performance of the imaging system, and also by geometric and shield integrity requirements in the target area[4].

IOP Conf. Series: Journal of Physics: Conf. Series **1021** (2018) 012082 doi:10.1088/1742-6596/1021/1/012082

The multi-wire grid, also supported by the PBIP, provides two 1-D profiles of the beam distribution. It is based on the reading of the current generated by the interaction of the beam with the wires[5]. The analysis of the profiles provides the projected current density, the position of the beam and the extent of the beam along each axis. This system provides redundancy to the imaging systems, but these two systems are not equivalent. Although the grid may be more sensitive than the imaging systems, it does not detect some errant beam conditions that could damage the target. Similar to the imaging system, the grid has to be positioned with ±0.5 mm accuracy. It can be interlocked to the machine protection system (MPS).

The aperture monitoring systems are located in three positions along the beam path: the first one is at the pivot point of the beam raster optics where the beam has to be in a fixed position and highly focused; the second one is in the PBW plug; and the third one is supported by the PBIP. These systems are based on two techniques to measure the current intersecting the defined aperture: metallic blades and thermocouples. The metallic blades measure the current induced by the intersecting beam. It is a fast response detector that can detect errant beam condition in the μs timescale. The thermocouples measure the current outside the defined aperture at the *s* time scale. The system detects errant beam conditions within a pulse, and also accumulation of small beam losses at the aperture. Errant beam conditions detected by these monitors can trigger the MPS.

In the following, we will focus on the design of the PBIP which supports the integration and the operation of the diagnostics suite.

2. Target environment

The PBIP is situated inside the Monolith in the ESS Target station (Fig. 1 and 2). It is positioned between the TW and the PBW, where it supports the beam instrumentation. The PBIP, and thus part of the instrumentation it supports, is situated in a high radiation environment. The PBIP and its components will receive high doses of secondary particles produced by the proton beam on the PBW, the target and the aperture. Therefore, materials of the PBIP and its components will be heated by energy deposited from these particles, and also degraded by accumulated radiation damage. The PBIP is located in the monolith shielding 1.5 m from the target and as a result, it will receive a higher dose at its bottom. Thus it will be experiencing a high temperature gradient, which will have to be controlled to manage the resulting mechanical deformation that could misalign the instrumentation components.

3. Design concept

The design of the plug should satisfy requirements issued from the diagnostics and from the target environment. The design should allow the diagnostics to be positioned with a typical accuracy of 0.5 mm, should minimize distortion and movement during operation, and should maintain the shielding integrity. The instrumentation lifetime is expected to be 55GWh [1]. However, in case of failure, instruments should be independently replaceable. Figure 2 shows a model of the PBIP. In the figure, one can see a solid base in which slices can be inserted. Each of these slices support a diagnostic instrument. The concept of slices permits each instrument to be maintained independently of the other ones. It is a single independent slice, so the alignment precision of the slice depends on the instrument it supports. For instance, for the optics, the precision should be in the range of 1 mm in position, but 1 mrad or less in angle. To achieve this, the mirrors are mounted on specially developed mounts that can maintain the mirror alignment. The slice is assembled horizontally in the loading bay before it is installed in the PBIP. The alignment precision at this stage is within the μm and μrad range. During lifting and installation of the slice into the PBIP, the rigidity of the mounts guarantee the that the pre-alignment is maintained. While being slowly lowered by the robotized crane above the PBIP, the solid base guides the slice into its position with 6 mrad accuracy.

[1] 2 years of operation at full power, 5500 h at 5 MW

IOP Conf. Series: Journal of Physics: Conf. Series **1021** (2018) 012082 doi:10.1088/1742-6596/1021/1/012082

In addition, the rigidity of the PBIP structure guarantees good alignment repeatability. This applies for the other slices too, where the grid and aperture monitors should be positioned to within 0.5 mm from the nominal beam axis.

The thermo-mechanical deformation of the PBIP structure has been studied, using MCNPx[2] and ANSYS[3] codes. The energy deposited into the structure, obtained from MCNPx, is used for the thermo-mechanical simulation studies with ANSYS. The initial results showed several degrees distortion to the structure. The induced angular motion to the first mirror of the optics assembly would reduce image intensity, degrade resolution, and add distortion beyond acceptable limits. With the addition of cooling in the PBIP, more acceptable results were achieved, which can be seen in Fig. 3. With an angular distortion of the order of 3 mrad, the degradation of the imaging system performance reaches its limit. More cooling can solve the problem.

3.1. Shielding integrity

The apertures in the plugs, from the point of view of the shielding, represent a hole in the shield and increase the environmental radiation level. As a result, the tendency in practice is to reduce the apertures, and add chicanes. This presents serious challenges to the optics design. To ensure the performance of the imaging system, a large aperture is needed, and the traditional 90 degree chicane with the rule of a 1:7 offset[4] is not easily achievable. In the current design, a large 100 mm aperture is provided in the PBIP optics slices[4]. Figure 4 depicts the element of the optics assembly in the PBIP that looks toward the TW. The other assembly that looks toward the PBW is similar. The MCNPx shielding calculation with the optics in the configuration shown here proved that the concept is acceptable. The additional radiation coming through the optics slices is attenuated by a passively cooled shielding block placed on the top of the PBIP. It is removable so the slices can be replaced.

3.2. Remote handling and waste management

In each extraction arm of the PBIP slices, remote handling features are included to be able to extract the component into a cask and transfer it to the Active Cells facility. Components are vertically extracted and can be exchanged individually. The whole slice is considered to be expendable after it has been removed from the PBIP. However, a possibility to design the slice in pieces is under study. The Monte Carlo simulation on the target model has shown that only the bottom part of the assembly needs to be considered as radioactive waste. Having the top part of the slices re-used could reduce the waste production by more than 50%. The radioactive waste management drives some aspect of the design of the PBIP. The immediate impact is a reduction of the overall cost, as long as the specifications for the instrumentation performance can be achieved. Also, the parts of the plug with the instrumentation and the cables should follow a recycling path. This includes the radioactive cooling time. The cost of storing these radioactive elements will be studied so that the real benefit can be evaluated – optimum waste management is included in the global performance of the facility.

4. Concluding remarks

The design for the PBIP has seen significant progress. The design of the plug which supports critical diagnostics for tuning and monitoring the beam on target has addressed most of the requirements driven by from the instruments' performance and by the radiation environment. Technical solutions are under evaluation to enable required performance of the instrumentation, and the large 100 mm aperture for the optics assembly has been shown to be compatible with the shielding requirements. The rigid structure of the plug and the alignment procedure is still under design consideration. However, critical alignment tolerances are well defined and drive the design.

[2] https://mcnpx.lanl.gov

[3] http://www.ansys.com

[4] 1 for the aperture 7 for the length after a 90 degree chicane

The waste management is also considered in the design. Producing less waste is one of the objectives. It reduces the running cost of the facility. At the same time, a recycling circuit for the upper parts of the plug, including the instruments and cables should be studied to fully evaluate the real benefit.

The design of the PBIP is mature, and it continues with few remaining issues to be addressed.

References

[1] Shea T J, Christian B, Cheymol B, Gallimore S, Hassanzadegan H, Pitcher E J and Thomsen H D 2013 *IBIC* pp 349–352
[2] Thomsen H D 2016 *HB* ed Eshraqi M, Trahern G and Schaa V R (JACOW) pp 427–432 ISBN 9783954501786
[3] Shea T, Pitcher E, Takibayev A, Zanini L, Andersen K, Celona L, Henry P, Mezei F and Thomsen H 2016 *ICANS* ed Oku T, Nakamura M, Sakai K, Teshigawara M, Hideki T, Yonemura M, Suzuki J and Arai M (JAEA)
[4] Ibison M G, Welsch C P, Adli E, Gjersdal H, Cour N D, Shea T and Thomas C A 2016 *IPAC* ed Petit-Jean-Genaz C, Kim D E, Kim K S, Ko I S, Kim K R and Schaa V R (JACOW) ISBN 9783954501472
[5] Lee Y, Shea T and Thomas C A 2016 *IPAC* ed Petit-Jean-Genaz C, Kim D E, Kim K S, Ko I S, Kim K R and Schaa V R (JACOW) pp 276–279

IOP Conf. Series: Journal of Physics: Conf. Series **1021** (2018) 012082 doi:10.1088/1742-6596/1021/1/012082

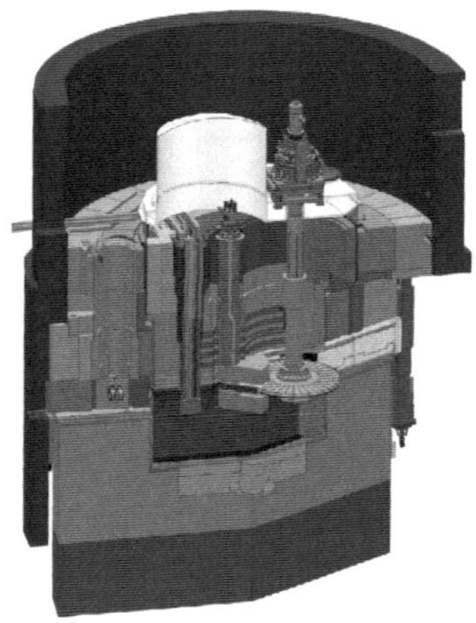

Figure 1. Cut through the Target Monolith.

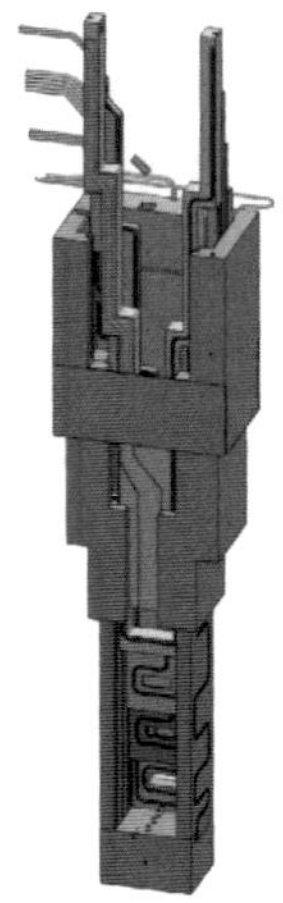

Figure 2. . PBIP design showing the slices supporting the instruments

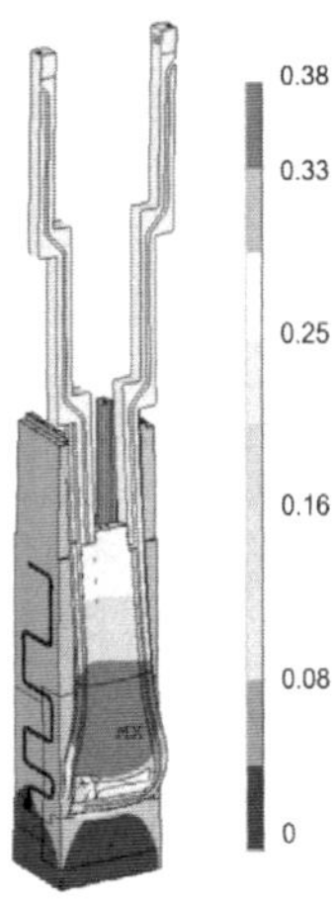

Figure 3. . PBIP thermal deformation at 5MW beam on target. Color scale is in mm

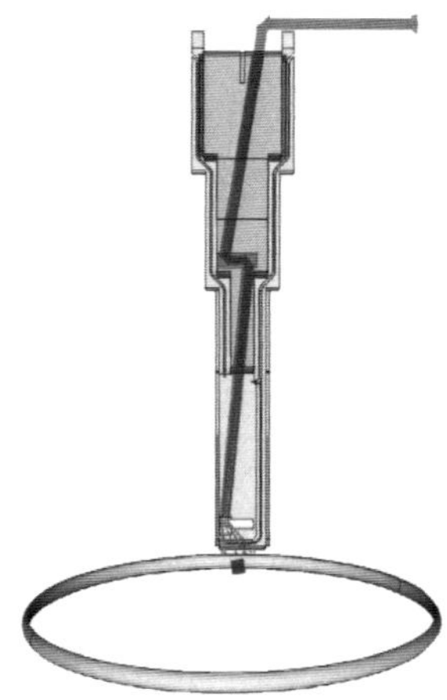

Figure 4. Part of the optics assembly supported by the PBIP.

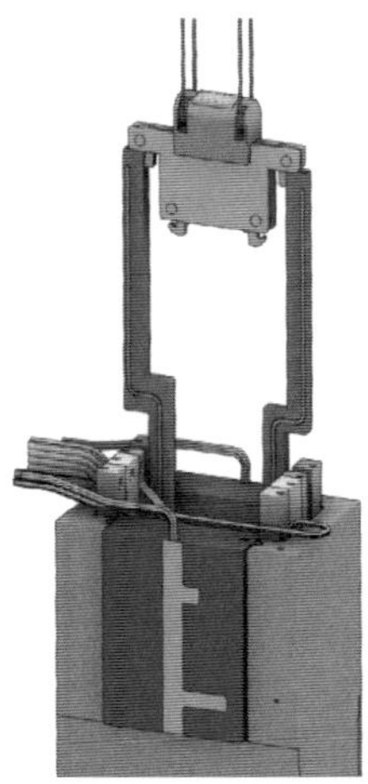

Figure 5. Handling tool crane to install or remove a PBIP slice.

IOP Conf. Series: Journal of Physics: Conf. Series **1021** (2018) 012083 doi:10.1088/1742-6596/1021/1/012083

Options for a very cold neutron source for the second target station at SNS

F. X. Gallmeier, T. Hügle, E. B. Iverson, W. Lu, I. Remec

Instrument and Source Division, Oak Ridge National Laboratory, Oak Ridge, TN 37831, USA

Email: gallmeierfz@ornl.gov

Abstract. Simulation work is underway to identify a very cold neutron source solution that would provide an order of magnitude higher neutron output than the conventional moderator suite. Moderator candidates such as water ice, beryllium, ortho-D_2, CH_4, and neon all solid at temperatures of 4-6 K were investigated. At present no moderator option was able to achieve the set goal; ortho-D_2 and CH_4 seem to be the most promising options.

1. Introduction

A second target station (STS) is in the planning in Oak Ridge to complement the existing sources, the HFIR reactor and the SNS target station for high-intensity cold neutron beams[1]. The new source will be operated at 700 kW power and 1.3 GeV proton energy and 15 Hz repetition rate by using every fourth pulse of the SNS accelerator in pulse-stealing mode. Two high-brightness coupled liquid hydrogen moderators and a decoupled moderator unit of ambient temperature water and liquid hydrogen will provide neutrons to 20+ beamlines. With three moderators on the target, there may be space for another moderator unit, envisioned to be a very cold neutron source optimized for delivering neutron beams of 10 Å wavelength and longer for novel scattering instruments and as a driver for a UCN source to serve fundamental physics experiments as suggested earlier by Carpenter et al. [2][3]. The requirements for a VCN are to provide neutron beams with wavelengths of 10 Å wavelength and longer at intensities an order of magnitude higher than any other moderator at STS with still well established neutron pulses. As the prime locations with proximity to the peak neutron production in the target is taken by the conventional moderator suite, the less than ideal remaining space around the target is being explored for hosting the additional VCN. In particular, the locations above and in front of the cylindrical coupled hydrogen moderator were considered as options.

This manuscript has been authored by UT-Battelle, LLC under Contract No. DE-AC05-00OR22725 with the U.S. Department of Energy. The United States Government retains and the publisher, by accepting the article for publication, acknowledges that the United States Government retains a non-exclusive, paid-up, irrevocable, world-wide license to publish or reproduce the published form of this manuscript, or allow others to do so, for United States Government purposes. The Department of Energy will provide public access to these results of federally sponsored research in accordance with the DOE Public Access Plan (http://energy.gov/downloads/doe-public-access-plan).

IOP Conf. Series: Journal of Physics: Conf. Series **1021** (2018) 012083 doi:10.1088/1742-6596/1021/1/012083

2. Probing moderator materials

MCNPX [4] calculations were performed probing the brightness on the viewed area for materials at cryogenic temperatures for which we had access to a scattering kernel such as for para-H_2[5], water ice[5], beryllium[5], ortho-D_2[6], CH_4[7], and neon[8] all solid at temperatures of 4-6 K. Configurations of box-shaped moderators of 50x52x88 mm^2 (the same size as box-type coupled hydrogen moderator) of materials containing hydrogen and beryllium were surrounded by ambient temperature light water, and were investigated with viewed areas of 50x50 mm^2. Moderators of deuterated materials and neon were sized to 140x140x200 mm^3 as those were deemed to require larger volumes for thermalizing neutrons. Their viewed area was scaled up to 100x100 mm^2. The moderators were positioned at about 15 cm distance from the target. Although being at single digit Kelvin temperatures, beryllium, water ice and para-hydrogen did not significantly cool the neutrons to lower temperatures. Solid CH_4 performed most promising as it was the only candidate that moved the spectral peak down to 1 meV. None of the configurations considered was able to approach the brightness of the box-type coupled moderator of the conventional moderator suite at the whole range of energies below 10 meV.

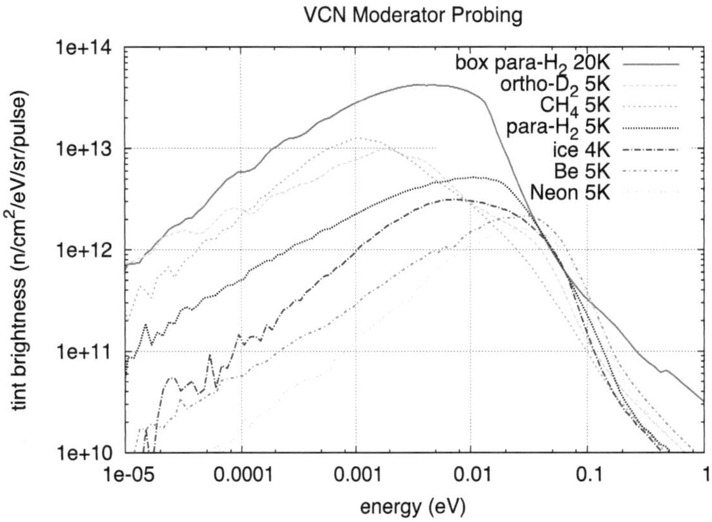

Figure 1: Comparison of brightness spectra of very cold neutron sources candidate materials compared to the performance of the box-type coupled hydrogen moderator (labeled B-H_2 in Figure 2).

3. Optimizing VCN moderators

The most competitive moderator materials, ortho-D_2 and CH_4, were entered into another round of investigations geared to optimize the size of the perspective moderators such that they deliver highest time-integrated pulse brightness[10]. Also, heavy or light water pre-moderators were optimized in dimensions. Optimizations were conducted for liquid para-H_2 and ortho-D_2 at 20K at the VCN locations for comparison to the 4-6 K moderators. Both locations above in front of the cylindrical coupled hydrogen moderator were being considered.

The optimized sizes of the VCN moderators varied vastly from a 20 mm thick slab for solid CH_4 with width and height of 110-120 mm to a 200mm cube for ortho-D_2 both for 5K and 20K. The optimizations of the VCNs of the same material arrived at very similar dimensions for the moderator dimensions for the top and front positions, however, the premoderator dimensions changed such that a moderator position closer to the target was favored. A reference liquid para-hydrogen moderator arrived at the width and height practically given by the dimensions of the viewed area and at a depth of 150 mm both for the front and top positions. Based on the time-integral brightness, none of the moderator options could overcome the disadvantage of larger distance to the target by cooling to 5K temperature compared to the box-typed conventional moderator at 20K except for the ortho-D_2 for energies below 0.04 meV as shown in Figure 3. Ortho-D_2 moderators deliver the neutrons in about a factor of five wider pulses compared to the para-H_2 and CH_4 moderators, which is shown both in the

IOP Conf. Series: Journal of Physics: Conf. Series **1021** (2018) 012083 doi:10.1088/1742-6596/1021/1/012083

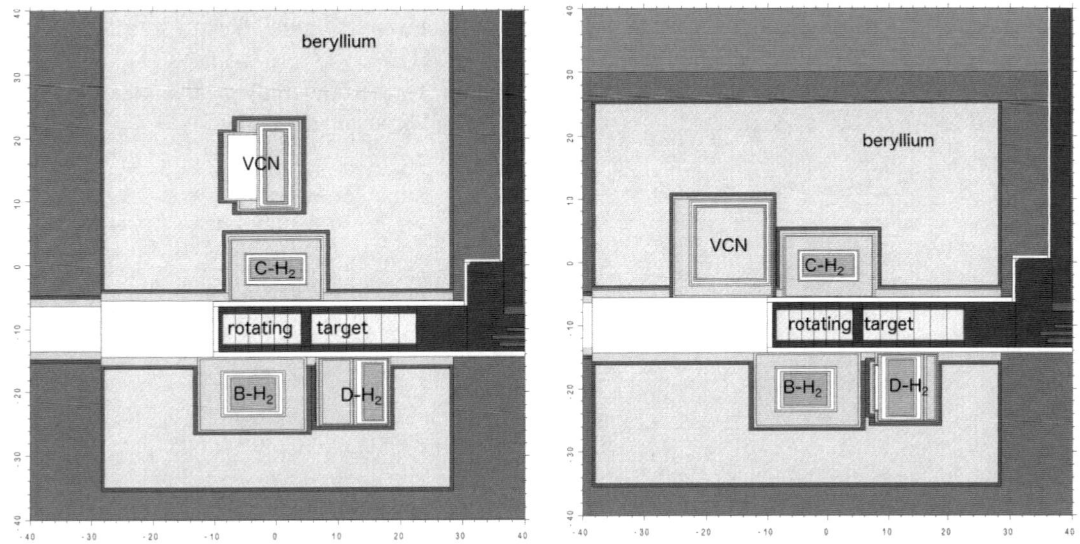

a) b)

Figure 2: Vertical cuts of STS MCNPX models with rotating target, and three conventional moderators including very cold neutron sources (VCN) of solid CH_4 at the front (a) and the top (b) of the conventional cylindrical para-hydrogen moderator (C-H_2).

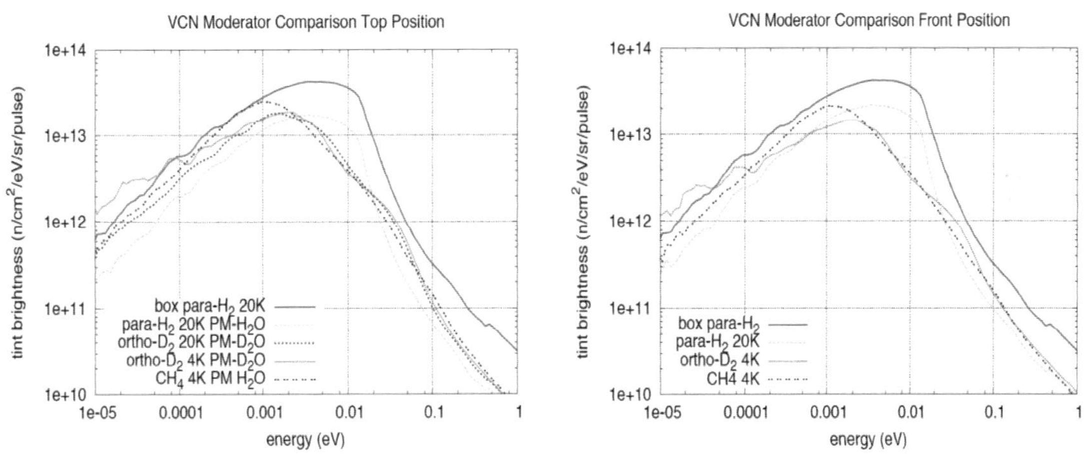

a) b)

Figure 3: Comparison of time-integrated brightness over a pulse of VCNs with various materials optimized in size compared to the performance of the of the box-type coupled hydrogen moderator (labeled B-H_2 in Figure 2): a) for VCNs positioned on top of C-H_2; b) for VCNs positioned in front of C-H_2.

full-width-at-half-maximum metric FWHM and the standard-deviation metric RMS as shown in Figure 4. In the energy range of cold neutrons below 5 meV, an substantial increase of the RMS metric compared to the box-type conventional moderator indicates the buildup of higher long-time tails for the off-ideal positioning of the VCN moderator.

Some of the CH_4 and ortho-D_2 calculations were repeated with alternative scattering kernels to check for consistency of the neutronics analyses. Analyses with ortho-D_2 kernels from Bernnat [5] confirmed the results obtained with the Granada kernel that was adopted for this study. The addition of the coherent scattering as provided by Bernnat seem to have little impact on the thermalization but comes at longer running times of the calculations. The use of the CH_4 kernel from Granada [11] reduces the very cold neutron output by a factor of 2. Differences between the hindered and free rotations of the crystalized CH_4 molecules seem to exist between the Shin and Granada approaches.

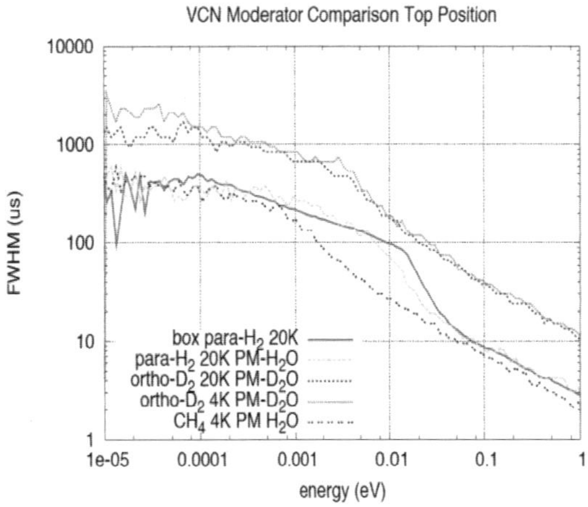

a)

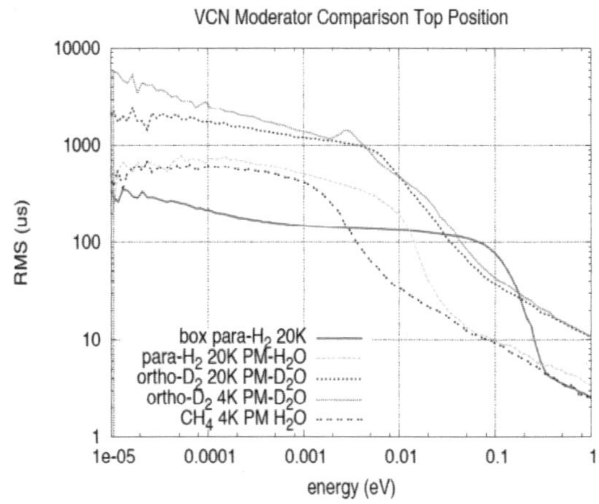

b)

Figure 4: Full-width-at-half maximum (FWHM) pulse width (a) and RMS standard deviation metric of the pulse time distribution (b) of neutron emission time distributions of various material VCNs positioned at the top of $C\text{-}H_2$.

4. Conclusion

The search for a VCN solution at STS that would provide an order of magnitude higher neuron output at energies below 1 meV (wavelengths above 10 Å) has so far not been successful. Also, a factor of 4 increase of the viewed area from 50x50 mm2 to 100x100 mm2 does not achieve the goal. Providing moderator material at single-digit Kelvin seem to be unable to compensate the disadvantage of a more remote position with regard to the conventional moderator suite. The reduction in physical temperature of the moderators studied is not accompanied by a reduction in the spectral temperature. In addition, the ortho-D_2 VCNs are penalized by a significant increase of the pulse width of a factor of 5 over hydrogen and CH_4 moderators.

Acknowledgement

This material is based upon work supported by the U.S. Department of Energy, Office of Science, Office of Basic Energy Sciences, under contract number DE-AC05-00OR22725.

References

[1] Galambos J. et al., 2015 Technical Design Report Second Target Station (ORNL/TM-2015/24).
[2] Carpenter J.M. and Mason T.E., Nucl. Inst. Meth. A, 545(1-2), pp 1-19 (2005).
[3] Micklich B.J. and Carpenter J.M (editors), Proceddings of the Workshop on Applications of a Very Cold Neutron Source, August 21-24, 2005, Argonne, Illinois, ANL-05/42, Argonne National Laboratory (December 2005).
[4] Pelowitz D. B., MCNPX User's Manual, Version 2.6.0, Report LA-CP-07-1473, Los Alamos National Laboratory, Los Alamos, New Mexico, 2008.
[5] Bernnat W. et al, in Proceedings of the 6th International Worksho[on Advanced Cold Moderators, Juelich, Sept 11-13, 2002, Matter and Materials, Volume 20, Juelich 2004, ISBN 3-89336-363-7.

IOP Conf. Series: Journal of Physics: Conf. Series **1021** (2018) 012083 doi:10.1088/1742-6596/1021/1/012083

[6] Granada J.R., EPL, 86 (2009) 66007, doi: 10.1209/0295-5075/86/66007
[7] Shin Y. et al., Nucl. Inst. Meth. A, 620, pp 382-390 (2010).
[8] Huegle T. et al, this proceedings.
[9] Chadwick M.B. et al., NDS, 107(12), pp2931-3060 (2006).
[10] Gallmeieer F.X. et al., Rev. Sci. Instrum., 87(6), 063304 (2016).
[11] Granada J.R., Nucl. Inst. Meth. B, 266, pp 164-172 (2008).

Neutronics Analyses for the ORNL's Spallation Neutron Source Second Target Station

I Remec[1], F X Gallmeier and M J Rennich

Oak Ridge National Laboratory, 1 Bethel Valley Road, Oak Ridge, TN37831-6476, USA

[1]E-mail: remeci@ornl.gov

Abstract. This paper presents an update of the status of the neutronics analyses performed for the Second Target Station (STS). The target station is driven with short (less than 1 micro second long) proton pulses at 15 *Hz* repetition rate and 700 *kW* proton beam power. The target will be optimized for high intensity and high resolution long wavelength neutron applications. The STS will accommodate 22 beamlines and will expand and complement the current national neutron scattering capabilities. The proton beam footprint as small as acceptable from the mechanical and heat removal aspects is planned to generate a compact-volume neutron production zone in the target, which is essential for tight coupling of the target and the moderators and for achieving high-intensity peak thermal and cold neutron fluxes. Present efforts to develop high fidelity engineering models for neutronics analyses with automatic CAD-to-MCNP conversion are described. Heating rates and radiation damage, which provide input in the engineering design are presented and the performance of the moderators is briefly addressed.

1. Introduction

The Spallation Neutron Source (SNS) and the High Flux Isotope Reactor (HFIR), two of the word-class neutron scattering facilities, are located at the Oak Ridge National Laboratory. The SNS and HFIR are funded by the U.S. Department of Energy, Office of Science, Office of Basic Energy Science, and are operated as user facilities, available to researchers from all over the world. Currently there are thirteen neutron scattering instruments in operation at the HFIR and twenty at the SNS First Target Station.

The SNS was designed from the beginning to allow addition of the Second Target Station (STS), and an upgrade of the accelerator power. At this time both advancements: the accelerator upgrade and the construction of the STS are in preparation.

Initially a stationary tungsten target was envisioned for the STS [1]; however, closer investigation of accident scenarios revealed that in the case of a loss of active cooling, decay heat alone could cause damage to the target and unacceptable consequences for the facility. For this reason, the current baseline STS design calls for a rotating tungsten target. The target station is driven with short (less than 1 micro second long) pulses of 1.3 *GeV* protons. Originally a 10 *Hz* pulse repetition rate and 467 *kW* beam power was proposed, while the current plan calls for a 15 *Hz* repetition rate and 700 *kW* beam power. The STS is optimized for high intensity and high resolution long wavelength neutron scattering applications. The proton beam footprint as small as acceptable from the mechanical and heat removal aspects is planned to generate a compact-volume neutron production zone in the target, which is essential for tight coupling of the target and the moderators and for achieving high-intensity peak thermal and cold neutron fluxes. The STS will accommodate 22 beamlines.

Recent neutronics analyses efforts were focused on transition to automatic conversion of computer-aided design (CAD) engineering models into high-fidelity models suitable for the analysis with Monte-Carlo (MC) radiation transport code MCNP6 [2].

2. Conversion of CAD to MC models

Two tools for the automatic conversion of CAD to MC models were tested: SuperMC developed by the FDS Team, China [3, 4] and Direct Accelerated Geometry Monte Carlo (DAGMC) Toolkit integrated into MCNP6, which is developed at University of Wisconsin-Madison [5].

Currently most of the work at SNS is performed with DAGMC. The CAD design starts in CREO [6]; in the next step the model is imported in SpaceClaim [7] where geometry checks and cleanups are performed. In the next step, the model is transferred in Cubit [8] as an ACIS file and additional checks of the geometry are performed, followed by imprinting and merging of surfaces and assigning materials to the volumes. Then the faceting is performed and the model is exported in the h5m file format. Finally, the "make_watertight" and "check_watertight" tools of the DAGMC toolkit are applied to assure the watertightness of the model. Several iterations may be required to develop acceptable MC model. A flowchart of the CAD to MC model conversion is shown in figure 1. The model developed with the automatic conversion is shown in Fig.2. The model is a 3 x 3 x 3 m cube intended to bound the actively cooled region of the core target region.

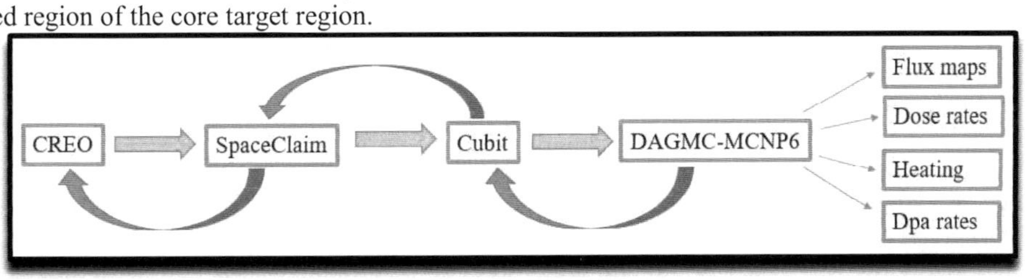

Figure 1. Flowchart of the CAD to MC model conversion.

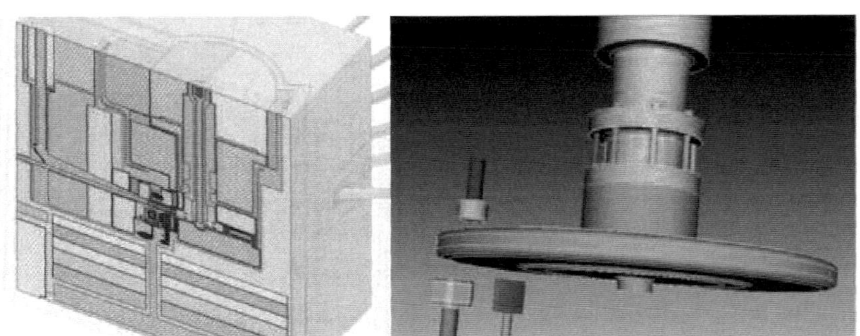

Figure 2. Vertical section through the MCNP6 model (left); detail of the rotating target and the moderators (right).

3. Results

Using model developed as described in the previous section heating rates and maps of displacements-per atom were calculated. Figure 3 shows heating rates around the rotating target. The isoline marking

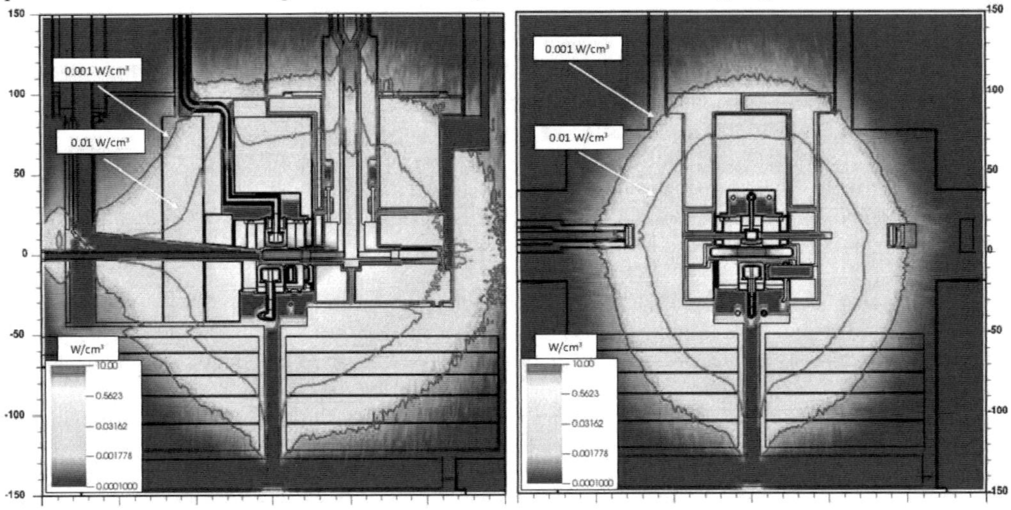

Figure 3. Heating rates around the rotating target: at left is the vertical section along the proton beam direction and on the right is vertical section perpendicular to the proton beam direction.

0.001 *W/cm*^{3} bounds the volume within which active cooling is typically required while the 0.01 *W/cm3* isoline defines the boundary of components requiring contact cooling. Figure 4 depicts displacement per atom (*dpa*) rates around the rotating target in aluminum and stainless steel; values shown are valid only for the static components. The *dpa/year* rates were calculated assuming 5000 hours of operation per year. The *dpa* rates can be used for determining the lifetime of certain components. This information is required for operational planning and making decisions regarding the segmentation of core components The high fidelity model also provides integrated heat loads in the core components. This information will be used to size cooling water flows, pipe sizes and cooling channels.

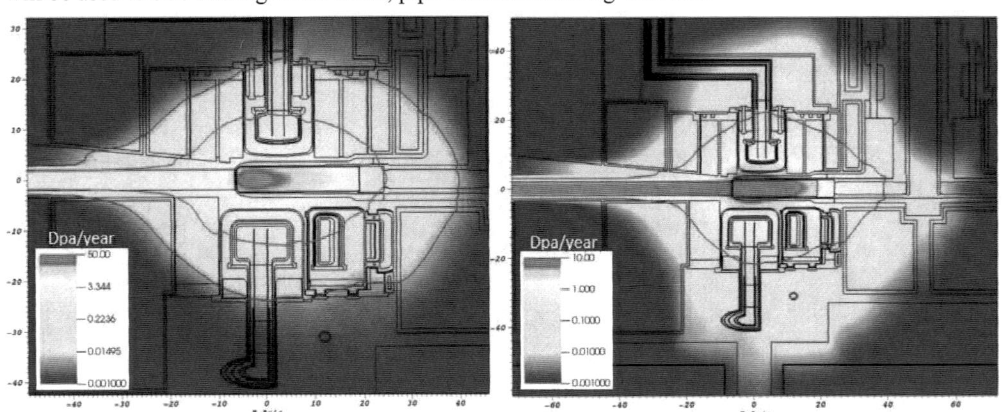

Figure 4 Radiation damage rates in *dpa/year* around the rotating target: in aluminum (at left) and in stainless steel (on the right). The pink isoline marks 1.0 *dpa/year* level (Al only), and the red isoline mark 0.25 *dpa/year* level.

The performance of the moderators has so far been calculated only with the simplified MC models, which allow application of the optimization procedures, but do not account for engineering details such as rounded moderator shapes and domes, and cooling pipe connections. The results therefore represent expected performance in optimal configuration. The coupled para-H_2 moderator peak brightness for the STS is about 10 to 14 times higher than the brightness for the FTS coupled moderators in the range below 10 *meV* (see figure 5). STS decoupled moderators are not optimally placed, but still exhibit gains in brightness by a factor of ~3 and ~4 for the para-H_2 and H_2O moderator faces at energies below ~1 *eV*, relative to the brightness of the FTS decoupled para-H_2 and water moderators (see figure 6). More discussion of moderators is provided in references [9] and [10].

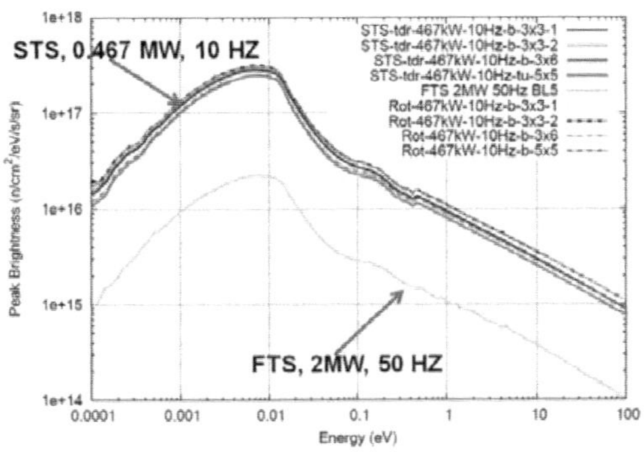

Figure 5. Coupled para-H2 moderators; peak brightness versus neutron energy, for the stationary STS (STS-tdr), rotating STS (Rot), and first target station (FTS).

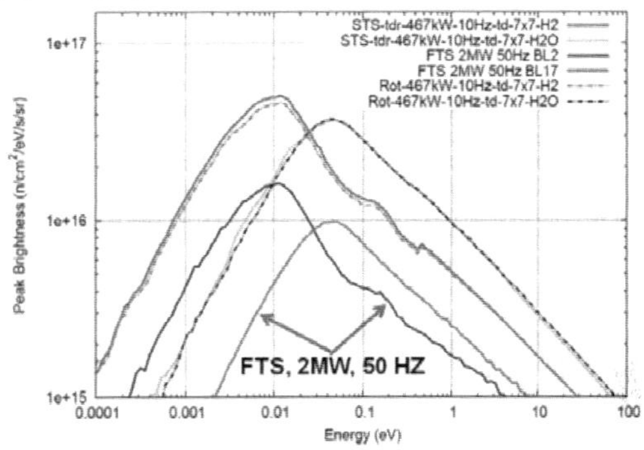

Figure 6. Decoupled para-H_2 moderator and ambient temperature H_2O moderator; peak brightness versus neutron energy, for the stationary STS (STS-tdr), rotating STS (Rot), and first target station (FTS).

4. Conclusions

Current work on neutronics analysis for the SNS STS was briefly reviewed with emphasis on newly adopted automatic conversion of CAD models to MC models with DAGMC toolkit which allows efficient creation of detailed MC models. First results obtained with detailed MC model were presented.

Acknowledgments
ORNL is managed by UT-Battelle, LLC, under contract DE-AC05-00OR22725 for the U.S. Department of Energy. This research was supported by the DOE Office of Science, Basic Energy Science, Scientific User Facilities.

References

[1] Galambos J *et al.* 2015 *Technical Design Report Second Target Station* (ORNL/TM-2015/24)
[2] Pelowitz D B *et al.* 2013 *MCNP6 User's Manual Version 1.0* (LA-CP 13-00634, Rev.0)
[3] Wu Y *et al.* 2009 CAD-based interface programs for fusion neutron transport simulation, *Fusion Eng. Des.* pp. 1987-1992
[4] Wu Y *et al.* 2015 CAD-Based Monte Carlo Program for Integrated Simulation of Nuclear System SuperMC, *Ann. Nucl.* 82 pp. 161–168
[5] DAGMC: http://svalinn.github.io/DAGMC/
[6] CREO: http://www.ptc.com/cad/creo
[7] SpaceClaim: http://www.spaceclaim.com/
[8] Cubit: https://cubit.sandia.gov/
[9] Gallmeier F X *et al.* 2015 *JAEA-Conf 2015-002*, pp. 117-124
[10] Remec I et al. 2015, *Proc. Int. Conf. on Math. and Comp. (Nashville)* vol 4, pp. 2713–2726

IOP Conf. Series: Journal of Physics: Conf. Series **1021** (2018) 012085 doi:10.1088/1742-6596/1021/1/012085

Recovery of helium refrigerator performance for cryogenic hydrogen system at J-PARC MLF

T Aso[1], M Teshigawara[1], S Hasegawa[1], H Muto[1], K Aoyagi[1], K Nomura[1], H Takada[1]

[1]Neutron source section, J-PARC center, 2-4, Ooaza-shitakata, Tokai-mura, Naka-gun, Ibaraki, Japan

E-mail: aso.tomokazu@jaea.go.jp

Abstract. At J-PARC's pulsed spallation neutron source, a cryogenic hydrogen system has been operated to provide liquid hydrogen (18K and 1.5 MPa) to the moderators since 2008. The pressure differences between No.1, No.2 heat exchangers (HXs) and an adsorber (ADS) in the helium refrigerator had begun to increase rapidly since the beginning of 2015, the refrigerator could not be operated continuously. The impurity in the refrigerator was measured by newly introduced quadrat mass spectrometer, but no significant impurities was observed. We suspected the oil contamination from the helium compressor, as it caused performance degradation of the cryogenic system in other facilities, such as RIKEN, CERN, etc. In the summer outage in 2016, we cleaned the HXs with Freon to remove the oil contamination, and replaced the activated charcoals of ADS and oil adsorber (OS-5). As a result, the performance of the helium refrigerator was recovered completely.

1. Introduction

In the Japan Proton Accelerator Research Complex (J-PARC), 1 MW pulsed spallation neutron source, called "Materials and Life science experimental facility, MLF", was constructed to promote cutting-edge materials researches [1]. A liquid hydrogen was selected as a cold neutron moderator from neutron irradiation point of view in a MW class source. A cryogenic hydrogen forced circulation system was applied to remove a nuclear high heat deposition (3.75 kW) [2] in the moderators. Furthermore, temperature increase in the moderator was designed to be less than 3 K in order to achieve the required neutronic performance.

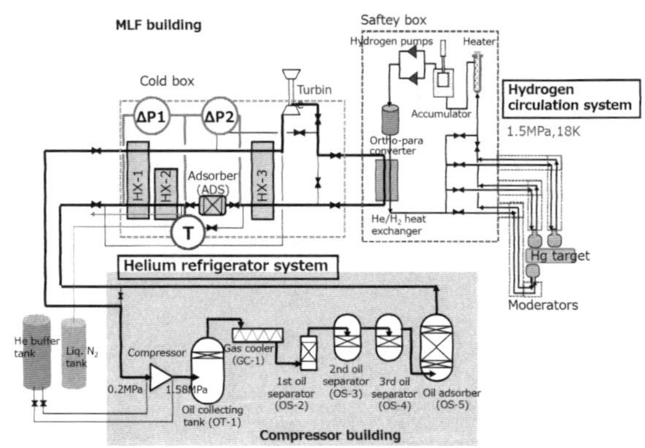

Fig.1 Overview of the cryogenic hydrogen system.

Figure 1 shows overview of the cryogenic hydrogen system, which consists of a hydrogen circulation system and a helium refrigerator system to produce the liquid hydrogen [3]. The refrigerator capacity was designed to be 6 kW in total. The helium flow rate of 280 g/s is used to provide the liquid hydrogen (18 K, 1.5 MPa and 180 g/s) to the moderators.

2. Performance degradation of the refrigerator

We have operated the cryogenic hydrogen system on good performance until December, 2014 with typical scheduled operation period of approximately 3 months. However, since January, 2015, the pressure differences at heat exchangers (HX-1 and HX-2) and the ADS increased gradually as shown in Fig. 2.

We suspected insufficient regeneration of activated charcoal of oil adsorver (OS-5) because about 100 ppm of nitrogen concentration in helium gas, which was sampled at closed ADS region, was detected after the operation from April to May, 2015 and the dew-point at heat exchangers (HX-1 and HX-2) was -45 °C at room temperature. The dew-point had been constant at about -80 °C under room temperature, and concentration of impurities, such as nitrogen, oxygen, methane and carbon oxide, which were measured by gas-chromatography, had been almost 0 ppm, respectively before January 2015.

In the 2015 summer shut-down period, therefore, we replaced the activated charcoal of OS-5 and regenerated it to remove the impurities by heated nitrogen gas (more than 100 °C) and then the nitrogen was removed by vacuum and filling helium gas. We restarted the refrigerator operation in October. However, the pressure difference increased again as shown in Fig. 2, and what was worse, after 21 days operation, the outlet temperature heat exchanger HX-2 began to increase from 78.5 K to 81 K, resulting in hydrogen temperature rise. Finally, we were forced to stop the operation, which meant interruption of scheduled beam user program.

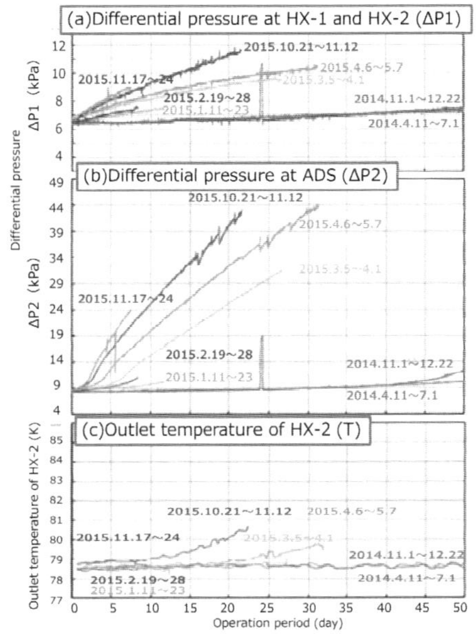

Fig.2 Trend of differential pressure at HX-1 - HX-2 (ΔP1) and ADS (ΔP2), and outlet temperature of HX-2 (T)

3. Investigation and countermeasure for the performance degradation

In December, in addition to regeneration of activated charcoal in OS-5, we regenerated activated charcoal in ADS and dried inside heat exchangers to remove the moisture. The dew points of OS-5 and ADS went down to -65 °C and -80 °C under heated condition, respectively, after one month. The heat exchangers were purged by nitrogen gas of 100 °C for 10 days. A quadrupole mass spectro-analyzer (Q-mass) with a wide range (1-200) of mass/charge ratio was newly applied to analyze the impurities in the gas, which was sampled from OS-5, ADS and the heat exchangers. We also observed inner piping of the cold box by a fiber scope and ultraviolet light. The debris, etc. were not found in the inlet filter, inlet of heat exchanger, and surfaces of some valves. The surface of inlet filter and inside pipe glistened slightly by ultraviolet light. It might be due to small amount of oil contamination.

After the maintenance, a test operation was started in January 2016. However, the pressure difference at the heat exchangers and the ADS began to increase in the same or somewhat faster changing rate than the previous one.

We surveyed several examples of the refrigerator performance degradation. The similar trouble was reported at RIKEN [4] and CERN [5] etc. due to large amount of oil accumulation (several kilograms) in cold box from helium compressor. If the oil contamination caused the performance degradation, we predicted that a major repair work, taking several months, was required to remove the oil. On the other hand, Beam user program was already scheduled until end of June. In order to keep

IOP Conf. Series: Journal of Physics: Conf. Series **1021** (2018) 012085 doi:10.1088/1742-6596/1021/1/012085

this schedule, we changed the operation cycle of cryogenic hydrogen system, such as 3 weeks operation (up to allowed maximum pressure difference) and 1 week shutdown, because we found that the pressure difference returned initial level for every restarting operation. We also reduced the helium flow by 25% to suppress oil exhaust from compressor to downstream.

4. Recovery measures and investigations at summer outage in 2016

In the 2016 summer maintenance period, we carried out cleaning of inside heat exchangers, replacement of ADS, and check of oil separators and exchange of the filters on the assumption that oil accumulation would cause the performance degradation.

4.1. Cleaning of inside heat exchangers

The heat exchangers (HX-1 and HX-2) with HX-3 were taken out from the cold box. They were transported to cleaning factory in Hyogo. The Freon, ASAHIKLIN AK-225G (c.a. 480 litters) was used to clean inside heat exchangers.

The cleaning procedure was as follows; filling Freon, circulating for 1 hour, circulating again with nitrogen gas bubbles for 1 hour, dipping overnight, circulating again, and then removing Freon. The extracted Freon (c.a. 350 ml) were analyzed to measure the oil content by NVR (Non-volatile residue) method. These were repeated three times. We found that total amount of accumulated oil was about 94 g inside heat exchangers. It seemed that the amount of measured oil was roughly equivalent to the predicted one from accumulation of oil contained in helium gas (10 ppb), which was design value.

4.2. ADS replacement

The ADS was replaced with a new one because most of the pressure difference was generated there. The ADS was removed from the cold box, and then new one with same design was installed at same position. The activated charcoal in new ADS was regenerated by heated nitrogen gas of 100 °C in order to remove moisture.

We observed inside of inlet and outlet area of used-ADS by fiber scope and ultraviolet light. The surfaces of inlet and outlet area were relatively clean. There was little glistening oil that was illuminated by ultraviolet light. The used-ADS was dismantled each component, such as felt and activated charcoal. We found that the just surface (3 mm in thickness) of upper felt, where helium gas flowed into firstly, was yellowed, like oil color in the visual observation. We conducted to measure the amount of oil in the felt. The oil concentration in the felt and charcoal was analyzed by solvent extraction and infrared spectroscopy. The upper and lower felts were separated eight layers and each thickness of felt was 4 cm. The analyzed results are shown in Fig. 3. Total amount of accumulated oil was 48.0 g in ADS, however, total accumulated oil in ADS was also equivalent to the predicted one from accumulation of oil contained in helium gas.Almost oil (45.6 g) was trapped in upper felt and even more surprising, 17.7 g of oil was accumulated in the first layer of the upper felt. It seemed that upper felt played a role of filter of oil. In generally, the felt was used not for filter of oil but for holding the activated charcoal. We considered that even if small amount of oil accumulation, accumulated oil like film on the surface of felt might cause pressure difference during cooling down process.

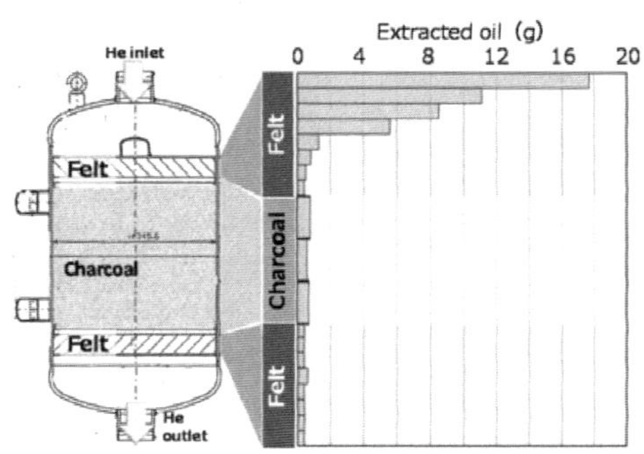

Fig.3 Result of oil analysis of used-ADS.

4.3. Check of oil separators and exchange of filters

We have three oil separators (OS-2, 3, 4) and one oil adsorber (OS-5) unit in order to separate the oil in helium gas after through screw compressor. In the summer outage, we confirmed that the OS-2, 3, 4, and 5 were normal situation without any clogging and breakage of filter. They were no problem from results of the check. Although not reaching the lifetime of each filter, these filters were exchange new ones in order to maximize the oil separation capability.

4.4. Recovery of the refrigerator performance after maintenance

After above measures at summer outage 2016, the refrigerator could be operated for over total fifty days (Nov. 1st to Dec. 20th 2016 and Jan. 12th to Mar. 8th 2017) without any problem. During the operation, the differential pressure at the heat exchangers (HX-1 and HX-2) and the ADS did not increase completely as shown in Fig. 4. The outlet temperature of heat exchanger HX-2 did not also increase.

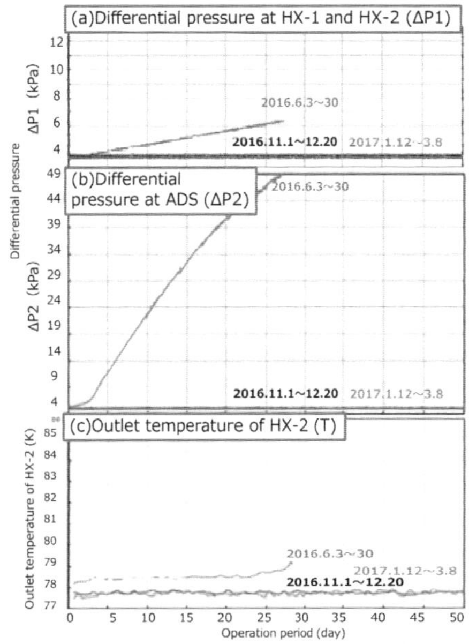

Fig.4 Trend of differential pressure at HX-1 - HX-2 (ΔP1) and ADS (ΔP2), and outlet temperature of HX-2 (T) after maintenance at summer outage.

5. Conclusion

Since Jan. 2015, we had performance degradation of the cryogenic hydrogen system, which is a critical function of the neutron moderator system at the pulsed spallation neutron source. The pressure difference rapidly increased between heat exchangers and ADS, resulting in hydrogen temperature increase.

The performance of helium refrigerator has been recovered by countermeasures, such as cleaning the heat exchangers with Freon and replacing ADS on the assumption that oil accumulation would cause the performance degradation. However, we have not yet identified the cause of degradation completely. We will carried out further investigation including any experiment to reveal the phenomena induced.

References

[1] Y Ikeda, 1MW Pulse Spallation Neutron Source (JSNS) under the High Intensity Proton Accelerator (J-PARK) Project, Proc. of the ICANS-XVI, 2003, pp.13-24.
[2] M. Harada, et al., Nuclear heating calculation for JSNS, Proc. of ICANS-XVIII, 2007, p.616-626.
[3] T Aso, et al., Design result of the cryogenic hydrogen circulation system for 1 MW pulse spallation neutron source (JSNS) in J-PARC, Advances in Cryogenic Engineering, 51A, 2005, p.763-770.
[4] H Okuno, Cryogenic cooling system for RIKEN superconducting ring cyclotron, Journal of Cryogenics and Superconductivity Society of Japan, Vol.50, No.12, 2015, pp.583-590.
[5] Caroline Fabre, Creaning and consolidation of the CMS helium refrigerator after hydrocarbon contamination, Workshop on Cryogenic Operations, 2016

IOP Conf. Series: Journal of Physics: Conf. Series **1021** (2018) 012086 doi:10.1088/1742-6596/1021/1/012086

A Comparison of Strains Induced by Manufacturing and Operational Conditions of a Tantalum Clad Tungsten Plate of an ISIS TS1 Solid Target

Yanling Ma[1,2], David Jenkins[1], Saurabh Kabra[1] and Shu Yan Zhang[1,2]

[1]ISIS Facility, Science and Technology Facilities Council, Rutherford Appleton Laboratory, Harwell Oxford, Didcot, Oxfordshire, OX11 0QX, UK

[2]Centre of Excellence for Advanced Materials (CEAM), 1 Libin Road, Dongguan, Guangdong Provence, China 523808

yanling.ma@stfc.ac.uk Yanling.Ma@ceamat.com

Abstract. This paper aims to reveal the relationship between two sets of residual strains. One set of the strains were induced by manufacturing when the Tantalum (Ta) cladding was diffusion bonded to the Tungsten (W) core by Hot Isostatic Pressing (HIP) process. Another set of strains was induced during target operations, i.e., when the HIPPED target plates were further bombarded by proton beams for neutron production. The former set was revealed by Neutron Diffraction Measurement (NDM) using the ISIS Engin-X instrument. The latter set was obtained via Finite Element Analysis (FEA) on the ISIS TS1 target assembly. The comparison and superposition of these two sets of strains demonstrated mainly an overall reduction of about 50 micro strains within the Ta cladding. Namely, the strains induced during the TS1 target operation could be beneficial for the specific region of target plate number five (P5) in this case.

1. Introduction

The ISIS TS1 target mainly consists 12 compound Tungsten and Tantalum plates as shown in Figure 1. The tantalum cladding of the target plates is diffusion bonded to the tungsten core by a Hot Isostatic Pressing (HIP) process, during which we believe that residual strains are induced in the HIPPED target plate. This is because W and Ta have different thermal expansion coefficients while they were cooled down from 500℃ to room temperature [1]. Neutron Diffraction Measurement (NDM) of the residual strains in a target plate was carried out using the ISIS Engin-X instrument. Residual strain distribution patterns and values within a sample target plate were revealed successfully [1]. This sample target plate of number five is indicated as P5 in red in Figure 1. The selection of this plate was because it is readily available. During the TS1 target operations, however, the target plates are further bombarded by proton beams for neutron productions. Such an operational condition is equivalent to about 100kW of heat load spatially deposited onto the solid target. As a

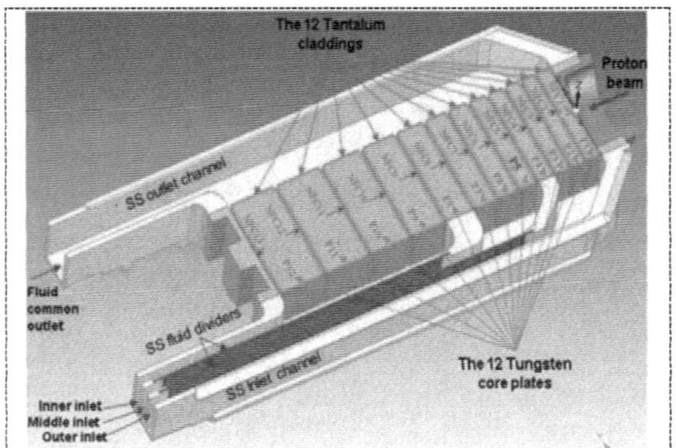

Figure 1. Details of a half (due to symmetry) simplified CAD model of TS1 Target used for CFD and FEA

result, the compound target plates experience elevated temperatures and thermally induced strains and stresses again.

Understanding the relationship between the above two sets of strains is necessary as it could guide the ISIS target safe operation. Namely, if the overall strains within the HIPPED target plate are decreased after the above two set of strains are superposed and compared, that could mean the operational condition is mechanically strengthening the HIPPED target plates. Extra assurance may be provided for operations. On the contrary, if the overall strains superposed within the HIPPED target plate are increased, this could indicate that the operational condition is mechanically weakening the HIPPED target plates. Extra care may need to be taken during operations. To reveal such information quantitatively, we use CFD results obtained previously [2] as input, and a thermal and structural analysis [3] was conducted subsequently. From the recent study, not only the temperature distributions and residual strains within each of the target components and the target assembly were obtained, but also the residual strains within the same area of the same target plate P5 in this case were revealed as planned. These findings may indicate that the strains induced during target operations could be beneficial as overall about 50 micro strains cancellation was produced from the superposition and comparison of the strains for the specific area of the specific target plate in this case. Some of the analysis results will be reported in this paper.

2. NDM of residual strains induced by manufacturing

A description of the NDM of residual strains of target plate No. 5 can be found in reference [2]. Details of the sample used in the NDM, including the measurement plane and location are shown in Figure 2 (a). A schematic diagram of the experimental set-up from the top view of the sample in cross section is given in Figure 2 (b), in which the specific measurement area and points are included.

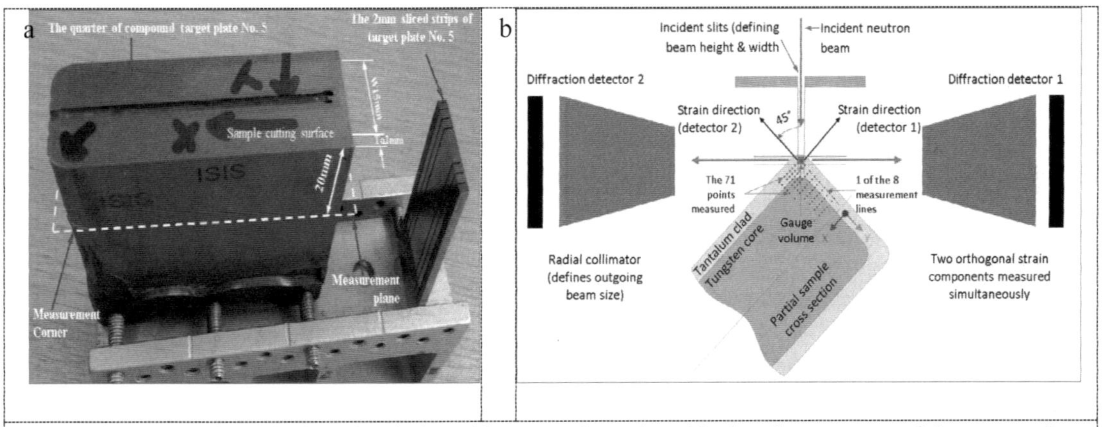

Figure 2. NDM of residual strains induced by manufacturing (a) details a quarter of the target plate No.5 and (b) schematic diagram of experiment set-up viewed from the top of the NDM plane

From this NDM, the residual strain distribution patterns induced by manufacturing (the HIPPING process) were revealed quantitatively for the issued area and points as presented in the following **Section 4** for comparison and discussions.

3. FEA of residual strains induced by operational conditions

To understand the residual strain distributions induced by operational conditions, an FEA was conducted, using CFD results [2] as input, on the ISIS TS1 target assembly (Figure 1). Owing to its symmetry, only a half of the target was considered in the thermal stress analysis. It also should be pointed out that the FEA used a worst-case scenario in terms of beam history, i.e., it assumed the full beam power of 800MeV, 200μA at 50Hz is constantly hitting the solid target. In addition, to accurately calculate residual stress, three residual strain components are needed, but since the NDM in this case only measured two components, corresponding FEA results were therefore post processed and used in this paper.

Some of the typical results are presented in Figure 3 for the residual strains along the FEA X direction. Figure 3 (a) is the strain map within the half target assembly and Figure 3 (b) is the strain distributions within the target plate No. 5. The measurement area indicated with a white square box on the NDM plane in Figure 3 (b) corresponds to the NDM area and points marked in Figure 2 (b) for the residual strains induced by manufacturing. The strains within this area revealed by FEA under operational conditions were further post processed with a view to comparing these two sets of strains.

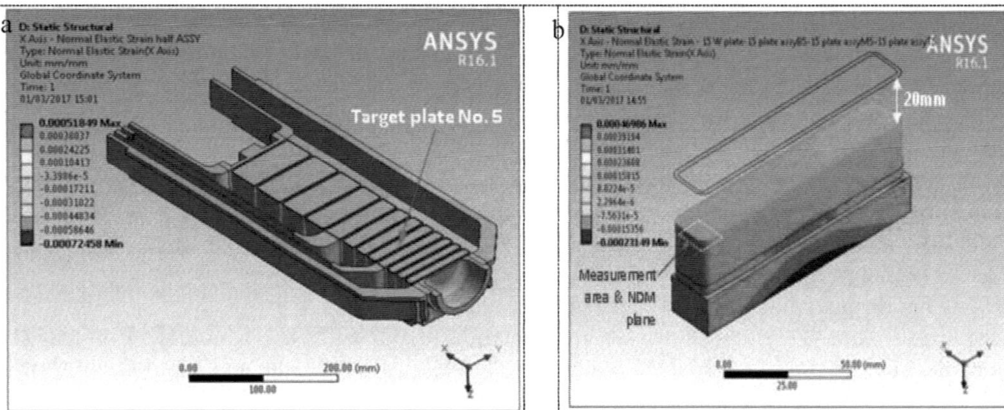

Figure 3. Residual strains induced by operational conditions along FEA X-direction (a) within half target assembly and (b) within target plate No. 5 along NDM plane and area

It should be noted that similar results for residual strains along the FEA Y direction were also obtained. They are not presented here due to limited space. However, the strain distributions of the FEA Y direction for the issued area and points were post processed in a similar manner to those for the FEA X direction as presented in the following **Section 4**.

4. Comparison of strains induced by manufacturing and operational conditions

From the above investigations, residual strain distributions along the NDM X direction within the specific area of plate No. 5 are shown in Figure 4, where TaVP stands for Tantalum cladding vertical part and TaHP stands for Tantalum cladding horizontal part. A superposition of the X direction strains within the TaHP is presented in Figure 5.

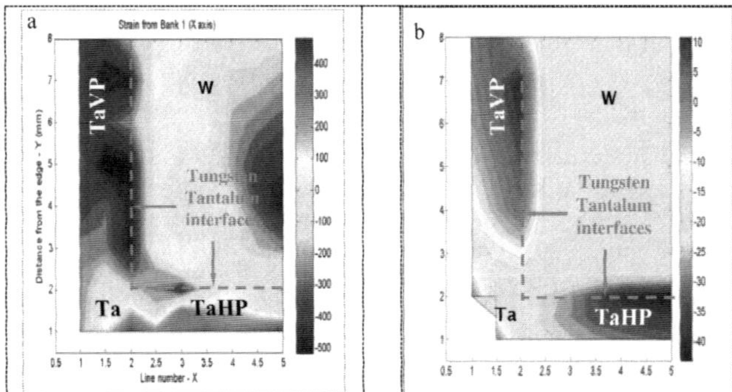

Figure 4. Residual strains distributions along the X-direction revealed (a) by NDM for manufacturing and

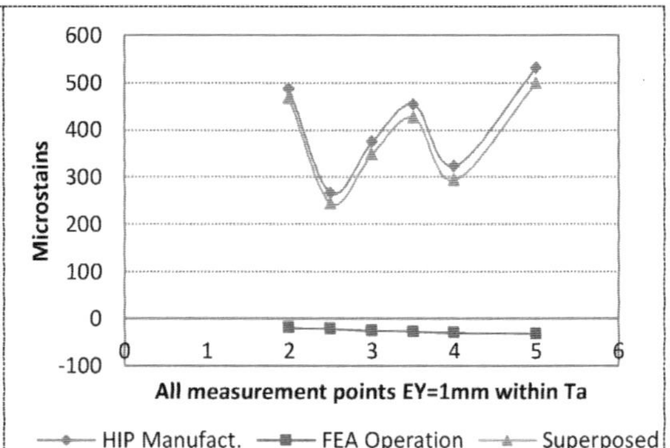

Figure 5. Superposition of residual strains along X direction within the Tantalum

(b) by FEA for operational conditions	cladding horizontal part

As shown in Figure 4 (a), when residual strains induced by manufacturing within the TaVP are mainly compressive, the corresponding ones induced by operation, Figure 4 (b), are mainly tensile, resulting in an overall cancellation of about 50 micro strains. Similarly, while the residual strains within the TaHP are mainly positive (Figure 4 (a)), mainly negative strains caused by operations can be seen within the TaHP in Figure 4 (b). In addition, a superposition of the X direction strains along the bottom line of the TaHP is given in Figure 5, from which a maximum of about 33 micro strain reduction is observed. Also the bottom line of the TaHP in Figure 4 (a) and (b) is the line that is 1mm in from the edge of the horizontal tantalum cladding.

Residual strains along the Y direction induced by manufacturing and by operations are similarly compared in Figure 6 with a superposition of them along the far left line of TaVP given in Figure 7.

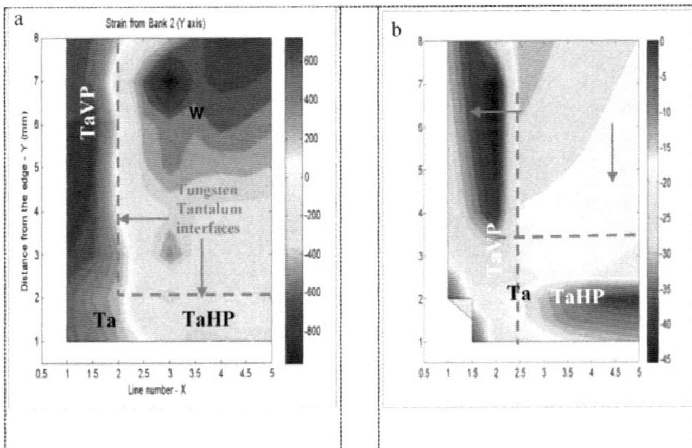

Figure 6. Residual strain distributions along theY-direction revealed by (a) NDM for manufacturing and (b) FEA for operational conditions

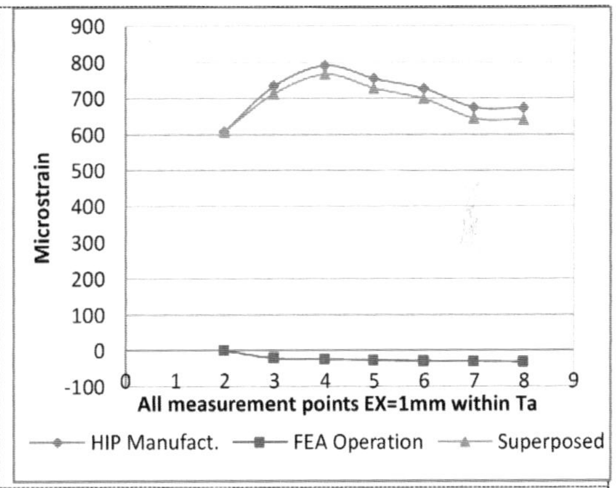

Figure 7. Superposition of residual strains along the NDM Y direction within the vertical Tantalum cladding

As depicted in Figure 6 for the Y direction residual strains, a similar relationship is revealed to those for the X direction ones (Figure 5). For instance, while the strains induced by manufacturing within the TaVP in Figure 6 (a) are mostly positive, mainly opposite sign strains were induced from the operational conditions within the TaVP in Figure 6 (b). This resulted in an overall cancellation of 45 micro strain. A similar reduction of about 32 micro strain was also seen from the superposition shown in Figure 7 along the far left line of the TaVPs in Figure 6 (a) and (b). This far left line within the TaVP is similarly the line that is 1mm in from the vertical edge of the vertical tantalum cladding in this case.

It also should be noted that there is an increase of about 19 micro strains within the TaHPs in Figure 6 for Y direction strains. This increase is however less than the overall maximum strain decrease of about 50 for the X direction and 45 for the Y direction, respectively. We may therefore say that the residual strains induced by operational conditions could be beneficial in this case.

5. Conclusions

The relationship between residual strains induced by manufacturing and by operational conditions has been revealed successfully. The latter could be beneficial in terms of mechanically strengthening the ISIS TS1 target as there is an overall strain cancellation within the Tantalum cladding when superposing the two sets of strains. The latter, however, is not beneficial within the Tungsten core as there are mostly strain additions from the same superposition of the two sets of strains. Further investigations of critical areas and critical plates may be needed as the findings from this study may only be applicable to the sample studied in this case. Topics to explore next should include the correlation between the residual strain with real beam history, accurate residual stress calculation, change of water channel dimensions, and radiation and fatigue damages.

Acknowledgments
The authors thank Stephen Gallimore for his support in preparing this paper and a poster for the ICANS XXII.

References

[1] Y. Ma, S. Y. Zhang, D. Jenkins and S. Kabra, "An Experiment of Using Neutron Diffraction to Investigate Residual Strain Distribution in a Hot Isostatic Pressed (HIPPED) Target Plate", Materials Today: Proceedings 2S, 2015, pp267-273.
[2] Y. Ma, D. M. Jenkins and L. Jones, "A CFD ANALYSIS OF A SOLID TARGET", NAFEMS World Congress, June 2015, pp. 194-201.
[3] Y. Ma. TS1 Thermal Stress Analysis, January, 2016, ISIS internal report.

Institute of Physics
Dirac House, Temple Back
Bristol BS1 6BE UK

ISSN: 1742-6588
ISBN 978-1-5108-6403-0

CURRAN ASSOCIATES INC.
proceedings.com

9th International Symposium on Steel Bridges 2018

IOP Conference Series: Materials Science and Engineering
Volume 419

Prague, Czech Republic
10 – 11 September 2018

IOP
Institute of Physics